Lecture Notes in Networks and Systems

Volume 93

The series "Lecture Notes in Networks and Systems" publishes the latest developments in Networks and Systems—quickly, informally and with high quality. Original research reported in proceedings and post-proceedings represents the core of LNNS.

Volumes published in LNNS embrace all aspects and subfields of, as well as new challenges in, Networks and Systems.

The series contains proceedings and edited volumes in systems and networks, spanning the areas of Cyber-Physical Systems, Autonomous Systems, Sensor Networks, Control Systems, Energy Systems, Automotive Systems, Biological Systems, Vehicular Networking and Connected Vehicles, Aerospace Systems, Automation, Manufacturing, Smart Grids, Nonlinear Systems, Power Systems, Robotics, Social Systems, Economic Systems and other. Of particular value to both the contributors and the readership are the short publication timeframe and the world-wide distribution and exposure which enable both a wide and rapid dissemination of research output.

The series covers the theory, applications, and perspectives on the state of the art and future developments relevant to systems and networks, decision making, control, complex processes and related areas, as embedded in the fields of interdisciplinary and applied sciences, engineering, computer science, physics, economics, social, and life sciences, as well as the paradigms and methodologies behind them.

**** Indexing: The books of this series are submitted to ISI Proceedings, SCOPUS, Google Scholar and Springerlink ****

More information about this series at http://www.springer.com/series/15179

Simon Fong · Nilanjan Dey ·
Amit Joshi
Editors

ICT Analysis and Applications

Proceedings of ICT4SD 2019, Volume 2

Springer

Editors
Simon Fong
University of Macau
Macau, China

Nilanjan Dey
Techno India College of Engineering
Kolkata, India

Amit Joshi
Ahmedabad, India

ISSN 2367-3370 ISSN 2367-3389 (electronic)
Lecture Notes in Networks and Systems
ISBN 978-981-15-0629-1 ISBN 978-981-15-0630-7 (eBook)
https://doi.org/10.1007/978-981-15-0630-7

This Springer imprint is published by the registered company Springer Nature Singapore Pte Ltd.
The registered company address is: 152 Beach Road, #21-01/04 Gateway East, Singapore 189721, Singapore

Organization

Organizing Chairs

Mr. Bharat Patel, Chairman, CEDB
Mr. Peter Kent, Chief Executive Officer (CEO), UKEI, UK

Organizing Co-chair

Mr. Nakul Sharedalal, Advisor, G R Foundation, India

Organizing Secretary

Amit Joshi, SITG Ahmedabad and Global Knowledge Research Foundation

Members

Dr. Vijay Singh Rathore, Department of CSE, JECRC, Jaipur, India
Mr. Aman Barot, G R Foundation, India
Dr. Mahipal Singh Deora, BN University, Udaipur, India
Dr. Nisheeth Joshi, Banasthali University, Rajasthan, India
Mr. Nilesh Vaghela, Electromech Corporation, Ahmedabad, India
Mr. Vinod Thummar, SITG, Gujarat, India
Dr. Chirag Thaker, L. D. College of Engineering, Gujarat, India
Prof. S. Rama Rao, Professor, Goa University, Goa, India
Dr. Nitika Vats Doohan, Indore, India

Dr. Parikshit Mahalle, Sinhgad Group, Pune, India
Dr. Priyanks Sharma, RSU, Ahmedabad
Dr. Nitika Vats Doohan, Indore
Dr. Mukesh Sharma, SFSU, Jaipur
Dr. Manuj Joshi, SGI, Udaipur, India
Dr. Bharat Singh Deora, JRNRV University, Udaipur
Mr. Ricky, P Tunes, Goa University, India
Prof. L. C. Bishnoi, GPC, Kota, India
Dr. Vikarant Bhateja, Lucknow, India
Dr. Satyen Parikh, Dean, Ganpat University, Ahmedabad, India
Dr. Puspendra Singh, JKLU, Jaipur, India
Dr. Aditya Patel, Ahmedabad University, Gujarat, India
Mr. Ajay Choudhary, IIT Roorkee, India

Technical Program Committee

Prof. Milan Tuba, Singidunum University, Serbia
Dr. Durgesh Kumar Mishra, Chairman, Div IV, CSI

Technical Program Committee Co-chairs

Dr. Mahesh Bundele, Dean Research, Poornima University, Jaipur
Dr. Nilanjay Dey, Techno India College of Engineering, Kolkata, India

Advisory Committee

Shri Nitin Kunkolienker, President, MAIT
Prof. S. Rama Rao, Professor, Goa University, Goa
Mr. P. N. Jain, Add. Sec., R&D, Government of Gujarat, India
Prof. J. Andrew Clark, Computer Science University of York, UK
Mr. Vivek Ogra, President, GESIA
Prof. Mustafizur Rahman, Endeavour Research Fellow, Australia
Dr. Kalpdrum Passi, Laurentian University, ON, Canada
Mr. Chandrashekhar Sahasrabudhe, ACM India
Dr. Pawan Lingras, Saint Mary University, Canada
Dr. S. C. Satapathy, Professor in CSE, AP, India
Mr. Niko Phillips, Active Learning, UK, India
Dr. Bhagyesh Soneji, Chairperson, Assocham Western Region
Dr. Dharm Singh, NUST, Windhoek, Namibia

Members

Prof. Abdul Rajak A. R., Department of Electronics and Communication Engineering Birla Institute of Technology and Sciences, Abu Dhabi
Prof. Hoang Pham, Rutgers University, Piscataway, NJ, USA
Prof. Shashidhar Ram Joshi, Institute of Engineering, Pulchowk, Nepal
Prof. Abrar A. Qureshi, Ph.D., University of Virginia, USA
Dr. Aynur Unal, Stanford University, USA
Dr. Manju Mandot, JRNRVU, Udaipur, India
Dr. Rajveer Shekawat, Manipal University, India
Prof. Ngai-Man Cheung, Assistant Professor, University of Technology and Design, Singapore
Prof. J. Andrew Clark, Computer Science University of York, UK
Prof. Yun-Bae Kim, Sungkyunkwan University, South Korea
Prof. Lorne Olfman, Claremont, California, USA
Prof. Louis M. Rose, Department of Computer Science, University of York
Nedia Smairi, CNAM Laboratory, France
Dr. Vishal Gaur, Government Engineering College, Bikaner, India
Dr. Aditya Patel, Ahmedabad University, Gujarat, India
Mr. Ajay Choudhary, IIT Roorkee, India
Prof. Prasun Sinha, Ohio State University Columbus, Columbus, OH, USA
Dr. Devesh Shrivastava, Manipal University, Jaipur, India
Prof. Xiaoyi Yu, National Laboratory of Pattern Recognition, Institute of Automation, Chinese Academy of Sciences, Beijing, China
Prof. R. S. Rao, New Delhi, India

Preface

The Fourth International Conference on ICT for Sustainable Development (ICT4SD 2019) targets theory, development, applications, experiences and evaluation of interaction sciences with fellow students, researchers and practitioners.

The conference may concern any topic within its scope. Workshops may be related to any topics within the scope of the conference. The conference is devoted to increase the understanding role of technology issues and how engineering has day by day evolved to prepare human-friendly technology. The conference will provide a platform for bringing forth significant research and literature across the field of ICT for Sustainable Development and provide an overview of the technologies awaiting unveiling. This interaction will be the focal point for leading experts to share their insights, provide guidance and address participant's questions and concerns.

The conference was held during 5–6 July 2019, at Hotel Vivanta By Taj, Panaji, Goa, India, and organized by Global Knowledge Research Foundation and supported by The Institution of Engineers (India), Supporting Partner InterYIT, International Federation for Information Processing, State Chamber Partner Goa Chamber of Commerce & Industry, and National Chamber Partner Knowledge Chamber of Commerce & Industry.

Research submissions in various advanced technology areas were received, and after a rigorous peer review with the help of program committee members and 56 external reviewers for 519 papers from 8 different countries including Algeria, USA, United Arab Emirates, Serbia, Qatar, Mauritius, Egypt, Saudi Arabia, Ethiopia and Oman, 113 were accepted with an acceptance ratio of 0.11.

Technology is the driving force of progress in this era of globalization. Information and Communication Technology (ICT) has become a functional requirement for the socio-economic growth and sustained development of any country. The influence of Information Communication Technology (ICT) in shaping the process of globalization, particularly in productivity, commercial and financial spheres, is widely recognized. The ICT sector is undergoing a revolution that has momentous implications for the current and future social and economic situation of all the countries in the world. ICT plays a pivotal role in empowering people for self-efficacy and how it can facilitate this mission to reach out to grassroots level. Finally, it is concluded that ICT is a significant contributor to the success of the ongoing initiative of Startup India.

In order to recognize and reward the extraordinary performance and achievements by ICT and allied sectors & promote universities, researchers and students through their research work adapting new scientific technologies and innovations, the two-day conference had presentations from the researchers, scientists, academia and students on the research works carried out by them in different sectors.

ICT4SD Summit is a flagship event of G R Foundation. This is the fourth edition. The summit was inaugurated by Dr. Pramod Sawant, Chief Minister of Goa, along with other eminent dignitaries including Shri Manguirsh Pai Raikar, Chairperson, Assocham's National Council for MSME; Shri. Prajyot Mainkar, Chairman, IT Committee of Goa Chamber of Commerce and Industry; Mike Hinchey, President, IFIP and Chair, IEEE, UK and Ireland; Milan Tuba, Vice Rector for International Relations, Singidunum University, Serbia; and Shri Amit Joshi, Director, G R Foundation.

Dr. Pramod Sawant shared his views and aim on creating a world-class IT Infrastructure and connectivity for e-governance in the state of Goa. Government is committed to create adequate infrastructure for the industry promotion of e-governance, e-education and streamlining of IT in Goa. He further stated that the infrastructure development and capacity building for promotion of IT is one of the main focus areas.

Shri Manguirsh Pai Raikar, Chairperson, Assocham's National Council for MSME, said that this summit is to provide a common platform and bringing forth significant industries, researches and literatures across the field of ICT for Sustainable Development and provide an overview of the technologies awaiting unveiling along with recognizing and rewarding the extraordinary performance as well achievements by ICT and allied industries, communally.

Shri Prajyot Mainkar, Chairman, IT Committee of Goa Chamber of Commerce and Industry, highlighted and encouraged the entrepreneurship and said that GCCI will continue to support such programmes for larger public benefits with great degree of excellence.

The international dignitaries including Mike Hinchey, President, IFIP and Chair, IEEE, UK and Ireland and Milan Tuba, Vice Rector for International Relations, Singidunum University, Serbia, also highlighted the issues and opportunities in the information processing and education sector.

Macau, China — Simon Fong
Kolkata, India — Nilanjan Dey
Ahmedabad, India — Amit Joshi

Contents

About the Editors

Simon Fong graduated from La Trobe University, Australia, with a 1st Class Honors B.Eng. Computer Systems degree and a Ph.D. Computer Science degree in 1993 and 1998, respectively. Simon is now working as an Associate Professor at the Computer and Information Science Department of the University of Macau, China. He is also one of the founding members of the Data Analytics and Collaborative Computing Research Group at the Faculty of Science and Technology. Prior to his academic career, Simon held various managerial and technical positions, such as systems engineer, IT consultant, and e-commerce director in Australia and Asia. Dr. Fong has published over 300 international conference and peer-reviewed journal papers, mostly in the areas of data mining and optimization algorithms.

Nilanjan Dey was born in Kolkata, India, in 1984. He received his B.Tech. degree in Information Technology from West Bengal University of Technology in 2005, M.Tech. in Information Technology in 2011 from the same University, and Ph.D. in digital image processing in 2015 from Jadavpur University, India. In 2011, he was appointed as an Assistant Professor in the Department of Information Technology at JIS College of Engineering, Kalyani, India, followed by Bengal College of Engineering and Technology, Durgapur, India, in 2014. He is now employed as an Assistant Professor in the Department of Information Technology, Techno India College of Technology, India. He is a visiting fellow of the University of Reading, UK. His research topics are signal processing, machine learning, and information security. Dr. Dey is an Associate Editor of IEEE ACCESS and is currently the Editor-in-Chief of the International Journal of Ambient Computing and Intelligence. He is the Series Co-Editor of Advances in Ubiquitous Sensing Applications for Healthcare (AUSAH), Elsevier, and Springer Tracts in Nature-Inspired Computing (STNIC).

Amit Joshi is a young entrepreneur and researcher who holds an M.Tech. in Computer Science and Engineering and is currently pursuing research in the areas of Cloud Computing and Cryptography. He has six years of academic and industrial experience at prestigious organizations in Udaipur and Ahmedabad. Currently, he is working as an Assistant Professor in the Department of Information Technology, Sabar Institute in Gujarat. He is an active member of the ACM, CSI, AMIE, IACSIT-Singapore, IDES, ACEEE, NPA, and many other professional societies. He also holds the post of Honorary Secretary of the CSI Udaipur Chapter and Secretary of the ACM Udaipur Chapter. He has presented and published more than 30 papers in national and international journals/conferences of the IEEE and ACM. He has edited three books, on Advances in Open Source Mobile Technologies, ICT for Integrated Rural Development, and ICT for Competitive Strategies. He has also organized more than 15 national and international conferences and workshops, including the International Conference ICTCS 2014 at Udaipur through the ACM-ICPS. In recognition of his contributions, he received the Appreciation Award from the Institution of Engineers (India) in 2014, and an award from the SIG-WNs Computer Society of India in 2012.

The Analytical CRM OLAP Analysis Tools and Data Mining

Sanjay Agal and Pooja Devija

Abstract This article focuses on Customer Relationship Management (CRM) or management customer relationship. We discussed this concept by focusing on its dimension analytical orientation justified by the choice of our subject, namely OLAP analysis tools and data mining. As we will see later, analytical CRM relies heavily on computing. Whether at the level of data production, storage or finally the analysis, useful also to optimize processes and decision of decision, IT is omnipresent. We have chosen to sometimes leave aside some aspects, particularly technical. That said, this work hopes to offer a vision most widely possible key issues related to analytical CRM in particular, as well as the management of customer relationships in general.

Keywords CRM · OLAP · Data mining

1 Introduction

Nowadays, technological developments, globalization of markets, and shorter product life cycles make competition ever rough. It becomes very difficult for a company to maintain its market share based only on prices and products.

The strong decline of mass advertising illustrates this difficulty to gain and maintain market share by focusing only on the product. Since the 80s, mass marketing techniques hardly bring positive results business, cede their places to direct marketing, customer-oriented; we must understand and comply with whom to communicate "directly" to optimize the future success of the company. CRM is clearly in this evolution and represents, in some way, the last direct marketing.

As we will see later in the work, the technology takes on a role essential in the CRM. It will allow extracting knowledge from data stored and managed in a data warehouse and then analyzed through OLAP and data mining tools.

S. Agal (✉) · P. Devija
Dr. V. R. Godhania College of Engineering & Technology, Porbandar, India
e-mail: sanjayagal@yahoo.com

P. Devija
e-mail: poojadevija@gmail.com

S. Fong et al. (eds.), *ICT Analysis and Applications*, Lecture Notes in Networks and Systems 93, https://doi.org/10.1007/978-981-15-0630-7_1

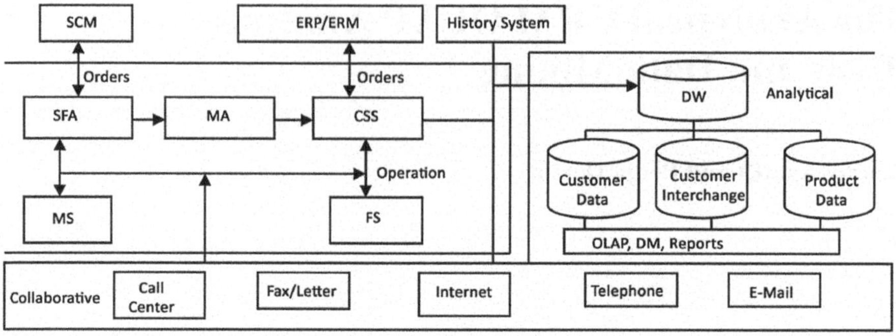

Fig. 1 The designing of CRM based on data warehouse

2 Dimensions of Customer Relationship Management

During our research, we found CRM frequently divided into three interrelated areas of activity: Operational CRM, CRM collaborative, and analytical CRM. The collaboration of these three parties aims to achieve the ultimate objective of the CRM business strategy, that is to say, development and optimization of customer relations. Figure 1 shows schematically the three elements of CRM and their interdependence.

3 Crossroads of Three CRM

Before going forward in analytical CRM in the definition and use of OLAP research tools and data mining, it is important to talk about the fundamental role that holds the data warehouse within the CRM.

Like in Fig. 2, talking about data warehouse or data warehouse is part of a broader concept which is that of systems integration heterogeneous information. The data warehouse refers in this case to integration by the data that are shared by autonomous systems. In observing the diagram, we can see that the data warehouse is literally pivot office between each of CRM elements.

Let's take an example. Whether we speak of the analyst, product manager, the marquetry, or the employee call center, everyone has access, through the warehouse data, the same data source, and to the same information.

The analyst uses its analytical tools to find in the "Customer Data Warehouse" the results of its petitions and new knowledge helps managers or engineers in their decisions as in optimization of the company's various business processes. The Head Product focuses his attention on sales of his product, by region, by period and, often via a data mart.

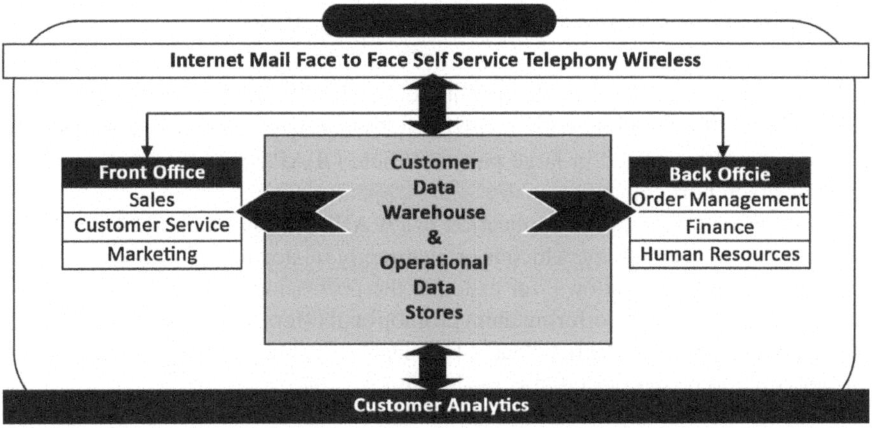

Fig. 2 Customer analytics making the difference in CRM

4 The Data Storage in the Data Warehouse

As noted above, the analytical CRM has the function of allowing the storing and analyzing of the data generated in particular by the steps of marketing, sales, customer service, and information collected on the client during telephone contacts, email, or through questionnaires.

As can be seen in Fig. 3, the step which precedes the analysis is which consists of extracting the data where it is and stored in a data warehouse or a data warehouse. This phase of the process is crucial and has some difficulties and risks which are important to state here.

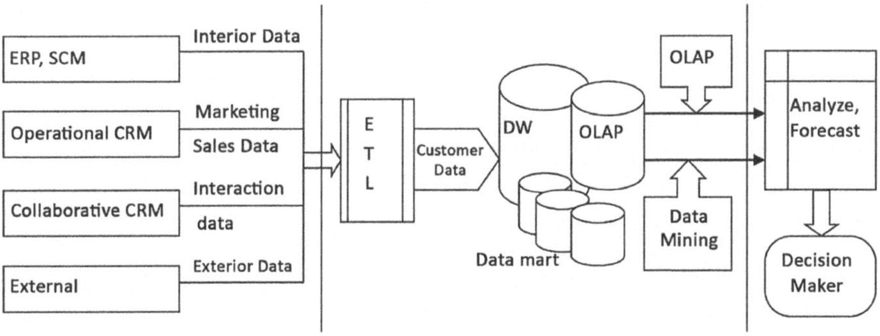

Fig. 3 The designing of CRM based on data warehouse

5 Application Examples OLAP Analysis Tools

For a more pragmatic view of the usefulness of OLAP tools, three practical cases can be exposed in which they are involved and what they can concretely serve. To do this, we will be guided in large part to Article OLAP applications available on the www.olapreport.com site.

Database marketing is the application of OLAP analysis tools, often combined with data mining technology, which is more directly related to customer relationship management. Indeed, it allows, for example the professional identify the best customers and retain them by offering them promotional offers. Conversely, the user can take the opportunity to identify "bad customers" or some customer profiles which an offer is not to be readjusted, it is to abandon it.

The analysis of sales is another essential function of OLAP tools. The latter may actually be profitable for knowledge companies, especially in the consumer goods industry and financial services, commercially details. Some of the questions to which these tools are able to provide response describe the whole point. They are as follows:

1. Are we able to achieve our sales targets by product and region in a given period?
2. Is the sales growth of a new product the same everywhere?
3. Do promotion actions have an effect on sales?, etc.

Analysis of the "click" of the Internet on a website can also be very useful, especially for companies devoted to e-business. Practically, OLAP tools make it possible. For example, demographic data collection, sometimes personal, on the client, identifying the path where he arrived on the company website or calculating the number of "Clicks" that it took to find on the website what he came there for.

6 Data Mining

Data mining literally means "data mining" or "data mining". This method, based on a series of algorithms or data mining models that we will not discuss in this work allows to extract information from data, information which, through analysis, are converted into knowledge. "Data mining is the analysis of a set of observations is intended to find unsuspected relationships and summarize data in a new way, so that they are understandable and useful to their holders" [1]. In other words, it consists in analyzing the information collected in data warehouses to identify relationships that would be a priori impossible to identify without this tool. This is an essential element in customer relations and a decision support system.

6.1 *The Tasks*

After defining what data mining is, it is appropriate to present the tasks it can perform. These are six in number; each assumes a specific function for analysis and is performed using different algorithms.

The importance of this task is to allow the analyst to interpret the results of a data mining model or an algorithm, the more transparent the more effective. "Thus, the results of data mining model should describe the characteristics clear that can lead to an interpretation and an intuitive explanation. Some methods of data mining are more suitable than others for a transparent interpretation" [2].

6.2 *Estimation*

After defining what data mining is, it is appropriate to present the tasks it can perform. These are six in number; each assumes a specific function for analysis and is performed using different algorithms.

6.3 *Segmentation*

Here, the target variable is not digital but categorical, such as income, which can be divided into three categories: low income, middle income, and high income. "The segmentation is to divide customers into homogeneous groups, that then be addressed by specific and appropriate means to characteristics and needs of each group. Members of the same group react the same way to marketing stimuli. They share a mode communication, purchasing behaviors and/or special needs" [3].

6.4 *Classification*

The classification is different from the segmentation in the sense that there is no target variable to segment. The classification will be interested in reunification data or observations in groups of similar objects. In other words, it will segment the data set to form homogeneous subgroups. These are called clusters, namely classes, which are groups in which data are similar to each other and different for definitions of the other groups.

6.5 Forecasting

The results of the forecast are, as the name suggests, in the future, which differs from the estimate. For the rest, it is similar to these two tasks explained above. The forecast allows for example to predict who will be winners of the Basketball Championship taking into account the comparison of results of each team or to predict what will be the death of decay rate on the road the following year, taking into account the increase of the limitations of speed and stricter measures of reprimand by the police in regard to drinking and driving.

6.6 The Association

This function of data mining to discover which variables go together, what are the rules that will allow to quantify the relationships between two or several variables.

For example, if we look at 500 customers coming to shop at a supermarket on a Friday night and we found that out of these 500 clients, 100 buy fruit and of those, 30 buy milk, and the rule of association is "if you buy fruit, so we buy milk," with a support measuring 100/500 = 20% and a confidence level of 30/100 = 33%.

7 Conclusion

The objective of this work was to define the analytical CRM and understand its issues by focusing on OLAP analysis tools and data mining. We have made throughout this work the Customer Relationship Management more like a theory on customer loyalty. Indeed, as the fundamental role that plays computing, CRM is extremely complex and difficult to define. Despite the difficulties, however, we are able to describe basic principles of customer relationship management starting from the subject we chose. We have defined CRM and showed its importance generally.

We then presented the concepts essential to the understanding of the analytical part of CRM to finally reach an example concrete in order to compare theory with practice. Although we have achieved most of our objectives, we felt that by researching and reflecting on the CRM, it seemed to us increasingly blur, vast and difficult to grasp. Indeed, the multiple ways of approaching CRM have prevented us in some ways to have a global vision. Outraged its computer dimension, its strategic interest, or in its place the evolution of marketing in general, it is the corporate culture we choose as the central element of customer relationship management. Whether in level of communication internally or outward, CRM is primarily a philosophy that should ideally help to redefine the set of processes and the company's values.

Without repeating the content of our work, we finally like to mention two important issues that appear in the literature that have haunted us throughout our research: The analytical CRM, defined as fundamental in theory, is it really in practice? Is the CRM not a mere fad promoted by companies wishing to sell their IT solutions?

References

1. Hand D, Heikki M, Padhraic S (2001) Principles of data mining. MIT Press, Cambridge, Massachusetts
2. Larose DT (2005) Discovering knowledge in data. Wiley, Hoboken, New Jersey
3. Peelen E, Montfort KV, Klerkx A, Beltman, R (2006) A study into the foundations of CRM success. NRG Working Paper, SSRN
4. Grenier C, Monk C (2003) Build the information system of the company. Editions Foucher, Paris
5. Lasserre L, Legrand B (2012) CRM, customer expectations. Global Village Press, Paris
6. Morgat P (2000) Build customer loyalty, strategies, tools and RCMP edition organisation, Paris
7. Lehu JM (1999) Customer loyalty. Organizational Edition, Paris
8. Tiwana A (2001) Knowledge management, CRM applications and e-business. CampusPress, Paris
9. Larose DT (2005) Des data to knowledge, an introduction to data mining. Vuibert Computing Editions, Paris
10. Brown S (2006) Customer Relationship Management. Pearson Education, France
11. Mayor P (2003) CRM study of the needs of the segment of business customers. The IIMT University Press, Fribourg/Switzerland

An Ensemble Framework for Flow-Based Application Layer DDoS Attack Detection Using Data Mining Techniques

K. Munivara Prasad, V. Samba Siva, J. Nagamuneiah and Siddaiah Nelaballi

Abstract The large number of requests flow exceeds the capacity of the target server drives to denial in the service to the legitimate users. Due to the server's oversized prospective, the flooding requests increase the server capacity generated by the malicious attackers from distributed environment defining the distributed denial of service attack. From the contemporary literature it is evident that applying the knowledge gained from the findings of previous request distributions is a suitable strategy to block the DDoS attacks. This strategy's key limitation is frisking to detect the new patterns of request flooding excavated by the attacker at the server from the previous knowledge on earlier attack distributions patterns. Therefore, this paper explains a novel trained ensemble classifier with new features which reflects in the traffic flow properties, so that, the traffic flow shows distribution diversity from each other which is considered and attached to individual classifiers. Ensemble classifier and AdaBoost are used to detect the flow by discovering the distribution resemblance involved in the multiple classifiers in the ensemble classification model. The experiment worked out on the voluminous traffic flow with visible distribution variety.

Keywords DDoS attack · Ensembles approach · K-S test and application layer DDoS attacks

1 Introduction

Nowadays the Internet plays a major role in human life in various activities and allows to do all the day to day activities online which attracts the attackers to compromise the network and user services. The Denial of Service (DoS) attack [1] is a malicious attempt by a single person to compromise the network resources which are not being accessed by the authorized person. If it is done by a group of people it is called as Distributed DoS attack (DDoS). One of the main threats to the internet applications

K. M. Prasad (✉) · V. S. Siva · J. Nagamuneiah
Chadalawada Ramanamma Engineering College, Tirupati, India
e-mail: prasadkmv27@gmail.com

S. Nelaballi
S R E S Group of Institutions, Tirupati, India

S. Fong et al. (eds.), *ICT Analysis and Applications*, Lecture Notes in Networks and Systems 93, https://doi.org/10.1007/978-981-15-0630-7_2

is the Application layer DDoS [2] attacks where all the user applications and services are targeted.

From the literature it is evident that several approaches were developed to detect the DDoS attacks, but each had its own drawbacks and advantages, but these methods failed to maintain consistent results when the traffic is from diversified traffic. Ensemble-based classifiers are used in this paper to maintain consistent results even though the traffic is from the diversified network.

2 Related Work

Though the detection method and defense measure have been widely researched the complexity of the DDoS attack is higher and the size of the DDoS attack is much larger than before. Paper [3] introduced several public datasets used in the recent years. Different types of DDoS attack datasets were presented in the paper.

The similarities of all the datasets were the large number of attributes and information, which posed a great challenge to detect the attacks among massive information. For better performance to process the huge amount of information data mining method has been researched to detect the DDoS attack.

In paper [4], two kinds of data mining methods, MLP and Rand forest method were applied to detect the DDoS attack. Both the methods were proven to detect the DDoS attacks while the consuming time and computing cost were high after experiment verification because of the high amount of dataset and lots of attributes used in this experiment.

To detect the DDoS attack with a huge amount of data, methods on reducing the amount of data and advanced method to improve the accuracy need to be researched. Different ranking methods, Info gain, gain ratio, and chi-squared were implemented in paper [4] in order to get more important attributes. The time taken in build model was saved and the detecting rate was improved after the one third selection of the voted ranking while the one third ranking whether can contain the whole Information need to be considered. And further improvement also needed to be done.

In paper [5], three different data mining methods Bagging, Rand forest, and k-NN were applied. The final result was voted among the three heterogeneous methods. Though the accuracy was improved according to the paper; the TNR was not the best compared with others. Normally, voting among different methods always leads to the middle value rather than the best which may lead to the detecting rate not being stable.

ARM was applied to select the important features in paper [6], and two datasets were experimented in this paper. It showed that accuracy to detect the attack was improved but the accuracy to identify the normal events was deceased. It makes sense in identifying the attack to some extent but still needed to improve the whole ability to identify both the normal and attack events.

The large amount of data needs to be processed in DDoS attack detection, but little error rate even means many attacks were incorrectly detected. Though some of them have contributed in improving the detection rate of DDoS attack to some extent, few paper majors in both improving the detection rate and reducing the amount of data at the same time. This paper aimed at improving the accuracy of DDoS detection by using ensemble data mining technology. The main target of this work is to reduce the unrelated data and improve the accuracy in detecting DDoS attacks at the same time.

3 Proposed Work

The proposed method includes the attack detection at the flow level rather than the request level. The dataset consists of attack and normal which is considered as the input for the process and each corpus is processed separately. The collection of normal requests from the input corpus is grouped as sessions with fixed time. The input dataset is now converted as session dataset. The sessions are grouped as the clusters using k-means cluster algorithm based on the session begin times. The clusters are grouped as the absolute time interval (ati) and the absolute time interval is defined as session begin intervals, session completion intervals, page access begin intervals, page completion intervals, and bandwidth consumption. The process is applied separately to the attack corpus and normal corpus separately and input dataset is converted from the request level to flow level where flow is defined as absolute time interval.

The absolute time intervals (ati) of attack and normal is considered for training, the collection of absolute time intervals (ati) are given as the input for the ensemble of classifiers for defining classifier pool for attack and normal independently. In the testing phase, the input corpus is again converted into absolute time interval (ati) that are validated through ensembles of classifiers. The Adaboost ensemble classifier with different classification algorithms in each level is used to validate the testing corpus as attack or normal.

3.1 The Absolute Time Interval (Ati) Is Defined Using the Following Parameters

- *Collection of Session begin intervals (CSBI)*: This parameter describes the time gap between begin times of the continuous two sessions in the absolute time interval.
- *Collection of Session completion intervals (CSCI)*: This parameter describes the time gap between end times of the continuous two sessions in the absolute time interval.
- *Collection of Page access begin intervals (CPBI)*: This parameter describes the time gap between the begin time of the page access requests in sequence in the absolute time intervals.

- *Collection of Page access completion interval (CPCI)*: This parameter describes the time gap between the completion time of the page access requests in sequence in the absolute time intervals.
- *Bandwidth consumption of Session (SBC)*: This parameter describes the bandwidth consumed by all the requests in each session of absolute time interval.

3.2 Feature Extraction from Dataset

Collection of Session begin intervals (CSBI) are defined as a set *sbi* (*Ci*) of size $|C_i| - 1$ related to specific cluster C_i contains the collection of absolute time interval (ati) $|C_i|$. The set *sbi(Ci)* of CSBI of the cluster C_i shown as:

$$\overset{|C_i|-1}{\underset{j=1}{\forall}} \left\{ sbi(C_i) \leftarrow \left(bt(s_{j+1}) - bt(s_j) \right) \right\}$$

Collection of Session completion intervals (CSCI) are defined as a set *sci* (*Ci*) of size $|C_i| - 1$ related to a specific cluster C_i includes the sessions of count $|C_i|$. The set *sci(Ci)* of CSCI of the cluster C_i is defined as:

$$\overset{|C_i|-1}{\underset{j=1}{\forall}} \left\{ sci(C_i) \leftarrow \left(abs\left(et(s_{j+1}) - et(s_j) \right) \right) \right\}$$

Collection of Page access begin intervals (CPBI) is stated as set $pbi(C_i)$ of size $|P(C_i)| - 1$ related to the collection of pages $P(C_i)$ which includes the pages in increasing order of session begin times. Let $|P(C_i)|$ represent the amount of pages available in every cluster C_i. The set $pbi(C_i)$ of CPBI of cluster C_i is defined as:

$$\overset{|P(C_i)|-1}{\underset{j=1}{\forall}} \left\{ pbi(C_i) \leftarrow \left(bt(p_{j+1}) - bt(p_j) \right) \right\}$$

Collection of Page access completion intervals (CPCI) is represented as a set $pci(C_i)$ of size $|P(C_i)| - 1$ related to the collection of pages $P(C_i)$ which includes the pages in increasing order of session end times. Let $|P(C_i)|$ represent the amount of pages available in every cluster C_i. The set $pci(C_i)$ of CPCI of cluster C_i is defined as:

$$\overset{|P(C_i)|-1}{\underset{j=1}{\forall}} \left\{ pci(C_i) \leftarrow \left(abs\left(et(p_{j+1}) - bt(p_j) \right) \right) \right\}$$

Bandwidth consumption of Session (SBC) related to a cluster C_i are defined as a set $bwc(C_i)$ of size $|C_i|$, Here $|C_i|$ defines the collection of sessions defined in cluster C_i. The amount of the bandwidth consumed by an individual request defined in cluster refers to the bandwidth in use. The set $bwc(C_i)$ of bandwidth occupied by every session in cluster C_i is shown as follows:

Step 1. $\overset{|C_i|}{\underset{j=1}{\forall}} \{s_j \exists s_j \in C_i\}$ Begin

Step 2. $bwc\ (C_i) \leftarrow \sum_{k=1}^{|s_j|} \{bw(p_k) \exists p_k \in s_j\}$ // Total bandwidth consumed $bwc\ (p_k)$ by each page p_k in session is moved to the set $bwc\ (C_i)$

Step 3. End //of Step 1

3.3 *Source Cluster Selection for Drift Detection*

Later the process of absolute time intervals (ati) grouping by their distribution similarity, the proposed model selects the cluster of absolute time intervals for training. Further, the selected clusters are used for training. The formulation of the cluster selection is as follows:

Let a set $CG = \{cg_1, cg_2 \ldots, cg_{|CG|}\}$ be the clusters defined and each cluster $\{cg_i \exists cg_i \in CG \wedge 1 \leq i \leq |CG|\}$ represents a set of absolute time intervals(ati) from each cluster-group which is depicted as follows:

Step 1. $\overset{|CG|}{\underset{i=1}{\forall}} \{cg_i \exists cg_i \in CG \wedge 1 \leq i \leq |CG|\}$ Begin // for each cluster-group cg_i depicted in set *CG*

Step 2. $csm = 0$

Step 3. $\overset{|cg_i|}{\underset{j=1}{\forall}} \{c_j \exists c_j \in cg_i \wedge 1 \leq j \leq |cg_i|\}$ Begin

Step 4. *if* $(csm < |c_j|)$ // if the number of ati $|c_j|$ in cluster c_j is greater than the value of *csm*

Step 5. $sc(cg_i) = c_j$ // selecting the cluster as source cluster of the cluster-group c_j, since it is having the maximum number of sessions than any of the clusters $c_1\ to\ c_{j-1}$ in cluster-group cg_i selected.

Step 6. $csm = |c_j|$ // considering the number of sessions $|c_j|$ in present cluster c_j as max sessions *csm* of the source cluster $sc(cg_i)$ of the cluster-group cg_i selected.

Step 7. End //of Step 4

Step 8. End //of Step 3

Step 9. End //of Step 1

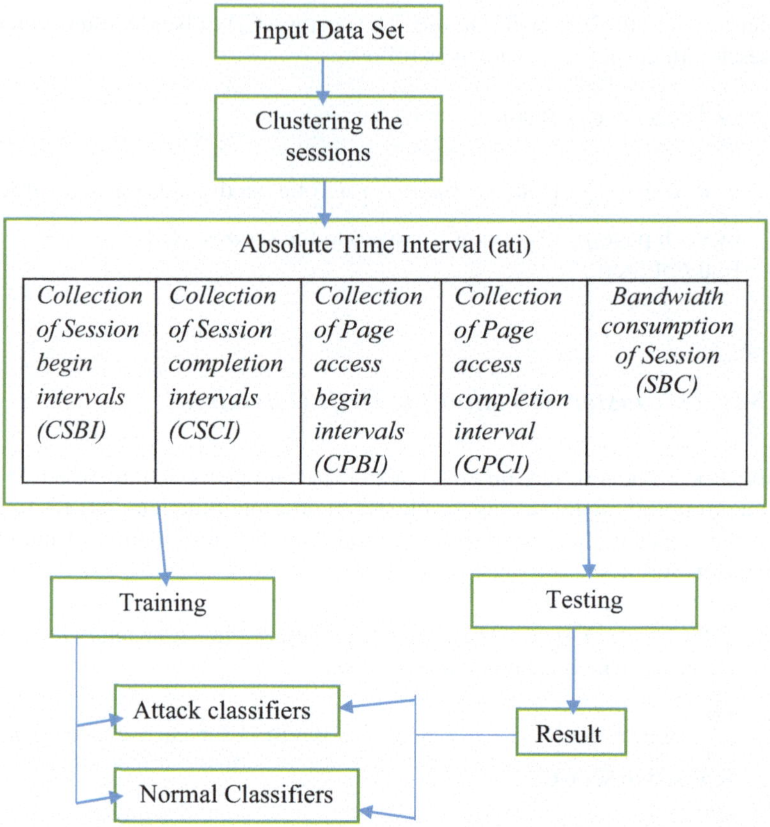

Fig. 1 Process diagram of the proposed model

3.4 Detection Model

As it is mentioned above, the large amount of data delivered during attack is the typical feature of the DDoS attack. The higher base number it is, the little—inaccuracy may lead to large error count, which is still the security problem that needs to be solved. In the detection part, the ensemble training is applied, which votes among the Bagging model, the boosting model, and the meta classifier. Bagging and Adaboost are both the improved model for the weak classifier with the sample training. In order to obtain the stable result of the DDoS detection, two ensemble models and the base classifier are combined.

3.4.1 Bagging

Bagging is a resample mode to training the weak classify which is present as Fig. 1. Sampling the k instances of the data with replacement is the key feature in Bagging model. After nth resample, n sub-datasets are selected as the figure shows. The n independent sub-datasets are trained to predict each own result. The voting procedure is conducted finally to obtain better results [7].

In this way, a new sample that represents the distribution of the original sample is rebuilt with few sample data. That means only few training datasets can also get high result of classification [8]. However, some training samples may be repeated or absent several times in a training session. Because the weight of each classifier is equal, the same mistake may be made in different classifiers. The accuracy of the result normally will increase with the number of resamples. But it may decrease when the resample times to some extent leads to an overfitting result.

3.5 Meta Classifier and Choose Reason

In order to verify the availability of the proposed ensemble framework, more than one classifier are designed to apply in the experiment. Two base classifiers Naive Bayes and J48 are implemented in the proposed design, respectively.

As for Naive Bayes, detecting the abnormal events depends on the probability of the different events [9]. It is used to predict the event as normal or attack by calculating the posterior probability that the event is an attack under the known features' probabilities [10]. That makes the detection rate of DDoS unsteady because the DDoS attack is more complex, which not only combines the different types but also contains different level attacks such as high speed and low speed. J48 predicts the event as normal or attack by calculating the entropy of every feature and divides the different groups by comparing the Information gain of each feature one by one until it identifies the event finally [11]. While J48 usually got the higher detection [12], however, the result may be limited in the size of the dataset and the number of attributes because of the large computing cost in training procedure. For this reason the J48 detection is also unstable with the volume of the DDoS attack dataset having increased. For the large volume of data that needs to be processed in DDoS attack and the various types of the DDoS attack launched nowadays, the two above base classifiers are applied respectively in the proposed voting model which combined the Bagging, Adaboost model, and the base classifier itself. Using the same base classifier separately, mainly because voting among the different detection methods usually lead to the middle result while voting for the different improved versions of one method can get a complementary result by the different sample methods and weighted result.

4 Experiment Configuration

Since the two parts feature selection and detection model were applied in this paper, the results also contained two parts the result of attributes' selection and the final result of detection. The DDoS attack dataset used for the experiment was NSL-KDD dataset which includes 41 attributes as given in Table 1. AS DDoS attack was featured for the high-volume data, detecting every instance with 41 attributes was time-consuming and the computing cost was high. The experiment was conducted on WEKA [13] which is an open platform for data mining. The parameter for ranking and searching is default.

Table 1 The original attributes of the NSL-KDD dataset

Attributes	Attributes
1. duration	22. is_guest_login
2. protocol_type	23. count
3. service	24. srv_count
4. flag	25. serror_rate
5. src_bytes	26. srv_serror_rate
6. dst_bytes	27. rerror_rate
7. land	28. srv_rerror_rate
8. wrong_fragment	29. same_srv_rate
9. urgent	30. diff_srv_rate
10. hot	31. srv_diff_host_rate
11. num_failed_logins	32. dst_host_count
12. logged_in	33. dst_host_srv_count
13. num_compromised	34. dst_host_same_srv_rate
14. root_shell	35. dst_host_diff_srv_rate
15. su_attempted	36. st_host_same_src_port_rate
16. num_root	37. dst_host_srv_diff_host_rate
17. num_file_creations	38. dst_host_serror_rate
18. num_shellsl	39. dst_host_srv_serror_rate
19. num_access_files	40. dst_host_rerror_rate
20. num_outbound_cmds	41. dst_host_srv_rerror_rate
21. is_host_login	

4.1 The Detection Result of Different Data Mining Procedures

In order to present the accuracy of the detection the confusion matrix was calculated. Confusion matrix is one of the most important metrics to evaluate the effectiveness of the attack detection, especially for the multiple attacks such as the sophisticated DDoS attack which contains more than one type of DDoS attack. To evaluate whether the result of the detection is reliable the confusion matrix is the key trait to compare.

It can be seen that True Positive (TP) is the number that attacks were correctly detected, and TN was the number that normal events correctly detected.

While False Negative (FN) meant that the attack instance was regarded as inaccurate, and FP represented the normal event that was regarded as attack. The detailed performances of the two base classifiers and the final result of the proposed model were compared in Table 2. In Table 2, eight parameters of performance are shown. Naive Bayes represented using Naive Bayes as the meta classifier of the RSV model and the same with J48. The As FP Rate was a metric of error rate, the less it was, the high performance the result was. As for other 5 parameters, the higher they were, the high performance the result was. It can be seen from Table 2, each FP Rate of the RSV models were decreased whether using the Naive Bayes or J48 as the base classifier. And every parameter of the two RSV-meta detection was better than base classifiers both Naïve Bayes. It proved that the performance of the RSV detection framework was all-round improved rather than just some aspects.

5 Conclusion

This paper contributes to how the DDoS attack id detected at flow level rather than the request level. From the contemporary literature researchers proposed many techniques to detect and defend the DDoS attacks particularly Application layer DDoS attacks, but nobody has addressed the detection in flow level. The detection accuracy and time is minimized in flow level attack detection rather than request level or session level. In this paper flow is defined with five attributes session begin intervals, session completion intervals, page access begin intervals, page completion intervals, and bandwidth consumption. The Input corpus is converted in terms of absolute time intervals which is known as flow. The ensemble classifiers are used to define multiple classifiers based on the diversity of the traffic, which increases the attack detection accuracy and minimizes the false alarms. In this paper Adaboost is used with different classifiers and validated that the detection accuracy is improved over the traditional and normal request level detection approaches. The overall process is experimented with KDD 99 cup dataset.

Table 2 The performance between the two meta classifiers and the RSV models

Classifier method	TP rate	FP rate	Precision	Recall	F-Measure	MCC	ROC area	PRC area	Class
Ada Boost with J48	0.967	0.03	0.961	0.967	0.964	0.936	0.995	0.994	Attack
	0.97	0.033	0.974	0.97	0.972	0.936	0.995	0.997	Normal
	0.969	0.032	0.969	0.969	0.969	0.936	0.995	0.995	Weighted Avg.
Ada Boost with Naïve Bayes	0.954	0.02	0.973	0.954	0.964	0.936	0.99	0.986	Attack
	0.98	0.046	0.965	0.98	0.972	0.936	0.99	0.992	Normal
	0.969	0.035	0.969	0.969	0.969	0.936	0.99	0.99	Weighted Avg.
Ada Boost with Random forest	0.98	0.03	0.962	0.98	0.971	0.948	0.995	0.99	Attack
	0.97	0.02	0.985	0.97	0.977	0.948	0.997	0.998	Normal
	0.974	0.024	0.974	0.974	0.974	0.948	0.996	0.994	weighted Avg.

References

1. Bhuyan M H, Bhattacharyya D K, Kalita J K (2015) An empirical evaluation of Information metrics for low-rate and high-rate DDoS attack detection. Pattern Recognit Lett 51(C):1–7. https://www.arbornetworks.com/
2. Bhuyan MH, Bhattacharyya DK, Kalita JK (2015) Towards generating real-life datasets for network intrusion detection. Int J Netw Secur 17(6):683–701
3. Alkasassbeh M, Al-Naymat G, Hassanat A et al (2016) Dataset-detecting distributed denial of service attacks using data mining techniques. Int J Adv Comput Sci Appl 7(1):1–10
4. Osanaiye O, Cai H, Choo KKR et al (2016) Ensemble-based multi-filter feature selection method for DdoS detection in cloud computing. Eurasip J Wirel Commun Netw 2016(1):1–10
5. Jia B, Huang X, Liu R et al (2017) A DDoS attack detection method based on hybrid heterogeneous multiclassifier ensemble learning. J Electr Comput Eng 2017(2):1–9
6. Moustafa N, Slay J (2017) The significant features of the UNSW-NB15 and the KDD99 data sets for network intrusion detection systems. In: International workshop on building analysis datasets and gathering experience returns for security. IEEE
7. Breiman Leo (1996) Bagging predictors. Mach Learn 24(2):123–140
8. Tumer K, Ghosh J (2015) Error correlation and error reduction in ensemble classifiers 8(3–4):385–404
9. John GH, Langley P (1995) Estimating continuous distributions in bayesian classifiers. In: Eleventh conference on uncertainty in artificial intelligence, San Mateo, pp. 338–345
10. Fouladi RF, Kayatas CE, Anarim E (2016) Frequency based DDoS attack detection approach using naive Bayes classification. In: International conference on telecommunications and signal processing. IEEE
11. Quinlan R (1993). C4.5: programs for machine learning. Morgan Kaufmann Publishers, San Mateo, CA
12. Patel J, Panchal K (2015) Effective intrusion detection system using data mining technique
13. Hall M, Frank E, Holmes G et al (2009) The WEKA data mining software: an update. ACM SIGKDD Explor Newsl 11(1):10–18

EM Design of Low RCS Proximity Coupled Patch Array

Avinash Singh, Deepa K. Sasidharan and Hema Singh

Abstract The low profile nature of a microstrip patch array is advantageous in stealth applications. However, the feed network of a typical corporate-fed patch array contributes considerably toward antenna scattering. A proximity coupled feeding technique can be adopted to overcome this problem. Since the feed network comes beneath the substrate, the radar cross-section (RCS) of the whole array can be reduced to an extent, with an added advantage of providing wider bandwidth. However, the multiple resonances that will be generated from proximity coupled arrays may contribute to higher RCS at corresponding frequencies. Incorporating slots in the ground plane helps in changing the path of surface current, thereby reducing the RCS peaks due to resonant modes. Further, results show that cutting slots in the ground plane of a proximity coupled patch array has aided in gain and bandwidth enhancement. A proximity coupled patch array with high impedance surface (HIS) layer and reduced slotted ground plane is proposed towards wideband structural RCS reduction from 11 to 30 GHz.

Keywords Proximity couple · High impedance surface · Structural RCS · Slotted ground plane

1 Introduction

Low profile microstrip patch arrays are often preferred in stealth technologies due to its conformal property, ease of fabrication, and light weight. One of the disadvantages of patch arrays with microstrip feed is the narrow bandwidth, normally limited to

A. Singh · D. K. Sasidharan · H. Singh (✉)
Centre for Electromagnetics CSIR- National Aerospace Laboratories Kodihalli, Bengaluru 560017, India
e-mail: hemasingh@nal.res.in

A. Singh
e-mail: avi.23110@gmail.com

D. K. Sasidharan
e-mail: deepasasidharan10@gmail.com

S. Fong et al. (eds.), *ICT Analysis and Applications*, Lecture Notes in Networks and Systems 93, https://doi.org/10.1007/978-981-15-0630-7_3

2–5%. Thicker substrates with lower dielectric constant can be used to improve the bandwidth. However, as the thickness is made larger, the surface waves and unwanted radiation from the feed network may increase. Further, a corporate feed network in patch arrays consumes comparatively larger area, when compared to series feed network.

Numerous techniques have been proposed to reduce the structural RCS of microstrip patch arrays. The use of high impedance surfaces (HIS) has proved to reduce the structural RCS of microstrip patch arrays over wider frequency band [1]. Apart from providing higher impedance bandwidth, aperture-coupled feed helps also in reducing structural RCS of a microstrip patch array [2]. This is due to the fact that the feed network is placed below a dielectric substrate layer, therefore the contribution of feedlines toward RCS will be minimized. It has been reported that aperture-coupled patch antenna with hybrid frequency selective surfaces (FSS)-based ground plane can provide wideband RCS reduction [3]. A chessboard configuration of two types of split-ring resonators and square patches have been used as the hybrid FSS.

In the present work, proximity coupled feeding technique has been employed to feed microstrip patch array. This has multiple advantages including higher impedance bandwidth, lower backward radiation, and minimal contribution to RCS. Appropriate overlapping of feedline and patch is required for enhancing fundamental resonance. Authors in [4] have demonstrated four configurations of proximity coupled patch array exhibiting dual resonance. However, the multiple resonances that will be generated from proximity coupled arrays may contribute to higher RCS at corresponding frequencies. Incorporating slots in the ground plane helps in changing the path of surface current, thereby reducing the RCS peaks due to resonant modes. Wu et al. [5] have proposed a technique to reduce the RCS of patch antenna by cutting slots of varying width in the ground plane. Along with cutting slots in ground plane, loading with chip resistor along the patch has been found to provide low RCS patch antenna [2]. A compact resonator consisting of metallic layer with vias connecting to ground plane can be used along the feedline of patch antenna to suppress the harmonic radiations [6]. The harmonic radiations can also be reduced by varying the dimension and position of the feed [7].

In this paper, the RCS reduction feature of HIS and proximity coupled feed network is combined to design a patch array with reduced RCS. A combination of JC-square patch elements arranged in a chessboard configuration has been used as the HIS layer. The EM design of a 4-element proximity coupled microstrip patch array with HIS layer resonating in the X-band is described and compared with the conventional patch array without any HIS layer. The array performance in terms of their radiation and scattering characteristics has been presented. The motivation behind the work being carried out is to achieve a significant reduction in structural RCS. Moreover, efforts have been put while designing, so that there is no degradation in the antenna radiation characteristics.

2 Proximity Coupling in Patch Array

Proximity coupled feed is a noncontact feeding technique employed in a patch antenna/array. Figure 1 shows the schematic of a proximity coupled patch antenna. The patch elements are electromagnetically fed through a feed network designed between the two substrates. In proximity coupling, the patch elements and the feed network are separated by a dielectric medium; the coupling between feed and patch is capacitive in nature. The dielectric constant of both the substrates can be either same or different, based on the specifications. Generally, dielectric material with lower permittivity is preferred for upper substrate to enhance radiation into free space and better efficiency. The bottom substrate is designed with thin dielectric with higher permittivity to minimize spurious radiation from feed [8]. When compared to coplanar feed, noncontact feed does not have patch-to-feedline discontinuities, hence facilitate minimizing of resulting unwanted radiation. The fundamental resonant frequency depends on the extent of overlapping of feedline with the patch. In order to obtain an optimum return loss, the ratio of feed-to-patch overlap (L_{ovl}) to patch length has to be greater than 0.55 [6].

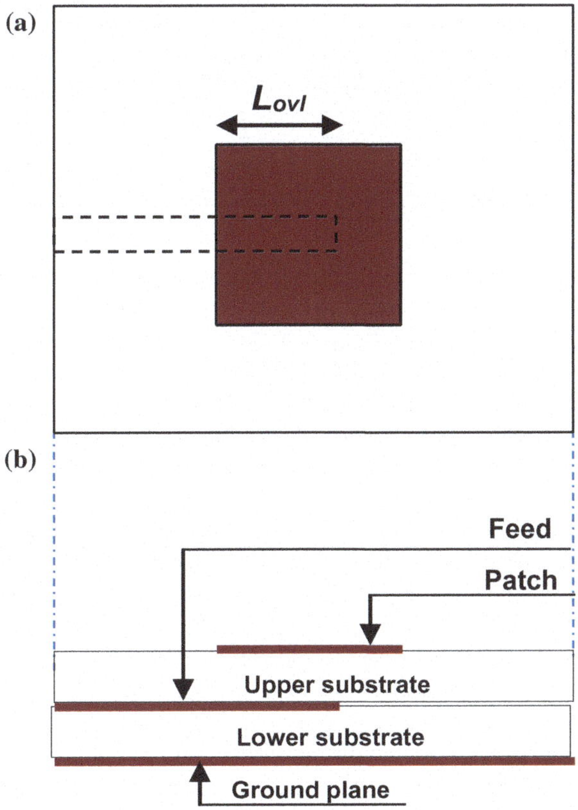

Fig. 1 A proximity coupled patch antenna—Schematic. **a** Top view **b** Side view

Proximity coupled feed provides maximum bandwidth of nearly 13%, when compared to coplanar and aperture-coupled feeding technique [8]. The radiation performance of a proximity coupled patch array can be improved by providing slots in the ground plane. The length of the slot is generally taken as $\lambda/2$, while the width to length ratio of the slot is maintained at 1:10. Slots with length comparable to $\lambda/2$ are called resonant slots, while smaller slots are called nonresonant. For maximum coupling, slots are to be placed beneath the center of the patch [6].

The deployment of slots in the ground plane also helps in reducing RCS of the patch array caused by the harmonic resonant modes. Incorporating slots causes change in the path of surface current, thereby suppressing the harmonic resonant modes [5]. Vertical slots cut off horizontally directed current, whereas horizontal slots cut off vertically directed current. In this way, the orientation of slots affects the array characteristics.

3 Radiation Performance of Proximity Coupled Patch Array

This section deals with the radiation characteristics (return loss, VSWR, and gain) of the proximity coupled 4-element patch array with conventional metallic ground plane, slotted ground plane and HIS layer with reduced slotted ground plane.

3.1 Conventional Metallic Ground Plane

The structure of a 4-element corporate-fed proximity coupled patch array with conventional ground plane, designed in X-band is shown in Fig. 2a. The top layer comprises of four metallic patch elements on a dielectric substrate (Fig. 2b). The patch dimensions are taken as 9.07 mm × 11.86 mm for 10 GHz resonant frequency. The substrate has a dimension of 49 mm × 90 mm. The patch elements are fed through electromagnetically coupled corporate-fed network designed as the middle layer (Fig. 2c). The overlapping distance of feedline from the edge of the patch (L_{ovl}) is 6 mm.

The feed network is sandwiched between two dielectric layers (Fig. 2c). Both the substrate layers are made of dielectric, RT Duroid ($\varepsilon_r = 2.2$, tan $\delta = 0.0009$) of the same thickness (1.58 mm). These layers are backed by a metallic ground plane. Copper ($\sigma = 5.8 \times 10^7$ S/m) with thickness 0.018 mm is chosen for all the metallic portions. Figure 3 shows the variation of return loss of the proximity coupled patch array with respect to the ratio L_{ovl}/L_p. Here, L_{ovl} represents the length of feed from edge of the patch to the open end of the feed, and L_p is the length of patch element. It can be observed that the proximity coupled patch array is resonant for L_{ovl}/L_p greater than 0.68.

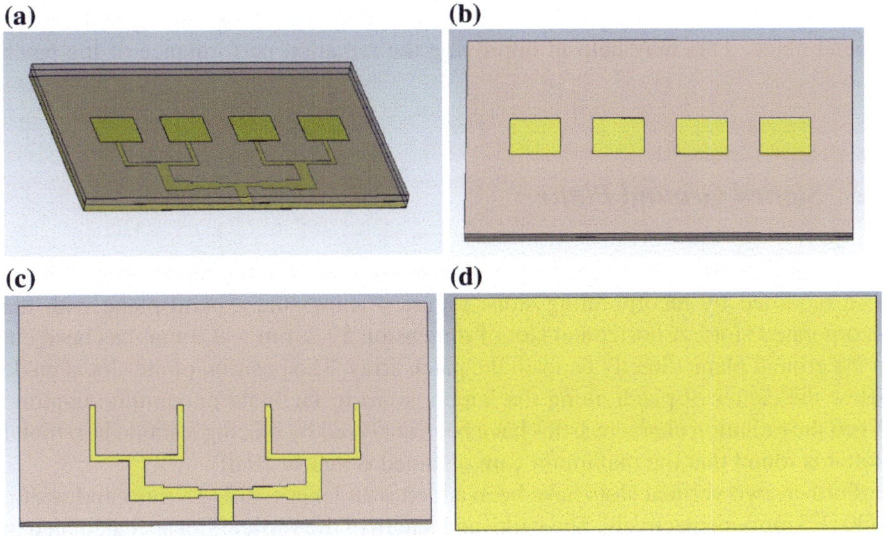

Fig. 2 A 4-element proximity coupled linear patch array with a conventional ground plane. **a** Perspective view **b** Top layer **c** Middle layer **d** Ground Plane

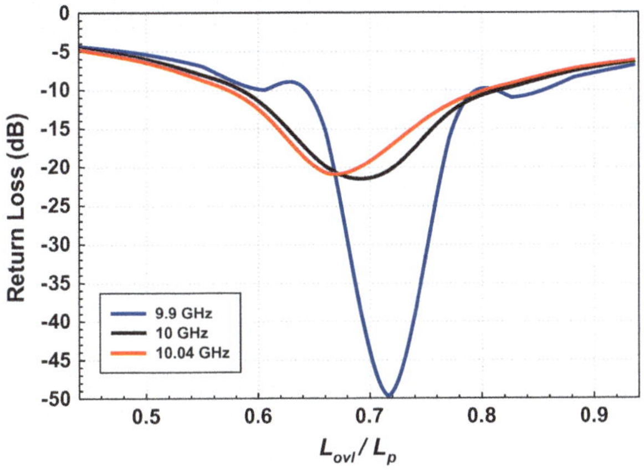

Fig. 3 Return loss of a proximity coupled 4-element linear patch array with respect to L_{ovl}/L_p overlap distance

While analyzing the radiation characteristics of the patch array, it can be observed that the array resonates at 10.04 GHz with a return loss of −20.76 dB. A bandwidth of 6% is obtained for the patch array. The VSWR obtained at the resonant frequency is 1.2. The gain of patch array along the specular direction is 7.58 dB, which is relatively small for a 4-element patch array. The side lobe level is −2.8 dB, and HPBW is 24.3°. This may be due to ineffective coupling of feed to the patch surface.

Coupling to the radiating patch elements can be improved by cutting slots in the ground plane. This may help in improving the radiation performance of the patch array.

3.2 Slotted Ground Plane

The ground plane of the proximity coupled patch array described in Sect. 3.1 has been modified by incorporating slots. Figure 4 shows the ground plane with the incorporated slots. A horizontal slot of dimension 59.2 mm × 1.5 mm has been cut in the ground plane directly beneath the patch array. The position of the slot is made below the center of patch along the length, so as to facilitate maximum coupling. When the radiation characteristics have been analyzed by placing a single horizontal slot, it is found that the maximum gain obtained is nearly 10 dB.

Further, two vertical slots have been added with length, $\lambda/2$ (15 mm) and width, $\lambda/20$ (1.5 mm), respectively. The width and length of the vertical slot are calculated in the ratio 1:10. The radiation characteristics of the proximity coupled patch array with slotted ground pane, illustrated in Fig. 4, are shown in Fig. 6a and b. It can be observed from Fig. 10a that the array resonates at 9.76 GHz, with a return loss of −33.33 dB. It can further be noted that the bandwidth has improved significantly to 12.94%, as compared to the proximity coupled patch array without slots (6% bandwidth). The VSWR is obtained as 1.04 (Fig. 10b). The gain of the array is 11.2 dB, which is 3.62 dB higher than the gain of proximity coupled patch array without slots. This clearly shows the effect of adding slots in increasing the coupling of input power to the patch surface. The array has a side lobe level of −6.7 dB, and HPBW of 20.8°.

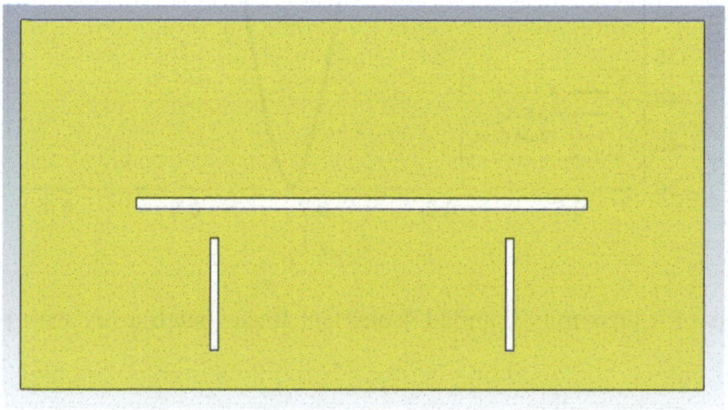

Fig. 4 Slotted ground plane of a 4-element proximity coupled linear patch array

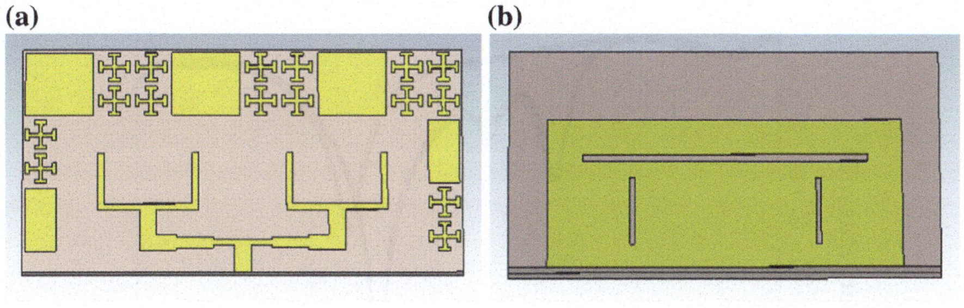

Fig. 5 A 4-element proximity coupled linear patch array with HIS layer with reduced ground plane **a** Middle layer **b** Slotted ground plane

3.3 HIS Layer with Reduced Slotted Ground Plane

The patch array design to be discussed in this section is a modification of the design already described in Sect. 3.2. The middle layer and the slotted ground plane of the proximity coupled patch array have been modified as shown in Fig. 5.

The HIS layer consists of alternately oriented 2 × 2 array of Jerusalem Cross (JC) elements and a metallic square patch. The radiation characteristics of the proximity coupled patch array with HIS layer-based slotted ground plane are analyzed. A comparison between these results with the radiation characteristics of proximity coupled patch array with conventional and slotted ground plane is illustrated in Fig. 5. It is evident that the radiation performance of patch array with slotted ground plane is not much affected by incorporating the HIS layer. It can be observed from Fig. 6a that the patch array with HIS layer and slotted ground plane resonates at the same frequency (i.e., 9.76 GHz) as that of conventional slotted ground plane. The return loss at the resonant frequency is obtained to be −34.16 GHz, and has a bandwidth of 11.3%. The radiating patch array has a minimum VSWR of 1.03 at the resonant frequency. The designed patch array shows a gain of 10.9 dB at $\theta = 0°$. The sidelobe level has reduced to −7.3 dB, and the HPBW is 19.5° (Fig. 6b).

4 Scattering Performance of Proximity Coupled Patch Array

The scattering characteristics of the patch array have been analyzed in terms of monostatic structural RCS. The simulated results of the structural RCS are illustrated in Fig. 7. The monostatic RCS pattern has been simulated at their respective resonant frequencies for $-90° \leq \theta \leq +90°$. It can be observed that the structural RCS value of the 4-element proximity coupled patch array with slotted ground plane and HIS layer

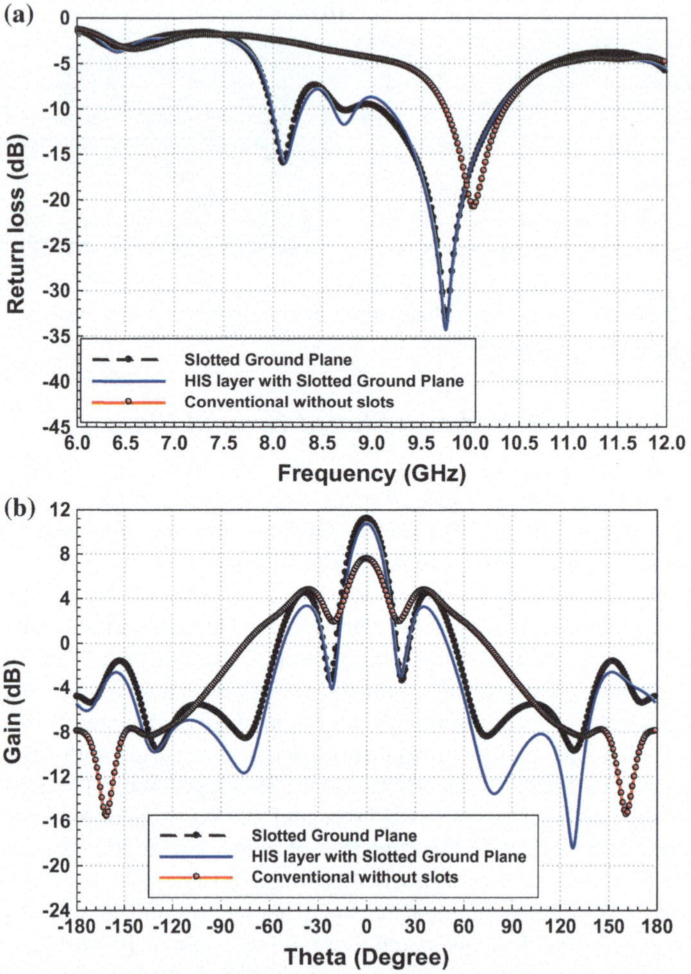

Fig. 6 Radiation characteristics and RCS of a 4-element proximity coupled patch array with different configurations. **a** Return loss **b** Gain

with reduced slotted ground plane is −7.46 dBsm and −7.85 dBsm, respectively. Further, the specular structural RCS of the proximity coupled patch array with HIS layer based reduced slotted ground plane is simulated with respect to frequency (Fig. 7b) and compared to that of conventional metallic ground plane (reference antenna). It can be observed that the specular structural RCS of the reference antenna increases with the frequency. It can be noted that a wideband reduction in RCS has been achieved from 11 to 30 GHz and beyond when compared to the reference antenna. A maximum reduction of 18.83 dBsm is observed at 24 GHz when compared to the reference antenna.

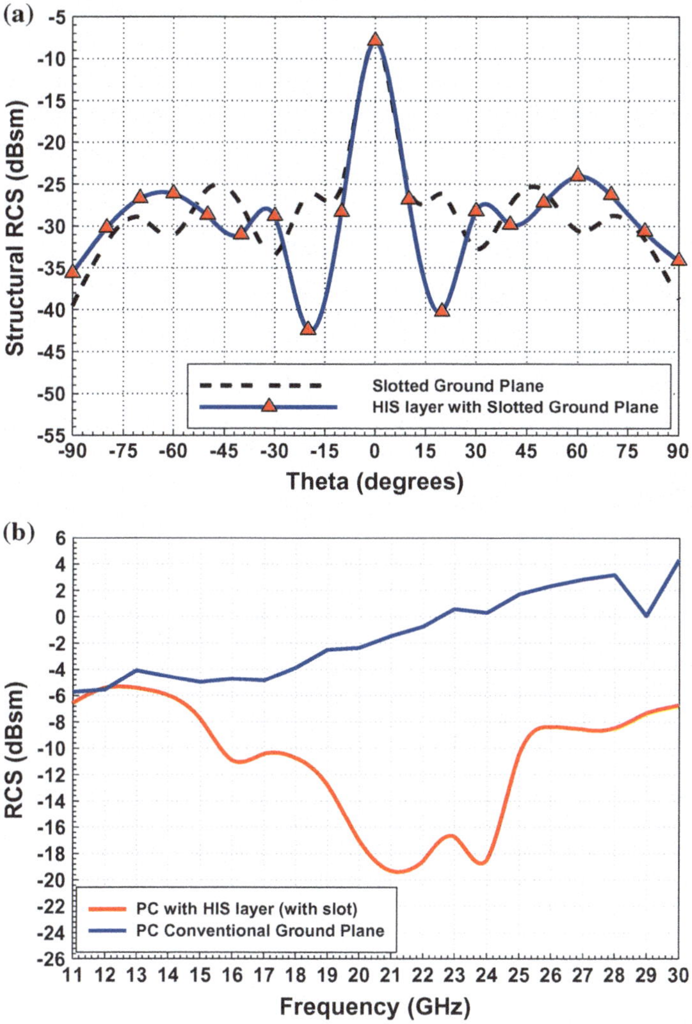

Fig. 7 Structural RCS of proximity coupled patch array. **a** Comparison of RCS pattern of slotted ground plane and HIS layer with reduced slotted ground plane **b** Comparison of specular structural RCS

5 Conclusion

The low profile nature of a microstrip patch array is advantageous in stealth applications. However, the feed network of a typical corporate-fed microstrip patch array will have more contribution towards antenna scattering. A proximity coupled feeding technique is adopted to overcome this problem. Since the feed network comes beneath the substrate, the RCS of the whole array can be reduced to an extent, with an added advantage of providing higher gain and bandwidth by cutting slots in the ground plane of a proximity coupled patch array. The patch array with slotted ground plane has shown a gain increment of 3.62 dB when compared to the conventional metallic ground plane. Moreover, the percentage bandwidth has doubled (6–12.94%) when slots were added to the ground plane of conventional proximity coupled patch array. The RCS reduction has been achieved by incorporating a HIS layer in the patch array. Specular RCS has been reduced significantly from 12 to 30 GHz and beyond, when compared to the other patch array configurations.

References

1. Singh A, Sasidharan DK, Singh H (2019) Estimation and control of total array RCS of microstrip patch array with hybrid HIS-based ground plane. In: URSI Asia-Pacific radio science conference (AP-RASC 2018), New Delhi
2. Zheng JH, Liu Y, Gong SX (2008) Aperture coupled microstrip antenna with low RCS. Prog Electromagn Res Lett 3:61–68
3. Cong L, Cao X, Song T (2017) Ultra-wideband RCS reduction and gain enhancement of aperture-coupled antenna based on hybrid-FSS. Radio Eng 26(4):1041–1047
4. Majidi N, Sobhani MR, Kilic B, Imeci M, Güngör OS, Imeci T (2018) Design and comparison of 4 types of dual resonance proximity coupled microstrip patch antennas. Appl Comput Electromagn Soc J 33(10):1135–1139
5. Wu T, Li Y, Gong SX, Liu Y (2008) A novel low RCS microstrip antenna using aperture coupled microstrip dipoles. J Electromagn Waves Appl 22:953–963
6. Sanchez LI, Roy JLV, Iglesias ER (2009) Proximity coupled microstrip patch antenna with reduced harmonic radiation. IEEE Trans Antennas Propag 57(1):27–32
7. Fistum D (2017) Efficient proximity coupled feed rectangular microstrip patch antenna with reduced harmonic radiation. Indones J Electr Eng Comput Sci 7(2):500–506
8. Balanis CA (2005) Antenna theory, analysis and design. Wiley, New Jersey

A Double-Weighted Parametric Model for Academic Software Project Effort Estimation

Jatinderkumar R. Saini and Vikas S. Chomal

Abstract The effort, cost, and time play a vital role in the success or failure of the software. The ratio of software project failure nowadays is growing like a storm in the world. One of the reasons behind this failure proportion is an imprecise and inappropriate estimate of required effort, cost, and budget for particular software project development. The motivation behind our research is to estimate the effort for software development accurately. Accurately estimate effort for software development is one of the most challenging tasks because from a very early stage it requires in-depth study as well as detailed statistics. Participation of members of the team is important during the development of the project along with various activities like domain area, sector, nature, software process methodology, and other resources. All these proofs do not appear in the initial phase of development, but on the other side, they discovered as software growth progresses. Many models proposed for estimating effort related to software development, but only a few are adaptable for effort estimation task in the academic software project. From the study of nearly 600 academic software projects and inputs from more than 30 Software Engineering experts, this research paper proposes an innovative model based on 8 parameters and 24 sub-parameters. Each of the sub-parameters is weighted twice, and the final effort estimation obtained by summation of the individual product of the weight of the parameter and weight of sub-parameter. We have coined the term "ASPEE units" (Academic Software Project Effort Estimation) for the estimated effort.

Keywords Educational perspective · Software engineering · Software estimation models · Estimate effort · Software project

---

J. R. Saini
Symbiosis Institute of Computer Studies and Research, Pune, Maharashtra, India
e-mail: saini.expert@gmail.com

V. S. Chomal (✉)
TMES Institute of Computer Studies, Mandvi, Surat, Gujarat, India
e-mail: vikschomal80@gmail.com

S. Fong et al. (eds.), *ICT Analysis and Applications*, Lecture Notes in Networks and Systems 93, https://doi.org/10.1007/978-981-15-0630-7_4

1 Introduction

One of the oldest definition of Software Engineering (SE) states that "it is a creation and utilization of comprehensive engineering principles to acquire software which is consistent, cost-effective as well as function competently on machines." Boehm states that "software engineering makes use of science and mathematics through which the competencies of computer equipment are provided to the user using computer programs, procedures and associated documentation" [1]. Haapio defines software engineering engagement of real people in real environments [2]. The software is shaped, conserved, and advanced by people. Apart from technical activities, SE also incorporates managerial activities such as defining roles and responsibilities, organizing the workflow, scheduling and tracking the progress of work down as well as synchronization of individuals working to fulfill a common objective [3]. The software managerial aspects focus on planning and estimation of software development time, resources required as well as cost and effort required. Software development transits through different phases as mentioned in the Software Development Life Cycle (SDLC) such exploration of requirements, design, code, test, documentation, and maintenance rigidly followed. Apart from these conventional stages, one of the most important as well as challenging tasks is to estimate the effort required in these various phases of SDLC accurately. If effort in software development is not measured correctly, calculated and followed, then it may result in quality failure, or even it may result in a complete failure of the software.

According to Haapio [2], the term effort in software development can define as the collective amount of time that is consumed by the project, and each project required a specific number of persons per hours, days, months, or years which is directly associated with the size of the project. Hence, the effort is the product of people and time which can be expressed as effort = people * time, whereas, estimation can be defined as a probabilistic valuation with a rational and precise value of the center of a range. Estimation can also be considered as a prediction. Therefore, all existing estimation model can be viewed as a prediction model also. For project scheduling, tracking, and controlling appropriate estimation is to be done and to make a real and effective estimation for any project is a challenging task. Estimation with software project development includes—(a) Estimation of size, (b) Estimation of Effort, (c) Estimation of Cost, and (d) Estimation of Schedule.

In effort estimation many existing models and techniques are available. Some of them are—(a) Empirical Parametric Estimation Model such as COCOMO, Putnam's Software Life Cycle Model (SLIM), (b) Empiric Non-parametric Estimation Model such as Optimized Set Reduction Technique (OSR), Decision-Making Trunk, (c) Expert Estimates such as Delphi technique, (d) Analogue Estimation Models such as ESTOR and ANGEL, (e) Downward Estimates, and (f) Upward Estimates. Software project development is a compulsory requisite for obtaining a master's degree in each computer science stream. The ultimate standard behind this is to make the students identify various factors that have a consequence on software development. During their software project development, students are required to focus on the following aspects of the software:

1. Requirement Analysis
2. System Design
3. Database Design
4. Structural and Object-Oriented Modeling Techniques
5. Coding
6. Testing

These six phases are extremely imperative as well as should be firmly and appropriately followed by students while developing their software within a stipulated time-bound. One of the collective opinions of academicians is that apart from regular monitoring and guidelines provided during the project work, the majority of students' projects remain incomplete. There may be many reasons behind this cause such as students fail to identify the nature and scope of software to be developed, requirements are not properly gathered, prioritized, and understood, having ambiguity in the functionality of the software and so on. It also perceived that students do not give crucial significance to recognize the effort required in every phase of software development. Looking at all these difficulties, an effort estimation model proposed which can utilize for estimating the effort in student's software project in the computer science curriculum. The rest of the paper is organized as follows. In Sect. 2, the related literature review presented, supported by systematic research methodology conducted during our study in Sect. 3. Section 4 represents experiments and results followed by the conclusion in Sect. 5.

2 Literature Review

Hathaichanok and Nakornthip proposed a framework which was used to estimate software cost and effort [1]. The proposed model based on the relational matrix utilizing the principle of multiple regression analysis and analogy method. J. Hill et al. studied the accuracy and effectiveness of experts' subjective estimates for the tasks that are carried out in software project development [4]. In their experiment, they concluded that the majority of functions were found to be overestimated. Further, Jovan et al. gave detail clarification as well classification of various effort estimation models and techniques used [5]. In their report, they accomplished that even though many prevailing models and methods then also effort estimation remains unpredictable. Jyoti and Vikas systematically represent a brief analysis of software effort and cost estimation [6]. They also conducted an experiment showing the performance of various estimation models and techniques. Kassem provided with multiple guidelines and checkpoints considered for effort and cost allocation and estimation for particular software development activities [7]. The existing requirements based estimation techniques used by the authors for conducting this study. Kjetil and Magne have proved that most of the software projects face either effort or schedule overruns [8]. During the study, they summarize that there is inadequate evidence of why this effort and schedule overruns occur. Lava in their research study chiefly considered effort estimation needed in testing the software been developed [3]. A survey and interview procedure was conducted, and the authors reveal that the majority of companies prepare separate estimation for testing phase and this estimation strictly associated with overall development effort. Jorgensen worked out with expert estimation technique and provided various guidelines based on expert estimation [9]. Mohammed and Rizwan presented an in-depth comparative study on software effort estimation models [10]. They also proposed a model that can be useful for information technology organizations for estimating efforts, budget, and in the decision-making process. Mudasir presented a review of various effort estimation and also suggested various advantages of models [11]. Muhammad et al. developed and provided with multiple checklists considering agile methodology and suggested various improvements made in effort estimation for agile methodology [12]. Paweł et al. proposed a course that was considered for understanding effort estimation in student's software project development [13]. Suri and Pallavi encapsulate various cost and effort estimation models and techniques [14]. They also implement simulators in their study and observed that no single technique best suited to given situations.

Poonam et al. focus on how various parametric and non-parametric estimation models used and what are their effects [15]. Putu and Sholiq considered Use Case Points popularly known as UCP in their exploration [16]. They made UCP as a base and premeditated amount of effort estimation required in a software development project. Reetuparna and Taniya described the various phase effort distribution patterns and causes of their variation. In their experiment, they found that the pattern shows consistency for phases like testing, as well as consistency in the effects of software size and team [17]. But a large amount of variation was observed in phases like design and requirement gathering. Basha and Dhavachelvan presented a detailed study of various software effort estimations [18]. Sangeetha and Pankaj investigated and presented a comparative study about different estimation techniques and also a detailed illustration of the model which makes use of Line of Code (LOC) and Function Point (FP) was presented [19]. Sarah and Maqbool proposed a model that can utilize to estimate effort for every task of SDLC [20]. Instead of relying on cost and line of code, they considered individual task as the primary factor while implementing their model. Thomas gave a domain-based effort distribution model that can help in estimating resource allocation for a particular domain [21]. Ursula and Shepperd conducted an experiment considering students' project with various checklists and group discussion [22]. The purpose was to improve the estimation. The result of the analysis shows that there was a significant improvement in estimation while applying these checklists and guidelines. Vishakha and Gaurav proposed a model for commercial software organizations [23]. Their model used to estimate and analyze various attributes of a software project like complexity, size, effort, development team as well as budget. This model can be used during the initial phase of software development to estimate these factors. Werner and Michel defined that effort is a term that consumed during software development considering the time duration of each phase of SDLC [24].

3 Methodology

The research initially commenced with an exploration of the literature review related to software effort estimation. The next phase includes exploration and investigation of various stages followed by students during software project development. For this investigation, we considered 589 software project documentations of final year students of the master's degree course. The duration time of these software project developments was six months. At the same time, all these projects were considered as full-time projects for obtaining a master's degree under computer science. The documentation was collected from the college library. After scrutinizing these massive software project documentations, we disclosed the following phases followed during their software project development. They are (a) Learning and exploring the technology which students utilized for software development, (b) Gathering and Analysis of software requirements, (c) Designing and Implementation of software, and (d) Testing. The various phases of the research task carried by us is presented in Fig. 1.

Further, the subsequent stage followed was listing and identification of parameters that are proportionally significant for estimating effort for software project development and explained in Table 1.

Further, these parameters are the deliberate perspective of estimating effort for software project development and not in the context of projects. After listing and clarification of these parameters, the next stage is to score these parameters. Scoring to these parameters was done based on the importance of each parameter in effort estimation for student software project development. It also followed that the total of these parameters should sum up to 100. For scoring these parameters, a survey was conducted with 31 academicians teaching in graduate and undergraduate colleges of the computer stream. All academicians fulfilled the criteria of holding at least the master's degree with first class and an experience of at least 10 years in the academic domain. Experiment and observations are presented in the next section.

4 Experiments and Results

The result of the survey is presented in Table 2 as well as graphically in Fig. 2a, b. The Column "CSA" in Table 2 specifies for the score provided by nth Computer Science Academicians with ranging from 1 to 31.

Based on the sum of parameters of the proposed effort estimation model values presented in Fig. 2a, it has been found that "Technology" scored the maximum sum (530) whereas "Experience" achieved the second highest score (475). The "Domain" was found to achieve the third highest score (450). Similarly, the minimum sum was

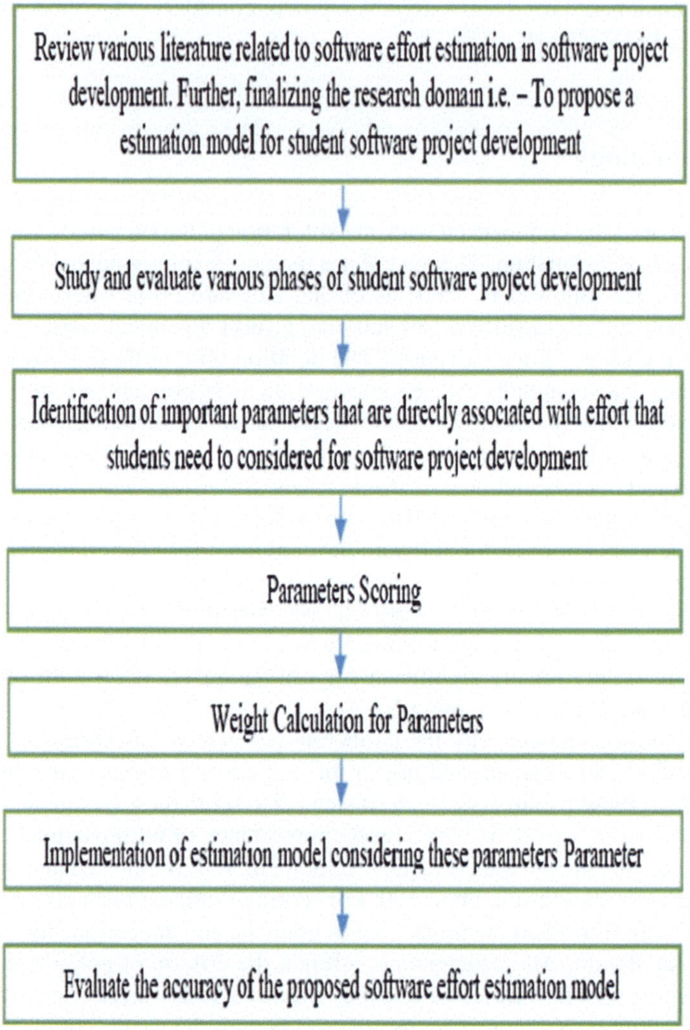

Fig. 1 Diagrammatic representation of research methodology

found to be assigned to "Team Size" (190) while the second lowest sum was found to be achieved by "Parallel Subject" (215).

Column Chart as depicted in Fig. 2b is used to represent clarification of the average values presented in Table 2, where on X-axis parameters of the proposed model are highlighted and on Y-axis average assigned are shown. From Fig. 2b it has been observed that maximum average was assigned to Technology (17.10%), while Experience with (15.32%) was found to have achieved the second highest whereas Domain was third highest with (14.52%). Similarly, the minimum average was found to be assigned to Team Size (6.29%) while the second lowest was found

Table 1 The parameters of the proposed model

Sr. no.	Parameters	Description
1	Project type	The type of project to be developed. For example, Desktop-based, Web-based portal
2	Technology	The technology used for developing software
3	Team size	The number of members involved in developing software
4	Domain	Application area for which software is developed. For example, Principle-based, Corporate- and Commercial-based, Decision-based, Network- or Security-based, Scientific-based, Social Networking-based
5	Project Duration	Time duration of software project development
6	Experience	The expertise of team members involved in software development. Having any prior knowledge of developing software
7	Functionality-based project complexity	The complexity of software to be developed
8	Parallel subject(s)	Any parallel subject to be learnt by the student along with software project development

to be achieved by Parallel Subject (6.93%). The next step is the calculation of weight for parameters identified and listed. The formula used for calculation of weight is:

$$\text{Weight} = (\text{Average } (\%)/100) \tag{1}$$

The value of variable Average (%) in formula (1) derived from Table 2. Further, the weight calculation for each parameter presented in tabular format in Table 3.

For better clarification regarding weight assignment to each parameter of the proposed model represented pictorial using a column chart in Fig. 3 where parameters plotted on the X-axis, and Weight was shown on Y-axis. From the Fig. 3, it is found that computer science academicians gave maximum weight to "Technology" (0.17), while "Experience" (0.15) was found to have achieved the second highest weight. Similarly, the minimum weight was found to be assigned to "Team Size" (0.062) while the second lowest weight was found to be achieved by "Parallel Subject" (0.069). Further, the next step was to provide weight to specific attributes of the main eight parameters listed in Table 1. The corresponding weightage presented in Table 4. At this stage, we are assigning weights to sub-parameters of parameters which have already assigned weights. Hence, we call our model a double-weighted

Table 2 Survey result

Sr. no.	Parameters	Respondent's response						Sum	Average (%)
		CSA1	CSA2	CSA3	CSA4	CSA5	…CSAn		
1	Project type	10	10	15	10	15	…..	415	13.39
2	Technology	20	20	15	10	5	…..	530	17.1
3	Team size	5	5	5	5	10	…..	195	6.29
4	Domain	20	20	10	15	10	…..	450	14.52
5	Project duration	5	5	10	15	20	…..	410	13.23
6	Experience	20	20	20	10	15	…..	475	15.32
7	Functionality-based project complexity	15	15	15	15	10	…..	410	13.23
8	Parallel subject(s)	5	5	10	20	15	…..	215	6.935
Total		100	100	100	100	100	…..	3100	100

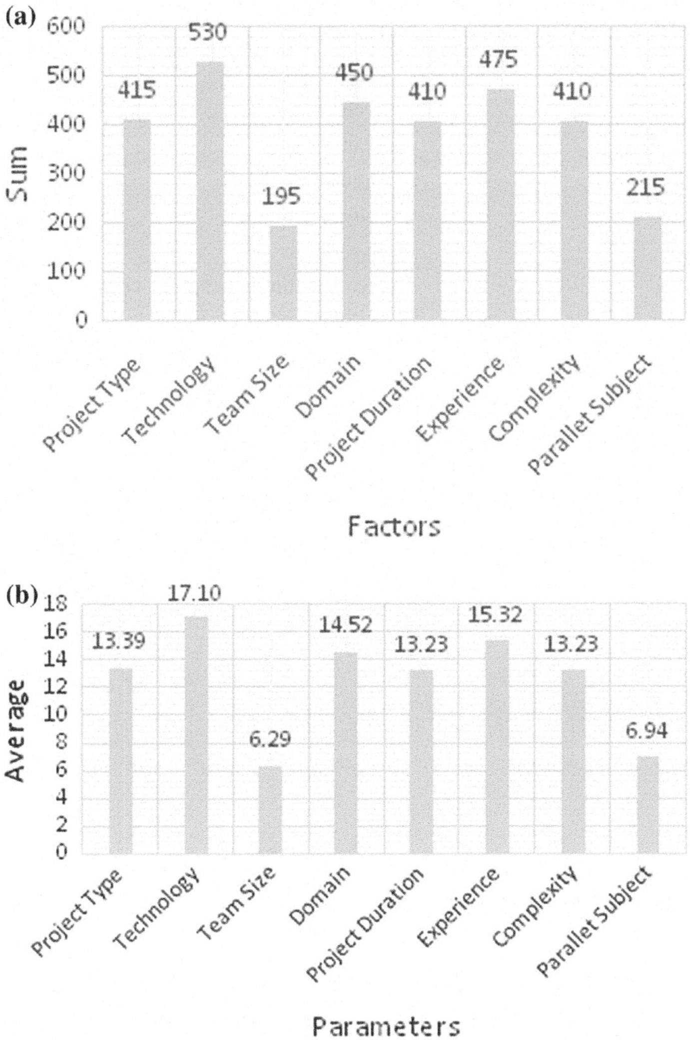

Fig. 2 **a** Sum of parameters of proposed effort estimation model. **b** Average of sum of parameters for proposed effort estimation model

parametric model. If the weight of a parameter is "n", then the product of the weight of its sub-parameter with "n" increases the weightage n-times.

In Table 4, the eight parameter(s) further characterized to sub-parameter (attribute/characteristics). For example, desktop, web, and portal are sub-parameter of "Project" type parameter. Our next step is to determine the impact of all these sub-parameters in estimating effort. In software development, the "attribute impact" consists of three group values with weight based on effect and complexity, e.g., (a). Low, Medium, and High with (1, 2, 3). (b). Yes, and No with (1, 2). (c). Lowest,

Table 3 Weight calculation

Sr. No.	Parameters	Weight
1	Project type	0.134
2	Technology	0.171
3	Team size	0.063
4	Domain	0.145
5	Project duration	0.132
6	Experience	0.153
7	Functionality-based project complexity	0.132
8	Parallel subject(s)	0.069

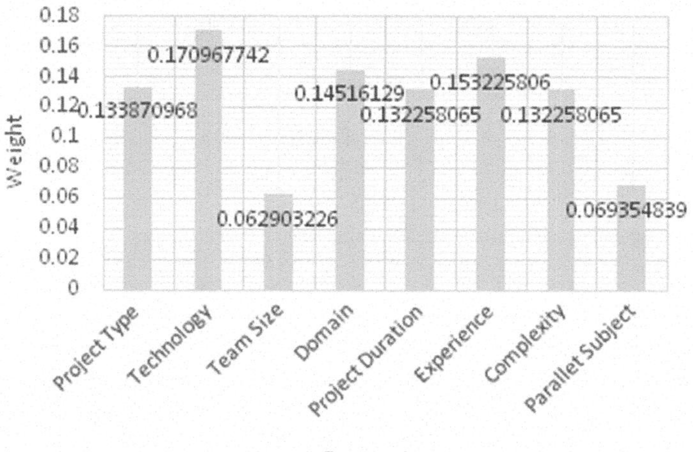

Fig. 3 Weight to factors of proposed effort estimation model

Lower, Low, High, Higher, and Highest with (1, 2, 3, 4, 5, 6). As each parameter and corresponding sub-parameters are unique and different from each other, it is noteworthy that this is repeated for each of the eight parameters under study. Now, the next step is to calculate the score for each parameter for estimating effort estimation. The formula mentioned below, and a unit considered in the formula termed as Academic Software Project Effort Estimation abbreviated as ASPEE (units), which is our coined term. For calculation of ASPPE units, the weight assigned to each parameter and attributes of parameters which represented in Tables 3 and 4, respectively are considered and mentioned below.

$$\text{Parameter(s)\_Weight} * \text{Attribute(s)\_Weight} \tag{2}$$

For example, let us consider project type as a parameter having weightage value (0.13) and desktop as parameter attributes which is having weight (1) since its effort

Table 4 Weight calculation for attributes of parameter(s)

Sr. no.	Parameter(s)	Attributes/Characteristics	Attribute Impact	Weight (Based on effort and complexity)
1	Project type	Desktop	Low	1
		Web	Medium	2
		Portal	High	3
2	Technology	Aware	Yes	1
		Not aware	No	2
3	Team size	1 Member	High	2
		2 Members	Low	1
		More than 2 Members	High	2
4	Domain	Scientific-based	Highest	6
		Principle-based	Higher	5
		Network or Security based	High	4
		Decision-based	Low	3
		Corporate and Commercial	Lower	2
		Social networking	Lowest	1
5	Project duration	6 Months	Low	1
		4 Months	Medium	2
		2 Months	High	3
6	Experience	Prior development experience	Yes	1
		No experience	No	2
7	Functionality-based project complexity	Less functionality required	Low	1
		Moderate functionality required	Medium	2
		Too much functionality required	High	3
8	Parallel subject(s)	No	Low	1
		Yes	High	2

and complexity is considered as low, hence the calculation of effort for this parameter will be:

$$0.13(\text{Parameter\_Weight}) * 1(\text{Attribute\_Weight}) = 0.13 \text{ ASPEE units} \quad (3)$$

Similarly, the calculation of effort for all 8 major parameters namely Project Type, Technology, Team Size, Domain, Project Duration, Experience, Functionality-based

Project Complexity, and Parallel Subject(s) as stated in Table 4 needs to be calculated. Finally, a summation of effort estimation for all parameters is done to yield the final effort estimate for the entire project. For experimentation of our proposed model, software project documentations of 10 groups were considered for calculating effort estimation. The experiment and relevant results presented in Table 5.

In Table 5, a total of ASPEE units are presented for each project. Further, if there is a change in the weights of the specified parameters such as technology, team size, and functionality-based project complexity or any other parameters, then there will be some deviations in total ASPEE units. Hence these parameters can also be termed as quantifiable parameters and comparison can be defined as a quantifiable comparison. The primary goal of formulating ASPEE units is to calculate and present the estimated effort for student's software project development.

5 Conclusion and Future Work

Today's digital world experiences swift variation in technologies and due to this software development in this period is at a challenging stage, and estimation of effort to be dedicated in the implementation of software remains complex as well as a challenging task. Further, the processes executed during software development and effort estimation are strongly cohesive. From the review, literature examined and observed that effort estimation in software development always remains a promising domain among the research community. During a period of the span, many diverse models and effort estimation methods were introduced and implemented. All these undoubtedly designate the awareness among the scholars of the need to improve effort estimation in software engineering. Practically, all IT companies globally consider effort estimation while developing a software project. But it is commonly not considered or unnoticed in developing academic software projects by students of computer science courses. Academic software projects form a mandated part of the curriculum for completion of a course. Often, as the project development progresses, various obstacles such as lack of technology-knowledge, the complexity of functionalities, pressure due to parallel subjects, etc. are experienced by the developer students. Considering the same, we believe that academic domain dealing with software project development oriented courses should focus, consider, and provide guidelines as well as approaches, models regarding effort estimation in software project development.

Our research also focuses on effort estimation with the significant objective of proposing a conceptual model for estimating the effort required for software project development in academic dominion. There are three primary stakeholders involved in the academic software project: Students, College/University Supervisor, and Company/Organization Supervisor. Given many projects and in lack of any quantified estimated value for expected effort, it often becomes difficult for either supervisor to assign a group of students to a specific project. This results, at times, in a difficult project assigned to a weak group of students. In any case, the project assignment is often done randomly and just based on intuitive guts rather than a quantified estimate

Table 5 Result of experimentation

Sr. no.	Project title	Parameters								Total (ASPEE Units)
		1	2	3	4	5	6	7	8	
		Estimate values								
1	Inventory management system	0.134	0.171	0.063	0.435	0.264	0.153	0.264	0.069	1.553
2	Online grocery store	0.268	0.171	0.126	0.29	0.264	0.153	0.396	0.069	1.737
3	Milk distribution	0.402	0.171	0.126	0.435	0.264	0.153	0.396	0.069	2.016
4	Lakshya blood bank	0.268	0.171	0.063	0.29	0.264	0.153	0.132	0.069	1.41
5	Network management system	0.268	0.171	0.126	0.58	0.264	0.153	0.396	0.069	2.027
6	Material management system	0.402	0.342	0.063	0.435	0.264	0.153	0.396	0.069	2.124
7	Accounting tracking system	0.268	0.171	0.126	0.725	0.264	0.153	0.396	0.069	2.172
8	Online food ordering system	0.268	0.171	0.063	0.29	0.264	0.153	0.132	0.069	1.41
9	A to Z GIS Map App	0.268	0.342	0.126	0.58	0.264	0.153	0.396	0.069	2.198
10	Online shoe store	0.268	0.171	0.126	0.29	0.264	0.306	0.132	0.069	1.626

for the effort. This research work has presented an effort estimation model for such academic software projects and claims that there will be a high success rate of project completions and completions on time and with fewer errors. In our research work, we identified 8 parameters and 24 sub-parameters. These individual sub-parameters are weighted twice, and the final effort estimation obtained by summation of the individual product of the weight of parameter and weight of sub-parameter. Further, we coined the term "ASPEE units" (Academic Software Project Effort Estimation) for the estimated effort. The fundamental goal of assigning double weightage to these parameters and to compute ASPEE units is to implement a model in the future that will be used to calculate effort estimation for academic software projects.

References

1. Suwanjang H, Prompoon N (2012) Framework for developing a software cost estimation model for software modification based on a relational matrix of project profile and software cost using an analogy estimation method. Int J Comput Commun Eng 1(2)
2. Haapio T (2011) Improving effort management in software development projects. Publications of the University of Eastern Finland Dissertations in Forestry and Natural Sciences
3. Kafle LP (2014) An empirical study on software test effort estimation. Int J Soft Comput Artif Intel 2(1). ISSN: 2321-404X
4. Hill J, Thomas LC, Allen DE (2000) Experts' estimates of task durations in software development projects. Int J Project Manag 18. www.elsevier.com/locate/ijproman
5. Zivadinovic J, Medic Z, Maksimovic D, Damnjanovic A, Vujcic S (2011) Methods of effort estimation in software engineering. In: International symposium engineering management and competitiveness 2011 (EMC2011) June 24–25, 2011, Zrenjanin, Serbia
6. Borade JG, Khalkar VR (2013) Software project effort and cost estimation techniques. Int J Adv Res Comput Sci Softw Eng 3(8). ISSN: 2277 128X
7. Saleh K (2011) Effort and cost allocation in medium to large software development projects. Int J Comput 5(1)
8. Molokken K, Jorgensen M (2003) A review of surveys on software effort estimation. In: ISESE'03 proceedings of the 2003 international symposium on empirical software engineering
9. Jorgensen M (2004) A review of studies on expert estimation of software development effort. J Syst Softw 70:37–60. www.elsevier.com/locate/jss
10. Aljohani M, Qureshi R (2017) Comparative study of software estimation techniques. Int J Softw Eng Appl (IJSEA) 8(6)
11. Kirmani MM (2017) Software effort estimation in early stages of software development: a review. Int J Adv Res Comput Sci 8(5). ISSN No. 0976-5697
12. Usman M, Petersen K, Börstler J, Neto PS (2018) Developing and using checklists to improve software effort estimation: a multi-case study. J Syst Softw. ISSN 0164-1212, E-ISSN 1873-1228
13. Skruch P, Długosz M, Mitkowski W (2017) Improving the success rate of student software projects through developing effort estimation practices. In: Springer International Publishing AG 2017, Trends in advanced intelligent control, optimization and automation, advances in intelligent systems and computing, p 577. https://doi.org/10.1007/978-3-319-60699-6_83
14. Suri PK, Ranjan P (2012) Comparative analysis of software effort estimation techniques. Int J Comput Appl 48(21) (0975–8887)
15. Rijwani P, Jain S, Santani D (2014) Software effort estimation: a comparison based perspective. Int J Appl Innov Eng Manag 3(12)

16. Primandari PL, Sholiq (2015) Effort distribution to estimate cost in small to medium software development project with use case points. In: The third information systems international conference, procedia computer science, vol 72, pp 78–85. www.sciencedirect.com
17. Mukherjee R, Gupta T, Thirugnanam M (2016) Review of effort distribution in IT companies. Int J Eng Techn 2(6)
18. Basha S, Ponnurangam D (2010) Analysis of empirical software effort estimation models. Int J Comput Sci Inf Secur 7(3)
19. Sangeetha K, Dalal P (2015) A review paper on software effort estimation methods. Int J Sci Eng Appl Sci (IJSEAS) 1(3). ISSN: 2395-3470
20. Safavi SA, Shaikh MU (2011) Effort estimation model for each phase of software development life cycle. www.igi-global.com/chapter/effort-estimation-model-each-phase/46269, https://doi.org/10.4018/978-1-61520-789-3
21. Tan T (2012) Domain-based effort distribution model for software cost estimation. A Dissertation Presented to the Faculty Of The Usc Graduate School University Of Southern California In Partial Fulfillment of the Requirements for the Degree Doctor of Philosophy (Computer Science)
22. Passing U, Shepperd M (2003) An experiment on software project size and effort estimation. In: IEEE Xplore conference: empirical software engineering. https://doi.org/10.1109/isese.2003.1237971
23. Sharma V, Garg G (2017) Usage and analysis of estimation techniques in software project management. Int J Innov Res Comput Commun Eng 5(5). ISSN (Online): 2320-9801 ISSN (Print): 2320-9798
24. Heijstek W, Chaudron MRV (2007) Effort distribution in model-based development. In: Proceedings of the 2nd workshop on model size metrics (Co-located with MODELS 2007) Nashville, TN, USA

Learner Performance and Preference Meter for Better Career Guidance and Holistic Growth

Nilay M. Vaidya and Kanubhai K. Patel

Abstract One of the biggest challenges for higher educational institutes is to increase the placement ratio. Another challenge is to increase the holistic development of the students. Looking at the global requirement, the companies require people not only excellent in the domain knowledge but required excellent in the soft skill too. Finding and predicting the performance factor of the student may help in improving the system and also give an indication to improve pedagogy being offered to students. Many tutoring systems and continuous evaluation patterns adopted by many institutes help in improving the performance of a student. As the trend changes toward holistic development of the students, focus is also upon the soft skills measurement factor. This encouraged us to have a model that helps predicting the holistic performance of a student based on the continuous evaluation as well as performance indicator of a student in other activities too. A gray-based decision-making theory helps assessing the required parameters that find the continuous performance measurement of a learner for each aspect. The multi-attribute situation decision-making theory helps in improving the criticality of the information system by recognizing the sensitivity of the criteria.

Keywords Employability · Grey decision-making · Multi-attribute situation decision-making · Tutoring system

1 Introduction and Related Work

With the revolution in the global market and the rapid development in Information Technology (IT) around, many companies started demanding the candidates with the good technical skill as well as soft skills. Soft skill development becomes one of the important components in good institutes nowadays. Institutes have started applying the different modules to build the soft skill measure among students. But

N. M. Vaidya (✉) · K. K. Patel
Charotar University of Science and Technology, Changa, Gujarat, India
e-mail: vaidyanilay@gmail.com

K. K. Patel
e-mail: kkpatel7@gmail.com

S. Fong et al. (eds.), *ICT Analysis and Applications*, Lecture Notes in Networks and Systems 93, https://doi.org/10.1007/978-981-15-0630-7_5

the challenge is to have a performance factor analysis that may help in combining the result of technical as well as soft skill measure to identify the holistic measure of the student. Now the biggest challenge is to find (i) basic competency of a student at entry time [1], (ii) continuous internal evaluation, and (iii) preferences and outcome in other activities. Entry time basic competency skill identification can be done by considering basic criteria like age, area, previous qualification, family background (parent's qualification), score in the last competitive exam and others. Model already been developed using Multi-Criteria Decision-Making theory (Multi-Objective Grey Situation Decision-Making Theory) that identifies basic entry time competency level of a student [1].

Major task then is to track the continuous internal evaluation [3, 6] and the preference of the learner for the holistic personality development measure. The task is to measure the holistic flavor or holistic factor of the student during the study period. According to the survey the best way to find the performance measure is to calibrate the performance indicators. To provide better effect measure, multi-criteria (multi-attribute grey) decision-making theory helps in finding the performance indicator after comparing the effect measure in the same environment. Finding or evaluating the intellectual ability and improving it becomes challenge in the current environment. The objective of our study is to find the preference meter and performance meter of the student to increase the employability factor.

2 Proposed Model

As shown the Fig. 1, key dimensions are Assessee (Learner/Student), Assessment System, and Global recruiters. The key aspect is to improve the intellectuality, awareness of core ethics, and to have more social touch that leads toward the understanding of the social quotient.

Decision-making process takes a number of steps: (i) problem identification, (ii) preference building, (iii) assessing substitutes, and (iv) finding best substitute option (Simon 1977; Keendy and Raidda 1993; Kleindorfer, Kunreuther, and Schoemaker 1993). In the decision-making process we focus upon the decision science and operation research rather than focusing upon the thinking or end user's research. So we can say the problem objective is to make use of prescriptive and normative analysis rather than descriptive analysis.

The problem we focus on considers the alternative criteria with performance rating; and not on the single criteria. This encourages in finding multiple criteria with alternative weights depending on the preferences, outcomes, and corelation among the criteria to be taken into consideration while finding the effective result. Expert agent helps in improving the implicit knowledge of the learner as well as provides the conditional content to the learners. The system helps in identifying the effectiveness and likeliness of the user and provides the content as well as selects the pedagogy accordingly [2]. Many agents nowadays are designed and implemented that helps in identifying the interest and basic understanding of the user to produce

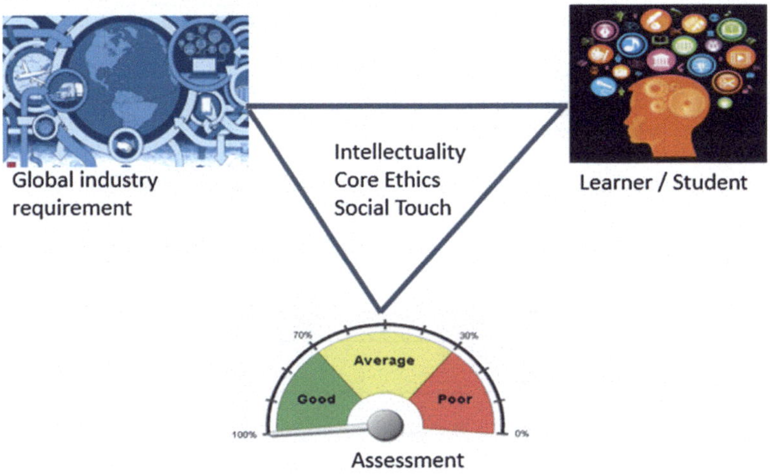

Fig. 1 Overall working of the model

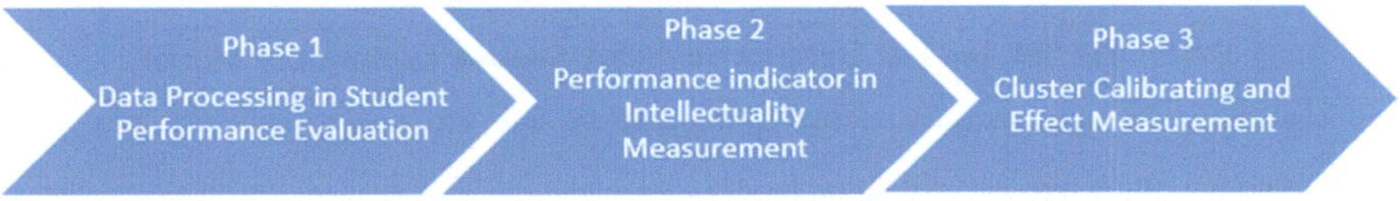

Fig. 2 Phases identified for the working model

the content. Now the challenge is to find the performance meter of the learner using an alternative approach that helps in giving the effective measure of the learner that helps tutor in knowing the exact indicator of the learner in the given cluster. This also helps in finding the performance matrix for the cluster that gives better indicator of the learner by calibrating the effect measures within the cluster.

Three-phase performance indicator meter identified that uses different criteria to generate the effect measures as shown in Fig. 2.

2.1 Phase 1

Data processing to evaluate the learning effectiveness in the domain knowledge using information systems by applying Multi-Objective Grey Situation Decision-Making theory. This makes the information system more adaptive by flavoring the intelligent agent onto it. The working flow is shown in Fig. 3.

The processing model collects and investigates data for the future prospecting that can be further utilized by alternative phase that helps in calibrating the multiple user effective measure grade. This phase also uses exploratory model that helps in

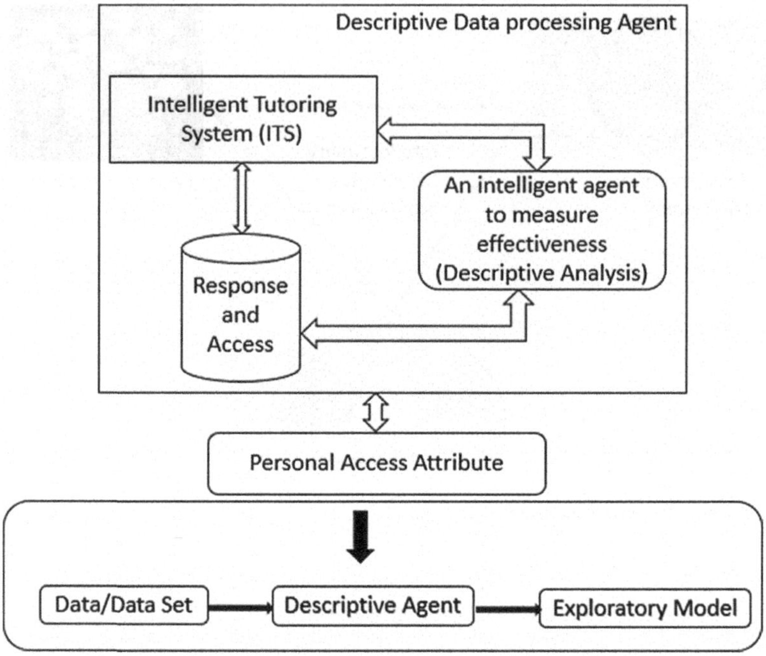

Fig. 3 Working of the internal evaluation model

identifying the correct learning pedagogy that suits the learning style and offers services accordingly.

2.2 *Phase 2*

Data processing with Multi-Objective Decision-Making (MODM) on the external environment and explicit feedback. This phase identifies several results as well as contributions of the assessee in terms of participation in the event or coordinating the event, competing in the several events, cracking or attempting the certification exams, feedback of the peer and the teacher or coordinator, and many more. Many of the criteria have fuzzy values and greyness associated with it. This fuzziness as well greyness needs to be evaluated in such a way that it helps in generating a better result. This leads to the normative analysis approach.

This phase also uses the scores of different aptitude tests being conducted by the internal evaluation information systems (phase 1). The scores are then being mapped with the criteria and also affect the weights associated with it.

General overall picture is shown in Fig. 4:

Each criterion has got the measuring scale from upper, lower, or central. The decision-making during the measurement of the performance factor includes four

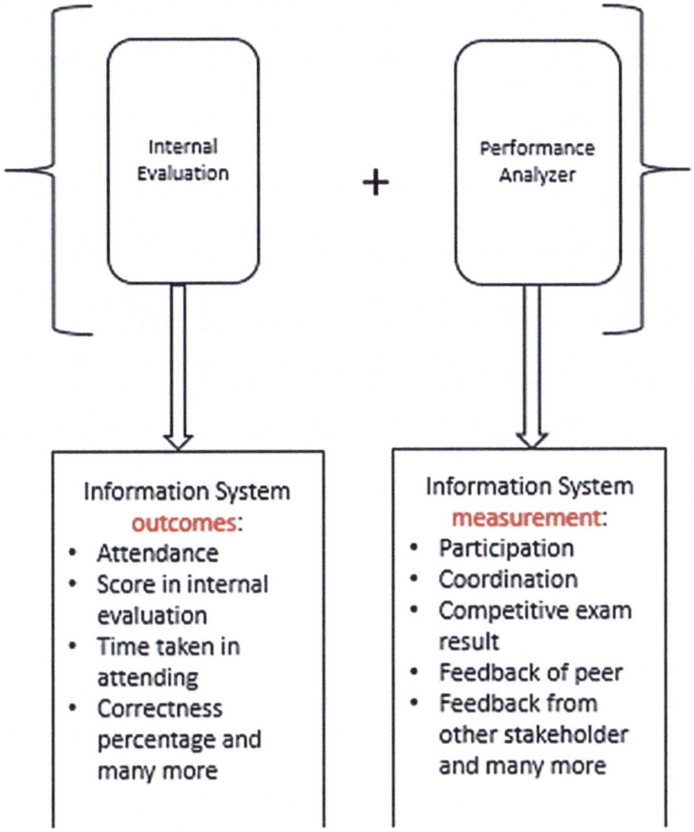

Fig. 4 Working of the continuous performance analyzer

major elements: (i) event, (ii) strategy, (iii) effect, and (iv) scale. In case of finding the performance factor of the student, here we considered the factors like participation in various curricular activities, co-curricular activities, certification exam cracked or appeared, events coordinated as head or member with team size and event type, direct as well as indirect feedback of the student from peer or teacher or counselor.

Depending upon the deviation factor of the criteria, the effect measure is classified into three levels: (i) upper, (ii) lower, and (iii) center. The criteria use of the effect measure is depended upon the deviation in size of the group, objective of the task, criticality defined for the task and many other criteria can be taken into consideration to have the effect measure for the criteria. To find the effect measure for different levels we need to fix the upper, lower, and central limits for the cluster to find the effect measures for the students. So the primary effect measure for the criteria with classified level as higher the better, i.e., upper level:

$$r_{ij} = \frac{u_{ij}}{u_{max}} \tag{1}$$

where u_{ij} is the actual effect measuring value and the u_{max} is the maximum data for the criteria $u_{ij} \leq u_{max} : r_{ij} \leq 1$. Similarly, for the effect measure with lower measure with the lower level:

$$r_{ij} = \frac{u_{min}}{u_{ij}} \tag{2}$$

where u_{ij} is the actual effect measuring value and the u_{min} is the minimum data for the criteria $u_{ij} \geq u_{min} : r_{ij} \geq 1$ [4, 5].

As there are several such objectives taken to find the performance meter, we need to find the comprehensive measure on the objectives, i.e., criteria defined which are called multi-objective situation decision-making. The effect measure for individual criteria for students need to be calculated as $s_{ij.}$ The decision factor then can be calculated as for the kth element as $\delta_i^{(k)}$ and the decision matrix which is prepared as

$$\begin{bmatrix} \frac{r_{11}^{(k)}}{s_{11}} \\ \frac{r_{21}^{(k)}}{s_{21}} \\ . \\ . \\ \frac{r_{n1}^{(k)}}{s_{n1}} \end{bmatrix} \tag{3}$$

Similarly, the comprehensive matrix then can be calculated as:

$$r_{ij}^{(\Sigma)} = \frac{1}{N} \sum_{k=1}^{n} r_{ij}^{(k)} \tag{4}$$

This comprehensive value being generated shall then be used to find the effectiveness of the student in the given environment. More the value is nearer to one more effective the result. The classification is shown in Fig. 5.

Every criterion has got the actual effect measure which is being calculated by the system once being inserted into the system. The greyness or fuzziness factors are converted into their equivalent crisp values before being processed by the algorithm. Each criterion has got the minimum effect measure or maximum effect measure or

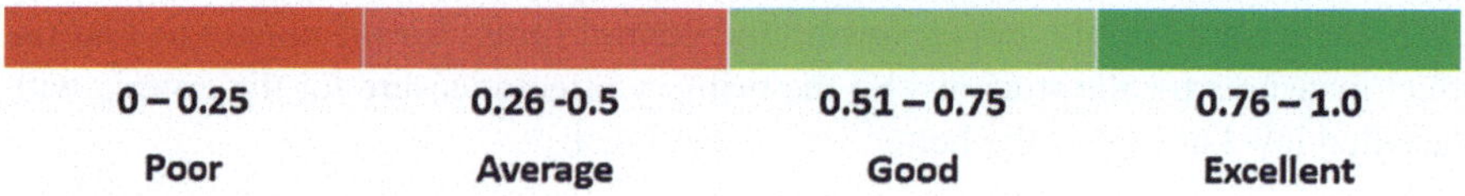

Fig. 5 Effect measure classification

Table 1 Sample data of students

	Participation in co-curricular events	μ1	Clearing any online course	μ2	Attending the workshop/seminar	μ3	Number of attempts to clear the aptitude test	μ4
S1	1	1	2	2	0	0	3	3
S2	4	4	5	5	1	1	1	1

The quantitative and qualitative assessment

central effect measure depending upon the deviation factor of the activity/criteria. These models use few criteria which are single level, few are multilevel, few are multistage. But all are dynamic and required comparison at all the levels.

Problem solving by the multi-objective situation decision-making theory for sample criteria for two students: Considering sample four parameters to find the effect measure for a student by considering the participation in co-curricular events (upper limit measure—minimum four per year), clearing any online course (higher limit measure—minimum five as one per course/subject), attending the workshop/seminar (higher limit measure—minimum two per year), number of attempts to clear the aptitude test (lower limit measure—out of three attempts offered) (Table 1).

The effect measure for different parameters:

$$r_{ij}^{(1)} = \left[\frac{1}{4}, \frac{4}{4}\right], r_{ij}^{(2)} = \left[\frac{2}{5}, \frac{5}{5}\right], r_{ij}^{(3)} = \left[\frac{0}{2}, \frac{1}{2}\right], r_{ij}^{(4)} = \left[\frac{3}{3}, \frac{1}{3}\right]$$

And the comprehensive effect measure shall be

$$r_{ij}^{(\Sigma)} = \left[r_{11}^{(\Sigma)}, r_{12}^{(\Sigma)}\right] = [0.4125,\ 0.7075]$$

As discussed above the effect measures calculated above shows that the competitiveness of student S1 is average and S2 is good.

2.3 *Phase 3*

Extending the comprehensive measure by comparing the effect measures of each student and standardizing will give the comprehensive analysis factor. This helps in finding and comparing the performance factors as well as the preference factor of the batch. Applying the aggregation gives better understanding of an individual in the group of assessees. In the previous example, taking the average of two comprehensive effect measures shall give us the aggregate of [0.4125, 0.7075], which is 0.56 which indicates that the overall classification of the batch is "Good" as per the effect measure classification shown in Fig. 5.

3 Conclusion

As the biggest challenge is to increase the employability of the student and to increase the ration and to make the students competitive we need to continuously observe the performance of the student. The proposed solution keeps on observing their details and finds the effect measure at the regular interval. The counselor or teacher is able to find overall personality performance meter that helps them to take corrective actions to improve the employability of the students and their holistic growth.

References

1. Vaidya N, Sajja P, Gor D (2015) Evaluating learning effectiveness in collaborative learning environment by using multi-objective grey situation decision making theory. Int J Sci Eng Res (IJSER) 6(8):41–45, ISSN 2229-5518
2. Vaidya N, Sajja P (2017) Agent based system for collaborative learning environment in an educational habitat. In: International conference on ICT in business industry & government (ICTBIG), 2016. https://doi.org/10.1109/ictbig.2016.7892644
3. Vaidya N, Sajja P (2018) Use of an adaptive agent in virtual collaborative learning environment. In: Satapathy S, Joshi A (eds) Information and communication technology for intelligent systems (ICTIS 2017), vol 2. ICTIS 2017. Smart innovation, systems and technologies, vol 84. Springer, Cham
4. Jiang D et al (2009) Study on the application of multi-objective grey situation decision-making theory for transportation mode choice, 978-0-7695-3583-8/09, IEEE. https://doi.org/10.1109/icmtma.2009
5. Wei W, Xuewu C (2007) Transportation planning, the people's transportation. Beijing
6. Vaidya N, Sajja P (2014) Intelligent virtual collaborative learning environment. In: National conference on computer science, information technology and management, Korti, Pandharpur (MS), India, 06–07 March 2014, ISSN: 2319-5010

Grading of Tuberculosis Bacilli Using Computer Vision Assisted Detection Method for ZN-Stained Images of Bright Field Microscopy

Vishakha Yadav and G. Thippeswamy

Abstract In the developing countries, tuberculosis (TB) is the primary and foremost cause of death in the infectious disease category. At present, the research for tuberculosis diagnostics tool development focuses on extemporizing microscopy procedure with a simpler technology for detecting smear-positive TB. In tandem, recent diagnostics are aiming at increasing the sensitivity or simplicity of diagnosing active TB disease. Smear microscopy of sputum is often the first TB test to be used in countries with a high rate of TB infection. Sputum smear microscopy examination is economical and modest, although the sensitivity is only about 50–60%. Automated detection of TB bacilli could accelerate diagnosis, enhance quantitative classification and reduce the manual errors pertaining to diagnosis. The bright-field microscopy screening can be assisted by latest computer vision technique for detection of TB using the minimalistic computer-assisted infrastructure setting at the rural health centers. In this paper we discuss the implementation of one of the computer vision techniques for grading of the Mycobacterium Bacilli for conventional ZN-stained images. The objective of the implementation is to detect the edges of the bacilli by considering the area and perimeter filtering of the given input images.

Keywords Bright field microscopy · ZN-stained · Canny edge

1 Introduction

Tuberculosis has affected mainly the population living in poor municipal and communal setups having lesser or negligible health infrastructure [1]. These setups lack access to the mainstream healthcare instruments and facilities.

Many patients are hesitant to visit the nearby health centers to give sputum samples due to various reasons. Economic and social factors influence the assessment process thereby depriving the continuity in the treatment even after a positive

V. Yadav (✉) · G. Thippeswamy
Department of CSE, BMS Institute of Technology and Management, Bengaluru 560064, India
e-mail: vishakhay@bmsit.in

G. Thippeswamy
e-mail: dean_academics@bmsit.in

S. Fong et al. (eds.), *ICT Analysis and Applications*, Lecture Notes in Networks and Systems 93, https://doi.org/10.1007/978-981-15-0630-7_6

diagnosis of the TB cases. The follow up of the diagnosis is very challenging in such cases [2]. The probabilistic diagnosis is a lot made on the basis of a combination of clinical symptoms, signs, and radiological findings. Nonetheless the sensitivity and specificity of the diagnostic approaches are inadequate in the endemic population grappled with HIV, systemic viral or bacterial infections, parasitic infections, and bacterial, viral, or atypical pneumonia [3]. Many times the patients specifically children are not documented in registers or informed to the government agencies who are offering various schemes and programs tracking the disease making it difficult to account for the actual beneficiaries [4]. Given the nature of the disease, the microbiological identification of tuberculosis is quite perplexing. The paucibacillary pattern of the bacilli may not always help in concluding the bacilli of TB. With the diagnostic advancements, it is still challenging to conclude about the presence of tuberculosis as the suspect may have very few bacilli [5]. Hence, the childhood tuberculosis has been neglected by various entities like the health workers and researchers, because children often have paucibacillary disease that are not thought to be infectious. Children are considered as least risk disease infection agents for propagation of TB disease within a community [6].

2 Literature Review

Given the nature of the disease, the microbiological identification of tuberculosis is quite perplexing. The paucibacillary pattern of the bacilli may not always help in concluding the bacilli of TB. With the diagnostic advancements, it is still challenging to conclude about the presence of tuberculosis as the suspect may have very few bacilli [5].

Many times the patients specifically children are not documented in registers or informed to the government agencies who are offering various schemes and programs tracking the disease. Making it difficult to account for the actual beneficiaries [4].

There is a worldwide resurgence of tuberculosis including India. An estimate of 14 million of the population is infected with radiologically active TB. These are roughly estimated figures as most of the affected don't visit nearby health centres. The public and private healthcare services suffer shortcomings like lack of accomplished experts, incorrect diagnosis and misjudged procedures. In the rural and urban slum population areas, specifically the children, women, and young adults are grappling with the contagious disease, due to malnutrition, unhygienic living and working conditions [7]. In the nascent stages, the probable disease affected people are often treated based on the initial symptoms of cough and fever. The actual diagnosis confirms the tuberculosis only after having done the sputum smear test. The suspect must undergo 2 sputum smear sample screening collected early in the morning. In the resource-constrained setting, especially rural areas the screening of these morning samples need expertise and precise modus operandi to confirm the tuberculosis.

The directly observed treatment, short course(DOTS) [8] and Dots Plus programs with an extended aim of extricating this disease has been conceptualized and rolled out across the globe. WHO has been primarily focusing on the African and Asian continents for the successful accomplishment of these programs over many years. Among the 5 components prescribed for DOTS program one of the critical constituent is case detection by sputum smear microscopy [9].

Culturing bacteria to test for TB is much more complex and expensive to carry out than sputum smear microscopy. This is because it requires specific equipment and laboratory facilities. Testing for and diagnosing TB using culture can also take weeks because of the slow growth of TB bacilli [10]. The increased community awareness leverages the successful implementation and wider reach out of DOTS program over the years. In a country as large as India, the complete eradication of Mycobacterium tuberculosis disease is plausible at low-cost infrastructure, administration, and governance. This is feasible by integrating the monitoring of DOTS program at the Rural Primary Health Centre with increased accountability and compliance [7].

Introductory screening and diagnosis of TB are the crucial phases. Primarily Fluorescent microscopy and Conventional (Brightfield) Microscope assisted techniques are used for screening of the bacteria (Mycobacterium). In fluorescent microscopy an acid-fast Fluorochrome (Auramine o) stain environment and the intense light source like halogen or high-pressure mercury vapor lamp are being used for the sputum samples, while the bright field microscopy uses ZN (Ziehl-Neelsen) staining technique on samples to find out sputum smears [11]. Thus the bright field microscopic examinations are way cheaper and accessible in countries like India. The fluorescent microscopes are not only costly but maintenance of equipment also has to be looked into and hence it is less preferred [12] in most of the countries of Asia Pacific region. Bright field/ZN-staining technique is the cheapest and widely used method of screening. In rural areas/remote locations, the ZN-stained microscopy test is the handy method for TB detection and monitoring. The milieu of Computer Assisted Diagnostic methods for well-organized and initial diagnosis of cancer (especially breast cancer), diabetic retinopathy, Chest X-Ray-based TB detection, chronic obstructive pulmonary disease (COPD), has fostered the development of algorithms and methods that will assist automatic detection of TB from ZN-stained slides in the resource-constrained setting [13].

Preemptive Research in the clinical morphology analysis of TB bacteria in India is being attempted by many researchers. Conventionally, the primitive screening and analysis of sputum smears are done manually constrained to various fatigue-related factors. In such scenarios assessment with precision becomes a crucial factor [14]. The programmed identification of TB Bacilli shall improve the accuracy in terms of sensitivity and specificity by lessening the burden of judgment load on the pathology staff in diagnosis [15].

Evolution of Computer Vision techniques for the laboratory diagnosis of tuberculosis assists in order to achieve quick, cheap, and more accurate results. Acid-fast staining and culture for mycobacteria remain at the core of any diagnostic algorithm [16]. The morphology of TB Bacillus is quite simple. These are a thin rod shape with a length of roughly 4 μm and a diameter of 0.5–1.0 μm. The waxy lipid in the cell wall of the bacilli appears red after acid-fast staining in high contrast to the blue background [17]. TB Diagnosis is specifically dependent on such morphology and color. The examination has confirmed 97.94% sensitivity and 83.65% specificity [18]. The performance of watershed segmentation method for detection and classification of bacilli from camera-enabled Smartphone microscopic images as tuberculosis bacilli positive or negative were 93.3% and 87%, respectively [9]. The sensitivity, specificity, and precision rate of watershed segmentation on different infection levels (Scanty, 1+, 2+, and 3+) provided a discordant rate of 3.73% with respect to tuberculosis positive and negative cases [19].

3 Methodology

The proposed system is to develop an automated system which will assist the concerned in counting the number of bacilli in each of the slides. This system is a web application which uses client server model for conveying the information. The important feature is its ability to predict the presence of bacilli in the ZN-stained slide image. In the given sample set of sputum smear images the detection of M.Tuberculosis bacilli is done using the two main features (a) Area and (b) Perimeter of bacilli. These features are processed after applying the morphological transformation using the canny edge detection method. The usability of this system is very high and can be used by primary healthcare center practitioners. The application may assist the technician with inadequate experience who are burdened with examining many samples throughout the day. Dependency and non-availability of an expert pathologist may be dodged. This application will grade the ZN slide based on the count of bacilli obtained from the system and analyze the effect of treatment based on difference in grading over the period of treatment. This system mainly assists the practitioners in counting the number of bacilli present in ZN slide and not to replace their human judgment in the procedure. The ZNSM-iDB [19] has been used for the development of the proposed system used in automated microscopy using computer vision techniques. In the developed system we take ZN slide images as the input and process them to obtain the bacilli count as the output.

The ZN slide image taken as the input and uploaded using Django framework. The transitional processing passes through multiple stages such as masking, morphological transformation, edge detection, area filtering, perimeter filtering, etc. (Fig. 1).

These phases are applied to remove the unwanted background noise. The uploaded image can be previewed on the home page. Once the image is uploaded and the processing is done, the result along with the intermediate results can be viewed. The result is shown by displaying the bacilli count and the grading for the obtained bacilli count. The user interface also displays the grading scale on the result screen for reference.

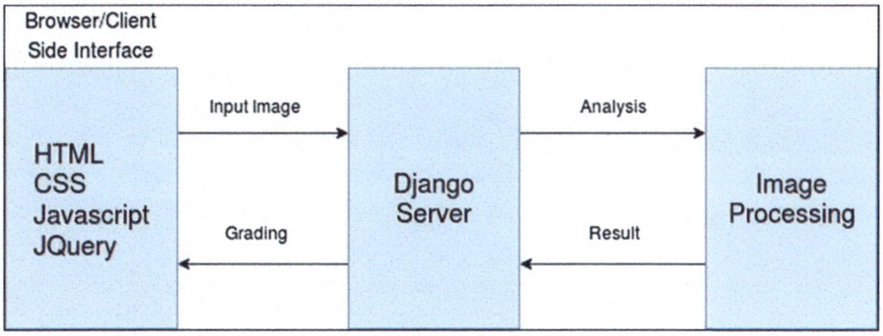

Fig. 1 Client server architecture used in the application

3.1 Implementation Framework

The application comprises of 3 methods to grade the cells,

- Divide and Conquer
- Masking and Tracing
- Canny Edge Detection.

3.1.1 Divide and Conquer

This is the first module implemented which divides the entire image (Fig. 2) into multiple segments similar to the divide and conquer method where a solution is found by recursively dividing a problem into its sub-problems and combining these results to obtain a final complete result of the original problem.

3.1.2 Masking and Tracing

In this module the whole image is masked and morphological transformations are applied to it. This provides us with a grayscale image of the ZN slide with bacilli in white on black background. Contours are applied on this to count the white segment which is present on the black background.

3.1.3 Canny Edge Detection

This functionality is used to find out the bacilli present in the image. This technique works in phases to detect edges in a given image. First, a Gaussian filter is used to reduce the noise in the image. In the second phase edges are detected horizontally, vertically, and diagonally by finding the intensity gradient of the image. This is done by applying a pair of convolutional masks. In the subsequent stage, the pixels which

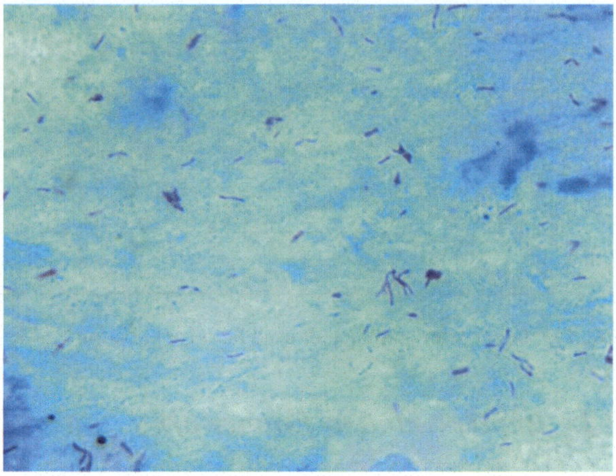

Fig. 2 Original image

are not edges are suppressed (or removed) from the image. Thus, only the candidate edges which constitute the edges will remain and move on to the next stage. In the last concluding stage, the algorithm uses two threshold values: upper threshold and lower threshold. If the pixel gradient is higher than the upper threshold, then it is accepted as an edge. If the pixel gradient is lower than lower threshold, then it is rejected. For the pixel gradient which lies between these two values, if the pixel is connected to a pixel with upper gradient, then it is accepted or else it is rejected. The developed system systematically utilizes all the above phases. It filters out unwanted background noise by methods such as area-based filtering (Fig. 3) and perimeter-based filtering (Fig. 4) which has a predefined threshold value and anything which was detected by Canny Edge Detection method which does not fall into the defined threshold value will be removed and will not be considered during the counting process (Fig. 5).

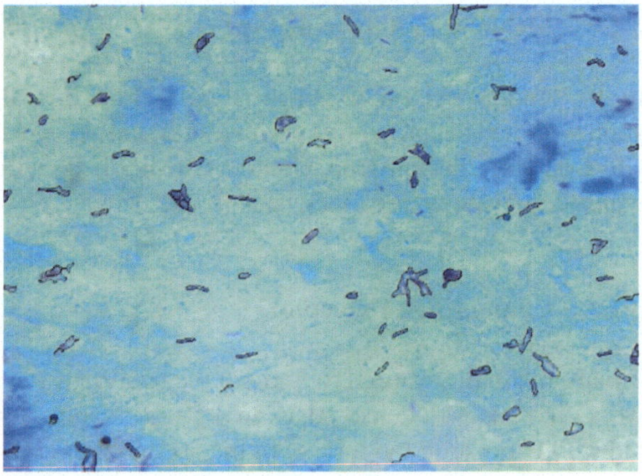

Fig. 3 Perimeter filtered image

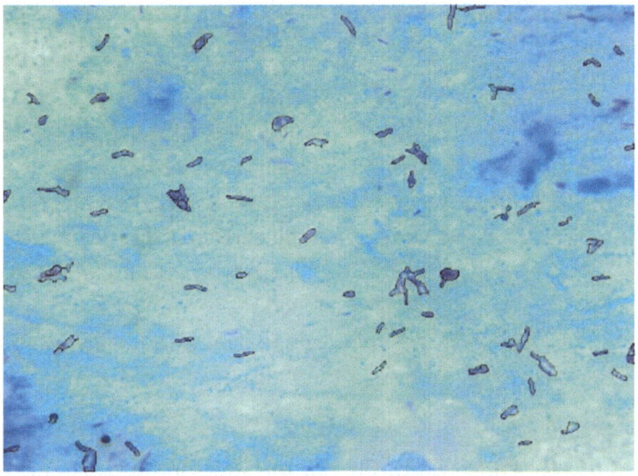

Fig. 4 Area filtered image

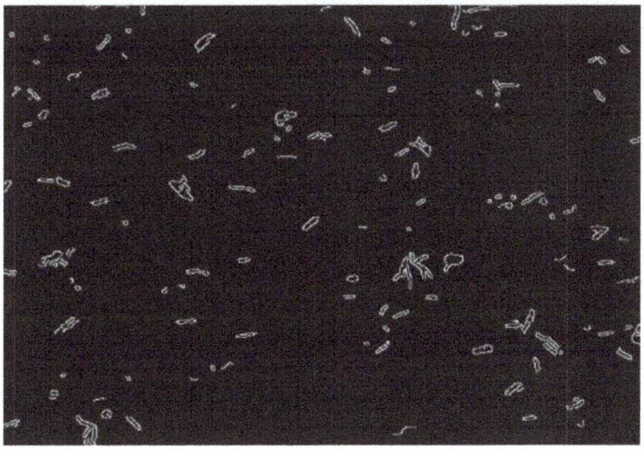

Fig. 5 Output of the canny edge detection method

3.2 Outcome

The number of bacilli seen in a smear reflects disease severity and patient infectivity. Therefore, it is important to record the number of bacilli seen on each smear. The grading of bacilli is done according to the grading scheme mentioned in the Table 1.

Here +++ (pronounced as three plus) is the most severe case of the disease. One positive specimen out of the two is enough to declare a patient as smear-positive TB. Sputum smear microscopy for bacilli is positive when there are at least 10,000 organisms present per ml of sputum. In the application 84 images were used as input from the ZNSM-iDB dataset. These images were picked up from Autofocusing Dataset, viewfields with overlapping (occluded) bacilli, viewfields with single or few bacilli, and viewfields without bacilli. All the images were put through the test and validation. The final result was taken based on the average of all the three aforementioned

Table 1 Grading of sputum smears

Number of bacilli	Result reported
No AFB per 100 oil immersion fields	0
1–9 AFB per 100 oil immersion fields	Scanty (or number of AFB seen)
10–99 AFB per 100 oil immersion fields	+ (1+)
1–10 AFB per oil immersion field	++ (2+)
>10 AFB per oil immersion field	+++ (3+)

methods. These methods mainly use OpenCV functions for analyzing and computing the results. Averaging is done since there might a loss of bacilli in one of the methods and a background noise may be considered as some additional bacilli in a different method. The grading application has shown the tolerance of up to 20% on all the assorted images. The tuberculosis bacilli were graded based on the 5-point grade scale mentioned in the above table. The system has classified the number of bacilli present with 80% accuracy.

4 Conclusion

All the present diagnostic methods like Serology, Microscopy, X-ray, and Culture-NNAT have their pros and cons. The usage of these methods is solely dependent on the resource demographics. The limitations of existing diagnostic methods used in most of the developing countries like India 1. Sputum Smear Microscopy, 2. Chest X-ray, 3. Genexpert/CB-NAAT, and 4. TrueNat paves advances in the automated TB diagnostic research. Genesis of upgradation in the rudimentary diagnostic technique may preempt the cost factor required in the infrastructure setup of various diagnostic routines. Quick implementation of new assessment techniques will address the growing requirements of affordable public health in developing countries. Keeping this in mind we are attempting at discovering an algorithm with an aid of computer vision technique. Manual diagnostics finding in tandem supported by automated tools may pull more and more rural patients toward continuation of the treatment of the disease. The developed application is designed at utilizing the open-source platforms which are cost-effective and shall help the government agencies in resource-laden countries.

References

1. Janssens J-P, Rieder HL (2008) An ecological analysis of incidence of tuberculosis and per capita gross domestic product. Eur Respir J 32:1415–1416
2. Zar HJ, Hanslo D, Apolles P, Swingler G, Hussey G (2005) Induced sputum versus gastric lavage for microbiological confirmation of pulmonary tuberculosis in infants and young children: a prospective study. Lancet 365:130–134
3. Hesseling AC, Schaaf HS, Gie RP, Starke JR, Beyers N (2002) A critical review of diagnostic approaches used in the diagnosis of childhood tuberculosis. Int J Tuberc Lung Dis 6:1038–1045
4. Du Preez K, Schaaf HS, Dunbar R et al (2011) Incomplete registration and reporting of culture-confirmed childhood tuberculosis diagnosed in hospital. Public Health Action 1:19–24
5. Dunn JJ, Starke JR, Paula AR (2016) Laboratory diagnosis of Mycobacterium tuberculosis infection and disease in children. J Clin Microbiol 54(6)
6. Wallgren AJ (1937) On the contagiousness of childhood tuberculosis. Acta Paediatr Scand 22:229–234
7. http://planningcommission.gov.in/reports/genrep/index.php?repts=b_repgen.htm
8. What is DOTS?. https://www.who.int/tb/publications/dots-who-guide/en
9. New Laboratory Diagnostic Tools For Tuberculosis Control (2009) Stop TB Partnership. http://apps.who.int/tdr/svc/publications/non-tdr-publications/
10. Schaaf S (2010) Tuberculosis—a comprehensive clinical reference, 1st edn. Elsevier Inc.
11. Balakrishna J, Shahapur PR, Chakradhar P, Saheb HS (2013) Comparative study of different staining techniques—ziehlneelsen stain, gabbet's stain, fluorochrome stain for detecting of mycobacterium tuberculosis in the sputum
12. Jan Z, Rafiq M, Muhammad H, Zada N (2015) Detection of tuberculosis bacteria in sputum slide image using morphological features. In: Kozielski S, Mrozek D, Kasprowski P, Małysiak-Mrozek B, Kostrzewa D (eds) Beyond databases, architectures and structures. BDAS 2015. Communications in computer and information science, vol 521. Springer, Cham
13. Yadav V, Thippeswamy G (2018) Efficacy of computer vision technique to identify and extrapolate the tuberculosis bacilli. In: 4th International conference on applied and theoretical computing and communication technology (iCATccT). Research Publishing, Mangalore, pp 198–201. 978-1-5386-7705-6/18/$31.00 c 2018
14. Kusworo A, Gernowo R, Sugiharto A, Pamungkas A, Bawono A, Mirnasri N (2013) Autothresholding segmentation for tuberculosis bacteria identification in the ziehl-neelsen sputum sample. In: The proceedings of the 7th ICTS, Bali, May 15th–16th (ISSN: 9772338185001)
15. Forero MG, Cristobal G, Borrego JA (2003) Automatic identification techniques of tuberculosis bacteria. SPIE proceedings of the applications of digital image processing XXVI, Vol. 5203. San Diego, CA, pp 7181, ISBN 0-8194-5076-6, Aug 2003. SPIE, Bellingham WA
16. Caulfield AJ, Wengenack NL (2016) Diagnosis of active tuberculosis disease: from microscopy to molecular technique. J Clin Tuberculos Mycobact Dis 4:33–43, 2405–5794
17. Svarna KS, Christopher L, Bancroft JD (2012) Bancroft's theory and practice of histological techniques. London, UK: Churchill Livingstone, pp 296–297
18. Xiong Y, Ba X, Hou A, Zhang K, Chen L, Li T (2018) Automatic detection of mycobacterium tuberculosis using artificial intelligence. J Thorac Dis 10(3):1936–1940. https://doi.org/10.21037/jtd.2018.01.91
19. Shah MI, Mishra S, Yadav VK, Chauhan, Sarkar M, Sharma SK, Rout C (2017) Ziehl–Neelsen sputum smear microscopy image database: a resource to facilitate automated bacilli detection for tuberculosis diagnosis. J Med Imag 4(2), 027503. https://doi.org/10.1117/1.jmi.4.2.027503
20. Shah MI et al (2016) Automatic detection and classification of tuberculosis bacilli from camera-enabled Smartphone microscopic images. In: Fourth international conference on parallel, distributed and grid computing (PDGC). IEEE, pp 287–290

Modeling the Dynamics of Carbon Dioxide Over an Educational Institute

Srikumar Sastry, Arnav Saha and Ranendu Ghosh

Abstract Carbon dioxide is a major contributor to climate change. It absorbs the outgoing longwave radiation, thereby increasing the temperature in the atmosphere. This study examines the variables which contribute to the flux of CO_2 over the 50-acre sprawling green campus of DA-IICT at Gandhinagar, Gujarat. The previous approach to this problem was to employ differential equations to model the CO_2 emissions. We believe that a compartment-based model that incorporates fossil fuels, electricity, human emissions, and a Light Use Efficiency (LUE) model would provide a better approximation. The LUE based model computes the total carbon that is sequestered by plants. It uses the Primary Productivity Capacity (ε) of plants and APAR (Absorbed Photosynthetically Active Radiation) to calculate the Gross Primary Productivity (GPP). Further, the Net Primary Productivity (NPP) is derived from the GPP. Three dedicated separate models using monthly MODIS NDVI, MODIS FPAR, and MODIS NPP time-series datasets were used to model this. To integrate the above, a Decision Tree-based algorithm was applied to compute the best fit curve and approximate it to the Keeling curve which is a graph of the accumulation of CO_2 in the Earth's atmosphere recorded at the Mauna Loa Observatory, Hawaii for all the three cases. The resultant curves indicated an MSE (Mean Square Error) close to zero and an upward trend was noticed for the future validation dataset.

Keywords CO_2 · LUE · FPAR · NDVI · GPP · Modeling

S. Sastry (✉) · A. Saha · R. Ghosh
Dhirubhai Ambani Institute of Information and Communication Technology, Gandhinagar, India
e-mail: 201601016@daiict.ac.in

A. Saha
e-mail: 201601022@daiict.ac.in

R. Ghosh
e-mail: ranendu_ghosh@daiict.ac.in

S. Fong et al. (eds.), *ICT Analysis and Applications*, Lecture Notes in Networks and Systems 93, https://doi.org/10.1007/978-981-15-0630-7_7

1 Introduction

Carbon dioxide is one of the major greenhouse gases. It occurs naturally as a trace gas. It is released mainly through human activities and natural processes. Carbon is one of the main components of CO_2. 99.96% of carbon is found in the Lithosphere and 0.00206% is found in soil and plants. It is interesting to note that 80% of a plant/tree volume is composed of moisture and the remaining 20% is just dry mass (biomass). Biomass has approximately 9% of carbon. Pertaining to DA-IICT campus, major carbon sources include burning of PNG, electricity consumption, and human respiration. Major carbon sinks include plants and soil.

A compartmental model approach [1] has been adhered to in which each sub-compartment has been allotted a factor that releases or absorbs CO_2 on campus. We incorporate the burning of PNG, electricity consumption, human respiration, and vegetation (plants and trees) into this model. Each of these have been explored in detail in the following sections. Methodology and various techniques applied in all of these four compartments have been discussed and a final composite of these have also been discussed in the sections ahead. Results obtained from this study are fitted with the actual CO_2 concentrations that are observed at the NOAA Laboratory, Mauna Loa [2].

2 Compartmental Model

We make use of a simple compartmental model [1] to study the dynamics of carbon dioxide in the institute campus.

$$\Delta CO_2 = \alpha \times \text{fossilfuel} + \beta \times \text{respiration} + \gamma \times \text{electricity} + \varepsilon \times \text{GPP} \times 0.5 \quad (1)$$

where α, β, γ, ε are model parameters.

- ΔCO_2 represents the flux of CO_2 on campus.
- fossilfuel represents the contribution to carbon dioxide from fossil fuel burning on campus.
- respiration represents the contribution to carbon dioxide from human respiration on campus.
- electricity represents the contribution from the electricity usage on campus.
- GPP represents the Gross Primary Productivity of the plants.

Note: All assumptions for the model and data sources are mentioned below.

3 Carbon Cycle

3.1 Carbon Dioxide Data

The carbon cycle is the key component of Earth Science. To quantify the trend in carbon cycle we use the data measured by NOAA observatory [2]. It has been recording the CO_2 concentrations from 1958 to date. The observatory is located in Hawaii which is a deciduous region and proves an ideal candidate for validation as Gandhinagar is also a deciduous region. Due to its location, Mauna Loa proves to be the least affected location from human activities [1]. But one thing to be noted is Mauna Loa could be the source of additional carbon dioxide from frequent volcanic eruptions. The yearly cycle can be observed with the slopes representing the seasons. A single year in the graph if isolated shows peaks for the winter months, November to January. A peak represents the low absorption of CO_2 by the vegetation during the nongrowing season. So, the overall concentration of CO_2 in the atmosphere increases. The graph also correctly shows a dip for CO_2 in the post-monsoon season, August to September, which is the onset of the growing season where the vegetation absorbs comparatively more CO_2. So, the overall concentration of CO_2 in the atmosphere decreases. There is an overall upward trend as we advance in the time scale which clearly shows that CO_2 levels are increasing yearly (Fig. 1).

3.2 Fossil Fuel Burning Data

There are mainly two sources of biomass burning within the campus—Piped Natural Gas (PNG) used in the canteen and burning of petrol in vehicles. Both of these data were collected in-situation through purchase bills and statistics, respectively.

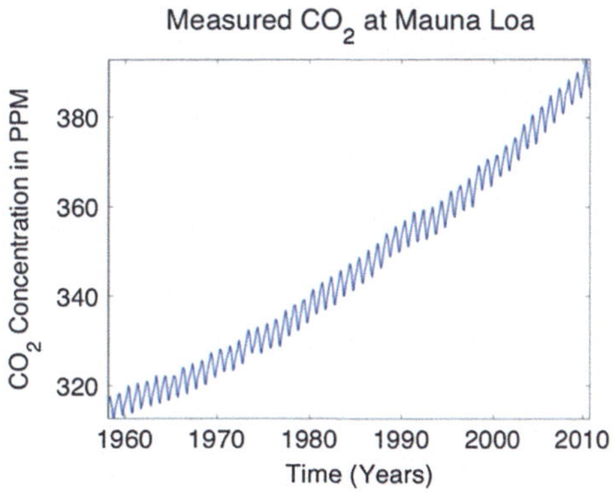

Fig. 1 Graph shows monthly CO_2 concentration in ppm from 1950 to 2010 at Mauna Loa, Hawaii

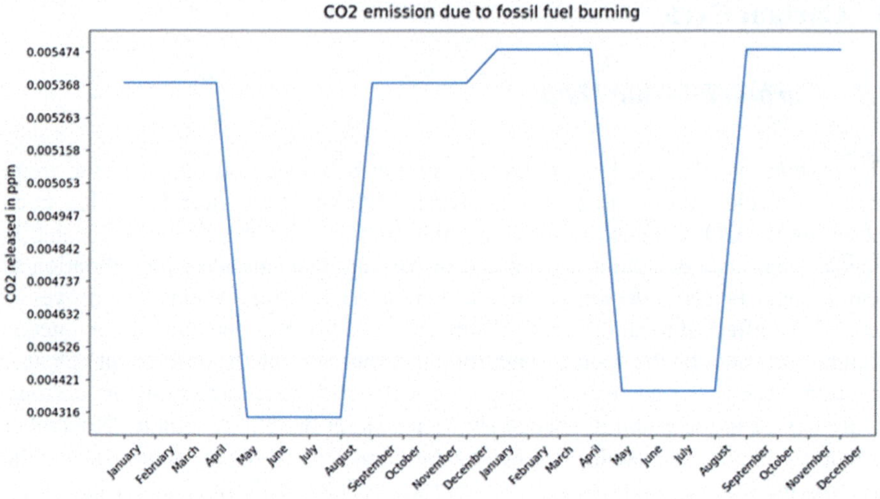

Fig. 2 Graph shows monthly CO_2 (ppm) emission due to fossil fuel burning. We can see a 2% increase in CO_2 emission between the year

We measure PNG in SCM (Standard Cubic Meter) unit and the relation between SCM and KG is: 1 kg = 1.423 SCM. PNG has a major component of Methane (CH_4) and 1 kg of CH_4 produces 2.75 kg CO_2. In total, 1608.54 kg of PNG gas was used in a normal academic month (Note: kg can be converted to ppm by dividing it by 1 million). Analyzing previous data, we assumed a 20% decrease in the PNG used in the nonacademic months (May to July). Also, we assumed a 2% yearly increase in the overall PNG use to incorporate the growing number of students on the campus.

Further, petrol consists of approximately 87% carbon, which is roughly 652 g (Note: 1 kg petrol weights approximately 750 g). Additionally, this carbon converts into CO_2 with the help of 1740 g of oxygen. So the total amount of CO_2 that is generated from 1 L of petrol is ($652 + 1740 = 2392$ g) or 2.392 kg. There are around 99 vehicles on the campus and each uses around 0.64 L of petrol per month. So, a total of $99 \times 0.64 = 63.36$ L of petrol. This gives a total of 151.557 kg of CO_2. Again, a 20% decrease was assumed for nonacademic months and 2% overall yearly increase (Fig. 2).

3.3 *Electricity Consumption*

DA-IICT uses electricity from the Torrent power plant located in Gandhinagar. It generates electricity from coal which eventually produces CO_2. Torrent uses bituminous coal which produces 0.3538 kg of CO_2 per unit of electricity. Total units of electricity consumed in a month are around 2,11,805. So, the total CO_2 produced due

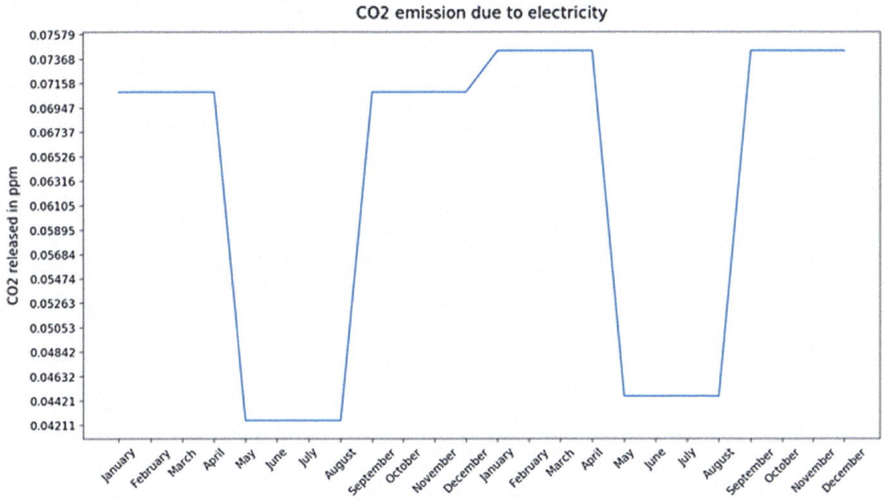

Fig. 3 Graph shows monthly CO_2 (ppm) emission due to electricity consumption

to electricity is around 84936.609 kg. From the previous statistics, A 40% decrease in nonacademic months and a 5% yearly increase was assumed (Fig. 3).

3.4 Human Respiration

An average adult produces 1.043 kg of CO_2 per day. There are approximately 1,400 people at DA-IICT. This gives a total of 43,806 kg of CO_2 monthly. The growing number of seats in the institute was noted and the number of students staying in the campus in nonacademic months was carefully examined, and thus a 40% decrease in nonacademic months and 10% yearly increase was assumed (Fig. 4).

3.5 Satellite Data

There are various biophysical variables which are associated with modeling environmental parameters such as type of vegetation, transpiration, carbon content, etc. Some of the important variables which are of our interest are the Normalized Difference Vegetation Index (NDVI), GPP, NPP, etc. We use NDVI and FPAR to quantify NPP and present three separate models. In simple terms, NDVI measures the amount of greenness of a vegetation canopy. For that, it uses the reflectance of the Near-Infrared (NIR) and Red bands (R) of the sun's incoming electromagnetic radiation. Vegetation reflects the incoming radiation as a function of wavelength. More in green wavelength due to chlorophyll followed by less reflected energy in the red wavelength and in NIR

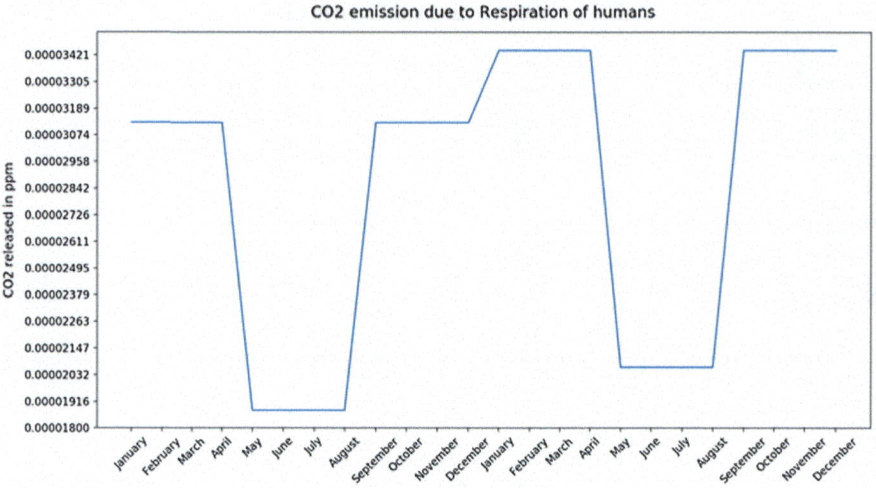

Fig. 4 Graph shows monthly CO_2 (ppm) emission due to respiration of humans

it is very highly reflected energy due to intercellular activity. So, more the difference between those reflected values, more is the NDVI, more is the chlorophyll content and healthier is the vegetation [3].

$$\text{NDVI} = \frac{\text{NIR} - R}{\text{NIR} + R} \tag{2}$$

The Moderate Resolution Imaging Spectroradiometer (MODIS) datasets are recorded by NASA's Terra platform on board two sun-synchronous satellites. It contains NDVI dataset at a spatial resolution of 250 m with 16-days composite [4]. It contains FPAR dataset at a spatial resolution of 500 m with 4-days composite [4]. Finally, it contains the NPP dataset at a spatial resolution of 500 m with 8-days composite [5]. PAR is calculated from solar radiation, in the short wave region which is obtained from IDAHO Terra Climate data (http://www.climatologylab.org/terraclimate.html). This dataset has a spatial resolution of 4 km and monthly temporal resolution. DA-IICT is at geolocation Latitude—23.184179° and Longitude—72.633213°. Monthly datasets for each of these were collected ranging from 2001 to 2018 (DA-IICT was established in 2001). The average value of the pixels falling inside DA-IICT campus will give a single value for these variables. Some of the values in the above datasets were null. These values were imputed to the mean of the rest of the data.

4 Light Use Efficiency (LUE) Based Model

On the land, the major exchange of carbon with the atmosphere results from photosynthesis and respiration [6]. Carbon sequestered into plants, also known as vegetation carbon comes from the Primary Productivity of plants using CO_2, sunlight, and moisture. The rate at which plants convert radiant energy into organic substances and other by-products is known as its Primary Productivity or Gross Primary Productivity (GPP). A fraction of this food reserve is used by the plants to participate in respiration, during this it releases back some of the CO_2 back into the atmosphere. Thus, the effective carbon sequestered by the plants is actually the respiration subtracted from the GPP. This end product is known as NPP. There are a lot of ways to measure the GPP of plants including allometric, Flux Tower method, Extrapolation, Light Use Efficiency-based model (LUE), and Process-based models.

To measure the primary productivity in the institute campus, we use the Light Use Efficiency model based on Monteith's Light Use Efficiency concept [7]. It indicates that the annual productivity of plants is directly linked to the Absorbed PAR (Photosynthetically absorbed radiation) (400–700 nm) and LUE. LUE (ε) is the efficiency with which plants convert the Absorbed PAR into dry matter.

$$\text{GPP} = \text{APAR} \times \varepsilon \tag{3}$$

APAR is expressed as the product of the PAR and the fraction of Absorbed Photosynthetically Active Radiation (FPAR).

$$\text{APAR} = \text{FPAR} \times \text{PAR} \tag{4}$$

PAR refers to that spectral range of solar radiation that can be fruitfully used by plants to participate in photosynthesis. The range is between 400 and 700 nm in the short infrared region. Similarly, FPAR is defined as the fraction of PAR that is actually absorbed by a vegetation canopy.

$$\text{GPP} = (\text{FPAR} \times \text{PAR}) \times \varepsilon \tag{5}$$

FPAR shows a linear relationship with NDVI [8]. By critical examination of this equation, we calculate the FPAR on campus.

$$\text{FPAR} = 1.164 \times \text{NDVI} - 0.143 \tag{6}$$

The SI unit of PAR is μmole m^2/s and the SI unit of Solar radiation (SR) is W/m^2. 1 $W/m^2 \approx 4.6$ μmole m^2/s [9]. But this approximation accounts solar radiation only for wavelengths from 400 to 700 nm. This region only represents about 45% of the entire spectrum. As a result:

$$\text{PAR} = 0.4762 \times \text{SR} \tag{7}$$

Combining all of the above equations

$$GPP = PAR \times FPAR \times LUE \tag{8}$$

If we factor out the energy given out during plant respiration over the same period, the Net Primary Productivity (NPP) is achieved.

$$\begin{aligned} NPP &= GPP - RE \\ RE &: \text{Respiration done by plants.} \end{aligned} \tag{9}$$

However, calculating the amount of energy released during respiration is very tough and mundane. Most remote sensing applications are driven by this equation:

$$NPP = 0.5 \times GPP \tag{10}$$

5 Decision Tree-Based Regression

The Decision tree is a supervised machine learning algorithm used for both classification and regression. The compartmental model uses a decision tree-based regressor to predict the CO_2 emission in the next month. The regressor uses a binary tree structure which is constructed on the basis of a loss function which is taken to be Mean Squared Error (MSE) in the model. Before training the model, the entire dataset is considered to be at the root. The dataset is then divided into two parts depending on the standard deviation and mean and this becomes the child of the root. This process continues until the max depth is reached which is taken to be 5 in this model. After the tree is constructed, the data is rearranged according to the MSE and the value at the leaves of the tree are used to predict the next value.

6 Results and Validation

The compartmental model was used for modeling all the obtained data. Data from 2001 to 2013 was used for training and the rest of the data served the purpose of validation and testing. All the three models (NDVI, FPAR, NPP) have been separately fitted using decision tree regression. The first model uses NDVI to predict NPP and is shown in Fig. 5. The second model uses FPAR data to predict NPP and is shown in Fig. 6. The final model uses NPP data directly and is shown in Fig. 7.

It can be quite clearly seen that the model using NPP gives the least validation error while the model using NDVI gives the largest validation error. This indicates NPP is quite a good quantifier to understand the dynamics of carbon Dioxide (Table 1).

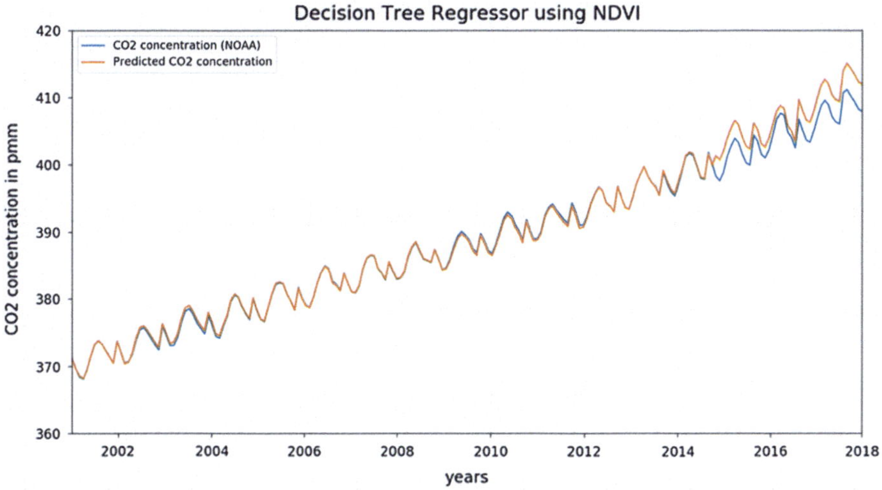

Fig. 5 Graph shows true CO_2 concentration against the concentration predicted by this model. The model uses NDVI for computing LUE. The training Mean Squared Error (MSE) was 2.918e−08 ppm. The validation MSE was 0.0003

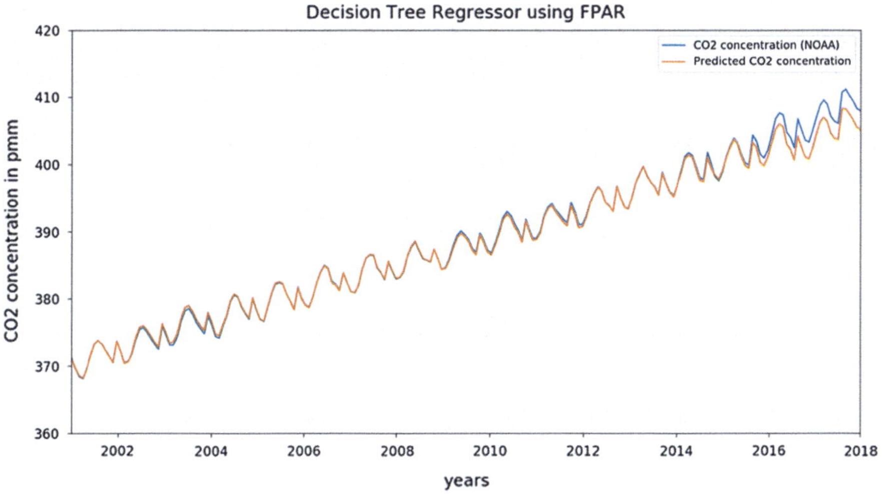

Fig. 6 Graph shows true CO_2 concentration against the concentration predicted by this model. The model uses FPAR for computing the LUE. The training Mean Squared Error (MSE) was 2.918e−08 ppm. The validation MSE was 0.000297

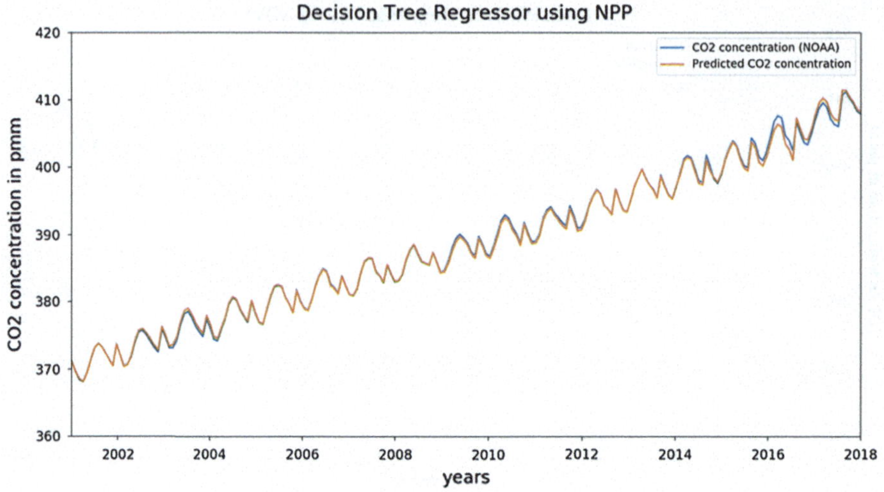

Fig. 7 Graph shows true CO_2 concentration against the concentration predicted by this model. The model uses FPAR for computing LUE. The training Mean Squared Error (MSE) was 2.918e−08 ppm. The validation MSE was 3.98e−05

Table 1 Table summarizing model using (NDVI, FPAR, and NPP) values

Model using	Training error (ppm)	Validation error (ppm)	Predicted CO_2 for January 2019 (ppm)
NDVI	3.937e−08	0.0003	411.140
FPAR	2.774e−08	0.000297	413.615
NPP	2.417e−08	0.0000398	413.87

Acknowledgements We would like to especially acknowledge the support that we have received from our institute. Also, MODIS and NOAA Observatory for providing the data necessary for this study.

References

1. Long WE, Ernstberger J (2013) Modeling atmospheric carbon dioxide over the United States. https://doi.org/10.1137/13s012352
2. Mauna Loa Observatory, noaa.gov, NOAA, n.d
3. Dash J, Curran PJ (2004) The MERIS terrestrial chlorophyll index. Int J Remote Sens 25(23):5403–5413. https://doi.org/10.1080/0143116042000274015
4. https://doi.org/10.5067/MODIS/MOD13Q1.006
5. https://doi.org/10.5067/MODIS/MOD17A3H.006
6. Rosati A, DeJong TM, Esparza G (2001) Physiological basis for light use efficiency models. https://doi.org/10.17660/actahortic.2002.584.9

7. Monteith J (1972) Solar radiation and productivity in tropical ecosystems. J Appl Ecol 9(3):747–766. https://doi.org/10.2307/2401901
8. Myneni RB, Williams DL (1994) On the relationship between FAPAR and NDVI. https://doi.org/10.1016/0034-4257(94)90016-7
9. Sager JC, Craig Mc Farlane J (2017) Radiation. In: Plant growth chamber handbook. https://www.controlledenvironments.org/wpcontent/uploads/sites/6/2017/06/Ch01.pdf

Orthogonalizing Weights in Capsule Network Architecture

Shubhranil Kundu and B. Gagana

Abstract Scalar neural network algorithms are limited in their ability to understand scale, rotational, or affine transformations within images and resort to average or max-pooling techniques which result in translational invariance. In an attempt to overcome these limitations, Hinton et al. introduced vectorized capsule network frameworks which support equivariance while capturing spatial relationships between data points, thus enhancing predictive capabilities of networks. However, experimenting with activation functions, hyperparameters, and optimizers have proven faster convergence and orthogonalizing weights within the layers of capsules enhance performance by slashing associated average error rates.

Keywords Weights · Orthogonality · Capsules

1 Introduction

1.1 Capsule Network Architecture

The vector based encoder–decoder neural architecture [1] mainly comprises of six layers namely convolutional 2D layer, primary capsule layer, digit capsule layer along with three fully connected layers while handling a total of 8,238,608 parameters during the process of image reconstruction.

The detailed architecture is as demonstrated (see Fig. 1).

A detailed breakdown of the layers of the capsule network is as follows.

Encoder. The encoder accepts the input image and predicts the set of structure variables corresponding to transformation matrices while dealing with margin as

S. Kundu (✉)
Madhav Institute of Technology, Gwalior, India
e-mail: neel.sk37@gmail.com

B. Gagana
PES University, Bangalore, India
e-mail: gaganab20496@gmail.com

S. Fong et al. (eds.), *ICT Analysis and Applications*, Lecture Notes in Networks and Systems 93, https://doi.org/10.1007/978-981-15-0630-7_8

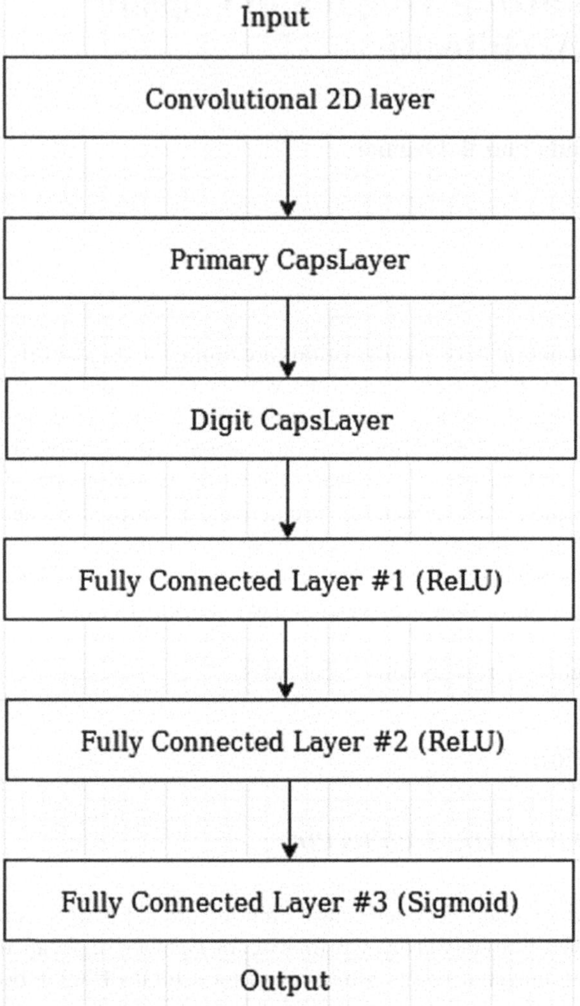

Fig. 1 Layer wise Capsule Network architecture

well as reconstruction loss. The functionality of each of the layers is as described below.

Convolutional 2D Layer. The layer with $9 \times 9 \times 1$ 2^8 kernels with stride 1, accepts the 28×28 input image and outputs a $20 \times 20 \times 256$ tensor and has 20,992 trainable parameters in all.

Primary Caps Layer. This layer which is composed of 32 primary capsule units, accepts the $20 \times 20 \times 256$ tensor from the previous layer, applies $9 \times 9 \times 256$ convolutions to the input volume and produces combinations of the already detected features. The output of this layer is a 6 × 6 × 8 tensor and has 5,308,672 trainable parameters.

Digit Caps Layer. The 10 digit capsules layer accepts the 6 × 6 × 328 dimensional tensor and maps the input to a 16D capsule output space. These coefficients are then used in routing [2]. The layer has 149,760 trainable parameters per capsule.

The loss of the system is defined by

$$\text{loss} = \text{margin\_loss} + \text{reconstruction\_loss} \tag{1}$$

where the reconstruction loss function is defined by a regularization factor which doesn't dominate over the marginal loss. The margin loss function for each category c is defined by a zero loss when correct prediction occurs with probability greater than 0.9 or when an incorrect prediction occurs having a probability less than 0.1 (non zero otherwise in either case). The same can be mathematically represented as follows:

$$L_c = T_c \max\left(0,\ m^+ - ||v_c||\right)^2 + \lambda\,(1 - T_c)\max\left(0,\ ||v_c|| - m^-\right)^2 \tag{2}$$

where λ is the numerical stability constant, and for a matching training label, there is only one correct DigitCaps capsule while the rest nine remain incorrect. The equation then retracts all values accordingly.

Decoder. The decoder part of the capsule system accepts the image format previously processed by the encoder and produces the final reconstructed output image after passing the input through three fully connected layers which also perform denoising thus, supporting routing by agreement. Each of the layers of the decoder accept processed input from the previous layers and pass the input to the next layer in a feed-forward manner while calculating number of parameters based on bias. The structure of each of the layers is as described below.

Fully Connected Layer #1. The layer with the ReLU activation function accepts the 16 × 10 input from the Digit CapsLayer and produces a 512 vector output while dealing with 82,432 parameters.

Fully Connected Layer #2. This ReLU activated layer accepts the 512 vector output from the previous fully connected layer and produces a 1024 vector output while dealing with 525,312 parameters.

Fully Connected Layer #3. The Sigmoid activated layer accepts the 2^{10} vector output from the previous fully connected layer and produces a 784 vector which is also the 28 × 28 reconstructed output while dealing with 803,600 parameters.

1.2 Orthogonality in Weights

As the training of neural networks is laborious due to vanishing or exploding gradients [3], proliferation or fluctuation in saddle points [4], and/or feature statistic shifts [5], introducing orthogonality in weights ensures faster and more stable convergence during training and is enforced through variants of Frobenius norm regularizer, mutual coherence or restricted isometric properties [6]. Orthogonality implies energy preservation and hence, stabilizes the distribution of activations over the concerned layer. The various methods for enforcement of orthogonalilty are as described below.

Double soft orthogonality regularizer. The soft orthogonality regularizer defined by:

$$\lambda||W^T W - I||_F{}^2 \tag{3}$$

is expanded to compensate for under and over complete cases using

$$\lambda\left(||W^T W - I||_F{}^2 + ||W W^T - I||_F{}^2\right) \tag{4}$$

where $W^T W = W W^T = I$ for an orthogonal W.

Mutual coherence regularizer. The mutual coherence regularization can be defined by:

$$\lambda(|| W^T W - I ||_\infty) \tag{5}$$

where the gradient can be solved with smoothing techniques applied to the l_∞ norm [7].

Spectral restricted isometry property regularizer. This regularization is defined by the following equation:

$$\lambda\sigma\left(W^T W - I\right) \tag{6}$$

where W is well-conditioned and regularizes computation cost from $O(n^3)$ to $O(mn^2)$

2 Related Work

2.1 Capsule Network Architecture

Fundamental research. While capsule networks are still in a nascent stage of research, experiments have been carried out around implementing additional inception blocks [8], and enabling sparsity in activation value distribution of capsules in primary capsule layer [9, 10].

Application research. Capsules have, in recent times, been employed in multitude of tasks including audio processing [11], mobility-on-demand network coordination [12], gait recognition [13], traffic classification for smart cities [14, 15], natural language tasks like sentiment analysis [16] and healthcare applications [17, 18].

2.2 *Weight Orthogonality*

Past research along the lines of weight orthogonality have been specifically focused toward reducing internal covariate shifts [19, 20], avoiding gradient explosions, exploring soft and hard constraints [21], using nonlinear dynamics to stabilize weight layer-wise activation distribution [20].

3 Methodology

The following section describes the architectural frameworks and the detailed experimental setup used to obtain the results described in Sect. 4.

A pytorch implementation of Capsule Networks was run on a tesla K40c GPU environment while implementing the same on the FashionMNIST [21] dataset. Since, under simplified assumptions [22], it's concluded that random initialisations produce similar convergence rates as in case of unsupervised pretraining, initial orthogonality wouldn't necessarily sustain throughout training and could breakdown if improperly regularized. With proper regularization, experimentation proves that adding orthogonality regularizations can impact accuracy as well as empirical convergence. However, the following regularizations and optimizations ensure faster convergence, and hence, have been executed for 20 epochs as the results are known to stabilize thereafter.

The modifications incorporated into the original architecture is as detailed below.

Activation or Transfer function. Activation functions are nonlinear complex functional mappings between incoming data and response variables. Capsule Networks by default employs the ReLU [23] activation function which can be mathematically represented as:

$$A(x) = \max(0, x) \tag{7}$$

However, our past experimentation [24] has shown that newer nonlinear transformations like Swish [22] outperforms standard ReLU. Ramachandran et al. proposed function can hereby be defined in terms of the sigmoid as:

$$f(x) = x.\text{sigmoid}(\beta x) \tag{8}$$

Table 1 Comparison between the original and the modified Capsule Network architecture in terms of accuracy

Architecture	Minimum accuracy value (%)	Maximum accuracy value (%)	Average accuracy value (%)
Original	48.7	93.8	82.8655
Modified	56.4	96.3	86.3420

Softmax optimization. In order to enhance the object recognition performance, additional softmax layers [25] are augmented into the decoder of the Capsule architecture and tend to have the same number of nodes as that of the previous layer [26]. The softmax layer computes the probability distributions for all involved classes and is mathematically:

$$P(y = j|x) = e_{w_j} T_{x+b_j} / \sum_{k \in K} e^{w} k^{T_x} + b_k \tag{9}$$

Optimizer. As optimizers like Adam [27] fail to stably converge at extreme learning rates, newer techniques like Adabound [28] tend to work better where the lower bound is η_t and upper bound is η_u with gradient clipping that enhances adaptive moment estimation with dynamic bounds.

Weight orthogonality. The capsnet system has two types of weights: (a) the weights between the primary and digit caps and (b) the dynamic weight connections. Applying the orthogonality equations to both of these ensures coherent and faster convergence.

4 Results and Discussions

The results of the experiments are as follows. The baseline is the original capsule network architecture which achieves maximum accuracy of 93.8% while the architecture with the above-suggested modifications achieves maximum accuracy of 96.3%. Hence, a relative increase of 3.4765% can be observed in the average accuracy values. The same are tabulated (see Table 1).

A visual representation of the same is as follows (see Fig. 2).

5 Conclusion and Future Work

It is evident from the above experimentation that implementation of weight orthogonality along with other optimizations and activation functions, ensure faster convergence and hence, lesser training time while enhancing reconstruction performance. Future work could entail aforementioned optimizations applied to more complex models or further enhancements to the architecture like experimenting with newer

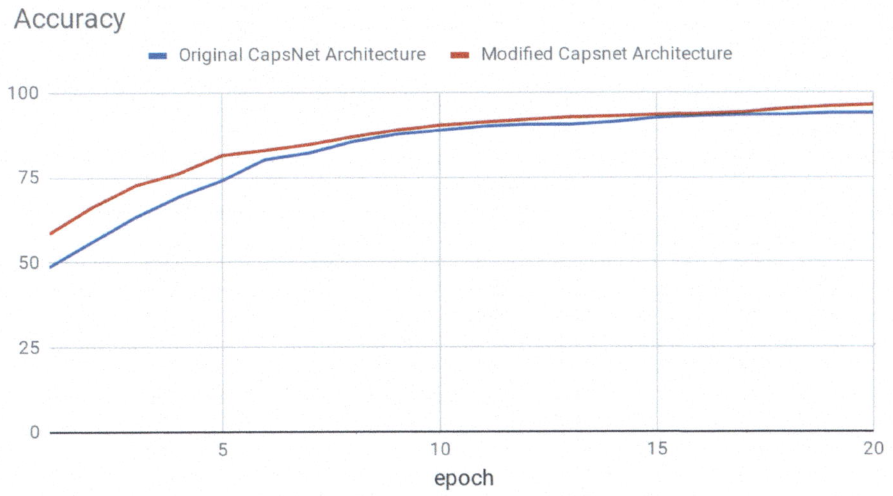

Fig. 2 Comparison between accuracy of original and modified Capsule Network architecture with reference to discussed dataset

techniques and methodologies. While the scientific community continues to explore capsule networks and other upcoming systems, the future could spur out challenging avenues to explore current and newer problems in interdisciplinary spaces as in case of nuclear variability in galaxies [29].

References

1. Sabour S, Frosst N, Hinton G (2017) Dynamic routing between capsules. In: Thirty first conference on neural information processing systems. arXiv:1710.09829
2. Hinton G, Sabour S, Frosst N (2018) Matrix capsules with EM routing. In: International conference on learning representations
3. Pascanu R, Mikolov T, Bengio Y (2013) On the difficulty of training recurrent neural networks. In: International conference on machine learning, pp 1310–1318
4. Dauphin Y et al (2014) Identifying and attacking the saddle point problem in high dimensional non convex optimization. In: Advances in neural information processing systems, pp 2933–2941
5. Ioffe S et al (2015) Batch normalization: accelerating deep network training by reducing internal covariance shift. In: International conference on machine learning, pp 448–456
6. Bansal N, et al (2018) Can we gain more from orthogonality regularizations in training Deep CNN's? In: Advances in neural information processing systems. arXiv:1810.09102
7. Balestriero R et al (2018) Mad max: affine spline insights into deep learning. arXiv:1805.06576
8. Zhu Z et al (2019) A convolutional neural network based on a capsule network with strong generalisation for bearing fault diagnosis. Neurocomputing 323: 62–75
9. Lin Z, Lu C, Li H (2015) Optimized projections for compressed sensing via direct mutual coherence minimization. arXiv:1508.03117
10. Zhong C (2019) Residual capsule networks. In: 11th annual undergraduate research symposium. Poster session. University of Wisconsin, Milwaukee

11. Jain R (2019) Improving performance and inference on audio classification tasks using capsule networks. arXiv:1902.05069
12. He S, Shin K (2019) Spatio-temporal capsule based reinforcement learning for mobility-on-demand network coordination. In: Association for computing machinery. ISBN: 978-1-4503-6674-8/19/05
13. Zhaopeng Xu, Wei Lu, Zhang Qqui, Yeung Yuileong, Chen Xin (2019) Gait recognition based on capsule network. J Vis Commun Image Represent 59:159–167
14. Yao H et al (2019) Capsule network assisted IoT traffic classification mechanism for smart cities. IEEE Internet Things J
15. Kim Y, Wang P, Zhu Y, Mihaylova L (2018) A capsule network for traffic speed prediction in complex road networks. In: Sensor data fusion: trends, solutions, applications. IEEE, Bonn, Germany. https://doi.org/10.1109/sdf.2018.8547068
16. Du Y, Zhao X, He M, Guo W (2019) A novel capsule based hybrid Neural network for sentiment classification. IEEE Access 7: 39321–39328. https://doi.org/10.1109/access.2019.2906398
17. Qiao K, Zhang C, Wang L, Yan B, Chen J, Zeng L, Tong L (2018) Accurate reconstruction of image stimuli from human fMRI based on the decoding model with capsule network architecture. arXiv:1801.00602
18. Mobiny A, Nguyer H (2018) Fast Capsnet for lung cancer screening. In: Frangi A, Schnabel J, Davatzikos C, Alberola Lopez C, Fichtinger G (eds) Medical image computing and computer assisted intervention (MICCAI). Lecture notes in Computer Science, vol 11071. Springer, Cham
19. Glorot X, Bengio Y (2010) Understanding the difficulty of training deep feed forward neural networks. In: Proceedings of thirteenth international conference on artificial intelligence and statistics, pp 249–256
20. He K, Zhang X, Ren S, Sun J (2015) Delving deep into rectifiers: surpassing human-level performance on image classification. In: Proceedings of IEEE international conference on Computer Vision, pp 1026–1034
21. Vorontsov E, Trabelsi C, Kadoury S, Pal C (2017) On orthogonality and learning recurrent networks with long term dependencies. arXiv:1702.00071
22. Saxe A et al (2013) Exact solutions to the non linear dynamics of learning in deep linear neural networks. arXiv:1312.6120
23. Rodriguez P, et al (2016) Regularizing CNN's with locally constrained decorrelations. arXiv: 1611.01967
24. Xiao H, Rasul K, Vollagraf R (2017) Fashion-MNIST: a novel image dataset for benchmarking machine learning algorithms. arXiv:1708.07747
25. Mass AL et al (2013) Rectified non linearities improve neural network acoustic models. In: International conference on machine learning, vol 30
26. Gagana B, Natarajan S (2019) Hyperparameter optimizations for capsule networks. In: EAI endorsed transactions on cloud systems
27. Gagana B, Athri U, Natarajan S (2018) Activation function optimizations for capsule networks. In: IEEE international conference on advances in computing, communications and informatics. ISBN: 978-1-5386-5314-2
28. Le QV et al (2017) Swish: a self gated activation function. In: Neural and evolutionary computing, computer vision and pattern recognition. arXiv:1710.05941
29. Liang M et al (2015) Recurrent convolutional neural network for object recognition. In: Conference on computer vision and pattern recognition
30. Kingma D, Ba J (2015) Adam: a method for stochastic optimisation. arXiv:1412.6980
31. Luo L, Xiong Y, Liu Y, Sun X (2019) Adaptive gradient methods with dynamic bound of learning rate. In: International conference on learning representations
32. Katebi R (2019) Examining extreme nuclear variability in the galaxies that host Active Galactic Nuclei. American Astronomical Society

Smart Billing Using Content-Based Recommender Systems Based on Fingerprint

Darshita Mittal, Sanyukta Shandilya, Dhruv Khirwar and Archana Bhise

Abstract A supermarket is a place where customers come to purchase items for their everyday needs. The long queues at the billing counter and manual billing makes way for a lot of errors and wastage of time. These problems can be overcome with the help of the smart shopping cart which is discussed in this paper. The smart cart will also recommend products based on the customers' purchase history which helps the shopping mart with increasing its sales and customers to remember the products they might forget. Finger print sensor has been used for registering/identifying customers, RFID technology for detection of products being added to the cart, and NodeMCU ESP8266 for Wi-Fi communication between the cart and the centralized system. Content-based filtering is used for implementation of Recommender systems [5].

Keywords Smart shopping cart system · RFID · Centralized system · Recommendation

1 Introduction

Individuals have constantly created innovation to bolster their requirements as from the start of humankind. The fundamental reason for development is that it leads to making our regular tasks simpler and speedier. Ever since the debut of wireless technology, electronic commerce has developed to such an extent to provide convenience, comfort, and efficiency in day-to-day life. One significant task that individuals invest

D. Mittal (✉) · S. Shandilya · D. Khirwar · A. Bhise
Mukesh Patel School of Technology Management and Engineering, SVKM's NMIMS University, Mumbai Campus, Mumbai 400056, Maharashtra, India
e-mail: darshita.mittal55@nmims.edu.in

S. Shandilya
e-mail: shandilya.sanyukta@gmail.com

D. Khirwar
e-mail: dhruvkhirwar1@gmail.com

A. Bhise
e-mail: archana.bhise@nmims.edu

S. Fong et al. (eds.), *ICT Analysis and Applications*, Lecture Notes in Networks and Systems 93, https://doi.org/10.1007/978-981-15-0630-7_9

maximum measure of energy is in shopping. Supermarket is a spot where individuals get their everyday necessities running from sustenance items, garments, electrical machines, and so forth [1]. Some of the time clients have issues with respect to the unspecific data about the item marked down and misuse of superfluous time at the counters. The main purpose of this paper is to provide centralized and automated billing system using RFID and to provide product recommendations to customers using Collaborative Recommender system of Machine Learning.

Radio frequency identification (RFID) is a rapidly growing technology that has the potential to make great economic impacts on many industries [2]. While RFID is a relatively old technology, more recent advancements in chip manufacturing technology are making RFID practical for new applications and settings, particularly consumer item level tagging. Supply-chain management, logistics, and inventory control are few sectors that are highly benefiting from these advancements. RFID tags are cards/tags attached to products that have transponders in them. When they are wirelessly interrogated by RFID readers, they send an ID that may be associated with arbitrary data records. RFID systems are a type of automatic identification system, just like optical bar codes. All products of supermarket will be supplied with an RFID tag, to identify the product and all its details like the price or brand. Every cart contains an RFID reader, Wi-Fi module, and a fingerprint sensor. There will also be a centralized database from which we can give product recommendations to the customer. We use the recommender system of machine learning. The fingerprint sensor takes the fingerprint of the customer relating it to all the products he bought and thus, the data is collected for the recommender system.

2 Related Work

In "Smart shopping cart with automatic billing system through RFID and ZigBee," the authors Mr. P. Chandrasekar and Ms. T. Sangeetha have implemented Smart Shopping cart using RFID for product detection and Zigbee for communication between carts and centralized system. Zigbee has a very low data rate and use of Wi-Fi module is a new technology that enables communication with higher data rate. This system does not add any extra features to the smart cart except for billing [3].

In "RFID based Smart Shopping: An Overview" by Zeeshan Ali and Reena Sonkusare. They have implemented an innovative concept of Intelligent Smart Shopping and Billing. This is a system implemented on RFID and Zigbee technology with no additional features [4].

In "Smart Cart with Automatic Billing, Product Information, Product Recommendation Using RFID and Zigbee with Anti-Theft," the authors Ankush Yewatkar, Faiz Inamdar, Raj Singh, Ayushya, and Amol Bandal have implemented Smart Cart with RFID and Zigbee and have added additional features of online payment system and anti-theft [5].

In "Electronic Shopping Cart for Effective Shopping based on RFID," authors Kalyani Dawkhar, Shradhha Dhomase, and Samruddhi Mahabaleshwarkar have implemented an electronic automatic cart system which automatically detects products being added to the cart and bills them on its own. The total value of the bill is displayed on an LCD screen [6].

Most existing projects have used Zigbee module for communication whereas we are using NodeMCU. Even though Zigbee is less expensive than a NodeMCU the range is less and a greater number of ZigBee modules will be required making the system more complex hence, in this project a NodeMCU is more efficient.

We are using recommender systems for product recommendations. Recommender system is the latest algorithm being used for personalized recommendations and is proven to be more accurate. It is also implemented by websites like Netflix.

The microcontroller being used is Arduino which is very simple to use and makes the system more affordable.

We are using fingerprint for customer identification making the processing faster rather than facial recognition which first of all the customers will not be very open to, plus it will make processing very complicated and hence slow.

3 Proposed Method

The approach that has been used in order to implement the idea can be elaborated using four steps which are stated below

1. Integration of fingerprint sensor, RFID reader module, and Wi-Fi module with the microprocessor (Arduino UNO) to form the PID
2. Creation of user dataset
3. Creation of product and purchase history dataset
4. Building the recommendation system.

Figure 1 shows the working of the project. The user first places their finger on the fingerprint sensor module attached to the PID which then identifies if the user is registered or not by checking the user database. If the user is registered, the cart is linked to the purchase history of the database and as new products are added the database keeps updating. Using this database, by analyzing the purchase history of

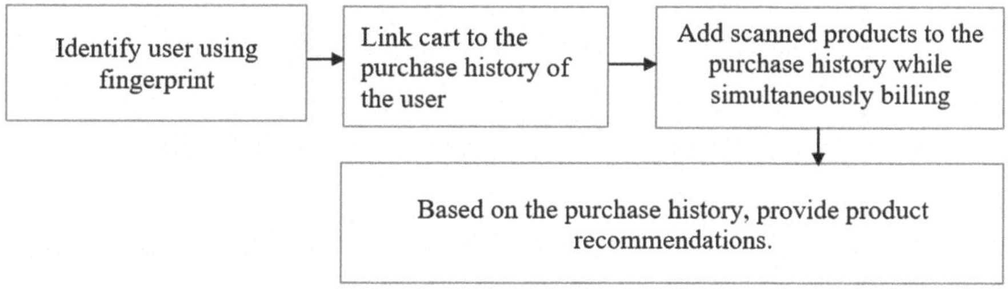

Fig. 1 Flowchart depicting the working of the project

users, we recommend other products that they have not previously purchased, to them.

As mentioned above, the detailed steps are:

A. Integration of fingerprint sensor, RFID reader module, and Wi-Fi module with the microprocessor (Arduino UNO) to form the PID

The microprocessor used here is Arduino UNO due to its cost-effectiveness and compatibility with most hardware. The components used here are—Fingerprint Sensor Module (FPS GT-511C3), RFID Reader Module (EM18), and Wi-Fi Module (NodeMCU ESP8266). The fingerprint sensor accepts the fingerprint, processes it to extract features to identify a user and assigns an ID to the user. EM18 reads the RFID on the products in the range of 10 cm and sends the ID on the RFID card which gives the system details of the product purchased. NodeMCU ESP8266 enables serial transfer of these ID's from Arduino to the central database.

B. Creation of user dataset

The user dataset has been created by us. It has a total of 35 users and their fingerprints. Every user has been assigned a unique ID by the fingerprint sensor itself. It is a simple database with just the user-fingerprint feature and the ID assigned to the user.

C. Creation of product and purchase history dataset

The product dataset is of size (48, 18), i.e., there is a total of 48 products falling in 18 categories. Each product falls into 1 or more categories of products. A 1 is assigned to the product for whichever categories it falls into, hence making a matrix of 1s and 0s (Fig. 2).

The purchase history database is of size (15, 50) with rows beings customer ID and columns being the list of products. For each user, the history of products purchased is mentioned with the quantity (Fig. 3).

These databases are further used by the recommendation engine to provide product recommendations to the users.

Product name	Price	Snacks	School supplies	Pet supplies	Frozen	Body care
Pedigree	0	0	0	1	0	0
Pringles	0	1	0	0	0	0
Saucepan	0	0	0	0	0	0
Nescafe	0	0	0	0	0	0
Pen	0	0	1	0	0	0
Nivea cream	0	0	0	0	0	1
Chicken Ham	0	0	0	0	1	0
Colin	0	0	0	0	0	0
Dal	0	0	0	0	0	0

Fig. 2 Snapshot of the product database

Customer ID	Customer Name	Pedigree	Pringles	Saucepan	Nescafe
12	Abhi	0	1	0	1
15	Anirudh	1	0	0	0
3	Dharmesh	0	1	0	2
6	Dhruv	2	0	0	1
4	Kruti	0	4	0	2
8	Mangesh	0	0	0	0
5	Neel	0	1	0	1

Fig. 3 Snapshot of the purchase history database

D. **Building the recommendation system**

Recommendation engines identify things that a user may like based on products that they've interacted with in the past. We have used TensorFlow to implement machine learning for this purpose [7, 8].

Figure 4 shows the algorithm used for building the recommendation system. Once we have the matrix of products and the user purchase history, the user quantity matrix, i.e., a list of products purchased by a user with quantities is multiplied with the product types to give a stack of users with their top product types.

Figure 5 shows the list created by multiplication of product types and user purchase history matrix.

The sum is taken across columns of this matrix and then dot product is calculated with the user products matrix. These values are then summed across rows to give the normalized values. These normalized values lie between 0 and 1 and show how similar the item is to the user's previous purchases. The product with the highest normalized value is thus recommended at the top by the recommendation engine followed by the products with decreasing values [8].

Figure 6 shows the snapshot of the simple GUI created to show the recommendations in an organized manner.

The GUI has been created by using PyQt5 library and its widgets (Fig. 7).

4 Results and Discussions

The smart shopping cart implements RFID and Wi-Fi technology along with Recommender systems using content-based filtering. Content-based filtering is based on comparing the history of the products purchased by the users and recommending products similar to the history or by comparing the history of users and recommending products from the history of similar users.

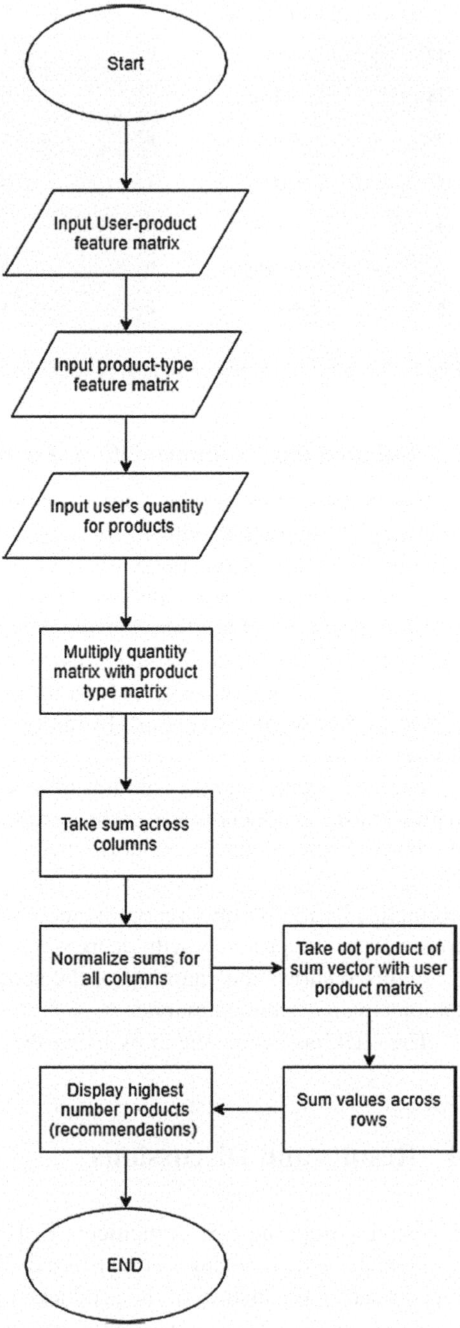

Fig. 4 Recommender systems algorithm

Key	Type	Size	Value
1	list	18	['Vegetables', 'Fruits', 'Household supplies', 'Spices/pulses', 'Snack ...
2	list	18	['Household supplies', 'Vegetables', 'Fruits', 'School supplies', 'Per ...
3	list	18	['Fruits', 'Household supplies', 'Vegetables', 'Snacks', 'Beverages', ...
4	list	18	['Snacks', 'Fruits', 'Vegetables', 'Household supplies', 'Spices/pulse ...
5	list	18	['Vegetables', 'Fruits', 'Snacks', 'Beverages', 'Household supplies', ...
6	list	18	['Snacks', 'Vegetables', 'Fruits', 'Spices/pulses', 'Cereals', 'Househ ...
7	list	18	['Snacks', 'Household supplies', 'Vegetables', 'Beverages', 'Personal ...
8	list	18	['Snacks', 'Beverages', 'Fruits', 'School supplies', 'Household suppli ...
9	list	18	['Snacks', 'Vegetables', 'Fruits', 'Spices/pulses',

Fig. 5 Lists of top user types

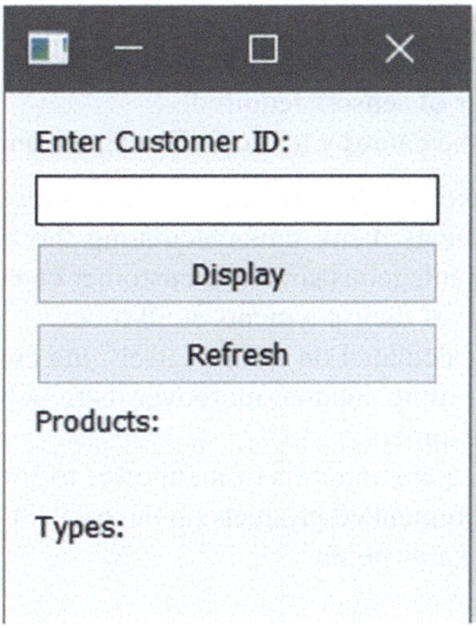

Fig. 6 Snapshot of the GUI

A more advanced algorithm for implementing recommender systems could have been collaborative filtering which takes many more factors into account while recommending products.

High-frequency or UHF RFID technology could have been used which has a better range. The availability of these modules is a problem and they are also not cost-effective [2].

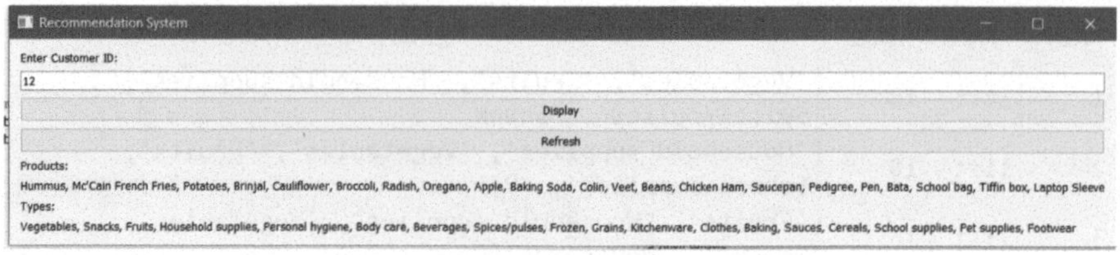

Fig. 7 Snapshot of GUI showing recommendations of user with ID 12

5 Conclusion

The smart cart system implemented was successful in generating databases of users and purchased products and give a customized product recommendation for each customer. The recommendation is displayed on the central computer where the recommendations are being generated. The data is communicated between the cart system (sensors attached along with the microcontroller to the cart) to the central computer via NodeMCU. This improves upon the existing smart shopping system implemented which requires multiple sensors in each aisle and relies heavily on image recognition for recognizing the user and the product being purchased. This allows for higher chances for technical error. The smart cart system implemented in this project requires the sensors to be attached to the cart instead on every aisle which reduces the number of sensors required.

The smart cart opens the gateway to the following advantages:

(1) The smart cart created will be used by the customers to put their products in, while the system detects them, thus calculating the bill, while the customer shops and this is advantageous since the customer can check their total paying amount and be aware of their payments at all times.
(2) With the bill being calculated on the cart itself, the customer is not needed to stand in lines at the billing counter, moreover there will be no billing counters.
(3) Saves time of the customers.
(4) Cuts costs of the company in terms of manpower to some extent.
(5) The customer is recommended products on the basis of his history and it makes it easier for them to know about.

References

1. Zhou RR, Soman D (2003) Looking back: exploring the psychology of queuing and the effect of the number of people behind. J Consum Res 29(4): 517–530. ISSN 0093-5301
2. Hin LK (2007) Radio-frequency identification for library: bookshelf. Department of electronic engineering. City University of Hong Kong
3. Chandrasekar P, Sangeetha T (2014) Smart shopping cart with automatic billing system through RFID and ZigBee. In: International conference on information communication and embedded systems (ICICES 2014)

4. Ali Z, Sonkusare R (2014) RFID based smart shopping. In: 2014 international conference on advances in communication and computing technologies (ICACACT 2014)
5. Yewatkar A, Inamdar F, Singh R, Bandal A (2016) SmartCart with automatic bolling, product information, product recommendation using RFID & Zigbee with anti-theft. In: 7th international conference on communication, computing and visualization
6. Dawkhar K, Dhomase S, Mahabaleshwarkar S (2015) Electronic shopping cart for effective shopping based on RFID. Int J Innov Res Electr Electron Instrum Control Eng 3
7. Badamasi YA (2014) The working principle of an Arduino. In: 11th international conference on information communication embedded systems electronics, computer and computation (ICECCO), pp 1–4
8. "Recommender Systems with TensorFlow on GCP" offered by Google Cloud. https://www.coursera.org/learn/recommendation-models-gcp
9. Gangwal U, Roy S, Bapat J (2013) Smart shopping cart for automated billing purpose using wireless sensor networks. SENSORCOMM 2013: the seventh international conference on sensor technologies and applications
10. Python for everybody, by Dr Charles Severance

Identifying Classification Technique for Medical Diagnosis

Potnuru Sai Nishant, Shashi Mehrotra, B. Gopesh Krishna Mohan and Gantakora Devaraju

Abstract The paper provides a comparative overview of machine learning techniques in medical diagnosis. We present a performance evaluation analysis of some of the state-of-the-art machine learning approaches applied for medical diagnosis. The research considered six machine learning classification algorithms: Naïve Bayes (NB), Logistic Regression (LR), Support Vector Machine (SVM), Linear Discriminant Analysis (LDA), and two decision trees classifying algorithms: C5.0 and Random Forest. UCI three medical data sets: Cleveland Heart Disease dataset, Wisconsin Diagnostic Breast Cancer dataset, and Pima Indians Diabetes Datasets. Our experimental results show the SVM classification algorithm has achieved the most promising result over all the three medical datasets.

Keywords Support vector machines · Naïve Bayes · Logistic regression · Random forest · Linear discriminant analysis

1 Introduction

Three severe diseases: heart disease, cancer, and diabetes are the main concern for the medical industry. These days, heart disease and cancer are the most common cause of death throughout the world. Breast cancer is one of the most common diseases among women. Many other problems develop if diabetics is not diagnosed or lack of proper treatment [15]. Utilization of machine learning techniques can be useful

P. S. Nishant (✉) · S. Mehrotra · B. G. K. Mohan · G. Devaraju
Department of Computer Science and Engineering, Koneru Lakshmaiah Education Foundation, Vaddeswaram, A.P, India
e-mail: sai.nishant8@gmail.com

S. Mehrotra
e-mail: sethshashi11@gmail.com

B. G. K. Mohan
e-mail: gopeshbhaskaruni@gmail.com

G. Devaraju
e-mail: g.devaraju0143@gmail.com

S. Fong et al. (eds.), *ICT Analysis and Applications*, Lecture Notes in Networks and Systems 93, https://doi.org/10.1007/978-981-15-0630-7_10

for identifying and predictions of these diseases at an early stage and may be a great help in medical assistance [1].

Cardiovascular diseases and attacks in people being one of the dangerous attacks all over the world. When a blood clot blocks blood flow to the heart, a heart attack usually occurs. Treatment ranges from changes in lifestyle. The Indian Heart Association reports that 50% of heart attacks occur in Indians under 50 years of age. Urban people are thrice at risk as rural populations to this disease [6, 16].

Another widespread medical problem among women is breast cancer. According to one of the reports by WHO, breast cancer is the most predominant cancer among women in the world. Indian women are indeed affected by this disease, as in the news, early discernment of this disease could probably increase chances of recovery [8]. Managing breast cancer among older women is indeed challenging due to many reasons, such as tolerance of medication, deformity of data from elderly women, many are undergoing inappropriate treatment which does not fit the patient's health. Mental health problems like depression, agoraphobia, terrible self-image, and the use of abnormal approaches to impersonation affect the life of a cancer patient. Although adequate treatments are available to cure breast cancer, there are some noticeable pains and suffering associated with the available treatment methods. Debilitating, severe pain acts as an addition to a person who is already distressed by many mental, social, and medical traumas.

Diabetes is a disease that occurs when your blood glucose is too high, also referred to as blood sugar. Blood glucose is the primary energy source, and it comes from the food they eat. Insulin, a pancreatic made hormone, helps food glucose get into their cells for energy use. The body sometimes does not produce enough insulin [12]. Diabetes is an uprising disease, primarily due to the type of nutrition we have these days and the different eating scheme and schedule we follow. Diabetes is mainly caused by obesity or high levels of glucose, and so on [4].

Long-term effects of diabetes can lead to stupor, arteria realis failure and retinal failure, pathological destruction of pancreatic islet cells, valvular heart dysfunction, arteriosclerosis, fat loss, infection and bacterial immunity, and carbohydrate levels and protein abnormalities in the body due to due to diminishing levels of insulin [17].

The paper is organized in the following way: Sect. 2 discusses related work. Sect. 3 is about a brief description of classification algorithms. Section 4 goes with the methodology which consists of datasets description, preprocessing of data, modeling and performance measures. Section V covers the experimental results and analysis. Section 5 is the conclusion and future research.

2 Related Work

Sisodia and Sisodia [15], discusses about the performances of three machine learning algorithms Decision tree, Support Vector Machines and Naïve Bayes for the prediction of Diabetes dataset which is the Pima Indian Diabetes Dataset (PIDD) which shows the results with accuracy of 76.30% with Naïve Bayes algorithm and they also evaluated other parameters like precision, recall, and F-measure.

Agarap [1], discusses the classification of breast cancer dataset using machine learning algorithms SVM, Linear Regression, Multilayer Perceptron, GRU-SVM, softmax regression, and nearest neighbor in which MLP stands out the best among them with an accuracy of 99.04%.

Zafiropoulos et al. [19], focuses on the implementation of SVM algorithm on Wisconsin Prognostic breast cancer dataset and Wisconsin Diagnostic breast cancer dataset which got an accuracy of 96.01% and 90%, respectively and evaluated other performance parameters: sensitivity, specificity, and precision.

Amin et al. [2], used classification techniques, K-Nearest Neighbor, and other few methods which we are using in our research, which was then compared with their newly proposed algorithm called Vote, a combination of Naïve Bayes and Logistic Regression which achieved an accuracy of 87.4% for the prediction of Cleveland heart disease dataset.

Aishwarya and Gayathri [3], uses Principal component analysis for preprocessing the Insulin Diabetes Dataset and prediction is done with SVM algorithm in machine learning, and performance analysis is validated using the ROC curves using the MATLAB 10.

Liu et al. [9], proposed a new system for the prediction of heart disease which is divided into two components, one is selection of feature selection, and the other is a classification system. The maximum accuracy obtained was 92.59% according to a jackknife cross-validation scheme and in their system, which was based on C4.5 classifier.

Ojha et al. [13], find out the significant features in the Wisconsin Prognostic Breast Cancer Dataset using the C5.0 algorithm and comparing the results with Naïve Bayes and k-means clustering for decision-making.

3 Methodology

3.1 Data Description

We collected three datasets from the UCI Machine learning repository for experiments.

Heart Disease Dataset. The heart disease was collected from Cleveland Clinic Foundation which consists of 303 observations with 14 features.

Diabetes Dataset. The Diabetes dataset, which is the Pima India Diabetes Dataset (PIDD) which consists of 768 instances with nine attributes.

Cancer Dataset. The cancer Dataset is taken from the UCI Machine Learning Repository, which is obtained from the University of Wisconsin Hospitals, consists of 683 observations with 32 attributes.

3.2 Preprocessing

In preprocessing, data cleaning and relevance analysis are performed, as all the features in the data set are not required for the study [10]. Properly selected feature eases the design process [11]. The Mice package is used for imputing missing values as our dataset instances are not very large and MICE is very efficient for small datasets. MICE is an acronym for Multivariate Imputation via Chained Equations [18]. For our datasets, MICE package will perform imputation based on the class of the variable. In our case, we use "polyreg" which is Bayesian polytomous regression as we have missing values in factor variables then this method is used.

3.3 Modeling and Experimental Results

After preprocessing of data, we implemented the following models; Naïve Bayes, Support Vector Machines (SVM), Random Forest, Logistic Regression, C5.0, linear discriminant analysis. The datasets are divided into a split ratio of 80–20 as training and testing data, respectively for evaluation. We experimented the models over all the three datasets using cross-validation k-fold method with metric accuracy. The k-fold method is a cross-validation technique which works on different sets of random training sets to get the best accuracy. The parameters accuracy, precision, and recall are used for the performance evaluation of the models.

Performance Measures. As we are dealing with classification, we calculate the accuracy, precision, and recall using the confusion matrix obtained after the prediction of testing data in R.

Table 1 presents, the negative and positive in the first row are the predicted ones whereas the negative and positive in the first column are the original ones.

We feel that False Positive and False Negative are very risky, as in the case of False Positive, if a person is healthy and is diagnosed with a disease, then a mental trauma will eventuate. In the case of False Negative, since he or she has been diagnosed healthy, a person will not receive proper medication.

Accuracy

Accuracy is the percentage of our model that shows how our model is good enough to predict with testing data.

$$\text{Accuracy} = (\text{TN} + \text{TP})/(\text{TN} + \text{FP} + \text{FN} + \text{TP}) \tag{1}$$

Table 1 Confusion matrix

	Negative	Positive
Negative	True Negative (TN)	False Positive (FP)
Positive	False Negative (FN)	True Positive (TP)

Precision

Precision deals with how accurate our model can predict out of those positive predictions.

$$\text{Precision} = \text{TN}/(\text{FP} + \text{TP}) \tag{2}$$

Recall

The recall is calculated for knowing the percentage of how many True Positive values are out of actual positives in the data.

$$\text{Recall} = \text{TP}/(\text{TP} + \text{FN}) \tag{3}$$

4 Results and Analysis

In this section, the performance parameters are tabulated, and graphs are shown for model wise performance for each of the diseases considered.

It can be observed from Table 2, Support Vector Machines (SVM) model has achieved the highest accuracy of 91.53%, 83.06%, and 99.12% for heart, diabetes, and cancer datasets, respectively.

Table 3 shows, the Support Vector Machines (SVM) model achieved the highest precision percentages of 93.55%, 83.04%, and 98.61% for heart, diabetes, and cancer datasets, respectively.

It can be noticed from Table 4, support vector machines, random forest, and Naïve Bayes models have achieved the same recall for the heart disease dataset. For diabetes dataset, SVM achieved the best recall rate. For breast cancer, Naïve Bayes and support vector machines obtained 100% recall.

Figures 1, 2, 3, 4, 5, and 6 depicts model wise performance using parameters accuracy, precision, and recall. Graphs are plotted using the ggplot2 package in R. The colors are according to their symbolic representation as in breast cancer goes with pink, diabetes with blue and heart with red color.

Table 2 Accuracy of the models over three datasets

Model	Heart (%)	Diabetes (%)	Breast cancer (%)
Naïve Bayes	89.33	77.78	97.35
Logistic	88.14	75.14	95.58
C5.0	86.44	76.47	92.04
Random forest	88.14	79.08	94.69
LDA	86.44	82.03	93.81
SVM	91.53	83.06	99.12

Table 3 Precision

Model	Heart (%)	Diabetes (%)	Breast cancer (%)
Naïve Bayes	86.11	80.56	95.95
Logistic	86.11	82.61	97.14
C5.0	83.33	80.19	95.59
Random forest	88.57	82.08	94.52
LDA	90.57	81.74	92.11
SVM	93.55	83.04	98.61

Table 4 Recall

Model	Heart (%)	Diabetes (%)	Breast cancer (%)
Naïve Bayes	96.88	95.00	100
Logistic	96.87	93.00	95.77
C5.0	93.75	85.00	91.55
Random forest	96.88	87.00	97.18
LDA	84.18	94.00	98.59
SVM	96.88	97.00	100

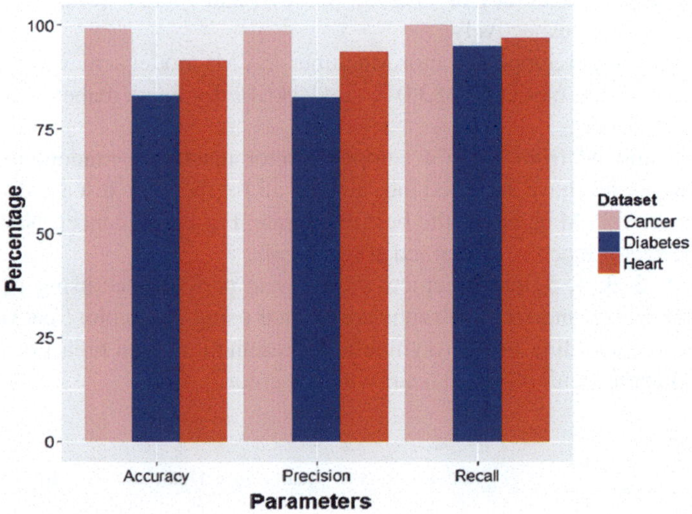

Fig. 1 SVM model performance

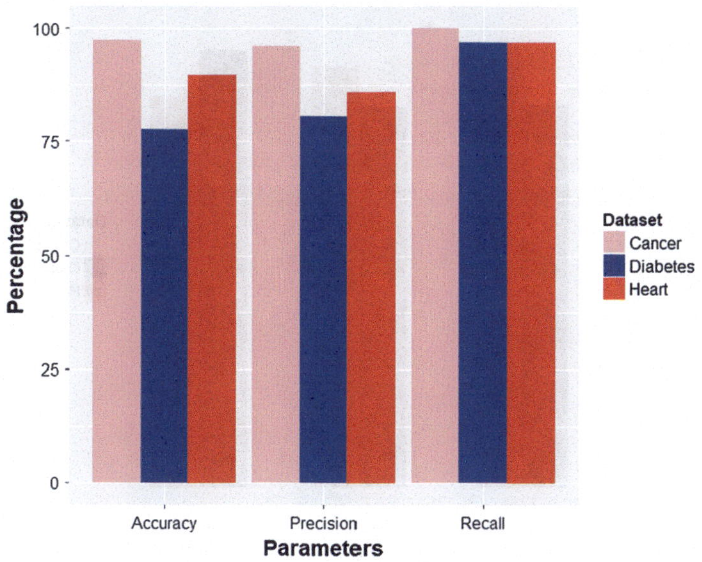

Fig. 2 Naïve Bayes model performance

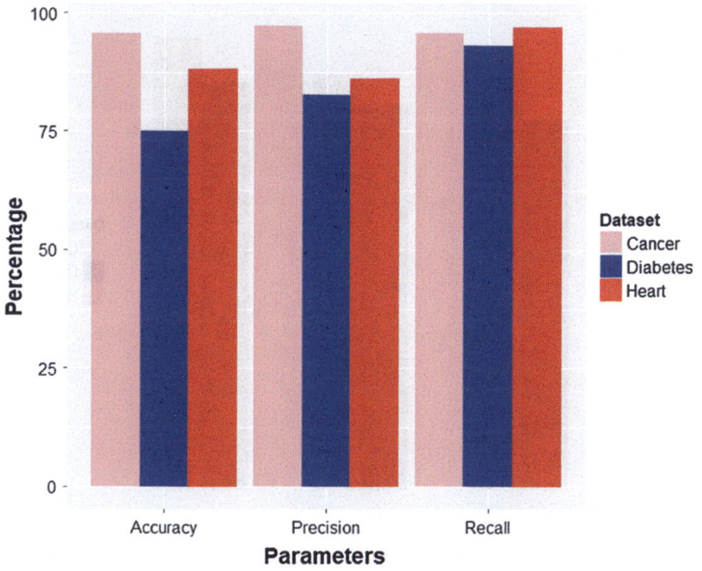

Fig. 3 Logistic regression model performance

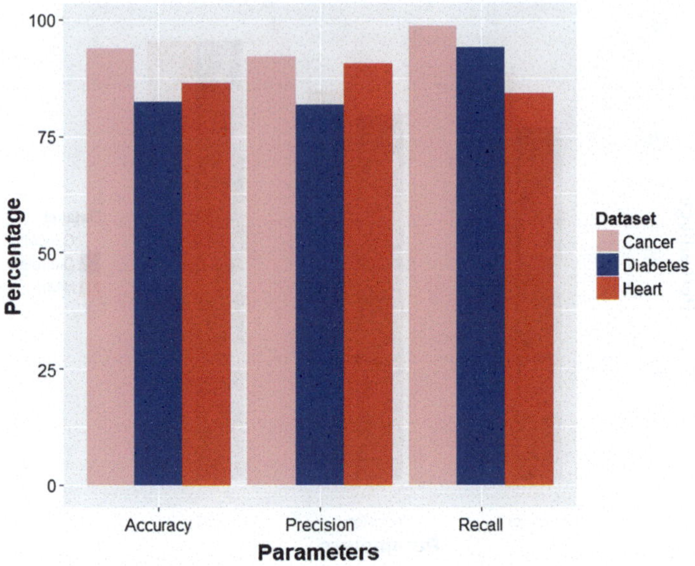

Fig. 4 LDA model performance

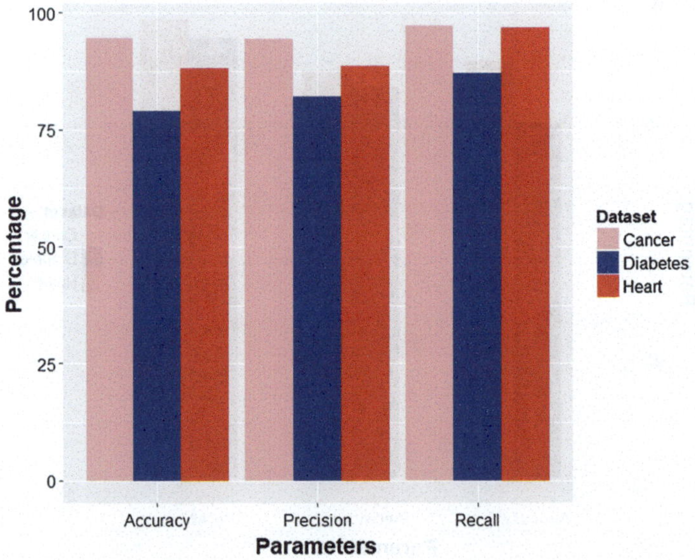

Fig. 5 C5.0 model performance

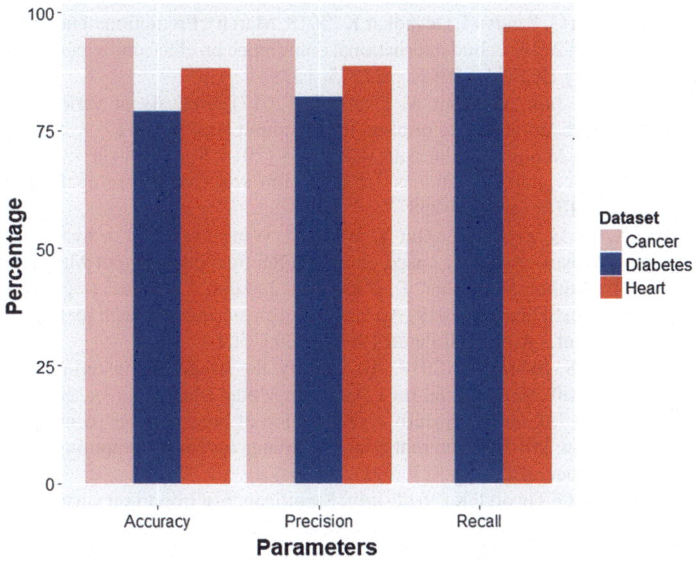

Fig. 6 Random forest model performance

5 Conclusion and Future Work

The paper presents a performance analysis of the six most common machine learning classification algorithms using heart disease, diabetes, and breast cancer data sets. The Support Vector Machine outperformed among all other algorithms for all the three datasets. This research can be broadened by performing these experiments on a large-scale real-world data sets. Our experimental results show SVM achieved an accuracy of 91.53% for heart, 99.12% for breast cancer, and 83.06% for diabetes. Support vector machines made a high percentage not only in accuracy but also wins this competition in other performance parameters like precision and recall.

References

1. Agarap AFM (2018, February) On breast cancer detection: an application of machine learning algorithms on the Wisconsin diagnostic dataset. In Proceedings of the 2nd international conference on machine learning and soft computing. ACM, pp 5–9
2. Amin MS, Chiam YK, Varathan KD (2019) Identification of significant features and data mining techniques in predicting heart disease. Telematics Inform 36:82–93
3. Aishwarya R, Gayathri P (2013) A method for classification using machine learning technique for diabetes
4. Dutta D, Paul D, Ghosh P (2018, November) Analyzing feature importances for diabetes prediction using machine learning. In 2018 IEEE 9th annual information technology, electronics and mobile communication conference (IEMCON). IEEE
5. Fratello M, Tagliaferri R (2018) Decision trees and random forests. Encyclopedia

6. Gavhane A, Kokkula G, Pandya I, Devadkar K (2018, March). Prediction of heart disease using machine learning. In: 2018 second international conference on electronics, communication and aerospace technology (ICECA). IEEE, pp 1275–1278
7. Gupta B, Rawat A, Jain A, Arora A, Dhami N (2017) Analysis of various decision tree algorithms for classification in data mining. Int J Comput Appl 8:15–19
8. Jacob DS, Viswan R, Manju V, PadmaSuresh L, Raj S (2018, March) A survey on breast cancer prediction using data mining techniques. In: 2018 conference on emerging devices and smart systems (ICEDSS). IEEE, pp 256–258
9. Liu X, Wang X, Su Q, Zhang M, Zhu Y, Wang Q, Wang Q (2017) A hybrid classification system for heart disease diagnosis based on the RFRS method. Comput Math Method Med, 2017. (Author, F., Author, S., Author, T.: Book title. 2nd edn. Publisher, Location (1999))
10. Mehrotra S, Kohli, S., Sharan, A (2019) An intelligent clustering approach for improving search result of a website. Int J Adv Intell Parad 12(3–4), 295–304
11. Mehrotra S, Kohli S, Sharan A (2018) To identify the usage of clustering techniques for improving search result of a website. Int J Data Min Model Manag 10(3):229–249
12. Ojha U, Goel S (2017, January) A study on prediction of breast cancer recurrence using data mining techniques. In: 2017 7th international conference on cloud computing, data science & engineering-confluence. IEEE, pp 527–530
13. Ojha U, Jain M, Jain G, Tiwari RK (2017, July) Significance of important attributes for decision making using C5. 0. In: 2017 8th international conference on computing, communication and networking technologies (ICCCNT). IEEE, pp 1–4
14. Sengur A (2008) An expert system based on linear discriminant analysis and adaptive neuro-fuzzy inference system to diagnosis heart valve diseases. Expert Syst Appl 35(1–2):214–222
15. Sisodia D, Sisodia DS (2018) Prediction of diabetes using classification algorithms. Procedia Comput Sci 132:1578–1585
16. Sowmiya C, Sumitra P (2017, March) Analytical study of heart disease diagnosis using classification techniques. In: 2017 IEEE international conference on intelligent techniques in control, optimization and signal processing (INCOS). IEEE, pp 1–5
17. Vaishali R, Sasikala R, Ramasubbareddy S, Remya S, Nalluri S (2017, October) Genetic algorithm-based feature selection and MOE Fuzzy classification algorithm on Pima Indians Diabetes dataset. In: 2017 international conference on computing networking and informatics (ICCNI). IEEE, pp 1–5
18. Yadav ML, Roy Choudhury B (2018) Handling missing values: a study of popular imputation packages in R. Knowl Based Syst 160:104–118
19. Zafiropoulos E, Maglogiannis I, Anagnostopoulos I (2006, June) A support vector machine approach to breast cancer diagnosis and prognosis. In: IFIP international conference on artificial intelligence applications and innovations. Springer, Boston, MA, pp 500–507

Character Recognition of MODI Script Using Distance Classifier Algorithms

Solley Joseph, Jossy P. George and Suhas Gaikwad

Abstract Machine simulation of human reading is an active research area since the introduction of digital computers. Optical character recognition aims at the recognition of printed or handwritten text from document images and converting the same into a machine-readable form. The focus of this work is handwritten character recognition of MODI Script. A proper recognition system for handwritten documents enables it to be conveniently viewed, edited, and shared via electronic means. The development of a character recognition system for some of the ancient script is still a challenging task due to the complex nature of the script. MODI script is one such script which is the shorthand form of the Devanagari script in which Marathi was written. Though at present MODI script is not an official script, there exists a huge collection of MODI documents in various libraries. In addition, it is observed that scholars and historians are taking serious effort to revive the script. The pursed study based on the implementation of two algorithms for the classification of handwritten MODI script. The algorithms use distance classifier method. The first experiment is done using Euclidean distance classifiers and the second one is with Manhattan distance classifier and the accuracy achieved is 99.28% & 94% respectively.

Keywords MODI script · Handwritten character recognition · Classification algorithms · Euclidean distance classifier · Manhattan distance classifier

S. Joseph (✉)
Christ (Deemed to be University), Bangalore, India
e-mail: solley.joseph@res.christuniversity.in

Carmel College for Women, Goa, India

J. P. George · S. Gaikwad
Christ (Deemed to be University), Lavasa, Pune, India
e-mail: frjossy@christuniversity.in

S. Gaikwad
e-mail: suhas.machhindra@cimlavasa.in

S. Fong et al. (eds.), *ICT Analysis and Applications*, Lecture Notes in Networks and Systems 93, https://doi.org/10.1007/978-981-15-0630-7_11

1 Introduction

MODI Script was initiated by Hemadri Pandit also known as Hemadpant, a famous leader of the Yadav dynasty, in the twelfth century. It was used as an official script for writing Marathi. This Script belongs to the family of the script known as Nagari Script.

MODI was easy to write and was commonly used for writing Marathi until 1950. The usage of the script was reduced and gradually stopped when printing technology was introduced in India as it was very difficult for type-setting the script. In addition to Marathi, there are other languages like Urdu, Kannada, Gujarati, Rajasthani, Hindi, and Tamil which used MODI script for writing [1]. It is reported that large volumes of MODI documents are preserved in India as well as other Asian and European countries. Denmark and other countries have also got some collection of MODI documents [2, 3]. Research and development on MODI Script will lead to the advancement of Devanagari and Marathi Script. The MODI script has 46 distinctive letters, of which 36 are consonants and 10 vowels. The basic character set is shown in Fig. 1. In this paper, we propose two algorithms for the classification of handwritten MODI script. The algorithms use distance classifier method. In the first experiment Euclidean classifier is used and in the second one uses Manhattan distance classifier, for the classification and recognition of MODI script. The results obtained from both algorithms are validated with each other. The remaining part of the paper is organized as follows. Section 2 deals with the literature review. Section 3 describes the methodology. The experimental study is illustrated in Sect. 4. The results are discussed in Sect. 5, and conclusion in Sect. 6.

Fig. 1 The basic character set of MODI script

2 Literature Survey

Research in the field of character recognition reveals that recognition accuracy depends heavily on the feature extraction and classification methods used in the recognition process. An analysis of different classification methodologies used in OCR of various languages is carried out in which the author discusses various classification algorithms used for pattern recognition [4]. An insight into different learning-based methods for character recognition is discussed in [5]. The use of KNN and Backpropagation Neural Network (BPNN) [6] for classification of MODI characters, in combination with structural similarity-based feature extraction, has resulted in 91–97% recognition rate. Otsus Binarization algorithm and Kohonen Neural Network-based classification method is used for MODI Script Character Recognizer [7] in which various collected datasets were tested and an overall character recognition rate of 72.6% is achieved. A recognition model based on image centroid and Chain Code has experimented for vowels in MODI script [8]. In the classification phase, a two-layer feed-forward Neural Network and SVM was used, and a recognition rate of 65.3–73.5% was reported. A theoretical analysis of MODI Script character recognition gives an insight into the basic features of the ancient script [9]. A comparison of the character recognition process of MODI, Devanagari, and Roman scripts was performed in this work and found that in the case of MODI Script structural features were difficult to extract due to various reasons such as the complex nature of the script, touching strokes in the character (due to hasty writing), presence of broken stroke primitives, etc. The authors conclude that the recognition process of MODI script is more difficult as compared to other Indian languages because of the factors such as cursive nature of the characters, inconsistency in writing style, shape similarity of some of the characters, etc. MODI character recognition and an attempt to convert it into corresponding English characters are carried out [10] in another study, where the authors list out different steps of the image processing which can be used for MODI script character recognition. Zernike and Zernike complex moments in combination with Euclidean distance classifier, experimented by Kulkarni et al. [11] have achieved 94.78% accuracy. In another experiment Euclidean distance classifier is implemented with the Hu's seven and Zernike features for the MODI numerals [12] as well as MODI characters [13]. In the case of MODI numerals, the recognition accuracy is 86.6%, and in the case of MODI script character, the recognition rate is 82.6%.

A comparative study of the feature extraction and classification techniques used in character recognition system is carried out in [14]. The authors have performed the study based on the various parameters such as sample size, data type, accuracy, etc. A method for freestyle Marathi online handwritten Character Recognition with the genetic algorithm approach is experimented in [15]. The authors also discuss some of the existing methods in their experiment. A survey of various preprocessing and segmentation techniques is performed in [16].

The authors have done a comparative study of different OCR techniques based on the data set, the methods used and the accuracy. Handwritten character recognition of Hindi characters using oriented gradients and Hu-geometric moments is experimented in [17]. They have used histogram oriented gradients and geometric moments in their experiments in combination with SVM classifier and achieved an accuracy of 96.8% in the recognition of Hindi Character data set.

3 Methodology

In the first step, vectorization is used to convert the raster image to the two-dimensional vector representation of the image. Vectorization is an import part of graphics recognition. It deals with converting the scanned image to a vector form that is appropriate for further processing and analysis. Many vectorization algorithms are available for this purpose [18].

The noise removal algorithms reduce or remove the visibility of noise by smoothing the entire image leaving areas near contrast boundaries [19].

The next step is classification. The classification stage is instrumental in mapping an unknown sample into a predefined class. Various methodologies of pattern recognition can be used at this stage for the classification purpose. BPNN, K-Nearest Neighbor, Support Vector Machine, and Euclidean distance classifier are the commonly used methods. The proposed experiment uses two algorithms [20], such as Euclidean distance classifier and Manhattan distance classifier.

4 Experimental Study

Data set used for character recognition consists of 100 samples each of 46 characters of MODI script. Thus the total data set comprises of 4600 MODI characters which are divided as train and test data set (70:30 ratio).

4.1 Vectorization

In the first step, vectorization is performed on the data. Then, the noise reduction technique is applied to the vectorized data.

4.2 Classification

In the classification process, two advanced algorithms are implemented using distance classifier. Euclidean distance and Manhattan distance classifier are used for classification purpose.

The first algorithm uses the Euclidean distance classifier. It has the following steps:

- Aggregation of MODI Script image data of test and train sample. (Load the image data of the test sample and train sample).
- Apply noise reduction filter on the test and train data, to remove the noise from the image (Stage One).
- Mean value computation of test and train sample (M). (Calculation of Standard deviation (Std)).
- Now the above value of two different data points is treated and subtracted to get the square and certainly square root of sample data

The same algorithm is also performed on the image after flipping it to 90°.

The second algorithm uses the Manhattan classifier. It has the following steps:

- Cramming MODI Script Image data (test and train sample).
- Using noise reduction filter removes the noise in the image (Stage 1).
- Forecasting of the mean value of the test and train sample (M).
- Access the probability value $P(A/B) = P(A)/P(B)$.
- Subtractions of probability $P(A/B)$ from the mean value (M) that is $P(A/B) - M$.
- Two distinct points required for subtraction and decisively mod of given value points.

The script image is flipped into 90° and performed the same procedure as described in the above algorithm. It is observed that the cluster formation of the image before and after rotating 90° forms identical clusters.

5 Results and Discussion

MODI script contains 36 vowels and 10 consonants. Each character is processed through the proposed algorithm and results are accomplished. The proposed algorithm extensively uses distance classifier that is Euclidean and Manhattan distance [19]. Both the distance classifier gives a relation between two distinct data point.

Table 1 consists of a comparison of two different algorithms. The first algorithm consists of Euclidean distance classifier. While the second algorithm uses Manhattan distance classifier as described above Sect. 4.

- In this experiment, we have used 3220 training samples 1380 test samples of handwritten MODI characters.

Table 1 The result set of the experiment using both the methods

Sr No	MODI Script	Method 1 Misclassification	Method 1 Time Complex	Method 2 Misclassification	Method 2 Time Complex
1	A	0.016733	0.523	0.0048578	0.421
2	AH	0.016401	0.523	0.0048613	0.421
3	DHA	0.01607	0.523	0.0048648	0.421
4	LA	0.015739	0.523	0.004868327	0.421
5	GA	0.015407	0.523	0.004871816	0.421
6	TA	0.015076	0.523	0.004877267	0.421
7	SHA	0.014745	0.523	0.004880756	0.421
8	THA	0.014747	0.523	0.004880756	0.421
9	RA	0.014413	0.523	0.004884246	0.421
10	NA	0.014082	0.523	0.004887735	0.421
11	DHA	0.01375	0.523	0.004891226	0.421
12	DA	0.013419	0.523	0.004894715	0.421
13	PHA	0.013088	0.523	0.004898205	0.421
14	CHA	0.012757	0.523	0.004901694	0.421
15	NA	0.012425	0.523	0.004905184	0.421
16	LA	0.012094	0.523	0.004908673	0.421
17	VA	0.011763	0.523	0.004912163	0.421
18	HA	0.011431	0.523	0.004915652	0.421
19	MA	0.0111	0.523	0.004921103	0.421
20	TA	0.010768	0.523	0.004924592	0.421
21	AM	0.010437	0.523	0.004928082	0.421
22	THA	0.010106	0.523	0.004931571	0.421
23	AI	0.0097746	0.523	0.004935061	0.421
24	E	0.0094432	0.523	0.00493855	0.421
25	DHA	0.0091118	0.523	0.00494204	0.421
26	JHA	0.0087804	0.523	0.00494553	0.421
27	NA	0.008449	0.523	0.00494902	0.421
28	NA	0.0081177	0.523	0.00495251	0.421
29	U	0.0077866	0.523	0.004955999	0.421
30	SA	0.0074552	0.523	0.00496145	0.421
31	DHA	0.0071239	0.523	0.004964939	0.421
32	CA	0.0067925	0.523	0.004968429	0.421
33	MA	0.0064611	0.523	0.004971918	0.421
34	SA	0.0061297	0.523	0.004975408	0.421
35	KA	0.0057983	0.523	0.004978897	0.421

(continued)

Table 1 (continued)

Sr No	MODI Script	Method 1 Misclassification	Method 1 Time Complex	Method 2 Misclassification	Method 2 Time Complex
36	YA	0.0054669	0.523	0.004982387	0.421
37	O	0.0051359	0.523	0.004985876	0.421
38	JA	0.0048045	0.523	0.004989366	0.421
39	BHA	0.0044731	0.523	0.004992855	0.421
40	AM	0.0041418	0.523	0.004996345	0.421
41	I	0.0038104	0.523	0.004999834	0.421
42	AU	0.003479	0.523	0.005005285	0.421
43	HA	0.0031476	0.523	0.005008774	0.421
44	BA	0.0028162	0.523	0.005012265	0.421
45	KHYA	0.0024852	0.523	0.005015754	0.421
46	DNYA	0.0018224	0.523	0.005019244	0.421

- Euclidean Distance based method fetched an average accuracy of 99.28% (with a time complexity of 6432.8 sec for 1380 test samples).
- The average accuracy achieved using Manhattan Distance based method is 94% (with a time complexity of 722 sec).
- It is observed that the reduction in accuracy rate in the case of Manhattan Distance Method is due to similar-looking characters.

Figure 2 shows the result of both methods for the recognition of each character of the MODI data set.

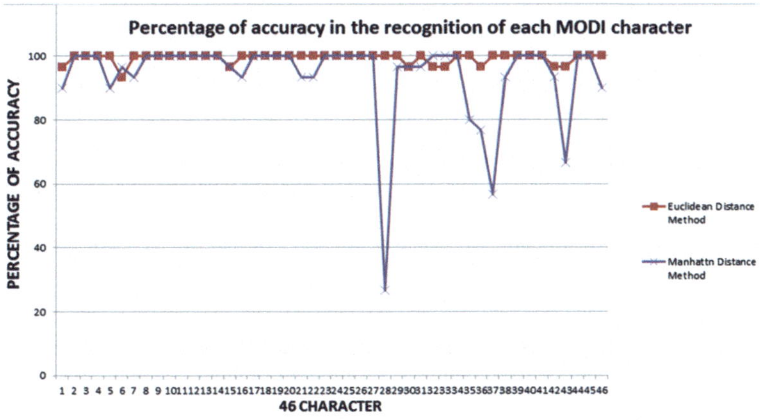

Fig. 2 Percentage of accuracy in the recognition of each character using the two methods

6 Conclusion and Future Work

For the optical character recognition of MODI script we have implemented two algorithms. The first implementation was using Euclidean distance classifier and the second one was done using Manhattan distance classifier and the accuracy achieved is 99.28% & 94% respectively. The experiment indicates that Euclidean distance gives comparatively better accuracy. It is also observed that Manhattan distance method performed better with respect to time complexity, though the accuracy was less compared to the other method.

MODI script character recognition is still in infancy and there is a scope for future work in this field. As a future plan we will be working with distance classifier advance algorithms and neural network for MODI script classification task. Cohort intelligence and the Genetic algorithm will also be explored for MODI character recognition. Cohort intelligence and Genetic algorithm are expected to give precise classification.

References

1. Pandey A(2011) Final proposal to encode the MODI script in ISO/IEC 10646, pp 26–27
2. Joseph S, George J (2019) Feature extraction and classification techniques of modi script character recognition. Pertanika J Sci Technol 27(4):1649–1669
3. Khillari R (2008) History of MODI script
4. Sharma P (2013) Classification in pattern recognition: a review. Int J Comput Eng Res (IJCER) 3(4):298–306
5. Liu CL, Fujisawa H (2008) Classification and learning methods for character recognition: advances and remaining problems. Stud Comput Intell, 1–23
6. Ramteke A, Katkar G (2013) Recognition of off-line MODI script: a structure similarity approach. Proc Int J ICT Manag 1(1):12–15
7. Anam S (2015) An approach for recognizing MODI Lipi using Otsu's binarization algorithm and Kohonen neural network. Int J Comput Appl 111(2):28–34
8. Besekar DN (2012) Special approach for recognition of handwritten MODI script's vowels. Int J Comput Appl 48:52
9. Besekar DN (2013) Study for theoretical analysis of handwritten MODI script-a recognition perspective. Int J Comput Appl 64(3):45–49
10. Rathi S (2015) recognition and conversion of handwritten MODI characters. Int J Tech Res Appl 111(2):28–34
11. Kulkarni S, Prashant LB, Manza R, Yannawar P (2015) Offline MODI character recognition using complex moments. Procedia Comput Sci 58
12. Kulkarni S, Borde P, Manza R, Yannawar P (2015) Analysis of orthogonal moments for recognition of handwritten MODI numerals. VNSGU J Sci Technol 4(1):36–43
13. Kulkarni S, Borde P, Manza R, Yannawar P (2015) Impact of zoning on Zernike moments for handwritten MODI character recognition. IEEE Int Conf Comput Commun Control
14. Malviya P, Ingle M (2018) Feature extraction and classification techniques in character recognition systems-a comparative study. In: Proceedings of international conference on recent advancement on computer and communication. Springer, Singapore, pp 527–538
15. Bagde PR, Gurjar AA (2016) A handwritten recognition for free style Marathi script using genetic algorithm. In: 2016 international conference on global trends in signal processing, information computing and communication (ICGTSPICC). IEEE, pp 43–48

16. Modi H, Parikh MC (2017) A review on optical character recognition techniques. Int J Comput Appl 160(6):20–24
17. Yadav M, Purwar RK (2018) Hindi handwritten character recognition using oriented gradients and Hu-geometric moments. J Electron Imaging 27(5):051216
18. Karabork H, Bildirici IO (2008) A neural network algorithm for vectorization of 2D maps. The international archives of the photogrammetry. Remote Sens Spat Inf Sci XXXVII(Part B2)
19. Hambal AM, Pei Z, Ishabailu FL (2017). Image noise reduction and filtering techniques. Int J Sci Res (IJSR) 6(3):2033–2038
20. Jain AK, Duin RPW, Mao J (2000) Statistical pattern recognition: a review. IEEE Trans Pattern Anal Mach Intell 22(1):4–37

Societal Transformations Through ICT as a Shared Public Infrastructure

Karan Rai Bahadur, Ojas Vyas and Prashant Mehra

Abstract Advancements in information technology have played a central role in the economic, political and cultural globalization of the world. While commercial ventures have thrived by leveraging information technology, governments and multilateral institutions might often view technology as a means for doing things as opposed to the way of doing things. Owing to the multitude of development initiatives being carried out at any given point in time development institutions face challenges in scaling the collective impacts of such interventions. In order to address these gaps, the technology division for social inclusion at Mindtree has developed a cloud-based 'Public Goods Platform'. The objective is to enable governments to provide digital platforms as a public service to its citizens, integrating multiple social development models, theories of market economics and scaling the impact to a larger section of the society especially the vulnerable sections. The core tenant of the Public Goods Platform is inspired by the concept of public goods—'it is non-excludable' and 'non-rivalrous'.

Keywords E-governance · Public goods platform · Technology platform · Government programs · Interoperability · Multi-domain · Sustainable development goals (SDGs)

1 Introduction

With India positioned to be one of the fastest growing economies of the world with a projected growth rate of 7.3% in 2019 [1]; about 20% of the country's population is still below the poverty line [2] and more than two-thirds of the adult population

K. R. Bahadur · O. Vyas (✉) · P. Mehra
Mindtree.org, Mindtree ltd, Bangalore, India
e-mail: ojas.vyas@mindtree.com

K. R. Bahadur
e-mail: karan.raibahadur@mindtree.com

P. Mehra
e-mail: prashant.mehra@mindtree.com

S. Fong et al. (eds.), *ICT Analysis and Applications*, Lecture Notes in Networks and Systems 93, https://doi.org/10.1007/978-981-15-0630-7_12

constitute the low-income group [3]. This dichotomy mirrors inequitable growth in which even the most basic amenities do not reach a large section of the population. This phenomenon is typical of not just India but of most developing economies such as China, Ethiopia, Rwanda, Bangladesh to name a few [2]. Moreover, in most of these developing economies, a significant section of the population works in the informal sector. For instance, in India, more than 80% of the economy is informal [4] making it very hard for the State and concerned institutions to ensure fair wages and social securities for such communities.

In a thriving democracy, by all means, the State becomes the most crucial agency for poverty alleviation and employment generation within diverse sectors. With a surge in job-seeking youth population, especially in countries like India, the rate at which formal jobs are created annually will need to increase manifold in order to bridge the existing economic divide. Since this divide is multidimensional; constituting of the constantly widening gap of urban and rural communities, gender and caste discrimination, the disparity in formal and informal sectors, and the digital divide; governments require a paradigm shift in the way economic growth is imagined. Our societies need more job creators than job seekers and technological innovations will need to power this transformation.

Information and communication technology is fundamentally redefining the way we live and operate as a global and a hyper-local community. In the past decades, technology innovation have altered our patterns of communication, work, consumption, formal education and the way we imagine our collective future. It has brought the global community closer than ever before. And while unfortunately those at the bottom of the socioeconomic pyramid haven't been able to reap the benefits of this digital transformation, the coming decades present an unprecedented opportunity to accelerate growth for all sections of the society. In a country such as India, with 88% of the Indian households already equipped with a mobile phone [5] and almost 500 million of the country's population having access to internet, [6] the possibilities for leveraging ICT are unprecedented. It offers the opportunity to scale up the connect between rural producers to urban consumers, job seekers from economically weak sections with potential employers, entrepreneurs at grassroots with incubating agencies. Technology platforms of the for-profit worlds have successfully demonstrated such multi-sided interactions connecting value creators with value consumers; thereby creating new marketplaces and enabling new economies to emerge.

In the domain of sustainable development, Mindtree.org's Public Goods Platform (PGP) aims to enable governments to make a transition from programmatic interventions to leveraging the capabilities of the digital platform to sustain and scale grassroots growth. The approach towards developing Public Goods Platform recognizes the State as the sole upholder and guardian of public goods, services and the communities that are creating or leveraging value. It is offered to government agencies to facilitate micro-interactions with diverse change agents in the concerned ecosystems. Public Goods Platform is designed with the underlying intent to hand over an instance of the principal platform to the government entity, as the primary custodian and enable socioeconomic interventions in various domains such as agriculture, education, skills development, solid waste management etc.

What makes platforms unique and powerful are the communities of stakeholders who can engage in meaningful transactions within the offered eco-systems. It's more about the communities and their shared values, and less about the product itself. The scope and the value offered only multiplies through the remarkable phenomenon of network effects as more and more people begin to leverage the platform's natural capabilities—thus becoming a source of endless possibilities.

2 Program to Platforms

In India, various developmental agencies have been working towards improving the socioeconomic outcomes for millions working in the informal sector and especially for those coming from vulnerable communities. These agencies include state and central government driven institutions, programs, schemes, and non-governmental actors such as multilateral organizations, social enterprises, Corporate Social Responsibility initiatives, foundations, philanthropy, etc. All such initiatives combined spend billions of dollars each year to drive the country towards sustainable development, driven by their specific individual charters, target groups and specific outcomes. While there is a wider consensus within the development community to approach sustainable development with a sense of urgency; there are certain fundamental challenges that prevent related initiatives to scaled and sustain.

2.1 Challenges in Sustaining Development

Program Approach. Most of the development initiatives take a programmatic approach to development. Which basically means that there is a set duration for the intervention focusing on a particular target group with a defined budget to drive it. The challenge is that these interventions often fail to sustain impact once the budget or the program execution period is over.

Data Gap. There is hardly any comprehensive and actionable real-time data available on socioeconomic development. The data is scattered in various forms and formats. Starting from development statistics and Personally Identifiable Information of beneficiaries stored in dusty paper files in remote villages to citizen data stored through high tech biometric systems. This not only makes data cross-sharing almost impossible but also causes years of delay before data can be consolidated to take any meaningful action.

Development Silos. Various government and non-governmental institutions which work towards sustainable development mostly carry out program planning and execution in silos. While there could be multiple parties engaged in each of these developmental programs, such interventions are not necessarily planned at a larger eco-system level. This leads to either duplication of efforts or multiple development interventions running in parallel losing out on the possibility of creating a larger collective impact.

In order to address these gaps in sustainable development, Mindtree's technology division for social inclusion—Mindtree.org has developed a digital platform called "Public Goods Platform". The primary approach is based on enabling sustainable development efforts to move from a program mode to a digitized platform approach.

2.2 Public Goods Platform (PGP) Architecture

The Public Goods Platform is a cloud-based solution. The platform's technology primarily constitutes of three layers:

1. **System of record**: This is the base layer in which all kind of data is stored. This would include people and organizations transacting on the platform, socioeconomic data, material movement data, financial transactions, geographical data etc.
2. **Layer of differentiation**: The data contained in the system of record is just raw data, the particular use case of this data is derived in the central technology infrastructure. This is called the layer of differentiation. This layer basically pulls out the relevant data from the system of records and performs the necessary computation to give a result based on the use case.
3. **Layer of engagement**: The user of the applications built on the Public Goods Platform primarily interacts with this layer. This interaction could be through a mobile device, tablet or a computer based interface. Based on the user's profile and the particular use case for the interaction, the inputs of the user is recorded in the layer of engagement, which then pulls out the relevant data from the system of records, puts it in the context of the use case at the layer of differentiation, does the necessary computation and displays the result/output back to the user (Fig. 1).

A host of digital interventions that are suitable for multiple social development domains emerge from the platform. These include but are not limited to applications for enterprise resource planning, digital payments, governance dashboards for monitoring and evaluation, tools for predictive analysis and solutions to facilitate interactions between different stakeholders who need to come together to drive sustainable development. The challenges and needs of users who deploy these solutions are unique and are dependent on the socio-cultural, geographical and political context. For example, a supply chain solution to tackle urban waste management might behave differently from a supply chain solution that connects marginal farmers directly to the market. However, since both solutions share the same technology and source code, they can represent one type of product offering. Likewise, a number of interrelated solutions can be bundled or unbundled as technology product offering and tweaked to suit the problem at hand. The layer of differentiation of the platform

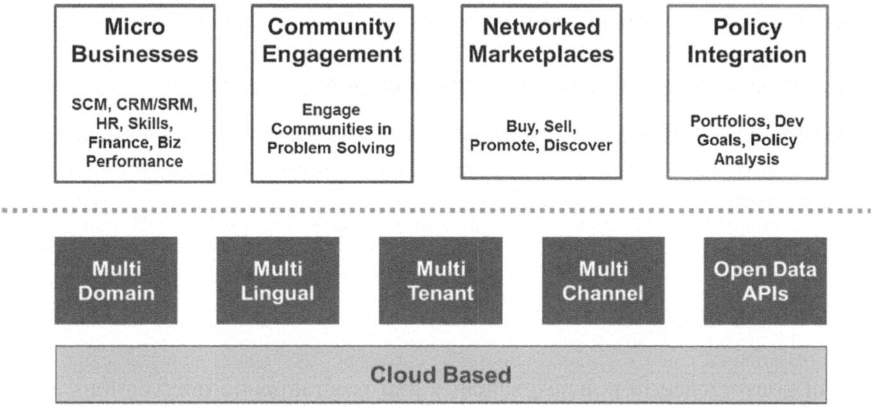

Fig. 1 Public goods platform architecture

helps codify the natural flow of interactions for the user and his larger context, barriers are being removed with each field testing experience and further understand the root cause. Further, the platform codifies people in unorganized sectors as independent micro-businesses and then proceeds to apply best practices and principals relevant to the hyper-local realities. The objective currently is to onboard different stakeholders on to the platform's marketplace in order to achieve a critical mass. This would then lead to a network effect as more and more stakeholders will find mutual value and the platform usage spreads across different sections in the society. From an architectural standpoint, it's a Multi-tenant, Multi-domain, Multi-lingual and Multi-channel platform. These are collectively termed as MULTI-X. The following section explains each briefly:

Multi-tenant. The platform leverages the principles of Multi-tenancy which primarily deals with security, privacy, subscriptions, customizations, metadata, analytics, and billing and metering. Every user uses the system as if they are the only users. Multi-tenancy is the basis for SaaS. Entities/Resources are qualified per tenant. However, we also have use cases for global data, cross-tenant data as well.

Multi-domain. It means that the platform has an ontology, white labeling, domain-mapping and multi-language. The solutions build on the platform are modular in nature and can be deployed in multiple domains. For example, the current modules of the platform are being used in domains such as education, skilling, solid waste management, agriculture, handicraft industry and community engagements.

Multi-lingual. The platform has multi-lingual enabled UI and domain labels; reference data and transaction data. For the Public Goods Platform, Multi-lingual support is the least common denominator support since languages and user preferences change from region to region in a diverse nation like ours.

Multi-channel. The Public Goods Platform is built with channel intelligence. It is channel aware and does not discriminate. It has tools and automation to build apps on any channel. This is further enabled through its App models.

Open Data APIs. The platform offers a rich API economy to open source developers, social entrepreneurs and other developmental agencies to build new solutions on top of the platform.

2.3 Interoperability

Lack of convergence in policies, schemes and programmatic interventions across government departments and institutions is often a key challenge in delivering sustainable development outcomes. And ICT can be a key enabler in overcoming these barriers by enabling meaningful multi-stakeholder and multi-process interactions. In many contexts governments and multilateral development agencies might end up using numerous ICT based solutions to address specific problems or to enable certain processes and interactions; however more often than not these technology solutions end-up being deployed and operated in silos without the attempt to cross-leverage technological capabilities by enabling information exchange and processes interaction between such solutions at a core architectural level.

While there is no limit to the kind of ICT applications that can be built on the Public Goods Platform, any government institution deploying Public Goods Platform can easily integrate its existing ICT solutions or new ones on the platform. This is made possible through the interoperability standards codified on the platform. Consider for example the domain of agriculture where a government institution may want to build a farm advisory solution on the Public Goods Platform for smallholder farmers in order to help improve farm productivity and reduce risk to crop failure. And say for instance it already has two technology solutions deployed on the field: one an IoT based system which captures data on soil and water health, and another a satellite based weather monitoring system. A farm advisory solution would need to collate and process information from these two systems and few others data sources in order to deliver an effective advisory service to the famer. It will not make much sense if these solutions provide independent outputs just in some data format. Under such a scenario, wherein multiple technology solutions are at play, the Public Goods Platform's interoperability standards would help integrate all such solutions enabling them to seamlessly interact with-each other and provide a meaningful outcome. Further if there are no such existing solutions currently deployed then the IoT based system and the weather monitoring solution can be directly built on the Public Goods Platform, using the platform's existing services and integrated into the overall farm advisory solution. With interoperability the possibilities are endless.

Figure 2 Illustrates how different isolated systems can interact with each other by leveraging the platforms interoperability standards.

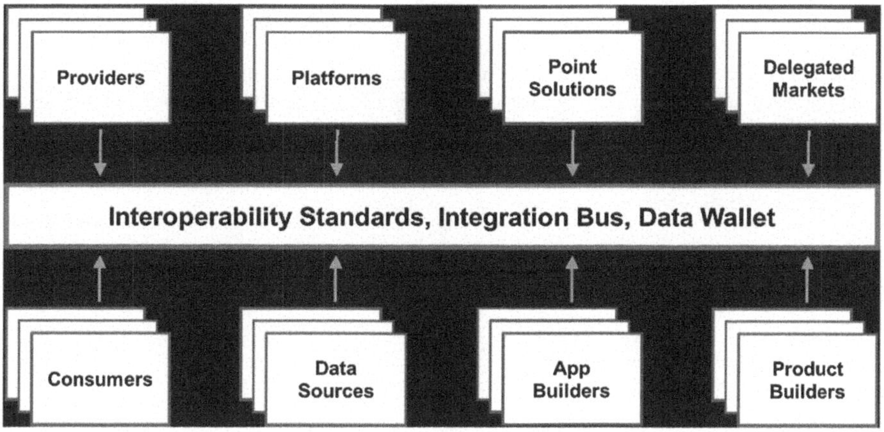

Fig. 2 Different isolated systems interact with each other by leveraging platforms interoperability standards

3 Conclusion

The Public Goods Platform has been designed for government institutions and multilateral development agencies to engage in socioeconomic transformation at various scale; be it at a hyperlocal level, national or global. Its architecture takes an eco-system approach to development interventions with the beneficiaries at the core and enabling multi-stakeholder interactions to create greater collective impact. The primary goals of the platform is to help socioeconomic development initiatives move away from a silo approach and towards a platform model enabled through information and communication technology. As a core principal the solutions built on the platform are made available free of cost to people below poverty line and many others in the informal sector who do not possess the means to leverage ICT in carrying out socioeconomic activities. The Public Goods Platform brings together public services, development eco-system stakeholders, strategic planning, monitoring and evaluation tools and open innovations to drive sustainable and measurable development.

References

1. Focus Economics (2019) The worlds fastest growing economies. https://www.focus-economics.com/blog/fastest-growing-economies-in-the-world
2. World Bank (2019) Data for India, lower middle income. https://data.worldbank.org/?locations=IN-XN
3. Livemint (2017) https://www.livemint.com/Money/HJkSWsigdz6Xvkg8wpr6jK/Where-are-you-in-Indias-wealth-distribution.html
4. The Wire (2018) https://thewire.in/labour/nearly-81-of-the-employed-in-india-are-in-the-informal-sector-ilo

5. Livemint (2016) https://www.livemint.com/Politics/kZ7j1NQf5614UvO6WURXfO/88-of-households-in-India-have-a-mobile-phone.html
6. The Economic Times (2018) https://economictimes.indiatimes.com/tech/internet/voice-video-and-vernacular-indias-internet-landscape-is-changing-to-tap-next-wave-of-users/articleshow/66102478.cms?from=mdr

Technology Based Self-learning—Case of Zucate

Pradnya Chitrao, Pravin Kumar Bhoyar and Rajiv Divekar

Abstract

Background Profound technological, economic and social transformations have marked the very beginning of the twenty first century. IT progress will greatly enrich possibilities in production, transportation, energy, commerce, education and health. Progress in information and communication technology has increased new opportunities for learning. Many colleges, schools, universities and organizations use the online-learning concept for enhancing in an effective and interactive manner the knowledge, abilities and skills of learners. Technology can expand what children learn by helping them to understand core concepts in subjects like math, science, and literacy.

Use of Technology by Enterprises for Improving Learning in India Today with both parents working, there is a profound need felt by both students as well as parents for resources that will take care of the learning needs. The 21st C has made available a lot of technology involving computers and internet that is being used in the learning aids being devised for students. One such enterprise is Zucate founded by Ms. Roli Pandey and Dr. Moitreyee Goswami.

Objective The paper studies Zucate, an enterprise that strives to create "a teacher-student-parent ecosystem with student at the center" (https://www.f6s.com/zucate). The objective is to examine how technology can be employed in a cost effective manner to enable students, especially school children to learn on their own post school hours.

Research Methodology The researchers have used the primary research technique of personal interview of the founders of the enterprise. They have also referred to secondary data.

P. Chitrao (✉)
Symbiosis Institute of Management Studies (SIMS), Constituent of Symbiosis International University, Pune, India
e-mail: pradnyac@sims.edu

P. K. Bhoyar · R. Divekar
SIMS, Pune, India

S. Fong et al. (eds.), *ICT Analysis and Applications*, Lecture Notes in Networks and Systems 93, https://doi.org/10.1007/978-981-15-0630-7_13

Significance of the Study The study is important as it gives insights into how technology can be used to create teaching and learning resources that will enable children and students to grasp concepts and do self-learning without taking recourse to costly, time bound tuitions or even tech savvy, expensive learning resources being offered by regular enterprises.

Keywords Technology · E-learning · Cost effective · Teacher-student-parent ecosystem

1 Introduction

Technology has been changing classroom practices and learning processes. Researchers have shown on the basis of the Second Information Technology in Education Study [5] that covered 28 countries in Africa, Asia, Europe, North America and South America, that technology is modifying classroom practices and learning processes. Gone are the days of mere chalk and talk by teachers. Curricula of schools has become very demanding going beyond reciting famous texts, recounting simple scientific facts, and solving basic arithmetic problems. In other words, students today are expected to learn a lot more complex concepts.

Today, more and more families are nuclear with both father and mother working. This leaves very little time for any of the parents to sit with the child and monitor studies after school hours. This has an adverse effect on the child's studies.

Technology can improve children's learning capability by supporting four fundamental characteristics of learning:

(a) Active engagement,
(b) Participation in groups,
(c) Frequent interaction and feedback, and
(d) Connections to real-world contexts

Johannessen [3] observes that we are increasingly using technology in all facets of our lives and we need to look at the question of whether the use of technology improves students' performance. Technology-mediated learning environments help students to search and analyze information, solve problems, communicate and collaborate; it equips them with a set of competencies that help them compete in the 21st century marketplace. Today, on account of technology, the role of the teacher has shifted from being the sole source of information to a more complex role of providing a varying degree of support for different students, monitoring students' progress, and encouraging reflection on classroom activities. Students also can, with the help of technology, are more engaged, and can make better connections between their previous learning experiences and the new concepts or principles being taught [4]. Tamim et al. [6], for example, found that learning aided by technology can have positive effects.

2 Literature Review

One of the biggest challenges of higher education reformation in developing solutions is facilitation and delivery of the right information and skills to the right learners at the right time. Virtual learning becomes a sub section of e-learning which describe internet based learning systems in an economical model rather than other conventional educational models.

E-learning is defined as "... any use of web and internet technologies to create learning experiences" [2]. It has also affected the education industry creating a paradigm shift in the way things are taught and learnt.

With the advent of mobile technologies, a new paradigm of teaching and learning with technology aid has emerged, that is mobile learning (M-Learning). Mobile technologies purvey opportunities to hold new and interesting methods of teaching and learning, both beyond and inside the classroom. The technological features of M-Learning are portability, immediacy, connectivity, ubiquity and adaptability [7]. It facilitates access to learning anytime and anywhere by enabling connectivity and the employ of multiple apps for educational purposes [1].

3 Objective

The research seeks to find out how some enterprises are using technology to bring about more effective self-learning among school children in a cost effective manner.

4 Research Methodology

The researchers focused on one case study of an enterprise called Zucate started by two women entrepreneurs for self-learning resources for school children. They conducted structured interviews of the founders. They also carried out secondary research.

5 Significance of the Study

The study is important as it gives insights into how technology can be utilized for creating teaching and learning resources that will enable children and students to grasp concepts and do self-learning without resorting to costly, time bound tuitions or even tech savvy, expensive learning resources being offered by regular enterprises. It also has the potential of helping students from low income groups to manage their studies in a cost effective manner.

6 Findings

6.1 Genesis

Zucate founders, Ms. Roli Pandey and Dr. Moitreyee Goswami are students of the EDP (2016 batch) conducted Symbiosis Institute of Management Studies (SIMS, Pune) for women. Dr. Moitreyee, post MBA in 2013 worked for a year for a company which was into Public Health. Ms. Roli worked as Business Development Manager for a startup in Healthcare. Both did not experience work satisfaction as they perceived their jobs as not giving any scope for creativity. So they signed up with an online learning resource company called Think Vidya (branded now as UrbanPro) as teachers. The subscription fee was Rs. 100, and for every student allotted Rs. 20 would get deducted. They both opted to teach BBA students, and charged Rs. 3,500/- for a subject (Rs. 18,000/- for all subjects of a semester). The students would come home and be taught on a one on one basis. They started dropping pamphlets of their classes in Vishrantwadi. They also gave medical stores Rs. 100 per month to display their posters. Also, by now their student network was growing. They then started teaching school students all subjects (except Marathi and French) on a one-on-one basis for Rs. 3,000/- per month. This business they called Enlightening Minds. It is still functional.

Both soon realized that there was a dire need to strengthen the foundation concepts of students in all the subjects. They found that elite schools were forcing parents to withdraw from CBSC and ICSC Boards and to enroll in open school boards namely National Institute of Open School University (NIOS). St. Helena's School in Pune is the center for NIOS schooling. The academic future of children passing from such boards, according to the founders, is not bright.

Ms. Roli contacted her relatives in Raipur who were into teaching children from poor families. There, children make aero planes out of the school books. They are also sent to school only for the free mid-day meal. The children had no access to proper schooling. At that time, many online educating businesses were starting. These businesses were offering each school subject for prices between Rs. 8000/- and Rs. 25,000/-. They offered videos with lots of animation and very less content. Some of them were offering school level. The two friends decided to start an online teaching platform of their own as by now they had a lot of content. They decided that their teaching platform would be holistic namely, study, revise, doubt clearing sessions, and exam preparation through mock tests all of which would be interactive, comprehensive, and multilingual.

They decided that the pricing would be subscription based with Rs. 100 as basic subscription fee and Rs. 10–15 per chapter being deducted from the basic registration fee. Their earlier business of teaching on a one-on-one basis is still functional and is called Enlightening Minds.

6.2 Recognition and Support

Potdar International School in an Indian Entrepreneurs' Meet in Mumbai suggested that they should use their content for training Aanganwadi teachers. Government will pay the subscription fees for training them. Since Aanganwadi centers are digitalized centers, this online training can be imparted. Their business is now one of the 15,000 startups that NASSCOM agreed to support.

F6S is a platform where multiple entrepreneurial bodies put their forms wherein they submit their business plans, financials, pitch (standardized form of presentation of 10–12 slides), marketing strategy, and product demos. The founders put up their business on F6S in December 2015, and got help from NASSCOM in March 2016. NASSCOM gave the technological backing i.e. tools worth a crore rupees free of cost for one year. The tools included technology development, HR accounting, marketing, e-mailers, cloud services, payment services, and toll free numbers. Through NASSCOM, they applied for Microsoft's BizSpark Plus Program which is an initiative for cloud services in association with Government of India. They got their subscription for one year worth Rs. 60 lakhs free of cost. NASSCOM gives every week training on business aspects in Kharadi, Pune. The founders attended the training in 2017.

The founders went to IIT, Powai (SINE, IIT) in June 2016 for a mentor matching event. They cleared the interviews and got their first mentor namely, Mr. Mahindra Kapadia, an ex-President of IIT, Powai Alumni Association. He hand holds the founders free of cost and the consultation is equity based (namely 1%). He also is part of an urban-rural special interest group that is into educating poor people from urban slums and from rural areas.

The founders also got Microsoft Big Spark in April 2016.

6.3 Business Costs

The business costs are under three headings:

(1) Technological which is about 10% and which comes to around Rs. 2000/- because of NASSCOM and Microsoft support.
(2) Marketing Cost which is organic that is word of mouth and through social media and which is about 30%.
(3) Administrative Costs which is around 60% i.e. around Rs. 12,000/- per year. Government compliances like ROC (Registrar of Companies), IT (Income Tax), annual returns are the main administrative costs.
(4) Total costs for first year will be around Rs. 20,000/-. The costs will subsequently increase each year as the ladies will start drawing salaries, employ people, and hire office premises. In 2017, the company was registered from Dr. Moitreyee's home in Kolkata. The firm is now profiled in World Disrupt June 2016 edition

6.4 Launch

The founders are launching content for the 10th standard in January 2018. Over a period of time, they will launch for other standards.

6.5 Premises

The founders have now taken premises on rent in Symbiosis Center for Entrepreneurship and Innovation (SCEI) campus at Lavale, Pune for Rs. 5000/- per month. They also pay 2% equity. The contract is for two years and began in August 2017.

6.6 Use of Technology for Self-learning Resources

The founder have prepared content that is in the form of a visually interactive video. For this they made use of artificial intelligence. This technology is handled by the founders themselves. Dr. Moitreyee is doing certification course in App Development from Cambridge University and Ms. Roli is a software engineer. The two of them have used artificial intelligence to make the learning platform a personalized one wherein it adapts to the child's pace of learning. Now the platform also includes cognitive learning. The platform takes the learner back to the previous level of learning, clears that concept, and then bring the learner back to the present level of learning. So the platform can take the student back to even the 3rd standard if he/she has forgotten some basic principle taught at that level. The platform thus utilizes a technology called Meta Cognitive Task Analysis. For example, for a compound interest problem in arithmetic, one needs to apply multiplication, division, and percentage principles. The platform identifies which task one is stuck with, and takes one back to the level where that particular concept is taught and refreshes the learning of that concept.

The founders themselves have done the coding for this App by using open source technology. They conducted two pilot studies with two hundred students of the platform from various education boards, one for the understanding of the content and the other for the ease of usability of the content. The results were 95% positive. Students performed better because of this platform. Another pilot study was done with teachers. They found that 88% teachers were willing to use Zucate as a medium of teaching.

The platform also uses Chat Bots. It also assists the students navigate smoothly without any adult supervision. It even saves the answers to the new queries and makes them available for future use by the next user who may have the same query. This reduces redundancy.

6.7 Awards and Recognition

Zucate won the Pune Pitch Fest in April 2017 organized by Spinta Global Accelerator, Silicon Valley, US. Post this, they got selected for Tech Innovation Fund by Kalaari Capital. The business also got selected on partial sponsorship to represent India in World Entrepreneurs' Investment Forum2017 organized by UNIDO namely, United Nations' Industrial Development Organization in Bahrain, Manama.

7 Conclusions and Recommendations

The findings clearly show that the enterprise facilitates self-learning at a reasonable cost. Established institutions (read schools, colleges and universities), more often than not, are resistant or reluctant to change or to adopt new paradigms of technology, stonewalling ed-tech startups. The case of Zucate shows that education sector will change forever and technology companies will bring about this change.

First and foremost artificial intelligence will personalize self-learning platforms. Education works through a nexus of Content, Coach and Community. Adaptive learning algorithms, targeted analytics and AI tools can help create the first 2 Cs, which can make education platforms connect with consumers. Gamification and social media connect can cater to the third C. This really is the future of education. We can well visualize in the near future a learning tool that can point and correct weak areas of a school kid even before the teacher or parents become aware of them. Secondly, with children today getting their first learnings from iPad and similar touch devices as also smart phones, early learning apps will soon become household phenomena and shape future generations. Established institutions will come around and realize the inevitability of technology intervention in education deliverance. The education industry will soon be massively impacted and revolutionized by technology intervention and positive disruption in the coming decade.

Government and venture capitalists should recognize similar enterprises like Zucate and support them. This will answer the problem of children's studies suffering on account of the present day trend and need of both parents working in a highly competitive corporate world and having very little energy for teaching and helping their children in their studies. Such technology driven platforms can also aid educators in teaching children concepts in a holistic manner. It will also enable low income group students access such self-learning platforms as it is cost effective.

References

1. Fundación Telefónica (2013) Mobile learning guide
2. Horton W (2003) E-learning tools and technologies: a consumer's guide for trainers, teachers, educators, and instructional designers. Wiley, Indianapolis, IN, USA, p 13

3. Johannessen O (2009) In search of evidence: the unbearable hunt for causality. Retrieved from the organization for economic co-operation and development. http://www.oecd.org/dataoecd/32/16/39458854.pdf
4. Kozma RB (2003) Technology and classroom practice: an international study. J Res Technol Educ 36(1):1–14
5. Law N, Pelgrum WJ, Plomp T (eds) (2008) Pedagogy and ICT use in schools around the world: findings from the IEA SITES 2006 study. Comparative Education Research Centre, Hong Kong
6. Tamim RM, Bernard RM, Borokhovski E, Abrami PC, Schmid RF (2011) What forty years of research says about the impact of technology on learning: a second order meta-analysis and validation study. Rev Educ Res 81(1):4–28
7. Valero CC, Redondo MR, Palacín AS (2012) Tendencias biomass in the use of devices moviles in Educación [Current trends in the use of mobile devices in education]. Educ Digit Mag 147:1–21

An Efficient Collaborative Recommender System for Removing Sparsity Problem

Avita Fuskele Jain, Santosh Kumar Vishwakarma and Prashant Jain

Abstract Recommender Systems is a special type of information filtering system which has become important in the information overloaded and strategic decision making environment. Recommender System is used to produce meaningful suggestions about new items for particular consumers. These recommendations may be based on the user profile or item ratings, facilitate the users to make decisions in multiple contexts, such as what items to buy, what online news to read or what music to listen. Recommender Systems helps their founders to increase profits by recommending items and attracting new consumers. Collaborative filtering technique recommends items basis of conclusion of opinions about various products by users of similar profile to the active user. This technique requires user-items-ratings matrix. Although this is the most mature and commonly implemented technique, it faces major problem of Data Sparsity problem. Sparsity Problem occurs as a result of lack of enough information when only a few of the total number of items are rated by the users. This produces a sparse user item matrix leads to weak recommendations. This paper presents a recommender system using collaborative filtering implemented with RapidMiner tool. The proposed recommendation system is designed with users' similarity calculated by Sequence and Set Similarity Measure (S^3M) with utilizing similarity upper approximation and a Singular Value Decomposition (SVD) model based technique used for recommending ratings for removing sparsity.

Keywords Collaborative filtering · Sequence and set similarity measure (S^3M) · Sparsity problem · Singular value decomposition (SVD)

A. F. Jain (✉) · S. K. Vishwakarma
Gyan Ganga Institute of Technology and Sciences, Jabalpur, India
e-mail: avitadocuments@gmail.com

S. K. Vishwakarma
e-mail: santoshscholar@gmail.com

P. Jain
Jabalpur Engineering College, Jabalpur, India
e-mail: pjain@jecjabalpur.ac.in

S. Fong et al. (eds.), *ICT Analysis and Applications*, Lecture Notes in Networks and Systems 93, https://doi.org/10.1007/978-981-15-0630-7_14

1 Introduction

The massive growth in digital information in this era of information technology was data is being generating with speed of thought. Availability of internet and glamour of online world have made people not only to overload information about them but also attracting e-commerce companies to understand the behavior of users so that attracting more customers by giving better customized products as well as recommending them for enhancing business profits or revenues of the company. Recommendation system are information filtering system that deal with the problem information overload [8], by filtering vital information fragment out of large amount of dynamically generated information according to user's preferences, interest, or observed behavior about item [12]. Since the recommendation system has the ability to predict the preference of any item for a user by understanding and finding similar "User-Item-Ratings" pairs.

The work we are following is a kind of Recommendation system with hybrid filtering where SVD model is applied on S^3M [9] metric which is Set Sequence Similarity Measure; a linear expression of weighted parameter "p" to emphasize on sequence similarity and relative variance to calculate Jaccard similarity. By varying the value of 'p' we compare the recommendation value. The contents we are experimenting is website ratings from dataset based on the websites visited by users. The major problem in collaborative Recommendation system is sparsity. The work we are performing to achieve sparsity reduction by suggesting recommendation ratings 'R_i'. A comparison is made on recommended rating outcome by varying weighted variable 'p' and 'q' used for quantitative and/or qualitative parts of similarity linear equation used in S^3M [9]. In this paper the work is focusing on web usage mining along with user item rating. The dataset used has been created based on user rating for websites.

S^3M similarly which is used to find similarity among users and then rough set based upper approximation for clustering is applied for forming soft clusters. The technique generate overlapping clusters contains common users which are common is showing interests/rating to multiple websites. We have compared our method performance with various conventional methods. The paper follows the work done on web recommendation by Mishra et al. [10]. The recommendation system is different from sequential pattern mining algorithms. In our work we have studied the variation in 'p' where we consider content (quantitative) not the sequential information (qualitative) as by Mishra et al. [10]. The users similarity is calculated by S^3M, SVD is used for sparsity removal and the R_i is calculated on basis of prediction quotient formula.

Our work has been initiated with the major issues of recommender system. One of the premier work is to remove sparsity problem in Recommender System with collaborative filtering. Mostly web users have a sequential approach for webpage accessing. The algorithm is motivational outcome of work done on web recommendation by Mishra et al. [10], which is actually web content mining and sequence mining based algorithm where the data is a sequential data. The prime work i.e.

the similarity calculation and prediction weight calculation both are adhere to the sequential data and the sequential behavior of web sessions.

The method proposed here is novel with respect to the user-item rating matrix as data. The similarity is calculated with partial consideration on sequential nature and more on the content by changing the value of 'p'. Also the prediction is made not only on the basis of cluster of similar users but also the calculation lies on the similar user with a specific rating given by new user for a particular website. The major contribution is that in order to remove sparsity the prediction vector is proportional to the rating value by similar user in the cluster. The selection of user for computing prediction quotient is also proportional to the rating. As the collaborative filtering primarily work on user-item ratings, above mentioned two levels of rating based prediction makes the work more promising to for accuracy.

The paper organization begins with introduction of the work with basic idea of proposed method, the motivation and contribution of our work, followed by the literature review with respect to understand the sparsity problem and the work done to alleviate this problem. Third section explains the methodology adopted and dataset. The paper concludes the performance of method in the last section.

2 Related Work

The evolution of Recommendation system begun with research paper on Recommendation system where it is defined as a decision making strategy for users under complex information environments [14]. Recommendation system also used as an E-commerce tool helps users filter through records of knowledge which is related to user's interests and preferences [16]. The phases of Recommendation system are information collection, learning or filtering phase, Recommendation phase. In the filtering phase approaches may be content based technique which predict on basis of user's information ignoring contribution of others users, collaborative filtering when user item rating are used from other similar users. Hybrid filtering approaches by harnessing benefits of both techniques [18]. The matrix is generated by user preferences or likes for items, finds similar users based on relevant interests. This approach is of two types memory based and model based [1, 5, 6].

Memory based collaborative filtering computers similarity between user item ratings. The algorithm of memory based systems is heuristics that make recommendations based on an entire collection of item already rated by users [4, 11, 15]. Model based collaborative filtering generates the descriptive model of the system, based on the user's preferences using various DM and ML techniques like Bayesian Model, Clustering Model etc.

Recommendation system faces some problem with respect to efficiency of recommendation system; recommendations quality is affected by cold start problem, data sparsity problem, scalability, synonymy and Matthew effect [2, 3, 13, 17]. These problems are reducing commercial benefits to an extent. The sparsity problem is lack of enough information when only a few items ratings by users in the matrix. Data sparsity problem directly affects the coverage of recommendation result [17]. This makes the matrix sparse which in turn disables to locate proper neighbours which finally leads to weak recommendations. Model based techniques solve the sparsity problem. This problem also exists in a user product based Product Attribute Model which is due to the subjectivity of product reviews since these reviews are not covering all aspects of product. The problem is resolved by Multiplication Convergence Rule and Constraint Condition equations to find the replacement of sparse values [19]. A web recommendation system which works on sequential mining and web mining also applies a weight calculation which adequately leads to substitute the next web page visit vector which is a sparse vector, has been proposed by Mishra et al. [10]. S^3M [9] is the similarity measure applied to set sequence similarity proposed by Kumar et al. [15] which is a linear equation based on the weighted parameter '*p*' Quality is determined by SetSeq(*A*,*B*) measure and quantity is calculated by SetSim(*A*,*B*) for content matching. S^3M is used to find similar users. Singular value decomposition is a model based technique. This deploys the previous ratings (user-item) to improve the performance of Collaborative Filtering Technique. SVD is used in Opt.space algorithm by Keshavan et al. [7] to deal with matrix completion problem.

In this paper we follow the work done on web recommendation by Mishra et al. [10]. The recommendation system is different from sequential pattern mining algorithms. The users similarity by S^3M metric and the SVD model is used for removing sparsity significantly and the R_i is calculated on basis of formula proposed by Mishra et al. [10].

3 Methodology

The work in the paper focuses on results which are recommendation generated using SVD model on soft clusters which are made on similarity of users by calculating S^3M similarity Mishra et al. [10]. Different values of '*p*' are taken and compared the result of "R_i" by the model proposed by Mishra et al. [10].

The methodology as shown in Fig. 1 is devised in three parts; Response Matrix '*A*' Generation, Prediction Quotient Q_{ij} Calculation and Recommendation Vector 'R_i' Generation.

In our work experiments performed on dataset which is generated manually from the survey done in the crowd of university UG program students regarding the websites they visit and asked them to rate. The generated dataset has ratings of websites,

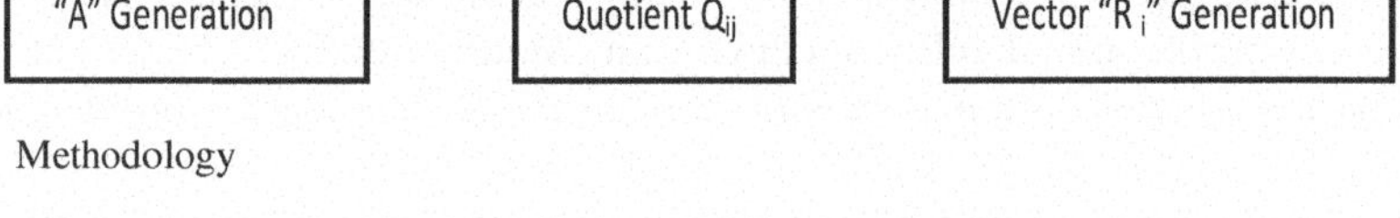

Fig. 1 Methodology

category of interested websites along with preferred website sequence partially The dataset is represented as matrix of user—website ratings. The user's ratings for websites ranges from 1 to 5 where 5 are highest and 1 is lowest rating order. The user—item-rating matrix is being developed where item a frequently visited websites is. The similarity of users from data set is found with full utilization of content similarity and partial consideration of order of Websites (a sequence) visited by user.

3.1 Response Matrix 'A' Generation

The work proceeds with formation of cluster which are soft clusters on the basis of website since a user may have multiple interest for which may belong to multiple clusters. A similarity upper approximation based clustering algorithm is used. The RS utilized a rough set based clustering approach. The similarity between users is calculated by similarity measure (metric). There exists many similarity metrics such as cosine, Jacquard etc. The metric used by Kumar et al. [15] is set sequence similarity measure. This measure enforces not only the similarity between two sets of data (vector or ordered set) but also considers the sequence (Fig. 2).

The algorithm of similarity upper approximation approach for cluster formation we are following is the same as per proposed by Mishra et al. [15]. The similarity of user is being measured here on the basis of S^3M [9]. The S^3M is Sequence Set Similarity Measure Kumar et al. [15] calculated as following:

$$S^3M[A, B] = p * \text{SeqSim}(A, B) + q * \text{Setsim}\ (A, B) \quad (1)$$

where 'p' is qualitative weight parameter of sequence similarly [9] and $q = (l - p)$ i.e. after sequence similarity content similarity is focused.

The following example explains similarity calculations. Let two users along with their preferred website rating sequence is shown as:

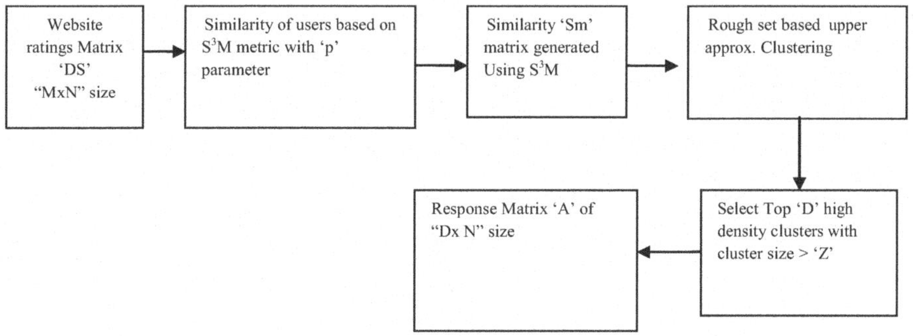

Fig. 2 Response matrix '*A*' generation

$U_A = \{1, 4, 18, 20, 11, 15, 6, 8, 5, 12\}$ and $U_B = \{4, 5, 11, 8, 1, 2, 3, 7, 9, 18, 15, 20\}$

$$\text{The length of sets}: L_A = |U_A| = 10, \quad L_B = |U_B| = 12$$

$$\text{And LLCS}(U_A, U_B) = \{4, 11, 8\} = 3$$

$$\begin{aligned}\text{So SeqSim}(U_A, U_B) &= \text{LLCS}/\text{Max}(|U_A|, |U_B|) \\ &= 3/12 = 0.25\end{aligned}$$

$U_A \cap U_B = \{1, 4, 5, 8, 11, 18, 15, 20\}$ and $U_A \cup U_B = \{1, 2, 3, 4, 5, 6, 7, 9, 11, 12, 15, 18, 20\}$
$\text{SetSim}(U_A U_B) = |U_A \cap U_B|/|U_A \cup U_B| = 8/14 = 0.57$

Let "Sm" be a similarity matrix (Table 1) such that Sm[i,j] $= \alpha_{ij}$ where α is the similarity measure value between users U_i and U_j. The value of $\alpha = 1$ for all $i = j$. The value of 'p' is 0.7 for calculation of α in the following Table 1 from the formula of S^3M Eq. (1).

The classification of web users Response matrix A is formed by selecting Top "D" clusters where we choose the higher density cluster such a way that the cluster density of selected clusters is 'Z' where "Z" > avg_cluster_density. The response matrix has clusters 'D' as rows and items as the columns will be the total 'N' (all) websites rated by users. The row vector of matrix 'A' will the average rating of respective website by the users of clusters. It is represented as $A_i = \{A_{ij}$: Average rating of W_j by all the users of $C_j\}$. The matrix A is of size $D \times N$ where 'D' is number of the top high density cluster and 'N' is the total websites rated.

3.2 Prediction Quotient '*Q*' Generation

After constructing response matrix, now it works to make a Prediction Quotient for new users who provide ratings for a few websites. A new user who provides a small pattern of ratings for some websites is the base for finding the similar users and to classify the respective cluster C_k. The prediction quotient is calculated by user ratings ratio/proportions. It is used for a new user who provides a small pattern of ratings for some websites. With the following calculations a recommended Prediction Quotient is calculated for a new user.

$$\text{The Prediction Quotient: } Q_{ij} = \frac{W_{ij}}{W_i} \tag{2}$$

where W_{ij}—total number of times ith website rated with "j" as the rating value in the cluster C_k.

W_i—total number of times ith website has been rated in the cluster C_k.

Table 1 Similarity matrix 'S_m' of users' of dataset

S_m	U_1	U_2	U_3	U_4	U_5	U_6	U_7	U_8	U_9	U_{10}
U_1	1	0	0	0	0.21	0.29	0	0	0	0
U_2	0	1	0	0.47	0.17	0.17	0.17	0	0.15	0.15
U_3	0	0	1	0	0	0.25	0.33	0.33	0	0.21
U_4	0	0.47	0	1	0.17	0	0.45	0.27	0.24	0.5
U_5	0.21	0.17	0	0.17	1	0.18	0	0	0	0
U_6	0.29	0.17	0.25	0	0.18	1	0.18	0.21	0	0.17
U_7	0	0.17	0.3	0.45	0	0.18	1	0.58	0.17	0.62
U_8	0	0	0.3	0.27	0	0.21	0.58	1	0	0.5
U_9	0	0.15	0	0.24	0	0	0.17	0	1	0.24
U_{10}	0	0.15	0.21	0.4	0	0.17	0.62	0.5	0.24	1

And if $W_i = 0$ then $Q_{ij} = 0$. A recommended Prediction Quotient vector is formed by placing Q_{ij} values in place of ith website rated by new user and the unknown values are filled with '×'. The length of the vector is 'N' the number of websites rated. Prediction Quotient vector is Ordered Set of Q_{ij} where ith website rated by new user and unknown website ratings = '× ', of 'N' elements. The vector is used with the output matrices of SVD applied on response matrix 'A'.

3.3 Recommendation Vector 'R' Generation

The response matrix 'A' is applied to Singular Value Decomposition (SVD) model which produces three matrices U, S, V^N. U is of size $D \times N$, S is of size $N \times N$, and the matrix V^N is also $N \times N$. The matrix S has diagonal elements as non-zero values other elements will be zero.

4 Result

The dataset has been created with the university students as described in Sect. 3.3 and consists of more than 5000 feedback records collected through their log files of web access. For experimentation purpose, RapidMiner the open source research data mining tool has been used. The unknown (sparse) ratings are the ratings of websites which have not been rated.

We have also compared performance results of our method with other collaborative recommendation system methods such as BMF, KNN and Slope-one. The experiments have been carried with the RapidMiner Data Mining tools in experiments. Our Method is very similar to Factor-Wise Matrix factorization method provided by RapidMiner FWMF operator performs. The following figures are the comparative diagram of the proposed method with the traditional methods.

The Figs. 3, 4, 5, and 6 are the comparative results of the proposed method with traditional methods. The tool RapidMiner uses following operators for experimenting the performance of proposed Recommender System Matrix factorization with factor-wise learning (FWMF) operator performs modeling Relationships at Multiple Scales to Improve Accuracy of Large Recommender Systems. The Bias Matrix Factorization (BMF) operator performs Matrix factorization with explicit user and item bias. This operator uses bold driver heuristics for learning rate adaption and supports Large-Scale Matrix Factorization with Distributed Stochastic Gradient Descent method.

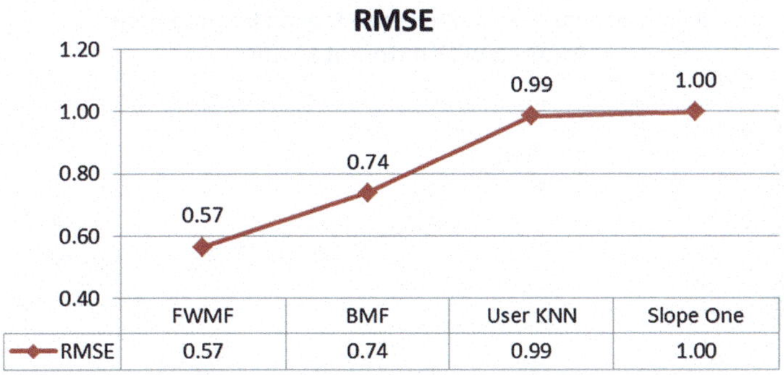

Fig. 3 Root mean square error

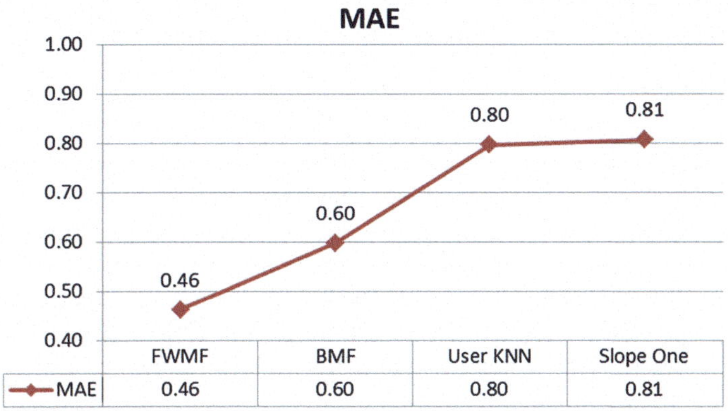

Fig. 4 Mean absolute error

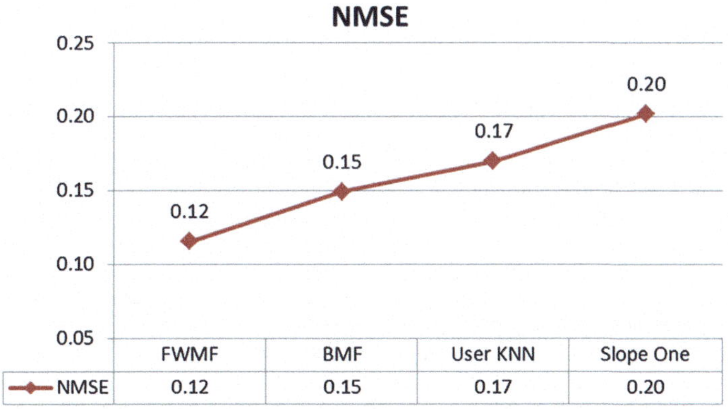

Fig. 5 Normalized mean square errors

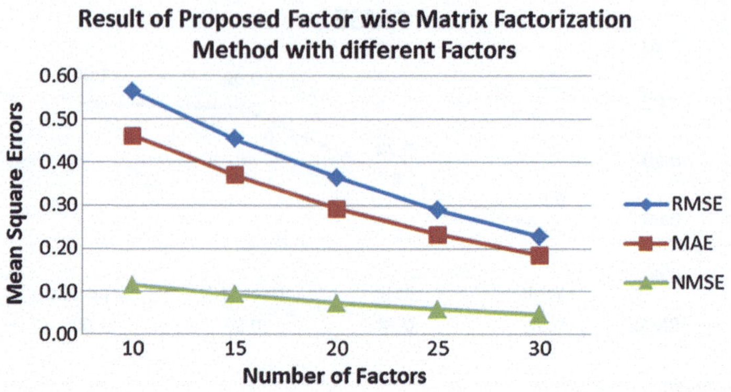

Fig. 6 Proposed method with different factors

5 Conclusion

This paper proposed a novel approach for collaborative recommender system based on S^3M approach of item and user ratings. The calculation for filling up zeros lies on the similar user with a specific rating given by new user for a particular website, is the foundation for generating the prediction vector makes it proportional to the rating at cluster level and as well at user level. The proposed method outperforms with all methods and gives the minimum errors with respect to RMSE, MAE, NMSE. The proposed method has been compared with different factors and it gives the minimum RMSE values for the ratings predictions.

References

1. Buder J, Schwind C (2012) Learning with personalized recommender systems: a psychological view. Comput Hum Behav
2. Burke R (2002) Hybrid recommender systems: survey and experiments—user modeling and user-adapted interaction. Springer
3. Burke R (2007) Hybrid web recommender systems—the adaptive web. Springer
4. Delgado J, Ishii N (1999) Memory-based weighted majority prediction—SIGIR Workshop Recomm. Syst …, 1999—pdfs.semanticscholar.org
5. Gadanho SC, Lhuillier N (2007) Addressing uncertainty in implicit preferences. In: Proceedings of the 2007 ACM conference, 2007—macs.hw.ac.uk
6. Herlocker JL, Kostan JA, Terveen LG, Riedl JT (2004) Evaluating collaborative filtering recommender systems. ACM Trans Inf Syst 22(1):553
7. Keshavan RH, Montanari A, Sewoong O (2010) Matrix completion from a few entries. IEEE Trans Inform Theory
8. Konstan JA, Riedl J (2012) Recommender systems: from algorithms to user experience—user modeling and user-adapted interaction. Springer
9. Kumar P, Raju BS, Krishna PR (2010) A new similarity metric for sequential data—warehousing and mining (IJDWM), 2010—igi-global.com

10. Mishra R, Kumar P, Bhasker B (2015) A web recommendation system considering sequential information. Decis Support Syst
11. Nakamura A, Abe N (1998) Collaborative filtering using weighted majority prediction algorithms. In: Proceedings 15th international conference. Machine Learning 1998
12. Pan C, Li W (2010) Research paper recommendation with topic analysis. In: International conference on computer …, 2010—ieeexplore.ieee.org
13. Park DH, Kim HK, Choi IY, Kim JK (2012) A literature review and classification of recommender systems research—expert systems with applications. Elsevier
14. Rashid AM, Albert I, Cosley D, Lam SK (2002) Getting to know you: learning new user preferences in recommender systems. In: Proceedings of the 7th …, 2002—dl.acm.org
15. Resnick P, Iakovou N, Sushak M, Bergstrom P, Riedl J (1994) An open architecture for collaborative filtering of netnews/PR GroupLens. In: Proceedings of the 1994 computer supported cooperative work …, 1994
16. Schafer JB, Konstan J, Riedl J (1999) Recommender systems in e-commerce. In: Proceedings of the 1st ACM conference on …, 1999—emunix.emich.edu
17. Wang H, Wang Z, Zhang W (2010) Quantitative analysis of Matthew effect and sparsity problem of recommender systems. In: 2018 IEEE 3rd International …, 2018—ieeexplore.ieee.org
18. Isinkaye FO, Folajimi YO, Ojokoh BA: Recommendation systems: principles, methods and evaluation. Cairo University Egyptian Informatics Journal, Elsevier
19. Yang X, Zhou S, Cao M (2019) An approach to alleviate the sparsity problem of hybrid collaborative filtering based recommendations: the product-attribute perspective from user reviews-mobile networks and applications. Springer

Exploring the Novice Approach to Orthorectification of Satellite Imagery

G. Mallikarjuna Rao, Ch. Mallikarjuna Rao, B. R. K. Reddy, D. V. Lalitha Parameswari and Mohammad Azeez

Abstract Orthorectification plays vital role in satellite image processing. This process imposes challenges due to the dynamism in capturing environment, capturing unit, satellite rotation, sensors parameters and overlying regions on earth. The required geometric modelling needs an accurate estimation of Ground Control Points (GCPs) and their processing. Most of the proposed models are computational intensive and use manual approach for locating GCPs. Further, GCPs co-ordinates are floating point numbers the computational capability of the system imposes the constraint on the accuracy and robustness of the respective models. In this paper we have studied orthorectification process and proposed instinctive processing framework for orthorectification of optical pushbroom sensor based satellite imagery. The frame work accompany metadata extraction, automatic ground control point (GCP) extraction using parallel processing, geometric modeling, orthorectification and image stitching processes. Experimental results with proposed framework confirmed the robustness of the technique and provided sub pixel accuracy on independent check points with positional accuracy around one pixel for orthoimage. Parallel SIFT features are extracted using SIMD architecture while performing image stitching.

Keywords Orthorectification · Generic geometric model · Ground control point (GCP) · Optical imagery · Random sample (RANSAC)

G. Mallikarjuna Rao (✉) · Ch. Mallikarjuna Rao · B. R. K. Reddy · M. Azeez
Gokaraju Rangaraju Institute of Engineering and Technology, Hyderabad, India
e-mail: Gmr_333@yahoo.com

Ch. Mallikarjuna Rao
e-mail: chmksharma@yahoo.com

D. V. Lalitha Parameswari
G. Narayanamma Institute of Technology and Science, Hyderabad, India

S. Fong et al. (eds.), *ICT Analysis and Applications*, Lecture Notes in Networks and Systems 93, https://doi.org/10.1007/978-981-15-0630-7_15

1 Introduction

The base map layer, a visual source, provided by satellite images is used as statistical analysis tool for disaster predictions and management, change detection and digital analysis. However some applications such as disaster prediction and man-agreement, require visual and accuracy factors of satellite image while the other applications such as military and navigation, needs location accuracy as main consideration. Radiometric (spectral) and geometric (spatial) corrections are performed to improve visual and accuracy properties. These operations require considerable computational effort and better resolution of satellite image.

1.1 Over View of the Development

The four basic components of orthorectification are (1) raw image (2) geometric model (3) DEM (digital elevation model) (4) GCP (ground control points). Orthorectification of satellite imagery imposes challenges due to the dynamism in capturing environment, capturing unit, satellite rotation, sensors parameters and overlying regions on earth. Further pixel positional displacements are common problems due to lens distortion and camera tilt. Orthorectification models [1–3, 9, 16] have taken care about positional displacement of pixels by employing real-ground co-ordinates (including x, y and z values). However these floating point co-ordinate vales have to be properly processed to provide pixel values of the orthoimage. Thus computational environment capability influences the accuracy. The geometric correction using geometric rectification and orthorectification is a common practice [6–8, 15] in the modern remote sensing (RS) to provide accurate orthoimage. Digital Elevation Model (DEM) along with GCPs (link between image and ground co-ordinate) is used for this purpose. However, the main challenge here is to capture the appropriate number of high quality GCPs and selection of correct quality of DEM. Some of the orthorectification models are:

(i) Polynomial rectification models: Polynomial function is applied to the surface adapt the polynomial to a number of check point is one the simplest image processing method to cancel the effects of tilts for both satellite images and aerial photographs. However the polynomial order is greatly influenced by number of GCPs and desired accuracy.
(ii) Rational Polynomial satellite sensor models: these models uses ratio of two cubic polynomial expressions and provide empirical mathematical models relating image coordinates (row and column position) to latitude and longitude using the terrain surface elevation. The coefficients of these two cubic polynomials (one for row and another for column) are computed by Satellite Company from satellite's orbital position, and orientation. Automatic Resampling Geometric relationship between object space and image space is provided by rational function sensor model. It relates object point coordinates (X,

Y, Z) to image pixel coordinates row and column (r, c) or vice versa using 78 rational polynomial coefficients (RPCs). Following Eqs. 1 and 2 provide the ground-to-image transformation:

$$r_n = \frac{p_1(X_n, Y_n, Z_n)}{p_2(X_n, Y_n, Z_n)} = \frac{\sum_{i=0}^{m1}\sum_{j=0}^{m2}\sum_{k=0}^{m3} a_{ijk}X_n^i Y_n^j Z_n^k}{\sum_{i=0}^{m1}\sum_{j=0}^{m2}\sum_{k=0}^{m3} b_{ijk}X_n^i Y_n^j Z_n^k} \tag{1}$$

$$c_n = \frac{p_1(X_n, Y_n, Z_n)}{p_2(X_n, Y_n, Z_n)} = \frac{\sum_{i=0}^{m1}\sum_{j=0}^{m2}\sum_{k=0}^{m3} c_{ijk}X_n^i Y_n^j Z_n^k}{\sum_{i=0}^{m1}\sum_{j=0}^{m2}\sum_{k=0}^{m3} d_{ijk}X_n^i Y_n^j Z_n^k} \tag{2}$$

In the above expressions $a_{ijk}, b_{ijk}, c_{ijk},$ and d_{ijk} denote Rational Function Coefficients, while (r_n, c_n) denote image space normalized row (line) and column (sample) index of pixels. Normalized coordinate values of object points in ground space are denoted by $X_n, Y_n,$ and Z_n. However, subjective procedure of selecting control points and associated RMS error of residue vales imposes limitation on achievable accuracy.

(iii) Projective rectification: It uses a geometric transformation between the image plane and the projective plane. At least four control points in the object plane are required to compute eight unknown coefficients, b_{ij}, using projective transformation. The row and column (r, c) are computed as per following expressions

$$r = \frac{b_{11}x + b_{21}y + b_{31}}{b_{12}x + b_{23}y + 1} \tag{3}$$

$$c = \frac{b_{12}x + b_{22}y + b_{31}}{b_{13}x + b_{23}y + 1}. \tag{4}$$

In general, orthorectifying low-to medium resolution (L-MR) satellite imagery Shuttle Radar Topography Mission (SRTM) 30 m DEM is sufficient while high-resolution (HR) and very high-resolution (VHR) satellite imagery require DEM with 15 m or less GSD (ground sample distance) is needed [4, 5, 15]. In our experimentation Ground Controls Points are manually collected with GeoExplorer 6000 series handheld (model: GeoXH 3.5G) receiver device and automatic extraction of GCP on earth surface was done using TerraSAR-1 with 1 m accuracy and TerraSAR-3 with 3 m accuracy. In our modest effort it is aimed to create an optimal accurate orthoimage, having smallest root mean square error (RMSE) and positional accuracy that could be used for benchmarking as well as reference source for quantifying created orthoimage set [7, 12, 14].

Inspiration: Pixel shift due to earth curvature is addressed by orthorectification while horizontal position of image positions is addressed by georeferencing and geotectification. For each pixel orhtorecification provide X, Y, Z ground coordinate values. In the rectification process the ground control points and associated surface

Table 1 Different DEMs and their resolution

DEM	Resolution
Shuttle radar topography mission (SRTM)	1 arc-second (30 m)
	3 arc-second (90 m)
Digital terrain elevation data (DTED)	Level 0: 900 m; Level 1: 90 m; Level 2: 30 m
Advanced Spaceborne thermal emission and reflection radiometer (ASTER)	30 m
Global digital elevation map (GEDM)	30 m

elevation reference plays a pivotal role and responsible for providing image and ground coordinate link. The following Table 1 gives some of the public products of DEM.

The main problem here, is qualitative GCPs selection. This may due to the dearth of available resources of GCP, expensiveness associated with collection of new GCPs and inaccessibility of the specific areas in some cases it is even impossible. Some of the limitations of manually collected GCPs are:

(1) ***Time overwhelming***: Covering vast area by physically visiting is not only time consuming but also difficult to reach due to environmental conditions and access restrictions.
(2) ***Monetary Impact***: To embark on such a project will have a high monetary impact on logistic facilities.
(3) ***Travelling constraints***: International immigration constraints have to be faced while collecting necessary information from other countries.
(4) ***Terrain restrictions***: Some terrain are not conducive to provide sufficient GCPs.

2 Experimentation

In general, orthorectification demand uniform distribution with complete image covering GCPs. However, exact number of required GCPs reliant on several aspects. Some of the factors deciding the count of GCPs [5, 9–11] are accuracy of usable GCPs, wide spread of GCPs, topographical characteristics and terrain variations. More than the number of GCPs, their quality and distribution will decide the efficiency of orthorectification process. In our experimentation we used the same base with varying number of GCPs and quality of DEM.

- **Data Gathering**: It associate with

  (i) Raw satellite Image.
  (ii) Reference data set for providing tie—points between raw image and reference image.

(iii) Digital Elevation Model data source as input reference for terrain height and slope estimation.
(iv) A geometric sensor model for correcting the various errors such as data recoding errors and systematic errors arise during the capture of satellite image.

- **Data Investigation**: Data analysis was performed by following two stages:

Stage 1: Nine experiments were performed to determine the accuracy of an ortho image by means of altering the number of GCPs and the quality of the DEMs. Three experiments each with 5, 13 and 25 GCPs with 30 m DEM, 12 m DTM and 2 m DTM respectively as shown in Fig. 1. Most precise orthoimage was identified form the experimentation and used as benchmarked one for evaluating outcomes of stage 2.

Stage 2: Scenarios to simulate the lack of GCPs that are irregularly distributed across an image scene are created using a vector road layer. Figure 2 shows created the five scenarios (West, East, North, and South and Random directions) for experimental simulation. These experiments were performed on the same image used during stage 1. A methodology is deduced from the results of stage1 and stage2. This methodology is used for improving geometric precision of orthorectification to the satellite image scene having irregularly distributed and insufficient GCPs. Accuracy, reliability and validity of the results are maintained by triangulation the data results.

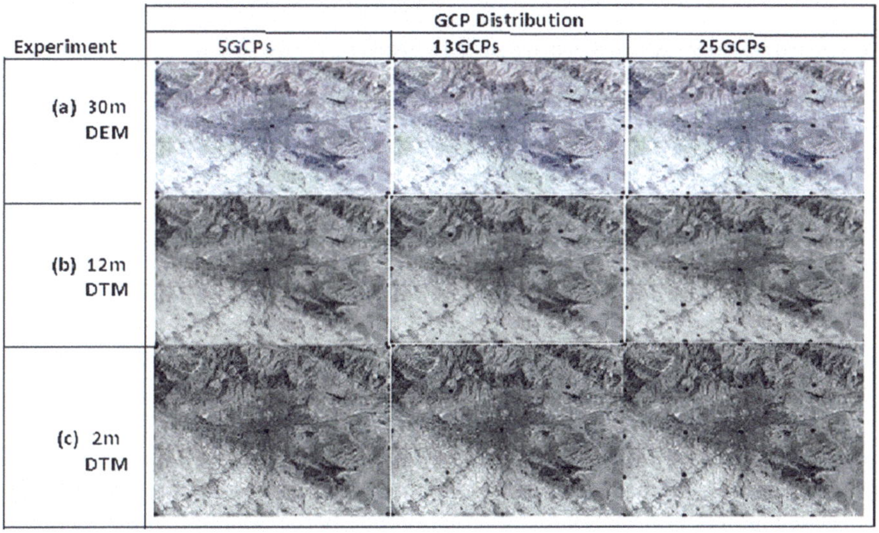

Fig. 1 Deficiency and unbalanced distribution of GCPs for several simulations

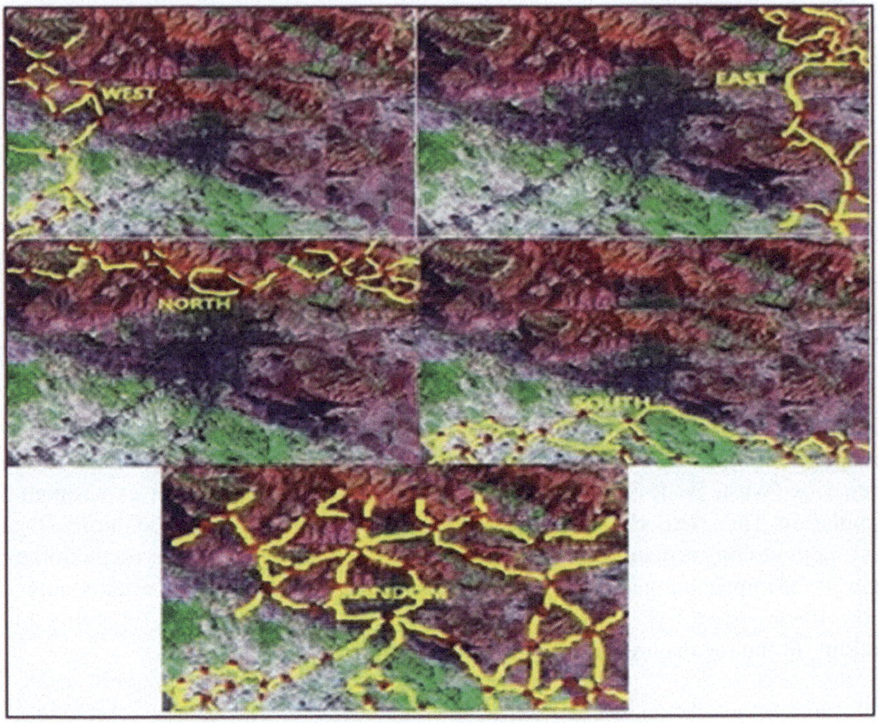

Fig. 2 DTM resolution upon varying uniformly distribute GCPs

3 Proposed Approach

The image data utilized to perform all orthorectification experiments was a Pléiades-1B primary panchromatic image. Flow diagram of the proposed frame work is given in Fig. 3.

3.1 Pseudo Code for Proposed Framework

Pseudo code used for the experimentation is given below. It is implemented ERDAS environment. Intel i7 quad core processor with 16 GB was used in the experimentation. Note that function names are given appropriately for the purpose they used and they are not directly refer any built in library function

```
 1. Input: data file, f.dat in DIMP v2 structure
2. Convert_Image(f,dat)  /* this function covert the data
file into '.img' format from DIMAP XML format ERDAS
IMAGEINE@ is used*/
3. Delelte_MAP_Projected_Inf()  /*Delete the MAP and Pro-
jected information as this information is emdedded in raw
image */
4. Select_Geometric_Model() /* Pléiades Orbital Pushbroom
geometric model */
5. Add_Elevation_Model()
6. Ref_Point_Select()
/* data sources for reference point maybe GPS points,
reference images, reference files, vector files or ASCII
files */
7. Choose_out_put_format()
/* JPEG200, TIFF, IMG etc */
8. Coose_Resample_Method() /* Resampling can be done us-
ing nearest neighborhood method, bilinear interpolation,
cubic convolution or bicubic spline   */
9 Run_Orthorectification()
10 Evaluate_orthimage()
/* assess the edges of the image for sign evaluation. Ir-
regular edges denote mountain terrain*/
```

The Fig. 4 represent extraction of GCPs for various cases. However, the sensor models and elevation data were altered to match the various experiment parameters. The Fig. 5 gives final orthorectifide image with TIN DEM.

3.2 Image Stitching

If the single orthoimage of the selected resolution is not covering entire scene area then set of orthoimages are generated by above procedure are stitched to generate single orthoimage [13]. Matlab environment is used for this purpose. SIFT algorithm is used to extract key point of the images to be stitched and SIFT descriptor is used for feature or key point matching. RANSAC (Random Sample Consensus) algorithm is used to remove outliers are wrong key points and finally alpha blending is used to provide sooth mosaic image. Figure 6 indicate image stitching.

Parallel Approach for SIFT feature extraction: Let convolved image $\boldsymbol{g}(\boldsymbol{x}, \boldsymbol{y})$, original image is $f(x, y)$ and $\boldsymbol{w}$ is 3×3 kernel convolved image is computed as per Eq. 5.

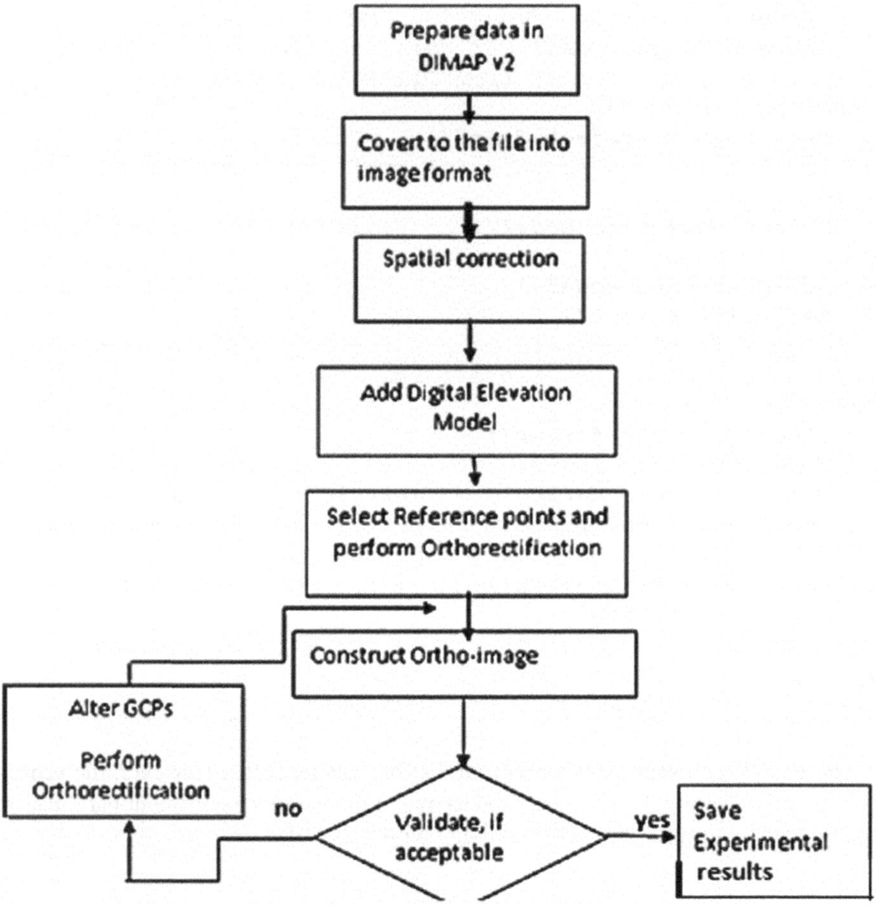

Fig. 3 Flow diagram of proposed framework

$$g(x, y) = w * f(x, y) = \sum_{s=-1}^{+1} \sum_{t=-1}^{+1} w(s, t) * f(x - s, y - t) \tag{5}$$

In our approach, each mage [1024, 756] to be processed is divided into 189, 64 × 64 sub images and one thread is proposed to each for computing associated Gaussian convolved sub image. Later all are merged to provide Single Gaussian Image. We have used quad core machine with **48** worker threads for each core. *SIMD* (Single Instruction Multiple Data) architecture was used for this purpose. Pseudo code for processing each sub image is given below.

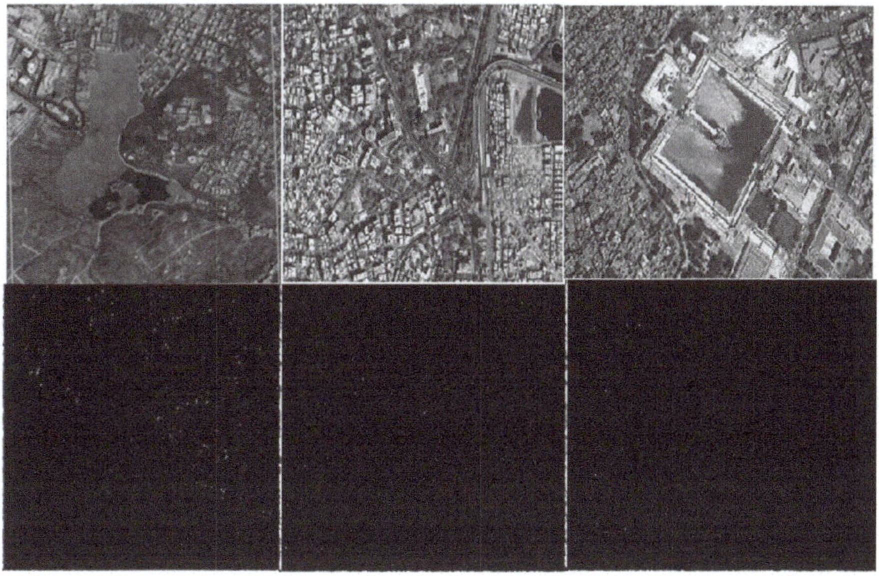

Fig. 4 GCPs for satellite images

Fig. 5 Orthorectifide image using TIN DEM

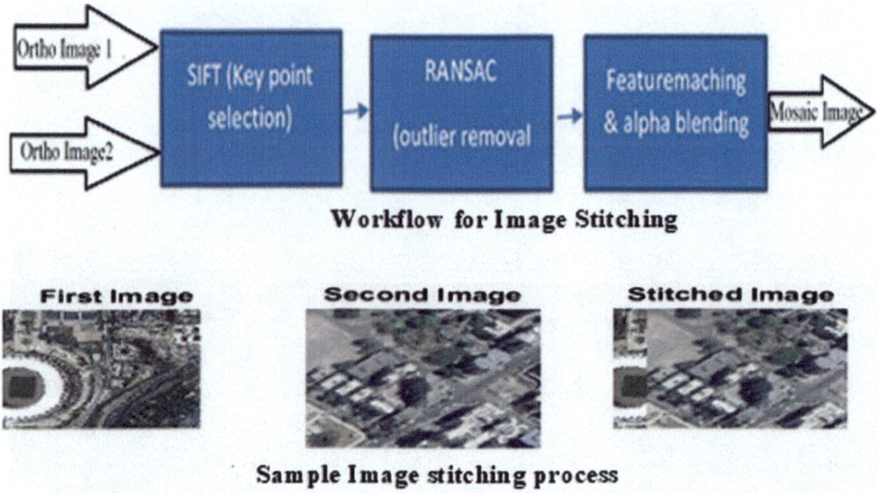

Fig. 6 Image stitching workflow

Pseudo code for Parallel Image stitching

```
for each Sub_ Image Si
  for each pixel(x,y) of Si

      Set accum_value =0;

              for each row element of kernel

                          for each column element of kernel

accum_value = accum_value + kernel(row,column)   *Si(x-row,y-coloumn)

      end for

   endfor

  set Gi(x,y)= accum_value     /* Gaussian valve sub image region location x,y */

endfor

endfor
```

4 Conclusions

The presence of variables such as sensor orbital data and elevation data make it difficult to achieve absolute accuracy by mere selecting input and reference GCPs while performing orthorectification. Further accuracy is restricted by operator involvement in this process who needs to manually select and place input points (GCPs) to corresponding reference points on the primary image data. HR and VHR satellite image processing tedious task even if the precise location of the reference points is known apriori. The operator needs to identify the pixel location by zooming into pixel scale of the primary image and associate the location to the input point that correspond to the specific reference point. In this paper we analyzed the influence of number of GCPs on the acceptable orthoimage. The GCP and pixel conversion are performed in parallel using SIMD architecture. Image stitching is performed for reconstructing large orthoimage covering vast area.

References

1. Auster HU, Richter I, Glassmeier KH, Berghofer G, Carr CM, Motschmann U (2010) Magnetic field investigations during ROSETTA's 2867 Šteins flyby. Planet Space Sci 58(9):1124–1128. https://doi.org/10.1016/j.pss.2010.01.006
2. Atif S, Salman N, Irfan AI, Zahir A (2015) Accuracy assessment of digital elevation model generated from pleiades tri stereo-pair. In: 7th International conference on recent advances in space technologies, pp 193–197
3. Bhardwaj A, Sam L, Martín-Torres FJ, Kumar R, et al (2016) Uavs as remote sensing platform in glaciology: Present applications and future prospects. Remote Sens Environ, 175–196
4. Parhami B (2009) Computer arithmetic: algorithms and hardware designs, 2nd edn. Oxford University Press, Inc, New York, NY, USA
5. Becker K, Detsis E, Nwosa CC, Palevska V, Rathnasabapathy M, Taheran M (2012). Space situation awareness. In: SSS Educational series 2012.Space generation advisory council
6. Chmiel J, Kay S, Spruyt P (2004) Orthorectification and geometric quality assessment of very high spatial resolution satellite imagery for common agricultural policy purposes. In: Proceedings of XXth ISPRS congress, pp 12–23
7. Costlow T (2014) The future of satellites: what are the options? Defense systems: knowledge technology and net-enabled warfare. http://defensesystems.com/articles/2014/03/21/satellite-communications-futureptions.aspx
8. De Caro D, Genovese M, Napoli E, Petra N, Strollo AG (2014) Accurate fixed-point logarithmic converter. EEE Trans Circuits Syst II: Express Briefs 61(7):526–530
9. Ranisavljević E, Devin F, Laffly D, Le Ni Y (2014) A dynamic and generic cloud computing model for glaciological image processing. Int J Appl Earth Obs Geoinf 27:109–115
10. Porikli F (2005) Integral histogram: a fast way to extract histograms in cartesian spaces. In: Computer vision and pattern recognition, 2005. CVPR 2005. In: IEEE computer society conference on, vol 1. IEEE, pp 829–836
11. Rieke M, Foerster T, Geipel J, Prinz T (2011) High-precision positioning and realtime data processing of UAV systems. Int Arch Photogramm, Remote Sens Spat Inf Sci 38:1–C22
12. Shahbazi M, Théau J, Ménard P (2014) Recent applications of unmanned aerial imagery in natural resource management. GiSci Remote Sens 51(4):339–365
13. Dey N, Bhatt C, Ashour AS (2019) Big data for remote sensing: visualization, analysis and interpretation digital earth and smart earth. Springer International Publishing AG, part of Springer Nature 2019, Springer, Cham

14. Yang W, Di L (2004) An accurate and automated approach to georectication of hdf-eos swath data. Photogr Eng Remote Sens 70(4):397–404
15. Wang S, Li W, Wang F (2017) Web-scale multidimensional visualization of big spatial data to support earth sciences—a case study with visualizing climate simulation data. Informatics 4(3). https://doi.org/10.3390/informatics4030017
16. Cao X, Wu C, Lan J, Yan P, Li X (2011) Vehicle detection and motion analysis in low-altitude airborne video under urban environment. IEEE Trans Circuits Syst Video Technol 21(10):1522–1533

Flower Pollination Algorithm for Test Case Prioritization in Regression Testing

Priyanka Dhareula and Anita Ganpati

Abstract Flower Pollination Algorithm (FPA) is a significant addition made to Nature Inspired Metaheuristic Optimization Algorithms (NIMOA). It is inspired by the pollination process of flowering plants. In this research, FPA is used for Test Case Prioritization (TCP) in Regression Testing (RT). The algorithm uses code coverage of test cases as the input. The algorithm has no prior information of faults covered by the test cases. This study deals with prioritizing (ordering) the test cases in such a way that only those test cases are executed that covers maximum faults in minimum time of execution. For validation of the results Average Percentage of Fault Detected (APFD) metrics is used. APFD values for different ordering of test cases is calculated for three applications written in Java. The empirical results of APFD metrics for FPA order (TS_1) and FPA order (TS_p) are better as compared to Random Order of Original Test Suite (TS_o) and Reverse Random Order of (TS_o). Therefore, this paper states that FPA for TCP gives efficient results in RT.

Keywords Regression testing · Test case prioritization · Flower pollination algorithm · APFD

1 Introduction

Regression Testing (RT) is foremost the crucial part of software life cycle. Whenever software changes RT is done in order to ensure the proper functioning of a software [13]. With the changes in a software, the size of test suite also varies. As the software evolves, the magnitude of test suite size becomes quite large. Rationally it is not feasible to execute all the test cases due to limited time and cost constraints. Therefore, it becomes important to deduce a mechanism to perform efficient RT with maximum number of faults revealing test cases in minimum time. So, test case optimization

P. Dhareula (✉) · A. Ganpati
Computer Science Department, Himachal Pradesh University,
Shimla 171005, Himachal Pradesh, India
e-mail: priyankarana.id@gmail.com

A. Ganpati
e-mail: anitaganpati@gmail.com

S. Fong et al. (eds.), *ICT Analysis and Applications*, Lecture Notes in Networks and Systems 93, https://doi.org/10.1007/978-981-15-0630-7_16

techniques are required to perform early fault detection in affordable time and cost [3]. Regression Test Case Optimization (RTCO) is further categorized into three main branches of test case optimization namely; Retest All [15], Test Case Selection [4] and Test Case Prioritization (TCP) [14]. In Retest All technique, all the test cases in original test suite are executed. This technique is time consuming and very expensive. Regression Test Case Selection executes subset of a test suite. Whereas, regression TCP techniques [6] deals with maximum and early fault detection by ordering the test cases based on priority given to them. Determining the faults covered by test case in advance is very difficult and time-consuming process, since each test case must be executed in advance to determine its fault coverage.

In the recent times Nature Inspired Metaheuristic Optimization Algorithms (NIMOA) have gained huge popularity [22]. Most of the NIMOA belong to the domain of Stochastic Algorithms (SA). SA are those algorithms whose outcome does not depend on its starting point as they follow a random search. Keeping in view the success of SA numerous algorithms for TCP in RT have been proposed in recent times [6, 7, 12, 21]. In 2012, a new algorithm namely, Flower Pollination Algorithm (FPA) was developed by Xin-She Yang [1]. FPA is inspired by the pollination process in flowering plants. Since, FPA is an optimization algorithm [8] therefore, in this paper TCP was done using FPA without any prior knowledge of faults covered by the test cases.

2 Related Work

Numerous studies have surfaced recently that adopted FPA in different domains like engineering, wireless sensor networking, image and signal processing, communication, structural design, computer gaming, software engineering and global function optimization [2, 11].

Alsewari et al. [1] proposed a strategy for Test Case Minimization based on FPA. Their strategy is termed as Test Generator Flower Pollination Strategy (TGFP). Their research is based on *t*-way testing. Where '*t*' represents the interaction strength between parameters. For different interaction strength their strategy performed well. Nasser et al. [10] proposed a hybridized FPA for generating *t*-way test suite. The main hybridized features used in their research are local search, mutation operator and elitism feature. It is shown in their work that elitism based FPA performed better than other variants of FPA.

Kabir et al. [5] proposed a technique to minimization test suite for *t*-way testing by using adaptive FPA. Their comparative analysis with existing techniques shows that their algorithm showed improved performance. Nasser et al. [9] proposed a Flower Strategy (FS) for generation of *t*-way test suite based on FPA. The research also critically performed the comparison of adoption Optimization Algorithm (OA). Their results show that FS outperforms OA based strategies.

Yadav and Dutta [21] used Genetic Algorithm (GA) to prioritize the test cases by using statement coverage as input to the algorithm. Average Percentage of Statements Covered (APSC) metrics is used to compute the efficiency of the algorithm. Their approach is compared with other techniques of prioritization viz., random prioritization and reverse prioritization. Their study gives optimum results.

Panwar et al. [12] proposed an improved version of Ant Colony Optimization (ACO) algorithm to prioritize the test cases. Triangle classification problem is used in this study. Their proposed technique is compared with different ordering of test cases viz., reverse ordering, random ordering, optimal ordering and original ordering. The antennas of ants are used to share the information for proposed approach. Their approach achieves better results as compared to other orderings of the test cases.

3 Flower Pollination Algorithm (FPA)

FPA was developed in the year 2012 by Xin-She Yang. There are four main rules for flower pollination process as discussed below [22].

I. Global pollination is achieved through biotic and cross-pollination process, and pollen- carrying vectors transfer pollens through Levy flight (Rule 1)
II. Local pollination is achieved through abiotic and self- pollination (Rule 2)
III. Reproduction probability is equivalent to Flower Constancy (FC) that is proportionate to the relationship between two flowers involved (Rule 3)
IV. Switch probability $p \varepsilon [0,1]$ controls the global and local pollination (Rule 4)

Figure 1 represents the FPA developed by Yang [22].

NIMOA are analyzed by the way they explore their search space [22]. In principle all algorithms have two main components i.e., exploitation and exploration which are also known as intensification and diversification respectively. Exploitation is a local search process which can leads to early convergence rate and may also get stuck at local optimum since the final solution depends on starting point. Whereas, exploration explore the global search space leading to a diverse set of solution. Exploration does not get stuck at local optimum [22]. FPA is one of the newest NIMOA [2]. Flower pollination is mainly related to the transfer of pollens, the transfer is achieved through pollinators viz., birds, insects, bats and other animals [23].

Pollination can be achieved in two ways biotic and abiotic. Nearly 90% flowering plants are a member of biotic pollination, the pollens are transferred with the help of vectors. Whereas 10% pollination belong to abiotic form where no vector is needed for transfer of pollens. The transfer of pollens take place with the help of water and wind. Various vectors develop FC [22]. FC can be defined as the affinity that is developed between the vector and the specific flower species. FC thus result in maximum transfer of pollens and reproduction of the same flower species. Biologically, the objective of flower pollination can be explained as the survival of the fittest and optimum reproduction of plants in number as well as fitness. Therefore, this natural process encourages this research to use FPA for TCP for RT.

Maximize or minimize the Fitness Function f(x) for all x_i where i= 1 to n
Create original population of n flower
Identify best flower from original population
Switch probability p ε [0,1] is defined
While (t<Maximum Generation)
For i=1 to n; where n is the number of flowers in the given population
If rand < p then perform global pollination
Perform Levy flight and draw L (step vector)
$x_i^{t+1} = x_i^t + \Upsilon L\,(g^* - x_i^t)$
Else
Draw ε from a uniform distribution in [0,1]
Perform local pollination as
$x_i^{t+1} = x_i^t + \varepsilon\,(x_j^t - x_k^t)$
End if
Calculate the fitness of new solution
If the fitness of new solution is better than previous solution then replace the previous solution
End for
Rank the solutions and keep the best solution found so far
End while

Fig. 1 Flower pollination algorithm (FPA)

4 FPA for TCP

This paper uses FPA for TCP to reduce overall time of execution in RT. FPA is used to prioritize test cases from the original test suite (TS_o) to give maximum fault coverage in minimum time of execution. The algorithm uses code coverage criteria as an input to prioritize test cases in TS_o. Original population of test cases is randomly generated and represented as $TS_o = (T_1, T_2, \ldots, T_n)$. Figure 2 shows FPA for TCP.

5 Explanation of FPA for TCP

Original population of 'n' test cases is generated and the test suite is identified as $TS_o = (T_1, T_2, \ldots\ldots, T_n)$. For each test case its code coverage is considered as the fitness function $f(x_i)$, where $x_i = (x_1, x_2, \ldots., x_n)$. The code coverage of test cases in TS_o is used to prioritize the test cases. Test case with highest coverage is maintained as g^*. If multiple test cases have same coverage, then First Come First Serve (FCFS) policy is adopted. A switch variable ($p = 0.8$) [23] is used to perform local (exploitation) or global (exploration) pollination [23]. Random number 'rand' (value ranging between 0 and 1) is used to determine whether local pollination will take place or global pollination will occur. If rand $< p$, global pollination is performed and the fittest reproduction is represented as g^* at tth iteration. In global pollination, the fittest reproduction is ensured through pollens that can travel to long distance. Therefore, for $t + 1$ iteration the pollens/test cases are picked from the tth iteration. Here the pollens are represented as test cases. In flower pollination vector has affinity

```
Set the Fitness Function f(x) for all x_i where i= 1 to n
Set the original population of n test case as random solutions
Set g* as test case with maximum coverage
Set a switch variable p= 0.8
While (t< stopping criteria)
For i=1: n (all test case)
If rand < p then perform global pollination
Perform Levy flight and draw L (step vector)
x_i^t+l = x_i^t + ϒL (g* - x_i^t)
Else
Draw ε from a uniform distribution in [0,1]
Perform local pollination as
x_i^t+l = x_i^t + ε (x_j^t - x_k^t)
End if
Calculate the fitness of new solution
If the fitness of new solution is better than previous solution
then replace the solution
End for
Rank the solutions and keep the best solution found so far
End while
```

Fig. 2 FPA for TCP

towards the pollens similarly test cases have affinity towards the fault. Those test cases with higher fault detection capability are of great value and are carried forward to the next generation of software. If we represent the fittest as $g*$. Then the FC and first rule (Rule 1) can be mathematically formulated as shown in Eq. (1).

$$X_i^{t+1} = X_i^t + \Upsilon L\left(g * - X_i^t\right) \tag{1}$$

In Eq. (1) X_i^{t+1} is the ith pollen/test case at $t + 1$ iteration, X_i^t symbolizes the ith pollen or the ith test case at iteration t, L is the strength of the pollination and has a certain probability of distribution. In this paper all test cases/pollens are considered to have FC or affinity toward the vector carrying them for pollination. Hence giving equal chance to all the test cases/pollens to participate in pollination process (TCP). For this reason, L (i.e., step size) is equated to one. To control the step size, Υ i.e., the scaling factor is used. It is assumed that uniform scaling is performed by vector to visit pollens. Therefore, Υ is equated to one. Local pollination described in (Rule 2) and FC can be formulated as shown in Eq. 2.

$$X_i^{t+1} = X_i^t + \varepsilon\left(X_j^t - X_k^t\right) \tag{2}$$

Pollens are transferred from different flowers, but these flowers belong to a single plant species. This simulates the FC in a small neighbourhood. Here $j = i$ and $k = i$

+ 1 in the *t*th iteration. This emulates the FC behaviour in small area. A local random walk variable ε is derived from a uniform distribution. This paper takes the value of ε as 0 or 1 representing either the test case is selected or not selected for further prioritization; since it may be the case that in a given local area two test case cover the same code. Therefore, redundant test cases are required to be discarded. Here redundant test cases are those test case which give same code coverage. Therefore, fifty percent of the chances of test cases with same coverage being selected is reduced. This result in a reduced and prioritized set of test cases represented as test suite TS_1. TS_1 is exercised upon the application under consideration to determine the number of faults covered by prioritized test suite TS_1. The test cases now in TS_1 are prioritized again to give test suite TS_p. TS_p gives maximum fault coverage in reduced time of execution by removing those test cases that give redundant fault coverage. Effectiveness of the final prioritized test suite TS_p is determined with the help of Average Percentage of Fault Detected Metrics (APFD). Let TS be the test suite having n test cases and let F be the set of m faults identified by TS. Let F_i be first test case in original test suite TS_o which identifies fault i. Therefore, APFD for TS_i is formulated as shown in Eq. (3).

$$\text{APFD} = 1 - \frac{F_1 + F_2 + F_3 + \ldots\ldots + F_m}{nm} + \frac{1}{2n} \tag{3}$$

6 Implementation

FPA for TCP is implemented in Java using Eclipse IDE. Three applications namely, Puzzle Game [16], Area and Perimeter [18] and Tritype [17] are used, which were written in Java programming language. Area and Perimeter calculates the area and perimeter of various mathematical shapes and tritype is a classic triangle classification problem. TestNG tools is used to design the test scripts for all the applications [19]. EclEmma tool is used to get the code coverage for applications used [20]. For the validation of results five versions of Puzzle Game, eight versions of Area and Perimeter and six versions of Tritype applications were created. Each version was seeded with a unique fault. Details of the applications used for experimental evaluation are shown in Table 1.

Table 1 Details of applications used

Sr. no.	Name of the applications	Size in LOC	Number of version/Unique fault	No. of test cases used
1.	Puzzle game	246	5	33
2.	Area and perimeter	916	8	113
3.	Tritype	106	6	45

7 Results

For the explanation of empirical evaluation only one application is considered namely, Puzzle Game. The original test suite (TS_o) for Puzzle Game contains 33 test cases. Test cases are designed in no order viz., $\{T_1, T_2, T_3, T_4, \ldots, T_{25}, T_{26}, \ldots, T_{33}\}$. There are five versions of Puzzle Game application with unique faults induced in each version viz., $\{F_1, F_2, \ldots, F_5\}$. TS_o for Puzzle Game is given as input to the FPA for TCP, the following sequence of test cases is given as the output from FPA i.e., $TS_l = \{T_3, T_5, T_8, T_{14}, T_{25}, T_{27}, T_{30}\}$. So, the output of FPA is a test suite TS_l consisting of seven test cases. TS_l was exercised on all the versions of Puzzle Game application to identify the faults covered by TS_l. Table 2 shows the faults identified by TS_l for Puzzle Game.

In Table 2, circled cells represent the test case that first identifies a given fault, which is also used to compute APFD metrics. It is evident from Table 2, that test case T_3 identifies maximum faults i.e., F_3, F_4, F_5. Hence, only T_3 test case is enough to identify faults F_3, F_4 and F_5. Test case T_3 is followed by T_5 and T_8 that identifies faults F_2 and F_1 respectively. Therefore, the prioritized sequence in final test suite i.e., $TS_p = \{T_3, T_5, T_8\}$ respectively. Table 3 shows the final test suite TS_p for Puzzle Game with the faults identified by TS_p.

Table 2 Faults identified by TS_l for puzzle game

↓Versions/Faults	Test Cases →	T_3	T_5	T_8	T_{14}	T_{25}	T_{27}	T_{30}
F_1 (Version 1)				(1)	1	1		
F_2 (Version 2)			(1)					1
F_3 (Version 3)		(1)					1	
F_4 (Version 4)		(1)	1				1	1
F_5 (Version 5)		(1)	1	1	1	1	1	1

Table 3 Faults identified by TS_p for puzzle game

↓ Versions/ Faults	Test Cases →	T_3	T_5	T_8
F_1 (Version 1)				(1)
F_2 (Version 2)			(1)	
F_3 (Version 3)		(1)		
F_4 (Version 4)		(1)	1	
F_5 (Version 5)		(1)	1	1

TS_p gives complete fault coverage as was given by TS_o. APFD metric value for $TS_1 = \{T_3, T_5, T_8, T_{14}, T_{25}, T_{27}, T_{30}\}$ for Puzzle Game application using FPA is shown in Eq. (4).

$$\text{APFD} = 1 - \frac{8+5+3+3+3}{33 \times 5} + \frac{1}{2 \times 33} \tag{4}$$

$$\text{APFD} = 1 - 0.13333 + 0.01515$$
$$\text{APFD} = 1 - 0.14848$$
$$\text{APFD} = 0.85152 \text{ in } 0.48 \text{ s}$$

Figure 3 shows the priority order and time of execution of TS_1. As it is evident from the Fig. 4 that all the test case for TS_1 failed. Hence, TS_1 was efficient in covering all the faults. Time of execution of TS_1 is 0.48 s. Similarly, for the Prioritized final test suite $TS_p = \{T_3, T_5, T_8\}$ its APFD metric value for Puzzle Game application is shown is Eq. (5).

$$\text{APFD} = 1 - \frac{8+5+3+3+3}{33 \times 5} + \frac{1}{2 \times 33} \tag{5}$$

$$\text{APFD} = 1 - 0.13333 + 0.01515$$
$$\text{APFD} = 1 - 0.14848$$
$$\text{APFD} = 0.85152 \text{ in } 0.286 \text{ s}$$

Figure 4 shows the priority order and time of execution for TS_p. Time of execution of TS_p is 0.268 s which is comparatively lower than the execution time of TS_1.

For the validation of the results APFD metric is computer for different ordering of test cases viz., FPA order (TS_1), FPA order (TS_p), Reverse FPA Order of (TS_p), Random Order of (TS_o) and Reverse Random Order of (TS_o) for Puzzle Game application. It is also evident from Fig. 4 that the Prioritized Test Suite (TS_p) covers all faults with reduced test suite size and reduced time of execution as compared to other ordering.

In Fig. 5 the random order of execution for original Test Suite (TS_o) in 0.94 s is shown.

In Table 4 APFD results and time of execution of all the application for different ordering is shown viz., FPA order (TS_1), FPA order (TS_p), Reverse FPA Order of (TS_p), Random Order of (TS_o) and Reverse Random Order of (TS_o). It is evident from Table 4 that the APFD results for FPA order (TS_p) and FPA order (TS_1) are better as compared to other ordering for all the three applications used in this study.

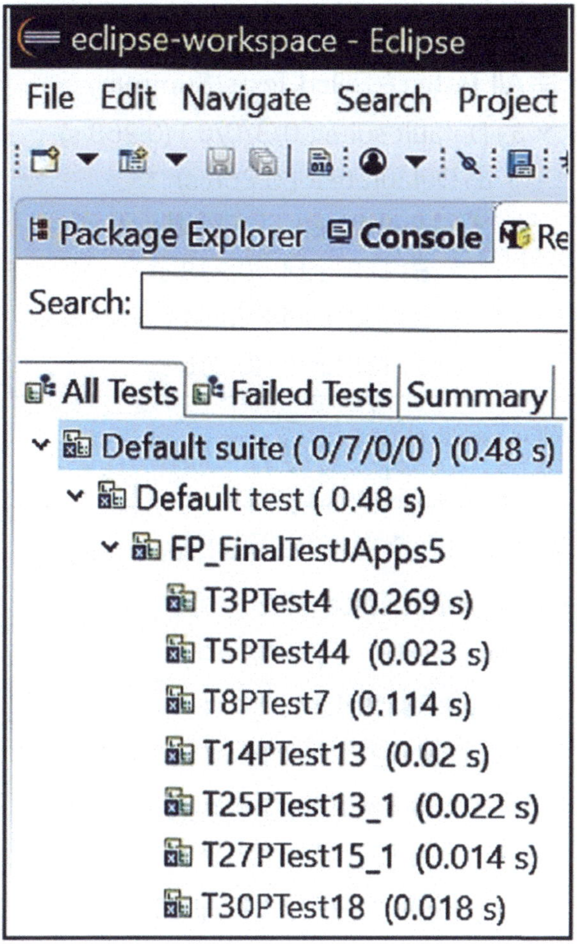

Fig. 3 Order and time of execution of TS_1

In Fig. 6 APFD results and time of execution of all the application for FPA order (TS_1), FPA order (TS_p), Reverse FPA Order of (TS_p), Random Order of (TS_o) and Reverse Random Order of (TS_o) is graphically represented.

8 Threat to Validity

The threat to validity for this study depends on various factors. Few aspects are that, FPA follows a random search mechanism which may result in an inefficient test suite giving minimal fault coverage and hence may not guarantee efficient results for TCP. In the industrial environment the size of applications and number of actual faults can

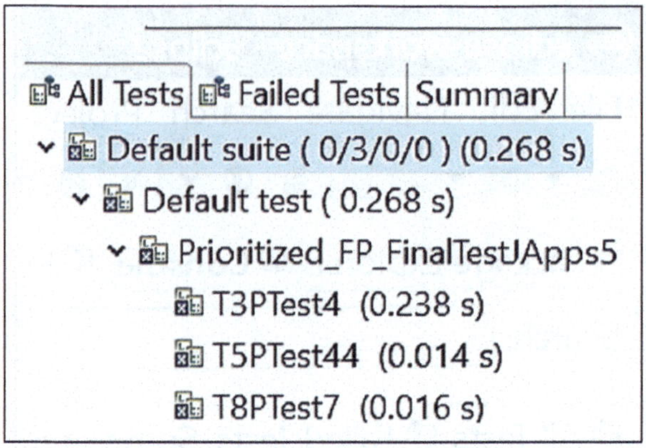

Fig. 4 Order and time of execution of TS_p

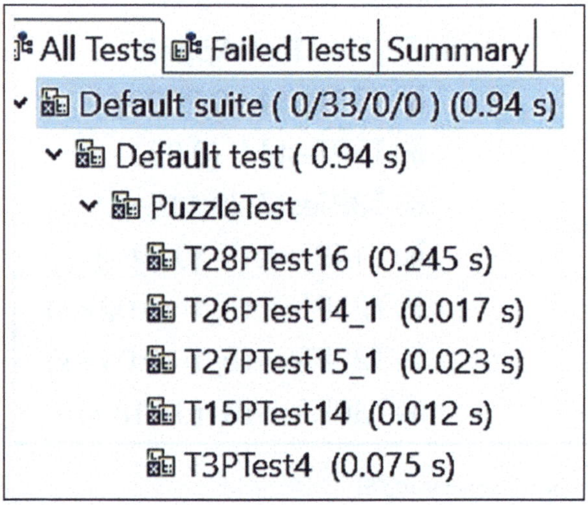

Fig. 5 Order and time of execution of random ordering of Tso

vary from large to very large including the variation in the operating environment, whereas in this study small size application with few seeded faults have been used. The value of control variables has been manipulated to suite the need of this study, which may have otherwise led to inefficient results.

Table 4 APFD results and time of execution for various ordering of test suites in all the applications

Sr. no.	Algorithm used	Puzzle game application			Area and perimeter application			Tritype application		
		Ordering	APFD results	Time (in s)	Ordering	APFD results	Time (in s)	Ordering	APFD results	Time (in s)
1.	FPA result (TS_1)	$\{T_3, T_5, T_8, T_{14}, T_{25}, T_{27}, T_{30}\}$	0.85152	0.48	$\{T_1, T_{43}, T_5, T_7, T_{11}, T_{27}, T_{35}, T_{39}, T_{40}, T_{42}, T_{44}, T_{51}, T_{59}, T_{60}, T_{68}, T_{74}, T_{77}, T_{81}, T_{88}, T_{99}, T_{106}, T_{108}\}$	0.44694	0.107	$T_1, T_{33}, T_2, T_6, T_9, T_{22}, T_{26}, T_{31}, T_{38}, T_{40}, T_{43}$	0.67039	0.05
2.	FPA result (TS_p)	$\{T_3, T_5, T_8\}$	0.85152	0.286	$\{T_1, T_{27}, T_{40}, T_{59}, T_{81}, T_{99}, T_{108}\}$	0.44698	0.088	T_1, T_2, T_6, T_{38}	0.67039	0.04
3.	Reverse order of FPA (TS_p)	$\{T_8, T_5, T_3\}$	0.8099	0.255	$\{T_{108}, T_{99}, T_{81}, T_{59}, T_{40}, T_{27}, T_1\}$	0.43696	0.078	T_{38}, T_6, T_2, T_1	03963	0.045
4.	Random ordering of original test suite (TS_0)	$\{T_{28}, T_{26}, T_{27}, T_{15}, T_3, T_2, T_{20}, T_{13}, T_{12}, T_{30}, T_{11}, T_7, T_{31}, T_8, T_5, T_{29}, T_{19}, T_{14}, T_{22}, T_{17}, T_{33}, T_1, T_{23}, T_4, T_{32}, T_{16}, T_{18}, T_{10}, T_{24}, T_{21}, T_9, T_{25}, T_6\}$	0.38485	0.94	$T_{24}, T_{79}, T_9, T_{85}, T_6, T_{90}, T_{13}, T_{101}, T_{83}, T_{77}, T_{105}, \ldots\ldots\ldots, T_{45}, T_{31}, \ldots\ldots\ldots, T_8, T_{21}, T_{37}, T_{58}, T_{14}, T_{44}, T_{23}, T_{89}$	0.42368	0.792	$T_{40}, T_{11}, T_{36}, T_{41}, T_{43}, T_{12}, T_{30}, T_{18}, T_{15}, T_{16}, T_4, T_{39}, T_{22}, T_{20}, T_{28}, T_2, T_{13}, T_{23}, \ldots\ldots\ldots, T_{44}, T_3, T_{35}, T_{34}, T_7, T_5, T_{42}, T_{29}$	0.23704	0.148
5.	Reverse random ordering test suite (TS_0)	$\{T_6, T_{25}, T_9, T_{21}, T_{24}, T_{10}, T_{18}, T_{16}, T_{32}, T_4, T_{23}, T_1, T_{33}, T_{17}, T_{22}, T_{14}, T_{19}, T_{29}, T_5, T_8, T_{31}, T_7, T_{11}, T_{30}, T_{12}, T_{13}, T_{20}, T_2, T_3, T_{15}, T_{27}, T_{26}, T_{28}\}$	0.27576	0.94	$T_{89}, T_{23}, T_{44}, T_{14}, T_{58}, T_{37}, T_{21}, T_8, \ldots\ldots, T_{31}, T_{45}, \ldots\ldots, T_{105}, T_{77}, T_{83}, T_{101}, T_{13}, T_{90}, T_6, T_{85}, T_9, T_{79}, T_{24}$	0.43363	0.793	$T_{29}, T_{42}, T_5, T_7, T_{34}, T_{35}, T_3, T_{44}, \ldots\ldots\ldots\ldots\ldots, T_{28}, T_{20}, T_{22}, T_{39}, T_4, T_{16}, T_{15}, T_{18}, T_{30}, T_{12}, T_{43}, T_{41}, T_{36}, T_{11}, T_{40}$	0.35186	0.148

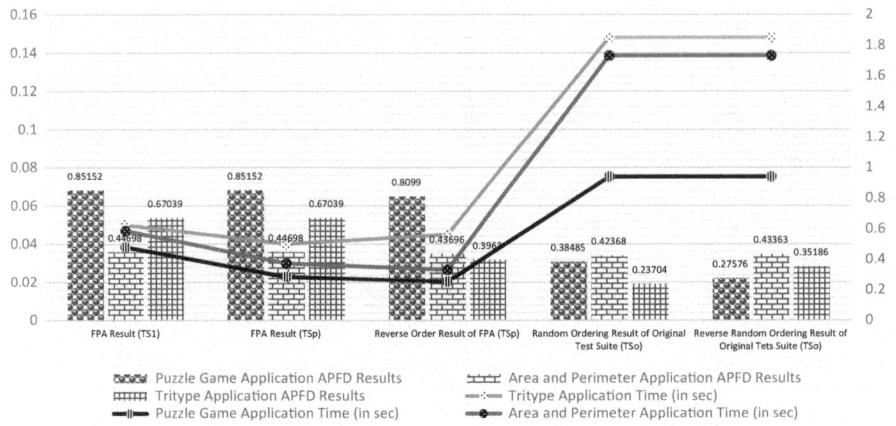

Fig. 6 APFD results for various ordering of test suites for all the applications

9 Conclusion and Future Work

In this research, FPA is used for TCP in RT. FPA uses code coverage of test cases without using any information of faults covered by the test cases. FPA is used to prioritize/order the test cases for RT to give maximum fault coverage in minimum time of execution. Three applications written in Java are used for empirical evaluation. Test scripts for all the applications were designed using TestNG tool. For all the applications three test suite ordering were considered namely, original test suite (TS_o) and two orderings generated by FPA (TS_1) and (TS_p). FPA for TCP initially generated prioritized test suite (TS_1) which was further reduced and prioritized (ordered) to get test Suite (TS_p). For validation of the results APFD metrics is used. APFD value for different orderings of test cases in three applications is calculated. The results of APFD metrics for FPA order (TS_1) and FPA order (TS_p) outperform the results given by Reverse FPA Order of (TS_p), Random Order of (TS_o) and Reverse Random Order of (TS_o). Since FPA for TCP converged early for all the applications used in this study, hence, only one iteration is used. Therefore, it can be stated that stopping criteria of FPA for TCP depends on the type and size of the application being used. For future validation of results an empirical evaluation of FPA for TCP will be performed with different NIMOA.

References

1. Alsewari AA, Har HC, Homaid AA, Nasser AB, Zamli KZ, Tairan NM (2017) Test cases minimization strategy based on flower pollination algorithm. In: International conference of reliable information and communication technology 2017 Apr 23. Springer, Cham, pp 505–512. https://doi.org/10.1007/978-3-319-59427-9_53

2. Chiroma H, Shuib NL, Muaz SA, Abubakar AI, Ila LB, Maitama JZ (2015) A review of the applications of bio-inspired flower pollination algorithm. Procedia Compu Sci 1(62):435–441. https://doi.org/10.1016/j.procs.2015.08.438
3. Harrold MJ, Gupta R, Soffa ML (1993) A methodology for controlling the size of a test suite. ACM Trans Sof Eng Method (TOSEM) 2(3):270–285. https://doi.org/10.1145/152388.152391
4. Jiang B, Wu Y, Zhang Y, Zhang Z, Chan WK (2018) ReTestDroid: towards safer regression test selection for android application. In: 2018 IEEE 42nd annual computer software and applications conference (COMPSAC) 2018 Jul 23, vol 1. IEEE, pp 235–244. https://doi.org/10.1109/compsac.2018.00037
5. Kabir MN, Ali J, Alsewari AA, Zamli KZ (2017) An adaptive flower pollination algorithm for software test suite minimization. In: 2017 3rd international conference on electrical information and communication technology (EICT) 2017 Dec 7. IEEE, pp 1–5. https://doi.org/10.1109/eict.2017.8275215
6. Khatibsyarbini M, Isa MA, Jawawi DN, Tumeng R (2018) Test case prioritization approaches in regression testing: a systematic literature review. Inf Softw Technol 1(93):74–93. https://doi.org/10.1016/j.infsof.2017.08.014
7. Luo Q, Moran K, Poshyvanyk D, Di Penta M (2018) Assessing test case prioritization on real faults and mutants. In: 2018 IEEE international conference on software maintenance and evolution (ICSME) 2018 Sep 23. IEEE, pp 240–251. https://doi.org/10.1109/icsme.2018.00033
8. Nabil E (2016) A modified flower pollination algorithm for global optimization. Expert Syst Appl 15(57):192–203. https://doi.org/10.1016/j.eswa.2016.03.047
9. Nasser AB, Sariera YA, Alsewari AA, Zamli KZ (2015) Assessing Optimization-based strategies for T-way test suite generation: the case for flower-based strategy. In: 2015 IEEE international conference on control system, computing and engineering (ICCSCE) 2015 Nov 27. IEEE, pp 150–155. https://doi.org/10.1109/iccsce.2015.7482175
10. Nasser AB, Zamli KZ, Alsewari AA, Ahmed BS (2018) Hybrid flower pollination algorithm strategies for T-way test suite generation. PLoS ONE 13(5):e0195187. https://doi.org/10.1371/journal.pone.0195187
11. Pant S, Kumar A, Ram M (2017) Flower pollination algorithm development: a state of art review. Int J Syst Assur Eng Manag 8(2):1858–1866. https://doi.org/10.1007/s13198-017-0623-7
12. Panwar D, Tomar P, Harsh H and Siddique MH (2018) Improved meta-heuristic technique for test case prioritization. In: Soft computing: theories and applications. Springer, Singapore, pp 647–664. https://doi.org/10.1007/978-981-10-5687-1_58
13. Pressman RS (2005) Software engineering: a practitioner's approach. Palgrave Macmillan
14. Rothermel G, Untch RH, Chu C, Harrold MJ (1999) Test case prioritization: an empirical study. In: Proceedings IEEE international conference on software maintenance-1999 (ICSM'99). Software Maintenance for Business Change (Cat. No. 99CB36360). IEEE, pp 179–188. https://doi.org/10.1109/icsm.1999.792604
15. Sivaji U, Shraban A, Varalaxmi V, Ashok M, Laxmi L (2019) Optimizing regression test suite reduction. In: First international conference on artificial intelligence and cognitive computing. Springer, Singapore, pp 187–192. https://doi.org/10.1007/978-981-13-1580-0_18
16. https://github.com/jkriti/JApps/blob/master/JApps.java
17. https://sir.csc.ncsu.edu/portal/bios/trityp.php
18. https://stackoverflow.com/questions/4479624/area-and-perimeter-calculation-of-various-shapes?rq=1
19. https://testng.org/doc/download.html
20. https://www.eclemma.org/download.html
21. Yadav DK and Dutta S (2017) Regression test case prioritization technique using genetic algorithm. In: Advances in computational intelligence 2017. Springer, Singapore, pp 133–140. https://doi.org/10.1007/978-981-10-2525-9_13
22. Yang XS (2014) Nature-inspired optimization algorithms. Elsevier
23. Yang XS, Karamanoglu M, He X (2014) Flower pollination algorithm: a novel approach for multiobjective optimization. Eng Optim 46(9):1222–1237. https://doi.org/10.1080/0305215X.2013.832237

A Study of the Effectiveness of Online Marketing Strategies of Packaged Health Food Brands

Amar Nath Gupta and Pradnya Chitrao

Abstract **Purpose**: The research study aims at exploring the effectiveness of online marketing strategies of packaged health food brands w.r.t. income of the customers. The study aims to investigate the relationship between buyer satisfaction and buyer recommendation w.r.t. the income of the customers. For the research, both primary data and secondary data were used. The researchers have conducted the pilot study on 105 respondents to investigate the effectiveness of online marketing strategies on packaged health food brands with respect to the income of the customers. The researchers have used a non-probabilistic convenience sampling method for the study. SPSS 21 version was used for the data analysis. Following hypotheses were proposed for the study:

H_1 There is some association between the awareness of packaged health food brands and the income of the customers.
H_2 There is some association between frequency of being online and income of the customers.
H_3 There is some association between frequency of online buying and income of the customers.

Findings: Results indicates that the higher the income level of the customers, the higher would be the awareness of health food brands among the customers.

Research Implications and Limitations: Practical implications—The marketers can apply the findings of this study in creating more effective online marketing strategies for the packaged health food brands with respect to the Income of the customers.

Keywords Income · Online marketing · Health food brands

A. N. Gupta (✉)
Symbiosis International University, Lavale, Pune, India
e-mail: amarg2460@gmail.com

P. Chitrao
SIMS, Khadki, Pune, India
e-mail: pradnyac@sims.edu

S. Fong et al. (eds.), *ICT Analysis and Applications*, Lecture Notes in Networks and Systems 93, https://doi.org/10.1007/978-981-15-0630-7_17

1 Introduction

The dynamic business environment and the technological advancement have redefined the consumption, ways of doing business and means of promotion for the Fast Moving Consumer Goods (FMCG) marketers. The twenty-first-century marketing has touched an entirely new scale with the penetration of the Internet among almost every household in India and it has resulted in the emergence of online marketing. The prominence of digitization and online presence has led the companies getting involved into a very cut throat competition for seeking the customer's attention.

The increasing awareness of consumers regarding health and nutrition issues has greatly helped the packaged health food brands to drive their demand in the market. The marketers are trying new and innovative ways to take an advantage of this growing packaged healthy food market. With the rise in the competition, the FMCG companies have now moved their attention from conventional approaches to modern approaches like E-commerce, due to the increasing mobile internet penetration globally.

Thus, keeping in view of the above all the tools available in the body of knowledge of marketing are highly utilized and innovation is always looked forward to build the brand association which results in repeated buying and boosting the overall profits. The marketers are now focussing more towards improving the quality of the packaged health food brands, which will create more brand loyalty and further help in crafting brand equity among the customers.

2 Literature Review

To identify the research gaps in the previous research studies, the literature review was carried out for drafting the Research objectives. Some of the referred research studies being conducted in past have been highlighted here below:

2.1 Jones and Reid [1]

Sandra et al. (2010) have explored how the leading Australian food organizations use the internet-based advertising techniques for their products. It was observed that most of the meals groups have internet web sites or sections of websites that are being targeted to the children and/or teenagers. The web sites also have the content for downloading which was aimed at direct marketing. The study was useful in understanding how the companies use the internet-based marketing tactics for marketing their products to children and teens.

2.2 *Kemp and Bui [2]*

Have investigated the variables which are vital in the course of brand building for the healthy food Brands. The study has examined the role of brand credibility in increasing the consumer confidence for the 'Healthy Brands'. The study has indicated that 'brand credibility, commitment and connection are very important for crafting the branding strategies'. The study is very helpful in identifying the positioning strategy of the selected 'Healthy Brands'.

2.3 *Nour et al. [3]*

Have carried out the research work to investigate how the "Ceramic and glass manufacturing companies in Jordan" make use of promotional strategies to influence on the buying decision making process of the consumers in Jordan. The authors state that the main aim of promotional mix is to reach its target customers and making them buy and rebuying the product again and again to be specific inducing them for repeat purchases. It was found that the degree of promotional mix elements used by the companies was reasonably high as the companies aimed at positively influencing their buying decision making process. This research has a constrained scope as it has been carried out only for the Ceramic and glass manufacturing corporations in Jordan. The future research can consider the applicability of findings of this research in different international locations with other Industries like FMCG etc.

3 Research Objectives and Scope

3.1 *The Following Research Objectives Have Been Proposed for the Study*

1. To explore the effectiveness of Online Marketing Strategies of Packaged Health Food Brands w.r.t. Income of the Customers.
2. To investigate if there exists any relationship between the awareness of packaged health food brands and the Income of the Customers.
3. To determine the association between frequency of Online Buying and the Income of the Customers.

3.2 Scope of the Study

The research is mainly focusing on the effectiveness of online marketing strategies of packaged health food brands. The products considered for this study are packaged health drinks, packaged healthy soups, packaged healthy breakfast cereals and meals, packaged healthy food supplements for adults and children, packaged dry fruits, and packaged sugar substitutes and packaged healthy oils. The study is further limited to study of effectiveness of online marketing strategies of packaged health food brands w.r.t. income of the customers. Other variables are not considered under this study.

4 Research Methodology

4.1 Research Design

The Descriptive Research Design was used for Conducting the Research.

4.2 Population

The Population for this research was the people of Pune City who are above 18 years age group.

4.3 Sources of Data

The researchers have used both primary and secondary data for the research. Primary data was mainly collected from the consumers of packaged health food brands from Pune region using the tools like personal observation, personal interviews and through the questionnaire. The secondary data was collected from the available sources like the print and electronic media along with the company's reports, research papers and internet websites.

4.4 Sample Design

For the Research, the researchers have used the Non-probabilistic Convenience Sampling technique.

4.5 Sampling Unit

The sample respondents were the customers buying food products from various retailers or grocery shops or through ordering online in Pune City.

4.6 Sample Size

Around 150 plus questionnaire were administered but only 105 responses were found to be completely filled for the purpose of research study. Therefore, the researchers have considered a Sample Size of 105 for the research study.

4.7 Sampling Area

The researcher has collected the samples from the Pune City between May and June, 2018.

5 Data Analysis

To check the reliability of the scale used in the questionnaire, the Cronbach's alpha was calculated. The value is shown in the Table 1.

Form the Table 2 it was found that since the Cronbach's alpha value is higher than 0.7; the scale used can be said reliable and can be used for further data collection process.

Table 1 showing the percentage of valid responses

Case processing summary			
		N	%
Cases	Valid	105	100.0
	Excluded[a]	0	0.0
	Total	105	100.0

[a]List wise deletion based on all variables in the procedure

Table 2 Showing the Cronbach's alpha value

Reliability statistics	
Cronbach's alpha	No of items
0.942	15

5.1 *Descriptive Statistics*

From above descriptive statistics it can be said that **Online Videos** of packaged health food provokes the customer to buy the packaged health food brands. Furthermore, it was also found that **Official and appealing website** of a packaged health food brand creates a good brand image and thus leads to positioning of the packaged health food brands. It was also observed that the respondents tend to buy Packaged Health food products if it is available on particular online retail stores' (Amazon, Flipkart etc.). It was also observed that, for the kids, the respondents are not so much dependant on online stores like Amazon, Flipkart, etc. for buying the packaged health food brands. Further, One Sample t-test is further done to know statistical difference between sample mean and a hypothesized value of the mean in the population. The t-test here is run to know the difference between satisfaction and recommendation (Tables 3, 4 and 5).

From the table we get the $p < 0.001$, which says that the satisfied consumers will further recommend the product to the other buyers. Further, a correlation test is run between the variable, which again shows that the variables are positively and significantly correlated. It means that if, satisfaction level of the consumer increases, the recommendation for the packaged foods will also increase as the satisfied consumers are likely do a positive word of mouth publicity for the product (Table 6).

6 Testing of Hypothesis

H_0 There is no association between the awareness of packaged health food brands and the Income.

H_1 There is some association between the awareness of packaged health food brands and the Income.

In the Chi-Square Table 7, we see that the Pearson chi-Square score is 90.362. Therefore, it proves that the two variables under the study are not related with each other.

Again, If the p value is less than or equal to the significance level, we reject the null hypothesis and conclude that there is a statistical significant relationship between the variables. Hence, we fail to reject the H_0 as the p value is less than 0.05. Therefore, the H_1 is accepted which states that there is a direct significant relationship between the awareness of packaged health food brands and the Income of the customers. It proves that higher the income level customers, the higher would be the awareness of health food brands.

H_0 There is no association between frequency of being Online and Income.

H_1 There is some association between Frequency of being Online and Income.

In line with the first hypothesis tested, the researchers have further tested the frequency of being online and Income. The results here show that, since the p value

Table 3 Showing the descriptive statistics of the variables

	N	Min.	Max.	Mean	Std. deviation
I am more likely to buy a particular packaged health food brand if I come across by an promotion **email** of that brand	105	1	5	3.25	0.8955
I am more likely to recommend a particular packaged health food brand if I come across by its testimonial through **social media**	105	2	5	3.46	0.6660
I am more likely to buy a particular packaged health food brand if I come across by its testimonial through **social media**	105	1	5	3.26	0.8994
I am more likely to recommend a particular packaged health food brand if I come across by an promotion **email** of that brand	105	1	5	3.42	0.9374
Online videos of packaged health food provoke me to buy that brand	105	1	5	**4.32**	0.9639
I buy (would buy) only those packaged health food brands for my kids which there are **good reviews are available online**	105	1	5	3.12	1.0277
Official and appealing website of a packaged health food brand creates a good brand image in my mind and thus I find that brand worth buying	105	2	5	**4.35**	0.9028
I am more likely to buy a particular packaged health food brand if it comes as one of the **online search results**	105	1	5	3.00	0.9094
Pop-up advertisement of packaged health food while browsing internet make me buy that particular brand	105	1	5	2.92	1.1966
I would recommend packaged health food brand if it is listed online retailer's (amazon, Flipkart etc.) website or mobile application	105	1	5	2.92	1.2046
I would more likely to recommend a particular packaged health food brand if it comes as one of the online search results	105	1	5	2.91	1.0295
I am more likely to buy particular packaged health food brand if it is available on particular online retail stores' (Amazon, Flipkart etc.) product portfolio	105	1	5	**4.21**	1.1131
I buy (would buy) only those packaged health food brands for my kids which are available on online stores	105	1	5	**2.63**	1.1804
I buy (would buy) only those packaged health food brands for my kids for which an official website is available	105	1	5	**4.41**	0.8933
Valid *N* (listwise)	105				

Table 4 Showing the Mean, Std. deviation and Std. error mean values of variables as calculated from One Sample t-test

One sample statistics				
	N	Mean	Std. deviation	Std. error mean
I am satisfied with the packaged health food brand which I am using currently	105	4.6857	0.95388	0.09309
I will recommend the brand of packaged health food to others	105	4.6190	1.00366	0.09795
I am ready to pay little higher price for good quality packaged health food brands as compared to local heath food (non-branded)	105	4.8095	1.02933	0.10045

Table 5 Showing the values from running One Sample t-test

One sample test						
	Test value = 4					
	t	*df*	Sig. (2-tailed)	Mean difference	95% confidence interval of the difference	
					Lower	Upper
I am satisfied with the packaged health food brand which I am using currently	7.366	104	0.000	0.68571	0.5011	0.8703
I will recommend the brand of packaged health food to others	6.320	104	0.000	0.61905	0.4248	0.8133
I am ready to pay little higher price for good quality packaged health food brands as compared to local heath food (non-branded)	8.059	104	0.000	0.80952	0.6103	1.0087

is less than 0.05, the null hypothesis (H_0) is rejected and the alternative hypothesis (H_1) is accepted. Here, the Pearson Chi-Square is 82.803 with a *df* of 16 so we can say that there is a direct significant relationship between frequency of being online and income. It proves that high income level customers are having high frequency of being online as compared to low income level customers (Table 8).

Table 6 Showing the correlation between outcome variables (satisfaction and recommendation)

Correlations				
		I am satisfied with the packaged health food brand which I am using currently	I will recommend the brand of packaged health food to others	I am ready to pay little higher price for good quality packaged health food brands as compared to local heath food (non-branded)
I am satisfied with the packaged health food brand which I am using currently	Pearson correlation	1	0.868[a]	0.810[a]
	Sig. (2-tailed)		0.000	0.000
	N	105	105	105
I will recommend the brand of packaged health food to others	Pearson correlation	0.868[a]	1	0.878[a]
	Sig. (2-tailed)	0.000		0.000
	N	105	105	105
I am ready to pay little higher price for good quality packaged health food brands as compared to local heath food (non-branded)	Pearson correlation	0.810[a]	0.878[a]	1
	Sig. (2-tailed)	0.000	0.000	
	N	105	105	105

[a]Correlation is significant at the 0.01 level (2-tailed)

Table 7 Showing the Pearson Chi-Square values

Chi-Square tests			
	Value	*df*	Asymp. sig. (2-sided)
Pearson Chi-Square	90.362[a]	16	0.000
Likelihood ratio	106.289	16	0.000
Linear-by-linear association	11.926	1	0.001
No of valid cases	105		

[a]16 cells (64.0%) have expected count less than 5. The minimum expected count is 0.23

Table 8 showing the Pearson Chi-Square values

Chi-Square tests			
	Value	*df*	Asymp. sig. (2-sided)
Pearson Chi-Square	82.803[a]	16	0.000
Likelihood Ratio	87.674	16	0.000
Linear-by-linear association	12.242	1	0.000
No of valid cases	105		

[a]17 cells (68.0%) have expected count less than 5. The minimum expected count is 0.23

Table 9 showing the Pearson Chi-Square values

Chi-Square tests			
	Value	*df*	Asymp. sig. (2-sided)
Pearson Chi-Square	72.399[a]	20	0.000
Likelihood ratio	71.278	20	0.000
Linear-by-linear association	6.081	1	0.014
No of valid cases	105		

[a]23 cells (76.7%) have expected count less than 5. The minimum expected count is 0.17

$\mathbf{H_0}$ There is no association between Frequency of Online Buying and Income.
$\mathbf{H_1}$ There is some association between Frequency of Online Buying and Income.

In continuation with the research study, the researchers have further investigated the frequency of online buying and the Income. The results here show that, since the p value is less than 0.05, the null hypothesis (H_0) is rejected and the alternative hypothesis (H_1) is accepted. Here, the Pearson Chi-Square is 72.399 with a *df* of 20 so we can say that there is a direct significant relationship between frequency of online buying and Income. It proves that high income group customers do more frequently online buying as compared to low income group customers (Table 9).

7 Findings

With the growing penetration of internet, more and more number of the customers are buying the products mostly online. Therefore, the marketers are now giving more importance to the online marketing strategies to have been more effective in terms of customer satisfaction for the packaged health food brands.

It was found that the consumers with high income level are more cautious about their health as compared to the consumers with low income level. It was also observed that the consumers with high income level generally buy branded packaged health food products as they do not want to compromise with the quality of the product. Furthermore, it was also observed that consumers with high income level tend to try new packaged health food brands due to their expectations of getting a better product. It was found that the customers generally pay higher price for good quality packaged health food brands as compared to local heath food as they don't want to take any risk of their health while buying such brands for which they are totally unaware.

It was also found that **Online Videos** of packaged health food provokes the customer to buy the packaged health food brands. Furthermore, it was also found that **Official and appealing website** of a packaged health food brand creates a positive brand image and thus leads to positioning of the packaged health food brands because the customers are more attracted towards such brands which have an official website and have a positive brand image. It was also observed that the respondents tend to buy packaged health food products if it is available on particular online retail stores' (Amazon, Flipkart etc.).

8 Conclusion

In conclusion, the research reveals that a significant positive relationship exists between frequency of being online and the income of the consumers. It has been concluded that the customers are more likely to buy a particular packaged health food brand if they come across by its testimonial through social media. The customers are more likely to recommend a particular packaged health food brand if they come across by a promotion email of that brand. It was further observed that satisfied customers will further recommend the packaged health food brand to the other buyers. It has been also found that the variables are positively and significantly correlated. It means that if, satisfaction level increases, the recommendation for the packaged foods will also increase. It has been observed that a significant positive relationship exists between the awareness of packaged health food brands and the income. it has been concluded that the online videos of packaged health food brands provoke the customers to buy that brand. Furthermore, it has been also concluded that a significant relationship exists between frequency of online buying and Income of the consumers.

9 Research Implications and Limitations

Practical implications—This research study would be useful to the marketers for crafting more effective online marketing strategies to promote the packaged health food brands online which will further add value to the overall offerings made to the customers. The research study highlights the growing importance of packaged health food brands among the customers and how packaged health food brands should amplify their online marketing efforts in crafting more consumer-centric online marketing mix strategies that could deliver more value to customers' with respect to their income levels. The limitation of this study is that it has been conducted on packaged health food brands. Therefore, future studies could be extended to other FMCG Products to generalize the findings of this study.

10 Discussion

The emergence and advancement of technology has directed the marketers to explore new possibilities to attract the customers of packaged health food brands. As more and more number of customers are now spending time on the Internet, packaged health food brands have amplified their online marketing efforts in crafting more consumer-centric content that could deliver more value to customers'. Most of the firms have embraced online marketing as means of communicating a desired message to their target market [4]. This research study attempts to study the different types of online marketing strategies which the marketers are implementing to attract the particular income group customers who are buying packaged health food brands online and to determine the gaps that exist in the previous research being done so as to help the marketers to subsequently design the online marketing mix strategies that can deliver the value to the customers at a profit. The study has helped in identifying the factors that contribute towards the effectiveness of online marketing strategies of packaged health food brands w.r.t. income.

References

1. Jones S, Reid A (2010) Children's magazines: reading resources or food marketing tools? Public Health Nutr 13(3):393–399. https://doi.org/10.1017/S1368980009991133
2. Kemp E, Bui M (2011) Healthy brands: establishing brand credibility, commitment and connection among consumers. J Consum Mark 28(6): 429–437. https://doi.org/10.1108/07363761111165949
3. Nour MI, Almahirah MS, Mohammed Said S, Freihat S (2014) The impact of promotional mix elements on consumers purchasing decisions. Int Bus Manag 8(2): 143–151. http://www.cscanada.net/index.php/ibm/article/view/4800, http://dx.doi.org/10.3968/4800
4. Kiang MY, Raghu TS, Shang KHM (2000) Marketing on the Internet-who can benefit from an online marketing approach? Decis Support Syst 27(4):383–393. https://doi.org/10.1016/S0167-9236(99)00062-7
5. Kathuria LM, Gill P (2013) Purchase of branded commodity food products: empirical evidence from India. Br Food J 115(9):1255–1280. https://doi.org/10.1108/BFJ-08-2011-0209
6. Clayton M, Heo J (2011) Effects of promotional-based advertising on brand associations. J Prod Brand Manag 20(4):309–315. https://doi.org/10.1108/10610421111148324

7. Shariq M, Khan BM, Rizvi AH (2014) An exploratory analysis of marketing elements and brand equity in the UAE FMCG category. Jindal J Bus Res 3(1–2): 121–136. http://journals.sagepub.com/doi/pdf/10.1177/2278682115593436
8. Mohan Kathuria L, Gill P (2013) Purchase of branded commodity food products: empirical evidence from India. Br Food J 115(9): 1255–1280. https://doi.org/10.1108/BFJ-08-2011-0209
9. Bui M, Kaltcheva VD, Patino A, Leventhal RC (2013) Front-of-package product labels: influences of varying nutritional food labels on parental decisions. J Prod Brand Manag 22(5/6): 352–361. https://doi.org/10.1108/JPBM-05-2013-0298
10. Chen MF (2009) Attitude toward organic foods among Taiwanese as related to health consciousness, environmental attitudes, and the mediating effects of a healthy lifestyle. Br Food J 111(2): 165–178. https://doi.org/10.1108/00070700910931986
11. Zeghache N (2014) The impact of packaging colour on children's brand name memorization (7–12 years old). Int J Retail Distrib Manag 42(11/12):1053–1068. https://doi.org/10.1108/IJRDM-08-2013-0168
12. Baglione SL, Tucci LA, Stanton JL (2012) Self-reported nutritional knowledge and the acceptance of health-related food benefit claims. Br Food J 114(4): 453–468. https://doi.org/10.1108/00070701211219496
13. Ali T, Alam A, Ali J (2015) Market structure analysis of health and wellness food products in India. Br Food J 117(7): 1859–1871. https://doi.org/10.1108/BFJ-12-2014-0438

ICT Intervention Challenges in Education in Rural India

Gopal Naik, K. N. Narasinga Rao and Ashwini Baje

Abstract People in rural India encounter greater paucity of facilities and services compared to their urban counterparts in scaling the opportunities created by technological advancement. Information and Communication Technologies (ICT) have the potential to provide much-needed succor by bridging the challenges of remoteness with satellite, web and mobile based applications. Education, being a critical requirement for social and economic well-being, and primarily a media-based service, lends itself well for technological interventions which aid in addressing the urban-rural divide. The case study that we have done shows that there are challenges in the eco-system that impact the smooth implementation of digital technologies in rural areas. Inadequacy in infrastructure both in terms of quality and quantity, inefficiencies and other systemic issues impact the ICT solution roll-outs causing delays and unmanageable cost escalations, thereby making solutions infeasible. The way forward is urgent, serious and needs concurrent efforts on many dimensions, which require strong collaboration of public as well as private partners.

Keywords ICT · Tele-education · Broadband · Interaction · Content · Education · Infrastructure · Power · Rural

1 Introduction

Education helps unleash the true potential of the human mind. It increases individuals' chances in the job market and helps them earn their living. Gary Becker has shown this empirically using the Human Capital Theory, which posits that education is an investment and provides a better payoff in terms of higher wages [1]. A study has also shown how GDP per capita is linked to learning levels [2, p. 2]. In addition to improving income distribution, education also brings in other benefits. It promotes innovation and entrepreneurship and raises people's productivity and quality of life. The educated population, generally tends to be well-informed about government

G. Naik (✉) · K. N. Narasinga Rao · A. Baje
IIM Bangalore, Bannerghatta Road, Bangalore 560054, Karnataka, India
e-mail: gopaln@iimb.ac.in

S. Fong et al. (eds.), *ICT Analysis and Applications*, Lecture Notes in Networks and Systems 93, https://doi.org/10.1007/978-981-15-0630-7_18

policies, lead a civic life and take decisions that benefit them individually and the society at large.

Despite considerable efforts by successive governments in India, the progress made has been inadequate compared to other countries. A comparison of how much prominent governments across the globe spent on education, during 2009–13, shows (Fig. 1) that the India's expenditure on education remained less than that of South Africa, Brazil, the European Union, the United States of America and even the World Average. In 2013, the Government of India spent less than 4% of its GDP, whereas, the other comparable developing countries, Brazil and South Africa spent about 6% of GDP on education.

According to a National Sample Survey Office (NSSO) study in 2014, rural India in comparison with urban India, performed poorly with respect to various indicators (Table 1) [3]. For instance, a huge gap of 20% in the Adult Literacy Rate was recorded in rural and urban India. Large disparities (of around 10–12%) also existed in the percentage of male and female population that completed graduation and above. The percentage of population (belonging to the age group of 5–29 years), who never enrolled in any educational institution, too was seen to be higher in rural India.

Poor learning outcomes among rural students could be attributed to a host of issues—not considering ground realities when framing education policy, limited

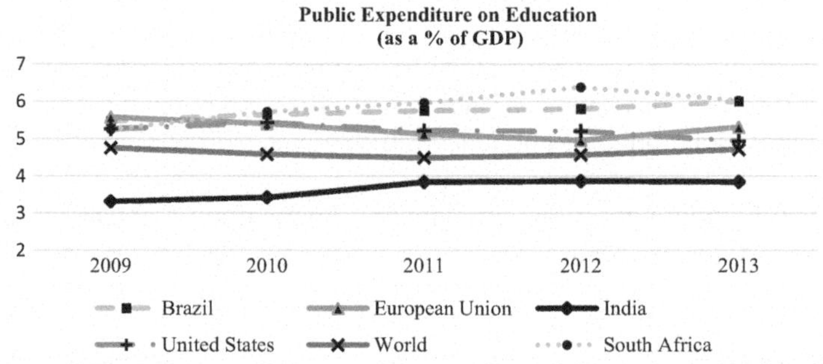

Fig. 1 Public expenditure on education. *Source* Created by authors based on World Bank Data (retrieved in March 2018)

Table 1 Educational disparities between Rural and Urban Populations in India (as of 2014)

Indicator	Rural (%)	Urban (%)
Adult literacy rate	64	84
Attained education level of 'graduation and above' (male)	4.5	17
Attained education level of 'graduation and above' (female)	2.2	13
Persons who never enrolled in any educational institution (age group 5–29)	11	6

Source Social Consumption: Education, conducted by the National Sample Survey Office (NSSO), 2014

administrative power for school headmasters (in terms of taking disciplinary actions, hiring or firing of teachers), overambitious curricula that must be completed in a given time, poor state of school infrastructure, lack of teachers and their training etc. [4]. Also, many rural students cannot afford private tuitions and have limited or no access to other knowledge sources. Their parents may not be educated enough to clarify doubts about various concepts taught in the school. Added to this is the poor state of educational infrastructure and manpower in schools. Another factor that impacts the rural students is the distance they have to travel to school. Only about 60% of rural students have a secondary school within a distance of 2 km, whereas about 91% of urban students can avail a secondary school within the same distance [3].

According to UNESCO report, primary schools in India require at least 3.7 lakh new teachers. This acute shortage of teachers has placed India second among world nations, next to only Nigeria in terms of teacher recruitment required to meet the existing education demand. It is projected that this number would snowball to 30 lakh requirement by the year 2030 [5].

Shortage of teachers implies higher Pupil-Teacher Ratio and overcrowded classrooms. World Bank Statistics indicate that despite a receding Pupil-Teacher Ratio in Primary Education in India, it is still higher than the world average and that of most other nations (Fig. 2).

The reasons for this shortage are many. In many schools, teachers and students are registered under pseudo-names. Fearing increase in expenditures upon hiring permanent staff, the government-run schools temporarily appoint teachers on an adhoc-basis. This leaves a lasting effect on the quality of education [5]. Another issue is difficulty in attracting qualified staff owing to factors like remoteness of village and low salaries.

This is a vicious circle, schools in rural India fail to attract talented staff because of their poor state of affairs, and this deteriorates the quality of teaching imparted. As a result, there are children who in spite of completing schooling, struggle to read basic text and perform simple arithmetic.

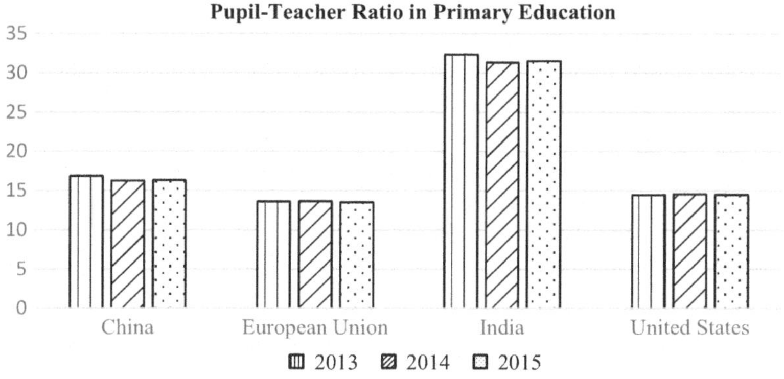

Fig. 2 Pupil teacher ratio in primary education. *Source* Created by authors based on World Bank Data (retrieved in May 2019)

2 ICT in Education

Information Communication Technology, which is often touted as the means for enabling knowledge transfer is today well-integrated with the educational system. ICT integration helps enhance the learning process of students [6, 7]. ICT fosters students' learning capabilities and retention power by making lessons more engaging and interesting. It also encourages students to undertake further reading that can help them enhance their knowledge.

With the goal of improving the quality, reach and cost-effectiveness of delivering instruction to students, integration of ICT is also being carried out in rural schools. In most rural schools, education through ICT is considered not as a replacement of teachers, but as a complementary teaching resource. It is mostly adopted as a distance education mode [8]. Use of ICT is particularly beneficial to students in rural and remote areas, who otherwise find it difficult to access courses or subjects of their interest due to geographical barriers. It can also help rural students become technologically better skilled and thus be confident and prepared for the new-age jobs.

To reap the full-benefits of ICT, it is critical that ICT is well-integrated into the rural education system. The process of integration of ICT is a continuous process [9] that needs constant support from the management, proper planning and appropriate policy making. Researchers contend that the development and success of ICT programs in the field of education in developing countries, depends on how well it is adapted to local social and economic factors [10]. It is critical to evaluate different aspects of ICT-based learning like "*usefulness, self-efficacy, willingness and challenges and readiness*" before adopting it [11, p. 154]. For any ICT project to be successful in rural areas, it would need to undergo a "trial-and-error" phase before the teachers and students are comfortable with its usage and are able to benefit from it [8]. The key is to find the right subject area that needs intervention (which involves complex concepts or skills), review of its curriculum and finding the right match of technology and pedagogy for these subjects and curriculum [8].

Studies have shown that teachers who have undergone ICT training are more effective and confident in using them when teaching than teachers who have not undergone any such training [12, 13]. Studies have found that students find it easier to grasp the subjects taught through the ICT projects, when done through local language [14] and used in conjunction with active pedagogical techniques. The quality of the content should be evaluated [15] for its distinctiveness, variety, volume [16], use of multimedia and ease of understanding. A study has shown that users' satisfaction and interest in a technology increases significantly, if it presents information in a vivid and interactive manner [17].

3 Approach of the Study

In this study, we use a case study to identify key challenges in reaching digital education in rural areas. The case study examines a project that provides tele-education in 1000 rural schools in Karnataka, India. This section covers the details of the case study.

Tele-education in 1000 Rural Schools in Karnataka
Owing to difficulties in achieving last mile connectivity in rural areas of India, innovators and policy makers have tried and experimented with various ICT models. One such model that has met considerable success is the SAMIE (Satellite & Advanced Multimedia Interactive Education) tele-education implemented in over 1000 schools of rural Karnataka. It was implemented in PPP model by an IIMB-led consortium.

The project made use of satellite communications for downstream communication and broadband internet for upstream communication to conduct live, multi-way interactive learning sessions for rural students. The classes were taught in Kannada by well-experienced subject matter experts, from a central studio. The teachers additionally made use of multimedia animation content to deliver the lessons.

First Phase—Pilot Project (2011–2012)
The pilot project (tele-education) was conceived by the Centre for Public Policy, Indian Institute of Management, Bangalore, led consortium. It was a satellite-based interactive educational program communicated live via the satellite EDUSAT. Phase 1 was completed by the consortium, with the tele-education technology led by Gumbi Software Pvt. Ltd. and multimedia animation content provided by Edutel Technologies Pvt. Ltd. The project involved roll-out in 14 government high schools located in the GP (Gram Panchayat) locations of Gubbi Taluk, Tumkur District, Karnataka State. The studio in the Government Department of State Education Research and Training (DSERT) was used for delivery of the classes. Quality training was provided to as many as 3112 students belonging to classes 8–10 in the subjects of Maths, English (Grammar) and Science. Significant improvement in passouts was observed as a result of the training. This hybrid technology was unique and involved the features provided by VSAT and terrestrial broadband connectivity. The technology enabled two-way voice and video. DSL broadband was used for reverse communication (student interaction). The hybrid model enabled 2-way communication with satellite channel for forward and broadband for reverse channel communication, as shown in Fig. 3.

Second Phase: SAMIE for a Thousand Schools
The Karnataka Department of Primary and Secondary Education approved a proposal to increase the number of schools from 14 to 1000, with a focus on backward areas. 18 such districts were identified; 2 talukas were randomly selected from each of those districts.

Department of State Education Research and Training (DSERT's) studio was used to conduct live classes which were beamed to schools through satellite for 30–35 minutes. The schools were provided with equipment for receiving the live stream

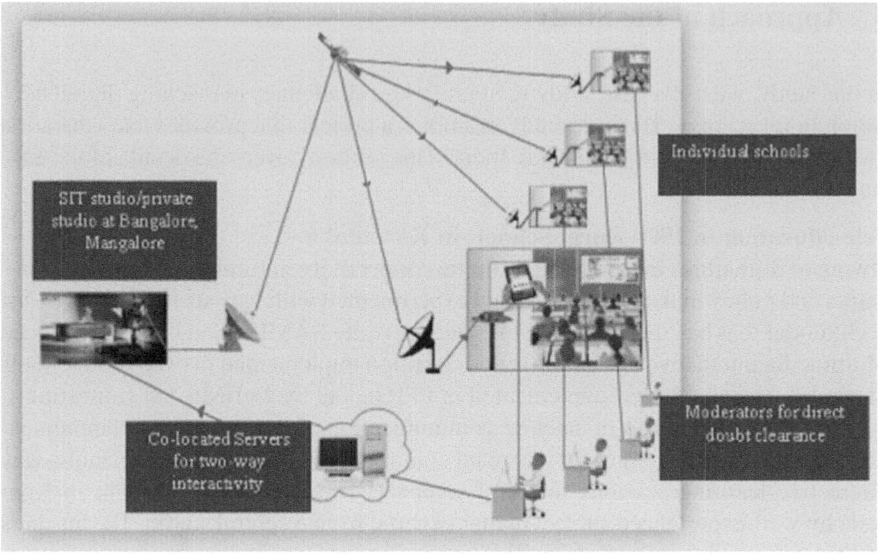

Fig. 3 Components of the SAMIE solution

sent. These were at managed Gram Panchayat levels. The course was designed to be run for classes 5–10th, for subjects English, Maths and Science. The duration was for 200 days and comprised a total of 1000 hours of classes in an academic year. The content created adhered to the state syllabus and provided methodical description for conveying the concepts, with animations where necessary. During the last 5 minutes of the class, interactions took place through broadband or cellular channels. Moderators were stationed to answer the question directly after the live classes. The tele-education classes were managed at the school level by one teacher in the projection room of the school. The content and process were arrived at by collaborating with the government department.

The consortium installed the technology components across 700 higher secondary schools and 300 upper primary schools. These schools were located across 36 taluks (18 districts). The training commenced with trial classes in October 2014, while the actual sessions commenced in November 2014. A total of about 2 lakh students belonging to classes from 5 to 10 participated in the project. The lower income group constituted in excess of 95% of the students. Nearly 95% of the Headmasters of the schools rated the program as excellent, as per a survey conducted.

Satellite and Multimedia Interactive Education (SAMIE) can be used to overcome the current deficiencies in the schools such as shortage of trained teachers and teacher absenteeism and provide educational content of high quality. The tests conducted showed that the project had a significant positive influence on the performance of students measured through a standardized evaluation for the three subjects.

The progress of the tele-education program was monitored through automatic updates from the remote end. Also, by building intelligence into the computing elements and mobile applications, the server got to know that a particular class was functioning. Progress of the project thus was collected, and consolidated on a dashboard so that it could be easily viewed by the key stakeholders. However, broadband availability was poor and therefore the interaction sessions had to be conducted through the cellular mobile alternative. The State Government, though responsible for providing broadband, could not organize it and instead decided to curtail the program as interaction could not be conducted via broadband in all the schools.

Key Factors Influencing the Outcome

The project was rolled out in November 2014 in 1000 schools. The lessons relayed through satellite for about 85% of the duration of the class, worked very well. However, the last 10 minutes of the scheduled interaction in each session faced constraints. The success of the project was impacted by various factors. An account of those factors is given below.

Technology Issues

Provision of reliable technology is a critical aspect of any ICT project. In several institutions, even though necessary infrastructure was present, internet connectivity required for interaction was not available. The challenges in technology can be best viewed against the aspects of connectivity and power supply availability with the required quality.

Data Connectivity

The choice of appropriate technology was a complex exercise. From an internet connectivity perspective, the service providers did not guarantee availability of connectivity in all the required schools. Despite inspection for feasibility before committing to the provision of connectivity, the actual rollout fell short significantly. Many locations, hence, had to face the possibility of being declared not feasible. Choosing alternative connectivity technologies, was complex and changed over time. The following three connectivity options were available.

1. Digital Subscriber Line (DSL) broadband connectivity
2. Cellular Mobile Data Connectivity
3. National Optical Fiber Network (NOFEN)

A depiction of data corresponding to DSL coverage in schools is given in Fig. 4.

When we looked at the reasons as to why the connectivity failed to function, the data collected reveals the following (given in Fig. 5). To start with, 1000 schools were selected based on the feasibility declared by the telecom service provider. However, when it came to provision of the connection, the feasibility got reduced to 900 schools, indicating a fall of 10%. Only 62% of the school locations declared feasible were enabled with broadband modems and connectivity. Out of this, only 34% were functioning satisfactorily. These inconsistencies ended up frustrating the user and posed difficulties in realizing the purpose and potential of the project. Also, this required continuous follow-ups and discussions that added up to the efforts and

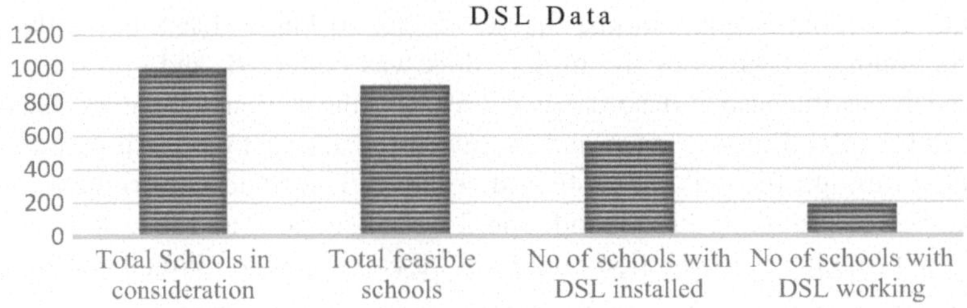

Fig. 4 DSL coverage data

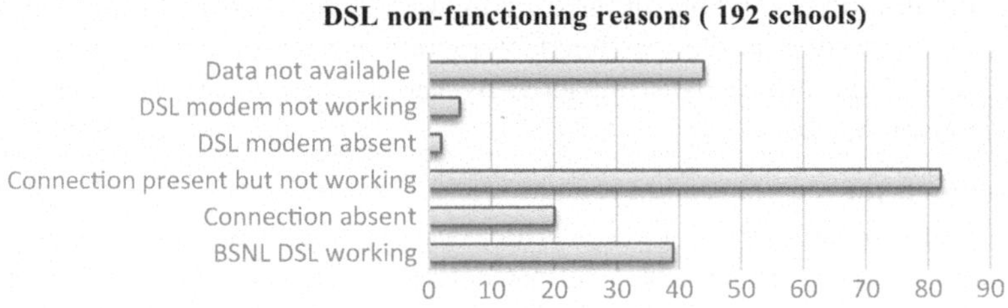

Fig. 5 Reasons for DSL not functioning

costs. Billing issues also cropped up where data mismatch was observed between those of the telecom service provider and that of the consortium partners.

From an analysis of 192 schools, valid data was available for 148 schools and only 39 of these schools had the DSL functioning to satisfaction (only 26%). About 82 schools had connections in place but internet was not working. In 20 schools, the wired connectivity from the exchange itself was missing. Five schools had modem malfunctioning, and in two schools, the modem itself was missing. Where the user expected 100% functioning of service, in reality only 26% of the 148 connections was working. This sort of issues impacts the confidence of user in offering any quality service in rural areas and also discourages the user from entering into any Service Level Agreement (SLA) based commitments that depend on this type of connectivity arrangement. Absence of competing telecom service providers in the rural areas, created further constraints to the user.

The consortium had to come up with a cellular alternative, which was not there in the original proposal of the project and differed from the Memorandum of Understanding with the Department. The Department itself was responsible for the broadband connectivity but could not do much with the chosen service provider. With respect to the data connectivity, the DSL broadband connectivity was not consistent in supporting real time interactive sessions. This made it difficult to commit to any SLA based on this connectivity. Even if data connectivity was present, the speeds were inadequate (<300 kbps) for real time video. This adversely impacted the 2-way real time moderator interactive sessions. Large scale deployment of broadband had challenges of operations and tracking. This resulted in increased manpower and escalated costs. Apart from data, even voice and SMS were not consistently available given the last mile reach challenges. The inconsistencies in SMS delivery, caused confusion as the automation of project tracking relied heavily on reliable SMS delivery. Telecom operators were not sufficiently equipped to attend to issues in the field leading to inconsistencies. The modem was not adequately provisioned by the service provider, this resulted in alternate purchase of modems and installations leading to cost and operational implications. Billing and tracking of problems were a significant issue given the scale and spread of locations. Inconsistencies in billing were observed which took a long time to verify and rectify. Lack of clarification in the bill-related queries on part of the telecom service provider, created confusion and unnecessary expenditure. Lack of consistency in data between different entities resulted in stalemate in decision making. Agencies came up with different sets of data (especially related to the number of DSL connections actually working) which gave rise to confusion and conflicts. Private telecom service provider's commitment to programs was also an issue and top leadership commitment for such critical responsibilities is absolutely necessary from the concerned organizations. In experiments such as in this project, there was a need to accommodate fine tuning of equipment parameters (like mobile antenna direction) so that the quality and quantity of data coverage could be improved. However, procedural delays at the operator's end resulted in the attempt becoming aborted. State-of-the-art technologies like NOFN also faced technological challenges and issues with co-ordination amongst stakeholders leading to inconsistent and unreliable service.

These observations indicate that there are serious issues with the internet service provider (ISP) in rural areas, emanating from a position of monopoly. We find that while the state-level officials of the ISP showed interest in the project, that interest did not translate into field-level action, requiring continuous and repetitive follow-up by the consortium on every small matter. This forced the consortium to look for alternative technology and service providers. While the alternative cellular technology was deployed smoothly, the Education Department initially agreed and later went back to say that DSL broadband was necessary, even though it was their own responsibility to provide broadband in schools. The Government ISP being the only service provider offering DSL, alternative DSL broadband ISPs were not available. Alternatively, cellular service providers were explored but the Education Department did not want to take private sector service provider's help even though it could provide better service at a lower cost. There was no enthusiasm on the part of the public service provider to address technology issues. Sending bills was the only prompt activity the public service provider did. Meetings and escalations to Divisional Heads of the public sector provider, did not help. It was an opportunity for the ISP to proactively participate in the consortium project and experiment on different solutions so as to encourage other users to follow. It called for proactive involvement of telecom engineers to be out in the field and study the problems, arrive at solutions given their end-to-end knowledge of last mile connectivity, network equipment configurations and their limitations, transport network configuration tweaking. But that was not to be, and the consortium had to work with the lineman whose knowledge and exposure was limited to basic installation and repairs.

The personnel deployed by the ISP in the last mile were shared between different locations. Getting their attention to attend to problems was an issue. Often, the connections would break due to line cuts and road cutting and other reasons. It was not clear if the ISP personnel followed any laid down standards for cable laying. In spite of the obvious challenges of the last mile cable laying in rural areas, it was expected that the ISP would ensure some certainty of connectivity. On many occasions, there was disagreement over the status of a particular connection whether working/not working. These conflicts never got resolved and led to those connections getting billed on a monthly basis even without the service being rendered. The disagreement over working and not working connections lingered on endlessly leading to frustrations as there was no solution in sight because there was no scope for any immediate and quick arbitration. The provision and choice of modem had compatibility issues. Those provided by the ISP were not compatible with the ICT solution installed in schools (no wifi capability). The consortium incurred expenditure in installing alternative vendor's modem to suit the scenario. Also, the fluctuations in bandwidth was common resulting in abrupt interruptions in streaming video. This perhaps necessitated fine tuning of configurations in upstream equipment in exchanges, in order to ensure proper share of upstream bandwidth. However, such experiments were never even entertained, leave alone doing it. Rigid policies on part of the ISP lacked the experimental spirit required for the success of such complex projects.

The alternate private mobile ISP did start off earnestly but got bogged down eventually as they could not bear the cost of experimentation in rural areas. With no government departmental assurance for business opportunity upon the success, the private mobile ISP did not show enough commitment. Even though the mobile data bandwidth offered was mostly in 2G/Edge range which would not suffice for video streaming, at least a consistent low speed connectivity would have helped the messaging between different software applications to work.

Eventually, with no alternatives, and with so much dependent on reliable data connectivity, the consortium struggled to progress on interaction in the project. The project got launched based on the premise that the DSL connectivity would be provided by the government departments and the ISP. When the ISP announces support for a service, it is assumed that its customers would bank on the assurance of service. However, it turned out to be a false sense of assurance. With the consortium focusing on the all-important purpose of creating multimedia animation content for 3 different subjects and 6 different classes (5–10), and providing good teachers, it was not a planned activity to invest time and money into experiments on providing connectivity.

In spite of the various challenges, the project was appreciated for delivery of multimedia content through satellite in the 3 subjects for classes 5–10. The project won the prestigious WSIS Prize (World Summit on Information Society) in 2015, by ITU, for the category of E-Learning. The ISP did not provide quality cellular service in a significant number of schools, as an alternative to its DSL solution. Alternative cellular service provider in those schools were private ISPs and the department did not want to engage a private cellular service provider. Had the ISP and the government department fulfilled their responsibilities, it would have been an extremely impactful experience in terms of takeaways from the project, effectiveness, costs and uncertainty. A video-based interaction supported by the application would have enabled better interaction and also allowed the students to have access to a record of the question asked and response given. This could have been part of a searchable library or FAQ. An alternative was indeed figured out with just a voice support; the student would give a missed call and would receive a call from any one of the free moderators. The alternative required not so trivial modifications to the software. This costed time and money which was at the expense of the consortium partner.

It cannot be expected that alternatives will be figured out all the time for all scenarios. Also, not all private companies will be ready to bear expenditure for lapses for which they are not responsible. Companies may go strictly by the MOUs and agreements without being flexible. This may result in conflicts leading to arbitration which again affects productivity and adds to costs. Even figuring out alternatives requires the staff to apply their mind, brainstorm and come out with a solution. All these efforts add up to unplanned costs. Hence it is better that, each member organization, be it government or private, delivers upon their responsibility, so that members can focus more intensely on their own areas of expertise. In the context of this case study, many experiments were conducted to discover broadband solutions, which were beyond the scope and responsibility of the consortium members (like the experiment on NOFN, and the trials with private mobile ISP).

Power Supply

Apart from reliable data connectivity, the provision of uninterrupted power supply is also a crucial need for the success of ICT projects. With rampant power cuts in rural areas, alternate standby arrangements are necessary. A typical power supply solution consists of the State Grid, Inverter, Battery and the Solar Panel. The ratings for the above components will depend upon various factors like—the load, the duration for which the power backup is needed, the extent of grid state supply available and the general sunlight conditions. The ratings are crucial as the investment on the solution is required to ensure that the purpose of uninterrupted power supply is satisfied.

The provision of quality power supply needs to factor in the following aspects.

1. Unpredictable state power supply
2. Lack of adequate sunlight hours, especially during monsoon
3. Quality of battery and its maintenance
4. Damage to externally mounted solar panels
5. Inadequate charging of battery due to premises supply being turned off daily
6. No proper indication of amount of battery backup available in order to ensure reliable service
7. Misuse of power backup for purposes other than what it is designed for.
8. Single phase power supply
9. Inadequate earthing results in equipment producing light shocks and also potentially can damage the equipment
10. During long vacations and holidays, the battery is not charged and it often goes dead. There is often very little understanding and concern by the existing system. This calls for a separate arrangement which add to the cost of power.

(Note: Reliable UPS systems with sensors for alerting failures are available, but the costs are high.)

The grid power provided by the government is not a 24/7 service and suffers from power cuts running into a few hours at least every day. This compelled the consortium to explore alternative solutions based on solar power involving solar panels, inverter and batteries. This had cost implications including that of regular maintenance. Also, the earthing provision was missing in many schools. This caused light shocks and also damage to equipment. The provision of earthing was not an easy exercise and involved significant efforts.

Infrastructure

Lack of adequate infrastructure is one of the key issues that rural schools in developing countries face when implementing ICT-based projects. This holds true both in terms of quantity and quality. The following issues were observed:

- Theft or damage to the infrastructure is seen in many institutions. This makes caretaking of the infrastructure an additional overhead and prohibits students from freely accessing the infrastructure.
- There is lack of classrooms to accommodate the students and thus they have to function from temporary facilities.
- Some students have to travel long distances to reach the schools. Dependency on public transports and their timings, constrains the students from being flexible with their timings in order to accommodate any special classes through tele-education.
- General infrastructure elements like—benches, lighting, ventilation have been observed to have issues in some schools.

Operational Issues

The disadvantage of remoteness is a major cause that impacts the smooth progress of projects in rural areas. For a private agency delivering services in rural areas, the need for being adequately organized becomes imperative. There are operational challenges. Owing to the distributed nature of institutions, reachability is an issue. Conveying information in a timely manner becomes difficult. Monitoring of progress of work is a challenge as the institutions are many and spread out. Depending upon the institutions that are being serviced, it requires a team of dedicated workers who need to follow-up. This increases the costs. Some areas may not be serviced by certain mobile operators, resulting in difficulty to reach the personnel in remote areas. Management of distributed workforce requires setting up of divisional and zonal hierarchy and setting up of offices, which in turn increase the costs. Payment of salaries and sundry expense claims to personnel becomes a challenge. Absence of banking facilities adds to the complexity.

Automation, web and mobile based applications, to some extent ease the operational challenges but the technology comes with its own challenges. The private organizations in the consortium seemed to struggle with operational issues and management of their human resources. Challenges of managing remote workers and tracking their performance imposed tough problems. Given the complex nature of the project, the staffing of skilled and motivated employees needed better funding and timely release from the government department. The private organization was already saddled with the responsibility of the last mile and setting up everything from scratch. The government departments with its administrative and operational presence in the last mile institutions, could have supported the consortium in a more positive manner.

Motivational Issues

The attitude of those involved in the project influences the outcome. Leaders who are passionate are found to positively impact the motivation of the entire project team, irrespective whether they are government or private. The support from top management of institutions, and policy makers, is also critical to the success of any education project. In many instances, stakeholders were required to be reminded several times of any responsibility to be fulfilled. It was an expectation by the stakeholders that the concerned party would follow-up on pending tasks. This compelled the agencies to invest in manpower to track and follow-up. Decision making was often delayed which adversely affected the project. Transfer of key persons responsible for the project, introduced intermittent delays. Some stakeholders were not interested in taking the project forward. They lacked the motivation and the drive to take initiatives and proactiveness was missing. Procedural lapses resulted in significant delay in payment to agencies by the government department. This resulted in salary delays to field staff which in turn led to loss of manpower severely affecting the project.

The project would have benefitted by a proactive involvement of the officials in government departments. They could have involved themselves in a more integrated manner given the important purpose of improving quality education. Even though at the school level many teachers were enthusiastic about the way the program was progressing, the officials at the government department level did not support the project adequately, and there was a disconnect with the overall purpose of the project itself. Sticking to strict academic plan was not a priority for the government department. Delays imposed on payments that were due to the consortium, resulted in many operational issues resulting in manpower loss and inconvenience to the staff of the consortium. As the consortium was burdened with the crucial responsibility of providing quality content and good teachers, many unplanned tasks resulted in loss of focus, also cost escalations and diversion of manpower. So, it can be observed that though the projects were bolstered by the presence and participation of the private organization, the significant dependencies on the government departments continued to persist in the rural areas.

4 Conclusion

Rural areas are disadvantaged in many aspects when compared to the urban developed parts. In spite of the ICT intervention efforts, to address this gap, the potential of ICT has hardly been exploited in rural areas. The challenges exist in terms of the inconsistencies in availability of quality power supply and data connectivity, which are the basic necessity for technology to work. The broadband availability is not satisfactory. There are many areas not served by the Digital Subscriber Line or other comparable technologies, which is required for stable bandwidth. The cellular wireless data access is inadequate to guarantee consistent and satisfactory access to content. Apart from technology challenges, the administrative setup which has to serve the people is also not efficient. Infrastructure woes exist in public organizations. Meager and inefficient administrative system imposes unexpected, unplanned costs on the parties who come forward to address the rural-urban divide. Slackness in decision making, unexplained delays, denial of services, and operational inefficiency, together make it very difficult, ultimately threatening to make the entire initiative infeasible. Estimates done before the commencement of the project, quickly become inadequate due to the unexpected costs. Provision for unplanned infrastructure related purchases, travel expenditures, escalation in manpower count, all contribute to the unmanageable expenditures affecting the feasibility of the project for private enterprises. Delays in decision making, monopoly of some services, arbitrariness in decisions, delays in payments and various dependencies on government, drive the private enterprises to resort to alternative options that again escalates the costs. This has a snowballing effect, that makes the private parties frustrated leading to gradual compromises and ultimately fall in quality and demise of the projects. The other outcome of this desperateness is the possibility of the private parties resorting to compromises that drastically impact the quality of the product and the service. This leads to possible unfair practices creeping in, which is a more serious aspect to worry about. Inability of the government agency to provide the required services all by itself or with the help of private partner is impacting the education service delivery for the rural citizens.

References

1. Patrinos HA (2016) Why education matters for economic development, 17 May 2016. http://blogs.worldbank.org/education/why-education-matters-economic-development. Accessed 3 May 2018
2. King EM (2011) Education is fundamental to development and growth, 10 January 2011. http://siteresources.worldbank.org/EDUCATION/Resources/278200-1295560712817/keynote_Beth_King-Ed_World_Forum.pdf. Accessed 3 May 2018
3. Ministry of Statistics & Programme Implementation (2015) Key indicators of social consumption in india on education show continued gender gap and rural urban differences, 30 June 2015. http://pib.nic.in/newsite/PrintRelease.aspx?relid=122881
4. ASER (2014) ASER 2014: annual status of education report. Pratham
5. Tripathi A (2016) Teacher shortage: affecting more than just the literacy levels? https://www.toppr.com/bytes/teacher-shortage-in-the-country/. Accessed 30 Dec 2018
6. Akpan JP (2002) Which comes first: computer simulation of dissection or a traditional laboratory practical method of dissection? Electron J Sci Educ 6(4)
7. Bork A (2003) Interactive learning: twenty years later. Contemp Issues Technol Teach Educ 2(4):608–614
8. Ghavifekr S, Rosdy WAW (2015) Teaching and learning with technology: effectiveness of ICT integration in schools. Int J Res Educ Sci 1(2):175–191

9. Young S (2003) Integrating ICT into second language education in a vocational high school. J Comput Assist Learn 19(4):447–461
10. Reijswoud VV (2009) Appropriate ICT as a tool to increase effectiveness in ICT4D: theoretical considerations and illustrating cases. Electron J Inf Syst Dev Ctries 38(9):1–18
11. Assar S, Amrani RE, Watson RT (2010) ICT and education: a critical role in human and social development. Inf Technol Dev 16(3):151–158
12. Winzenried A, Dalgarno B, Tinkler J (2010) The Interactive Whiteboard: a transitional technology supporting diverse teaching practices. Australas J Educ Technol 26(4):534–552
13. Hennessy S, Ruthven K, Brindley S (2005) Teacher perspectives on integrating ICT into subject teaching: commitment, constraints, caution, and change. J Curric Stud 37(2):155–192
14. Masperi P, Hollow D (2009) An evaluation of the use of ICT within primary education in Malawi. In: 2009 international conference on information and communication technologies and development (ICTD), Doha, 2009
15. Kim G-M, Ong SM (2005) An exploratory study of factors influencing M-learning success. J Comput Inf Syst 46(1):92–97
16. Douglas D, Vyver GVD (2004) Effectiveness of E-learning course materials for learning database management systems: an experimental investigation. J Comput Inf Syst 44(4):41–48
17. Nicholson J, Nicholson D, Valacich JS (2008) Examining the effects of technology attributes on learning: a contingency perspective. J Inf Technol Educ 7:184–204

ODDMS: Online Distributed Dynamic Meeting Scheduler

Archana Nigam and Sanjay Srivastava

Abstract With advent of social media, need for scheduling very large meetings while proving a degree of privacy to the participants has become an important problem. Existing solutions based on a global calendar expose individuals data to the calendar provider and thus are unsuitable for open meetings with quorum constraints. We propose an online distributed dynamic meeting scheduler (ODDMS). It is able to efficiently schedule meetings involving a large number of participants, without having complete knowledge of individual participants and their preferences, thus preserving privacy. The algorithm uses a modified negotiation-based distributed schedule that resolves the problem of deadlock and contention using hidden naive Bayes learning method. We compare our work with a baseline centralized algorithm and two existing algorithms based on voting mechanism and naive Bayesian methods. Simulation studies show that ODDMS performs similar to baseline centralized algorithm under light load condition and significantly outperforms the existing distributed algorithms under heavy load condition.

Keywords Distributed system · Meeting scheduling · Knowledge base · Deadlock handling · Hidden naive Bayes · Contention avoidance

1 Introduction

Scheduling a meeting is a naturally distributed task which can be defined as a method of finding the suitable and optimal time slot based on preferences and priorities of both host and participants [6]. If participants involves are huge in number, i.e., in thousands, for example, social media-based meeting, alumni meetup,

A. Nigam (✉) · S. Srivastava
Dhirubhai Ambani Institute of Information and Communication Technology, Gandhinagar, India
e-mail: archana_nigam@daiict.ac.in

S. Srivastava
e-mail: sanjay_srivastava@daiict.ac.in

S. Fong et al. (eds.), *ICT Analysis and Applications*, Lecture Notes in Networks and Systems 93, https://doi.org/10.1007/978-981-15-0630-7_19

event schedules, etc., then such problem involves more complexity, as many meeting being scheduled at the same time considering several constraints [8]. Till now most of the work had been done for organization-level meetings; this motivates us for a more realistic case where meeting scheduling for non-organizational setup is involved. Realistic distributed meeting scheduling is an online problem in which a new meeting arrives while the number of the meeting is being scheduled concurrently; therefore, the status of the local and non-local resources may change over time [13]. In this paper, we are considering a multi-agent environment. There are two types of agents in our system: host agent (HA) and participant agent (PA). HA is responsible for scheduling the meeting on the users behalf. PA responds on behalf of users who are in the guest list of the meeting. In centralized meeting scheduling, a central host has all the information regarding preference and availability of participants; therefore, the host takes a decision and schedules a meeting on their behalf. However, this method breaches the privacy constraint as each participant and central host can see the data of other participants and also participants remain isolated from decision-making process. Whereas in distributed meeting scheduling, information is kept private by sharing the limited information among entities which are involved in a meeting; therefore, entities have partial information about each other [2]. This information transfer is taken place through negotiation where an agent negotiates with other agents to select a common schedule considering the user's preferences and calendar. Another approach in distributed meeting scheduling is a voting-based method, HA releases a set of date, and all the PAs are asked to adapt date set according to their constraint and rank the date set accordingly. Negotiation-based method gives more freedom to the individual participant to take part in decision-making. In the distributed system because of partial information, meeting scheduling requires a large number of negotiation rounds and to make it converge learning method is adopted. In an online distributed system, if negotiation will not converge in every round, then with an increase in a number of pending meeting requests deadlock condition may arise. In this paper, we are solving a contention and deadlock problem in distributed meeting scheduling system using a hidden naive Bayes learning method. In this paper, we propose an online distributed dynamic meeting scheduler (ODDMS) which efficiently schedules meetings involving a large number of participants, without having complete knowledge of individual participants preferences, thus preserving privacy. To make our system more realistic, we introduce two important factors in our system: quorum and overlapped criteria. Quorum is defined as the number of participants mandatory for the meeting to be held, and overlapped criteria are defined as the number of participants common in more than one meeting. Quorum is important as it signifies the minimum number of participants to successfully schedule a meeting, failing which meeting cannot be scheduled. We compare ODDMS with techniques found in literature and the result shows that ODDMS perform better than most of the distributed meeting scheduling system. As per best of our knowledge under this setting, no work has been done in the literature. The paper is organized as follows: In Sect. 2, we discuss the literature review. In Sect. 3, the problem statement is given. Section 4 discusses the proposed architecture and algorithm. Section 5 discusses simulation results; we conclude in Sect. 6.

2 Literature Survey

Doodle [5] is a well-known service that allows users to find all common availability. However, it does not provide privacy as each user and doodle server can see the busy or free state of every other user. To solve the distributed meeting scheduling problem, two types of approaches were discussed in the literature: negotiation-based and voting-based. In [10], the author solves distributed meeting scheduling problem using negotiation technique, where a host agent communicate with all other agents in order to schedule a meeting and feasible schedule is searched based on participants calendar information passed during negotiation. This approach does not takes into account participant's preferences, whereas in [11] participant's preferences and calendar information is communicated during negotiation, therefore, privacy is not warranted. The main objective of using a learning method in distributed meeting scheduling is to reduce the burden of negotiation by calculating the probability of acceptance of the meeting based on past knowledge [7]. In [7], the author uses naive Bayes classifier for calculating the probability of meeting acceptance. But the naive Bayes classifier works for independent attributes, i.e., it considers only participants availability, not considering other attributes such as preferences and priority. Also, it does not use negotiation result for knowledge updating. In [13], Bayesian network is used for learning users' preference and this system learn from negotiation result. But learning an optimal Bayesian network has been proved to be NP-hard [3]. In fact, the most time-consuming step in learning a Bayesian network is learning the structure [12].

Another approach in distributed meeting scheduling is voting-based meeting scheduling system. In [2], host transfers a set of dates based on user preferences and priority, and all the participants are asked to adapt the date set according to their constraint and rank the date set accordingly. Date with a high number of votes is chosen as a date for a meeting. Here only date attribute is taken into consideration, other attributes such as time, day, duration are also important. Other concern with negotiation-based and voting-based methods is that they will suffer in the highly loaded scenario where a number of participants are common in more than one meeting for the same time interval and if a quorum is also high. In online distributed system, if negotiation will not converge in every round then with increase in number of pending meeting requests deadlock condition may arise.

3 Problem Statement

A meeting schedule is defined as a set of meetings for a group of participants. Given a set of n meetings and j participants, a scheduling problem is represented as $MS = (A, M)$, where $A = \{1, 2, \ldots, j\}$ is the set of participant and $M = \{m_1, m_2, \ldots, m_n\}$ is the set of n meetings to be scheduled. A time slot is represented as a day, date, hour tuple $<Da, Dt, H>$. A set of contiguous time slots is called a time interval. A meeting is represented by a tuple:

$$m_u = (A, h, dr, q, \tau)$$

where A is a set of participants of the meeting; h is host agent of the meeting; dr is the duration of the meeting in hours; and q is the quorum value.

τ is the time interval for which the meeting m_u is finally scheduled and is represented by an ordered set $\{<Da_u, Dt_u, H_u>, <Da_u, Dt_u, H_u + 1>, \ldots, <Da_u, Dt_u, H_u + dr_u - 1>\}$ (here Da_u gives the day, Dt_u gives the date, and H_u gives the starting hour for which meeting m_u is scheduled) if the meeting could be scheduled, and by null set otherwise.

3.1 Criteria for System Evaluation

In our system if multiple meeting requests arrive for a same time interval where participants are common and quorum is high, then meeting scheduling will be difficult. Therefore, offered load in our system is defined as

$$L = \begin{cases} L = \widetilde{Q}\widetilde{O}, & \text{if } \tau_i = 1 \\ 0, & \text{otherwise} \end{cases} \tag{1}$$

If n meeting request arrives correspond to ith time interval, $\widetilde{Q}$ and $\widetilde{O}$ are the average quorum and overlapped value, respectively, for n meetings. If there is no overlap then the load is null.

We evaluate our system in terms of two first-order metrics, namely, efficiency and satisfaction index. Efficiency reflects the degree to which system succeeds in scheduling the important meetings. Satisfaction index reflects how much satisfy the participants are corresponding to a particular schedule.

Efficiency is defined by the following equation:

$$\eta_i = \frac{|m|}{|M|} \tag{2}$$

where M is total meeting request arrived; m is meeting finally scheduled in i time interval.

Satisfaction index (SI) is defined as a meeting that is scheduled in the first round itself corresponds to optimal time interval choice. Each round of negotiation needed to schedule a meeting comes at a cost of lowered satisfaction. Thus, we define SI for meeting in a time interval as a weighted average, with weight $w = \frac{1}{r}$.

$$\mathrm{SI}_i = \frac{1}{|m|}\sum_{i=1}^{r_{\max}} \frac{|mr_i|}{i}, \qquad |m| = \sum mr_i \tag{3}$$

$r_{\max}$ is the maximum number of negotiation round required for meeting to schedule, $|mr|$ is the number of meeting scheduled in each negotiation round.

If there is total k, then $\widetilde{SI}$ is the average SI, and $\widetilde{\eta}$ is the average fraction of meeting schedule for total k time interval. The primary objective function is to maximize efficiency and satisfaction index simultaneously.

4 Online Distributed Dynamic Meeting Scheduler

In our proposed system, we are considering a multi-agent environment. There are two types of agents in our system, host agent (HA) and participant agent (PA). HA is responsible for scheduling the meeting on the user's behalf. PA responds on behalf of users who are in guest list of the meetings. HA calculates the probability of acceptance of a meeting proposal based on past data. If the probability is high, the proposal is sent to all the PAs. PA can accept the proposal if time interval is free or send a pending message. Finally, if quorum criteria are met, then meeting is confirmed else canceled. At most three rounds of negotiation are involved in our system. Under such setting negotiation-based method works perfectly as compare to voting-based method because negotiation-based method gives more freedom to the individual PA to take part in decision-making. In distributed system due to partial information, proposing a time interval suitable to all is a difficult task and it require large number of negotiation rounds, thus there will be no guarantee that negotiation method will converge. If there are n PA and meeting scheduling require r rounds of negotiation, if r is large then the cost it incurred will be $O(rn)$. To reduce the cost of subsequent round and to make negotiation method converge, learning method is adopted. Therefore to solve this problem, we are using a hidden naive Bayes learning method (HNB) [12]. In online distributed system, if negotiation will not converge in every round then with increase in number of pending requests deadlock condition may arise. Another condition of deadlock which is common in distributed system is hold and wait condition. We are solving the problem of deadlock using weighted priority and time out mechanism. The architecture of both host and participant agents is shown in Fig. 1. In HA architecture, a meeting request is given as input to contention avoidance module, details will be discussed in Sect. 4.1. In contention avoidance module, analysis function calculates the probability of acceptance of meeting request. If probability of acceptance is higher, meeting request is send to all PA. On receiving the message from PA, offer analyser module will check whether quorum criteria are met or not. If met, changes are made into calendar and confirmation is send to PA. On confirmation or rejection of meeting knowledge base is updated. Algorithm 1 describes the working of host agent, and Algorithm 2 describes the working of participant agent. In PA architecture, on receiving a meeting request for the first time

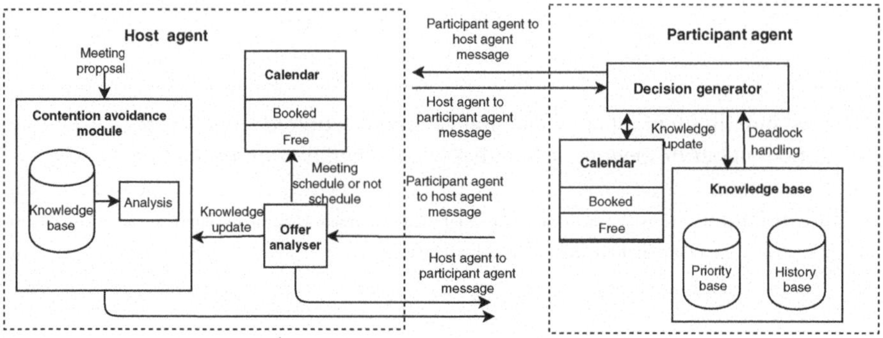

Fig. 1 Host agent and participant agent architecture

decision generator module will check the participant's calendar. If meeting is already booked reject message will be send, if time interval is free but there is request for same time interval from other HA then priority of both the HAs is calculated.

Algorithm 1 Host agent

Input: Meeting request
procedure CONTENTION AVOIDANCE
  Input: Set D of past history data
  Output: Probability of success or fail
  if probability of success is high **then**
    Send: meeting proposal to all participant, Call Offer analyser
  end if
end procedure
procedure OFFER ANALYSER
  if receive: $accept \geq q$ **then**
    Send: Meeting Confirm;
  else
    call Meeting Pending;
  end if
end procedure
procedure MEETING PENDING
  if $round \leq 2$ **then**
    negotiation round, call Offer analyser
  else
    Compute time out, call Offer analyser
  end if
end procedure

PA send acceptance message to higher priority HA and pending to other. If time interval is free and no conflict, then PA send acceptance to HA. If message is received for the second round then there might be deadlock between two requests for same time interval. Therefore, decision generator performs deadlock handling by calculating the weighted priority of both the HAs; accept message will be send to one with higher weighted priority and reject to other. If both the HAs have same weighted priority, then PA will send nothing. Both HA will wait for time out. Knowledge base is updated by decision generator and corresponding changes are made into the calendar of participant.

4.1 Contention Avoidance with Hidden Naive Bayes

In distributed system, there is no guarantee that random negotiation method will converge. Also the cost it incurred will be $O(rn)$. Therefore to solve this problem we are using a hidden naive Bayes learning method (HNB) [12]. The advantage of HNB is as it consider various attributes like preferences, priority, availability during learning. Instead of becoming slow when more situation is learned, it is more responsive and improves its precision by learning from negotiation. Therefore, the probability of success in organizing a meeting will increase. In HNB learning method, prior knowledge is combined with observed data to determine the final probability of a hypothesis. It creates a hidden parent for each attribute, which combines the influences from all other attributes [12]. $\{A_1, A_2, \ldots, A_n\}$ are n attributes and C is the class node. The joint distribution represented by an HNB is defined as follows:

Algorithm 2 Participant agent

```
procedure DECISION GENERATOR
    Input: meeting proposal receive
    if round == 1 then
        Check calendar availability.
        if available and no contention then
            send: accepted
            if available and contention then
                send: pending, after comparing priority
            else
                send : reject
            end if
        end if
    end if
    if meeting confirm then
        Update Calendar
    end if
    if round == 2 then
        call: Deadlock Handling.
    end if
end procedure
procedure DEADLOCK HANDLING
    compute weighted priority to break tie
    Update History
end procedure
```

$$P(A_1, \ldots, A_n, C) = P(C) \prod_{i=1}^{n} P(A_i | A_{hpi}, C) \tag{4}$$

Here in our case class is either success or fail, and attributes are a participant agent, date, time, day, and duration. For example, $P(\text{Agent} = A, \text{date} = 10/5/2019, \text{time} = 10\,\text{am}, \text{day} = \text{Mon}, \text{duration} = 2\,\text{h}, C = \text{success})$. We used the Laplace estimation to avoid the zero-frequency problem [12].

4.2 Deadlock Handling

In online distributed system, if negotiation will not converge in every round then with increase in number of pending meeting requests deadlock condition may arise. Another condition of deadlock which is common in distributed system is hold and wait condition. The problem of deadlock is solved by giving the preference to high priority host agent to break the tie [4]. But in the worst case, if all the host agent has the same priority then deadlock problem will remain the same. Therefore to solve this issue, we use weighted priority between priority list defined by participants' agent as well as priority list derived from the past meeting. To calculate weighted priority we have a smoothing factor β.

$$W_\text{p} = \beta p + (1 - \beta)h, \quad 0 \le \beta \ge 1 \tag{5}$$

where W_p is weighted priority, p is priority from priority database, and h is derived priority from history database.

History database. To calculate h, we use naive Bayes classifier [9]. Host agent with high h has done more number of meeting with the participant agent. In worst case if weighted priority is again same for both the host agents, then we use timer to drop particular host agent.

$$T_o = 2 \times \text{RTT}_a + \frac{\text{acc}_i}{q_i} \tag{6}$$

where acc_i is total number of accept received for a meeting i, and q_i is quorum defined for meeting i. RTT_a is the time of accept message reaching from participant agent to host agent. Therefore, we are dropping the host agent whose work done is less compare to other host agent.

5 Simulation Result

In this section, we simulate our design using JADE (Java Agent Development) [1] framework. In our simulation test-bed, we have 1000 meetings spread across 20 time interval; multiple meetings are overlapped randomly in a single time interval.

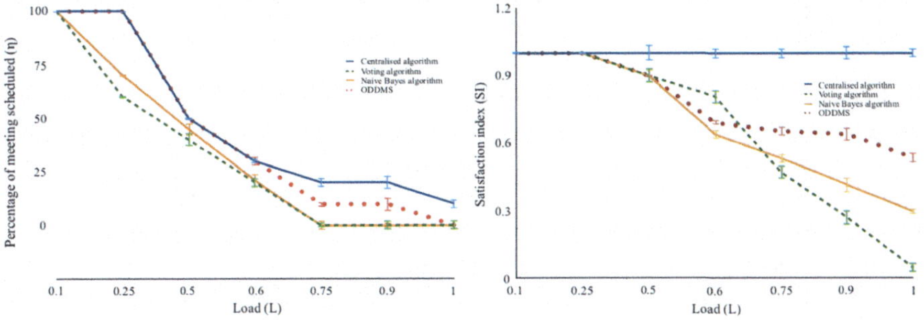

(a) Percentage of meeting scheduled as a function of offered load

(b) Satisfaction index as a function of offered load

Fig. 2 Performance and satisfaction index as a function of offered load

Each meeting involves number of participants between 500 and 2000. Performance of ODDMS is measured using η and SI as a function of offered load. We compare ODDMS with centralized algorithm, learning-based system [7] and voting-based system [2].

In our simulation results centralized algorithm is used as a benchmark for a comparison, this is the optimal result that can be achieved. In Fig. 2a, we are comparing η as a function of offered load between all the four algorithm. Under low load condition, ODDMS perform similar to centralized algorithm because learning method that we use is more restrictive, so it will send proposal to those who have a high chance of acceptance. Under high load condition, voting-based method gives poor performance because of increase in disagreement which increases the contention in the system and leads to deadlock; similarly in naive Bayes learning algorithm, there is no mechanism to detect and resolve deadlock. Therefore in both of these methods, the number of meeting schedule comes to zero. ODDMS performance is better because we are handling deadlock and contention using learning method. In Fig. 2b, we are comparing the satisfaction index of the four algorithm. In classical centralized algorithm, host has all the preference and priority data of all agents; therefore, it has complete knowledge of when to schedule meeting such that most of the agent will be satisfied. In voting-based method, during moderate-high load condition number of disagreement increases and this disagreement continues till meeting rejected or accepted thus poor satisfaction index. In naive Bayes method, due to contention negotiation round increases which degrade the performance of algorithm. In ODDMS, the number of negotiation round is three; in worst case much less compare to voting and naive Bayes because better scheduling decision is made based on the past knowledge considering all the dependent attributes.

6 Conclusion

We have proposed a distributed algorithm to solve the online meeting scheduling problem. ODDMS is able to efficiently schedule meetings involving large number of participants, without having complete knowledge of individual participants preferences, thus preserving privacy. We introduce the notion of quorum and overlapped criteria. We compare our ODDMS with voting and naive Bayes algorithm. Our simulation result shows that ODDMS is able to schedule about 43% more meeting under low load condition and about 25% under high load condition. Similarly, quality of the scheduled meeting is measured by satisfaction index (SI). SI is about 60% more than the voting and naive Bayes distributed algorithm when offered load reaches to 1.

References

1. Bellifemine F, Caire G, Poggi A, Rimassa G (2003) JADE a white paper. Telecom Ital EXP Mag 3(3). http://www.cis.umassd.edu/~hxu/courses/cis481/references/WhitePaperJADEEXP.pdf
2. BenHassine A, Ho TB (2007) An agent-based approach to solve dynamic meeting scheduling problems with preferences. Eng Appl Artif Intell 20(6): 857–873. https://doi.org/10.1016/j.engappai.2006.10.004
3. Chickering DM (1996) Learning Bayesian networks is NP-complete. In: Fisher D, Lenz H (eds) Learning from data: artificial intelligence and statistics V. Springer, p 121–130. https://doi.org/10.1007/978-1-4612-2404-4_12
4. Crawford E, Veloso M (2005) Learning dynamic preferences in multi-agent meeting scheduling. In: IEEE/WIC/ACM international conference on intelligent agent technology. https://doi.org/10.1109%2Fiat.2005.94
5. Doodle scheduling service website (2016). http://www.doodle.com. Accessed 14 Sep 2016
6. Garrido-Luna L, Sycara K (1996) Towards a totally distributed meeting scheduling system. In: Annual conference on artificial intelligence. Springer, Berlin. https://doi.org/10.1007/3-540-61708-6_50
7. Hossain SMM, Shakshuki EM (2013) A deliberative agent for meeting scheduling. In: 2013 IEEE 27th international conference on advanced information networking and applications (AINA). IEEE. https://doi.org/10.1109/AINA.2013.87
8. Mooney EL, Rardin RL, Parmenter WJ (1996) Large-scale classroom scheduling. IIE Trans 28(5): 369–378. https://doi.org/10.1080/07408179608966284
9. Murphy KP (2006) Naive bayes classifiers, vol 18. University of British Columbia, p 60. https://datajobsboard.com/wp-content/uploads/2017/01/Naive-Bayes-Kevin-Murphy.pdf
10. Sen S, Durfee EH (1991) A formal study of distributed meeting scheduling: preliminary results. In: ACM conference on organizational computing systems. https://doi.org/10.1023/A:1008639617029
11. Sen S, Durfee EH (1995) Unsupervised surrogate agents and search bias change in flexible distributed scheduling. In: First international conference on multi-agents systems. https://www.aaai.org/Papers/ICMAS/1995/ICMAS95-045.pdf
12. Zhang H, Jiang L, Su J (2005) Hidden naive bayes. In: AAAI. http://dl.acm.org/citation.cfm?id=1619410.1619480
13. Zunino A, Campo M (2009) Chronos: a multi-agent system for distributed automatic meeting scheduling. Expert Syst Appl 36(3): 7011–7018. https://doi.org/10.1016/j.eswa.2008.08.024

Digital Readiness Index—Empowering the Nation

Sitadevi Bharatula and B. S. Murthy

Abstract Digital Readiness Index (DRI) also referred to as Networked Readiness Index or Technology Index indicates the status and growth of information and communications technologies (ICT) and how it is effectively used to achieve maximum benefits to the country and its citizens. DRI explains the status of digital sovereignty of a nation and is used to monitor and compare developments in ICT domain. The Index is designed to understand the ICT Development Index (IDI) of a country, by which it can be ascertained the level of 'digital readiness' of the nation. The importance of ICT for digital India and E-Governance has been well brought out in the National Digital Communication Policy (NDCP) 2018 by Government of India. Development of Digital Readiness Index has been envisaged in the NDCP document. In this paper, important aspects of NDCP for making India digitally sovereign have been explained. The government policy on E-governance in ICT sector for making the country as 'digital India' has been explained in this article. The parameters of DRI and their significance for empowering the Nation has been brought out in this article. Methods by which these parameters could be measured and the difficulties involved therein are elaborated. Further, the importance of developing DRI and socio-economic impacts of measuring the index are explained in this article. The benefits to the people, to the States and the nation as a whole, by bringing out DRI of the States are brought out in this paper.

Keywords Digital readiness index · Information and communication technology · ICT development index · National digital communication policy · Networked readiness index · E-Governance · Telecom service provider

---

S. Bharatula (✉)
Aarupadai Veedu Institute of Technology, Paiyanoor, Chennai, India
e-mail: sitabharatula@gmail.com

B. S. Murthy
Department of Telecommunications, New Delhi, India
e-mail: budaraju.sm69@nic.in

S. Fong et al. (eds.), *ICT Analysis and Applications*, Lecture Notes in Networks and Systems 93, https://doi.org/10.1007/978-981-15-0630-7_20

1 Introduction

The Digital Readiness Index (DRI) measures the capacity of countries to use ICTs for increased competitiveness and well-being. It indicates the level of performance of the countries in ICT sector. The significance of DRI is that several basic infrastructure facilities, policy frameworks etc, are to be built to achieve the benefit of digital revolution. The DRI measures the level of preparedness of countries on this front.

With over two decades of commercial access to the Internet, India has come a long way from dial-up connections to super-fast broadband speeds. The total number of Internet subscribers in India stood at about 450 million of which over 95% is wireless. The pace of progress indicates that there would be 2 bn networked devices and 6.5 exabytes of data used per month in India by 2021. Much of this growth will be fueled by massive consumer adoption of smartphones, Internet of Things (IoT) and use of machine-to-machine (M2M) connections [1].

The increased connectivity across devices is expected to enhance efficiency for all sectors of the economy including healthcare, agriculture, manufacturing, retail, transportation as well as citizen services. The 'Digital India' programme championed by the government resonates with this vision. Connecting rural areas with high-speed Internet networks and improving digital literacy is also a deep-rooted objective of this national program. In the last couple of years, government is putting massive efforts towards inclusive growth in electronic provision of services, through initiatives like the JanDhan, Aadhaar and Mobile (JAM) trinity. India stands at 91st rank in 2016 in a global ranking exercise of Network Readiness Index of 144 countries published in the Global Information Technology Report [2]. According to International Institute for Management Development (IMD)'s World Digital Competitiveness Rankings 2018 [3], India's position climbed up to 48th rank in 2018 from 51st rank in 2017. Also, as per World Economic Forum report 2018 on Global Competitiveness Index, India's position improved by 5 places to 58th rank in 2018 when compared to 2017 [4].

The DRI studies political and business environment, readiness, usage, economic and social impact of ICT in the country [5]. It may be seen that different studies are being carried out by different forums for the assessment of digital readiness, network readiness, ICT development index (IDI), etc. of a country. However, for a large country like India, it is necessary to measure digital readiness index of each individual State, in order to assess the industry/investment friendliness, IDI, digital infrastructure availability of the State and also to identify the gaps and challenges. In this paper, measurement and implementation of DRI at State level and its benefits to the investors, State administration and general public have been explained. The government policy and steps taken by the government towards implementation of E-Governance and empowering the nation to make 'Digital India' a reality is brought out in this article. The objectives of National Digital Communication Policy 2018 of Government of India are outlined in Sect. 2. Section 3 presents the Digital Readiness Index framework and parameters selected for measuring DRI and its implementation. Results and conclusion are given in Sect. 4.

2 National Digital Communication Policy 2018

The National Digital Communications Policy (NDCP) 2018 aims to fulfil the ICT needs of industries and citizens and to support the country's transition to a digitally empowered economy by setting up of robust Digital Communication Technologies (DCT) infrastructure and services [6]. Some of the important targets envisaged in NDCP document are (see Fig. 1):

- to provide 50 Mbps broadband connectivity to all
- to connect all Gram Panchayats with bandwidth of the order of Gbps
- to provide mobile/data connectivity to the remotest parts of the country
- to attract investments in DCT sector of the order of $ 100 bn
- to impart skill training in ICT sector to 1 mn manpower
- to reach 5 bn IoT devices target
- to ensure data protection and security

The policy envisages implementation of National Building Code 2016 and Common Rights of Way (RoW) policy in all States for ease of doing business with simplified procedures. The Policy gives thrust in the following areas:

(i) Fibre connectivity to all office/residential complexes. NDCP targets w.r.to number of towers and fabrication of towers is shown in Fig. 2.
(ii) Simplified procedures and approvals for telecom infrastructure.
(iii) Provide speedy approvals for Rights of Way permissions.
(iv) Encourage and facilitate sharing of active infrastructure.
(v) Incentivizing the use of renewable energy technologies.
(vi) According Telecom Infrastructure the status of Critical and Essential Infrastructure at par with other sectors.
(vii) Rationalizing taxes and levies on DCT equipment and infrastructure.

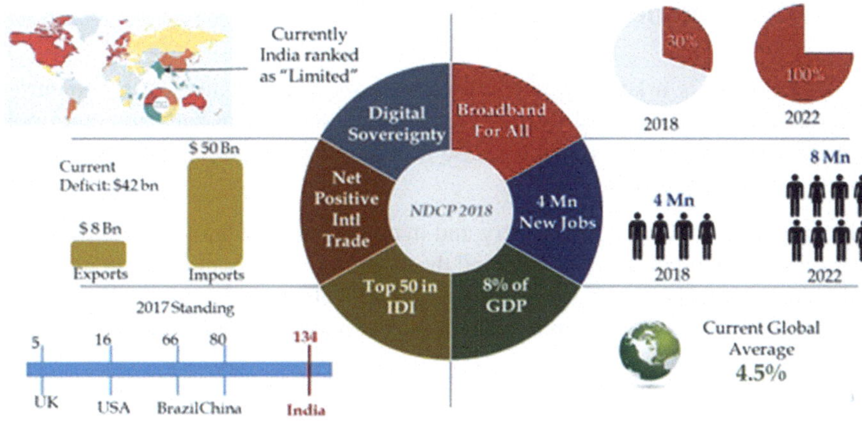

Fig. 1 NDCP 2018 strategic objectives

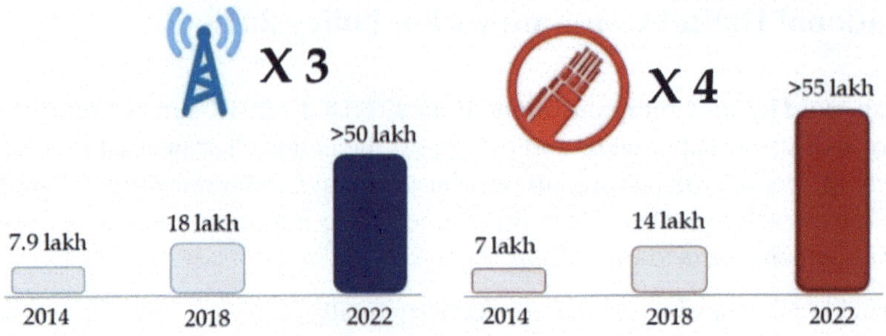

Fig. 2 Targets w.r.to. No. of towers & Ferberization

(viii) Addressing security issues across layers.
(ix) Connecting all towers with optical fibre cable.

In order to implement NDCP effectively, States have been asked to take necessary steps in the following areas:

(i) Alignment of all States Telecom Infrastructure Policy with Central RoW Rules, November 2016, i.e. Uniform State Policy with One-time nominal fee, Single-window clearance in a time-bound manner etc.
(ii) No restriction to be imposed on locations of mobile towers.
(iii) Availability of Government Land & Buildings for installation of Mobile towers.
(iv) Electricity connections to be provided on priority at industrial tariffs.
(v) No coercive action by State Government against Telecom Infrastructure without prior approval from State Nodal Authority.
(vi) Need for a Common Duct policy under 'Dig Once' umbrella.

3 Digital Readiness Index (DRI)

Digital readiness is a key indicator to measure the performance of the countries in digital domain. It is a tool for helping countries to identify gaps, structure policy dialogue, catalyze action and increase transparency and accountability [7]. It is measured to understand the level of:

- the digital divide within the country and in comparison, with other countries;
- the development potential of ICTs and the extent to which countries can make use of them to enhance growth and development in ICT sector;
- availability of necessary digital technologies to make a country 'digitally ready'.

3.1 DRI Parameters and Its Implementation

DRI parameters are divided into three categories, viz. digital environment, digital readiness and usage. The proposed index is envisaged to accomplish the following objectives:

- Create Robust and High-Quality Digital Communications (DC) infrastructure
- Attract Investments in creating next generation DC infrastructure
- Simplification of compliance procedures
- Ensure implementation of RoW policy by all States

DRI is measured in 6 areas as enumerated below:

(i) Micro and macro landscapes of administrative issues for developing Telecom Service Provider (TSP) infrastructure in States
(ii) Current Gaps and Challenges in developing infrastructure RoW issues, Power, Coordination challenges
(iii) Resource support from partner institutions
(iv) Leveraging existing data sources such as Tarang Sanchar
(v) Public Private Partnership in creating institutional mechanism for managing the DRI
(vi) To integrate relevant data requirements of other indices such as IDI

The details of DRI parameters vis-a-vis the provision in the NDCP are shown in Table 1.

3.2 Measurement of Parameters Through Tarang Sanchar

Tarang Sanchar portal is a set of web services for all stakeholders, Ministry of Communications, Telecom Service Providers for Electro Magnetic Field (EMF) compliance of telecom towers in India and general public [8]. In order to know the EMF radiation details of telecom towers and to have better accuracy, timely sharing of information amongst operators and eliminating other in-efficiency of self certificate submission, an exhaustive and comprehensive database of all the EMF emitting sources in mobile communication has been made available on Tarang Sanchar portal. The portal has potential to provide details of telecom infrastructure such as fibre connectivity to towers, availability of grid supply, mobile network coverage, etc. The portal will be suitably modified to collect information about the Digital Readiness Index parameters as mentioned in Table 1 above.

Table 1 Digital readiness index parameters

Sl. No.	NDCP provision	DRI parameter	Weightage (%)
1	State policy on RoW facilitation	Availability of State Policy on RoW and Towers	5
2	According accelerated RoW permissions	ROW cases (%) permission given within 60 days of the first application	7.5
3	To ensure implementation of RoW policy by all States	Availability of Centralized IT Portal for ROW clearances across all Government land and building owning authorities	2.5
4	To adopt National Building Code and provide cabling and in-building solutions in all office and residential complexes	Adoption of National Building Code 2016	5
5	To connect towers with optical fibre cable (OFC)	% of Mobile Towers with fibre	5
6	Fibre First Initiative: Network Infrastructure	No. of Fibre kms/sq. KM or per capita or per 100 households (HH)	5
7	Fibre First Initiative: Connecting public places	% of no. of important public institutions/offices connected by fibre (Hospitals including PHCs, Police stations, Schools and CSCs)	5
8	Enabling Infra: Power	% of Towers/sites with defined grid supply to (Duration: Urban 20 h; Rural 12 h)	7.5
9	Setting up of one mn public Wi-Fi access points	Public Wi-Fi Access Points (Penetration per capita)	7.5
10	ICT Development Index	Households (HH) (%) using computer/laptop with internet connection	7.5
11	Broadband availability	Households HH (%) with Fixed broadband connection	5
12	Internet users	Internet users as % of population	7.5
13	Mobile penetration	Individuals who own a mobile phone	7.5
14	Smart Phone penetration (an indication of BB use)	No. of Smart Phones/100 population	7.5
15	Digitally Enabled Households	Percentage of households with at least one member digitally literate	7.5
17	Mobile BB Coverage	Number of BTS per capita (3G & 4G)	7.5

4 Results and Conclusion

Telecom tower complete details, viz. type of BTS, radiated power levels, EMF compliance, etc. are obtained through Tarang Sanchar portal (see Fig. 3). The portal also provides information about optical fibre connectivity, grid supply availability to telecom towers, etc. With the help of data collected from the States and Tarang Sanchar portal, the Digital Readiness Index is measured for each individual State. For example, the current status of RoW policy implementation in the country is shown in Fig. 4.

The rapid changes in digital technologies which bring digital transformation in a country, if managed effectively, will bring good results to the countries in terms of development of infrastructure which in turn bring its benefits to the general public. In this work, the framework of digital readiness index of a country is explained. Further, the DRI framework is extended to State level and parameters to measure DRI of a State along with its implementation is given in this work. The development of Digital Readiness Index for States will attract investors and lead to industry promotion. This will bring competitiveness among the States and help them to align their policies industry friendly and take steps for the overall development of the State to make it digitally ready. E-Governance activities of the government along with measurement

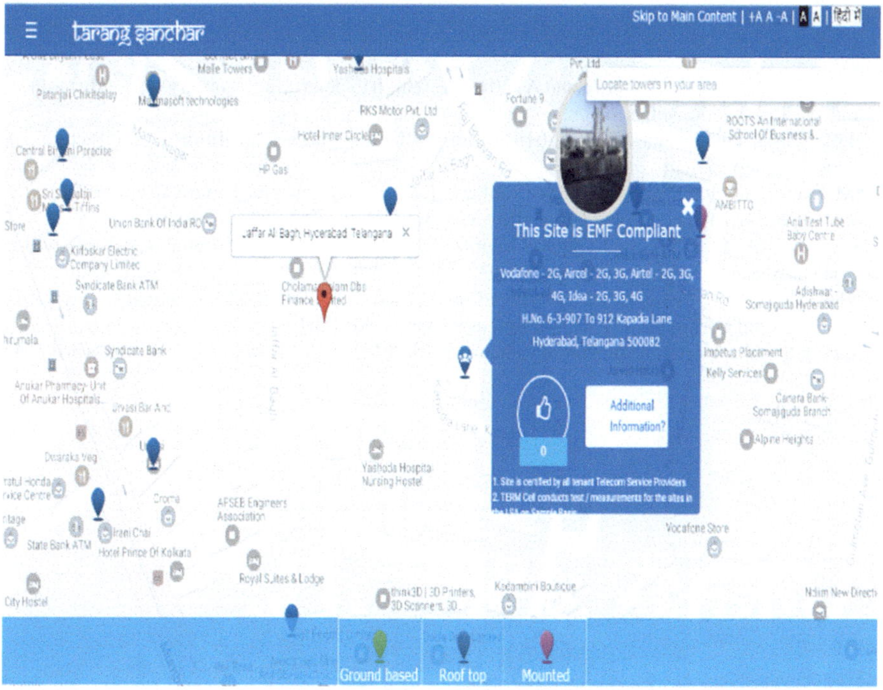

Fig. 3 Information about telecom towers in Tarang Sanchar portal

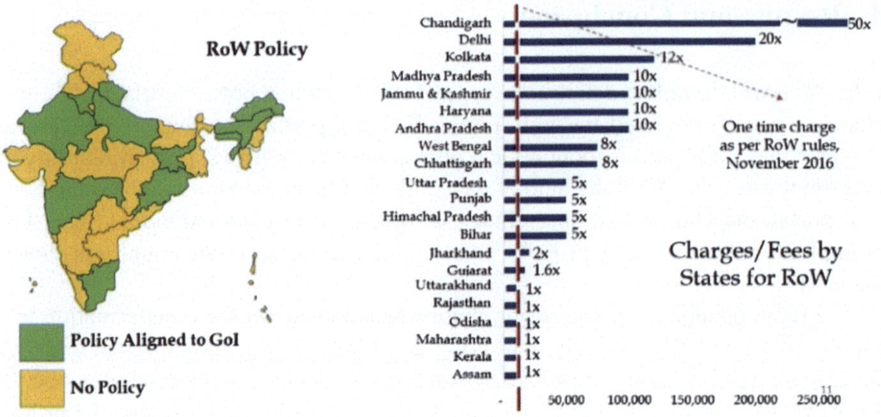

Fig. 4 Status w.r.to RoW policy implementation

of Digital Readiness Index of the States will empower the citizens and also encourage them to be digitally literate and contribute towards making the nation 'Digital India.

References

1. Yoo T, De Wysocki M, Cumberland A (2018) Research to determine a country's digital readiness and key interventions. Cisco
2. Baller Silja, Di Battista Attilio (2016) Soumitra Dutta. The Global Information Technology Report, Bruno Lanvin
3. IMD World Digital Competitiveness Ranking, IMD World Competitiveness Centre (2018)
4. Schwab K (2018) The Global Competitiveness Report. World Economic Forum
5. Sanchez MA, Ines Zuntini J (2018) Organizational readiness for the digital transformation: a case study research. https://doi.org/10.20397/21776652/2018.v18i2.1316 ResearchGate
6. NDCP, http://dot.gov.in/sites/default/files/Final%20NDCP-2018_0.pdf. Accessed 30 April 2019
7. ITU Regional Forum on ICT Measurement Report, UAE (2016)
8. Tarang Sanchar portal, https://tarangsanchar.gov.in/EMFPortal/Home. Accessed 15 April 2019

Real-Time Lane Detection for Autonomous Vehicle Using Video Processing

Chinmay Hasabnis, Sanjay Dhaygude and Sachin Ruikar

Abstract This paper presents the development of lane detection techniques using video processing which works in real time on highway and various similar roads. The Canny edge detection algorithm is used for detection of edges of lanes on the road which provides robust output. Hysteresis thresholding is used to identify the pixel that belongs to real edge pixels in the captured image. To get clear identification of lane markings Hough transform is used on respective edges.

Keywords Lane detection · Autonomous · Driving

1 Introduction

Nowadays road accidents are a major issue in the developed world. Recent studies show that one of the major reasons for accidents is quick change in lane on a fast driving road, as well as changes in the speed of vehicle [1]. In the past some years, there is formidable increase in the electronic products in machines used for transportation such as vehicle, where perceptive driver assistance plays an important role. The various electronic systems such as advanced driver assistance system have played an important role towards the intelligent assistance system for driver of vehicle [2]. In an intelligent world, a driving assistance system is used nowadays for unmanned vehicles. The new technology provides assistance to the vehicle as an automatic control unit in the unmanned vehicle.

The next-generation cars are being introduced with so many advanced electronics systems so as to make journey towards autonomous driving [3]. This autonomous feature reduces the risk of accidents, improves safety and performance for drivers. With rapidly changing computer and information processing technology, vehicle is finding a way towards autonomous driving capability. Vehicle automation used to reduce the chance of accidents, increase safety and capacity, reduce fuel consumption and increase comfort and performance for drivers [4]. More automated automobiles relieve the driver from many undesirable routines of the driving task.

C. Hasabnis (✉) · S. Dhaygude · S. Ruikar
Walchand College of Engineering, Sangli, India
e-mail: chinmayhasabnis@gmail.com

S. Fong et al. (eds.), *ICT Analysis and Applications*, Lecture Notes in Networks and Systems 93, https://doi.org/10.1007/978-981-15-0630-7_21

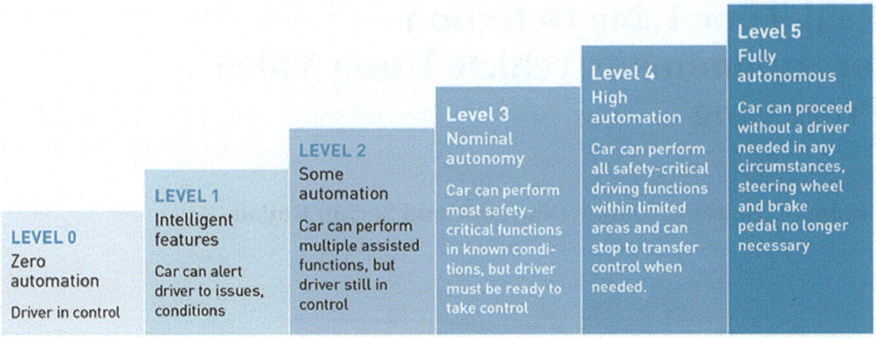

Fig. 1 Levels of autonomous vehicles (*Source* https://www.google.com/5 levels of car autonomy)

It is essential to develop systems that can assist the driver while navigating on the road. Lane detection system can immensely help in this context so as to follow certain predefined lane for particular speed. As shown in Fig. 1, vehicles are being developed from level 1 to level 5 for autonomous driving. Level 1 of autonomy is advanced driver assistance system [5]. It consists of features like cruise control and lane detection using edge detection algorithm.

The lane detection algorithm works on a series of images of the road conditions from the input video sequence. Then lane markings are highlighted on each image and finally output video is created from processed frames. Python IDE, OpenCV platform is used as convenient tool for image processing.

2 Literature Survey

Wu et al. [1] have proposed a lane detection system with the help of escape warning system based on an ultralow complexity block. In this the lane markings are distributed based on the area which is vicinity to the vehicle to determine the bounded region of interest (ROI).

Nguyen et al. [2] have proposed a method for lanes and vehicles detection simultaneously. The proposed method combines both lane detection algorithm and vehicle detection algorithm. Kodeeswari et al. [3] have proposed to identify the lane lines using image processing tools on the hilly road in transform domain. It shows that the Hough transform is suitable for lane detection. It works on the real-time video captured by camera with MATLAB simulation tool to obtain the marking position of lane. The system consists of an algorithm to detect the lane lines available on the road at specific distance. He et al. [4] proposed a prototype for Lane Departure Warning System using lane detection. The Canny edge detector algorithm is used to identify edge on the road and Hough transform is used to detect the lane markings accurately on the road. The region of interest (ROI) is obtained in real time, to further reduce the noise and to increase the accuracy. It is also useful to improve the execution period.

This lane detection study provides the extraction of lane information from real-time road images in real time efficiently and very accurately. Somasundaram et al. [5] presented a prototype of a robust lane detection method. It consists of the human visual properties such as far–near adaptation, lateral inhibition which is joined the mutual support to extract the feature. Chee and Lau [6] developed a prototype of Lane Departure Warning System on Raspberry Pi which consists of (1) a camera module, (2) a processing module and (3) an alarm module. The implemented system, on the road, are able to identify lane of the road base on the road mark (processing and camera module), and alert (warning module) to the user as the driver, at the time of departure of vehicle from the lane, in real time. Li et al. [7] have developed lane detection technique using two-stage Hough transform (HT) algorithm. Experimental tests done with a series of optimization in programming shows that the proposed algorithm gives better performance despite severe noise environment.

3 Methodology

Step 1: RGB to Grayscale conversion of image:
Input taken from a camera mounted on vehicle is used to produce frames of images. Then these images are sent for RGB to Grayscale conversion. It converts RGB image data to Grayscale data with the mathematical expression consists of the weighted sum of R, G and B components values.

$$\text{Gray} = 0.30 * \text{R} + 0.59 * \text{G} + 0.11 * \text{B} \tag{1}$$

Step 2: Filtering Technique:
Gaussian blur filter is mostly preferred to reduce the noise available in the images at the time of acquiring or any other noise. Each pixel of the images is transformed into a new pixel using Gaussian function as a process of the image smoothing technique [6] (Fig. 2).

Step 3: Edge Detection:
In edge detection technique the edges obtained with their discontinuities present in the image. The Laplacian, Sobel, Robert, Prewitt and Canny are the various edge detection technique [7]. In uses a Canny edge detection technique to detect the edges. This edge detection technique extracts the information from the present objects as well as it reduces time to process data in the image [8].

The procedure for Canny edge detection is described below.

(a) *Noise Reduction*

The edge detection technique is more prone to noise in the image. The noise is removed in the image using Gaussian filtering technique. It also helps to remove high-frequency information from the image. Therefore Gaussian filter also known as a low pass filter is used to remove high-frequency information as a noise [9].

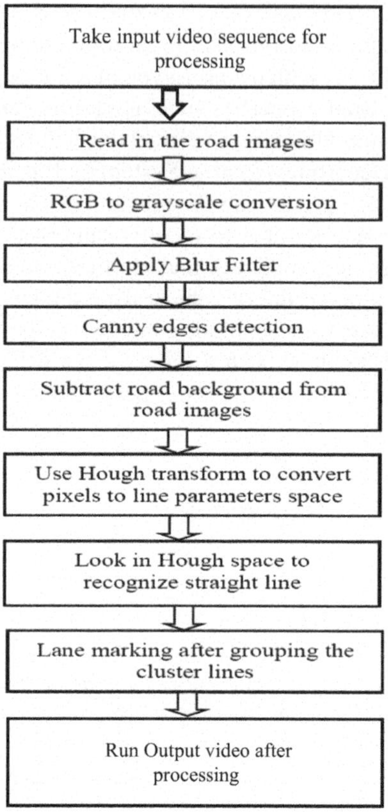

Fig. 2 Flowchart of lane detection algorithm

Gaussian kernel is given by

$$G_\sigma = \frac{1}{2\pi\sigma^2} e^{-\frac{(x^2+y^2)}{2\sigma^2}} \tag{2}$$

where x and y are the positions in the coordinates of the pixel value of the kernel according to the centre of the kernel

(b) *Finding Intensity Gradient of the Image*

After application of the noise reduction technique, image is smoothened. It is further filtered using Sobel kernel in each direction. The horizontal and vertical directions are used to get the first derivative in horizontal direction (Gx) and vertical direction (Gy), respectively.

It is helpful to obtain gradient of edge and provides the direction for each pixel as below (Fig. 3).

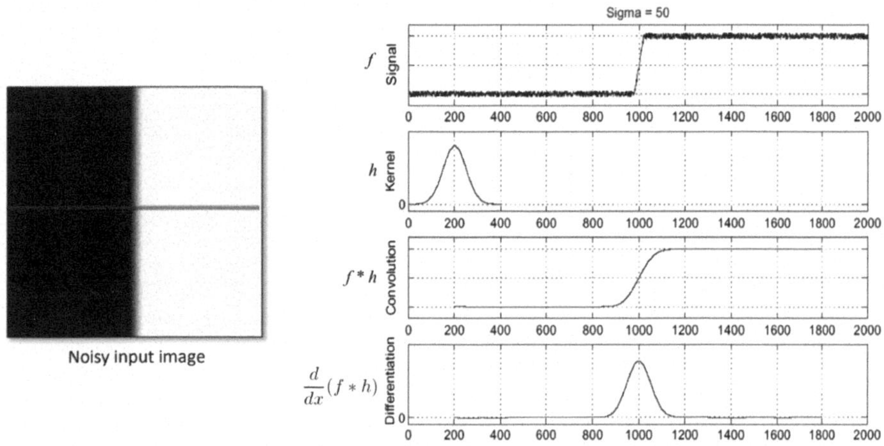

Fig. 3 Intensity gradient of the image (*Source* https://www.google.com/Canny edge detection)

To find edges, look for peaks in

$$\frac{d}{dx}(f * h) \tag{3}$$

Finding intensity gradient of the image given by

$$\nabla f = \begin{bmatrix} g_x \\ g_y \end{bmatrix} = \begin{bmatrix} \frac{\partial f}{\partial x} \\ \frac{\partial f}{\partial y} \end{bmatrix} \tag{4}$$

where
$\frac{\partial f}{\partial x}$ is the gradient in the x direction, i.e. horizontal
$\frac{\partial f}{\partial y}$ is the gradient in the y direction, i.e. vertical
The gradient direction obtained using

$$\theta = \tan^{-1}\left[\frac{g_y}{g_x}\right] \tag{5}$$

and the magnitude is given by:

$$\sqrt{g_y^2 + g_x^2} \tag{6}$$

(c) *Non maximum Suppression*

The non-maximal suppression technique uses gradient method to suppress any unwanted pixels in an image since they do not contribute to the part of the edge [10]. Each pixel is rigorously checked in the direction of gradient to obtain result as a part of a local maximum with their neighbourhood (Fig. 4).

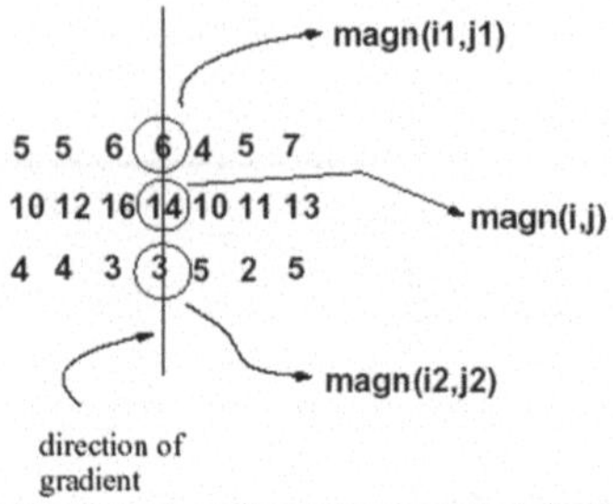

For each pixel (x,y) do:

if $magn(i, j) < magn(i_1, j_1)$ or $magn(i, j) < magn(i_2, j_2)$
  then $I_N(i, j) = 0$
else $I_N(i, j) = magn(i, j)$

Fig. 4 Non maximum suppression

(d) *Hysteresis Thresholding*

This stage is used to find real edges. This works on two threshold function as minVal and maxVal treated as the external threshold. The intensity gradient of edges having higher value than the maxVal are considered as part of the edges and if the value is minimum than the minVal are supposed to be non-edges part of the image [11]. The points identified between these two thresholds values which is useful to classify them as edges or non-edges part of the image depending on their connectivity. The pixels connected which is part of the edge pixels to obtain the edges detected output, else they are removed

Step 4: Haugh Transform:
Hough transform is a convenient tool to identify lines and shape of an object [12, 13]. Edge detectors like Canny edge detector or Sobel edge detector defines edges of an object. Hough transform is used to fill pixels between edges to produce shape of object. Hough transform is used to efficiently reconstruct the lines onto an output video stream by using Canny edge detector [14].

Step 5: Reconstruct final video output:
This process carried out on edge detection produces output video by combining all frames which are processed with the above algorithm.

4 Results

Samples of images from input video captured at rate 30 frames per second are shown below (Figs. 5, 6 and 7).

For each frame in video, Canny edge detection algorithm is applied.

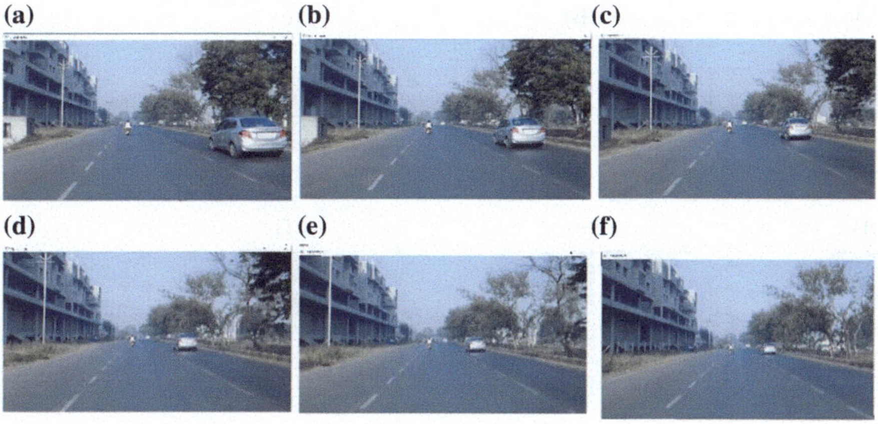

Fig. 5 Input frames from video for processing algorithm

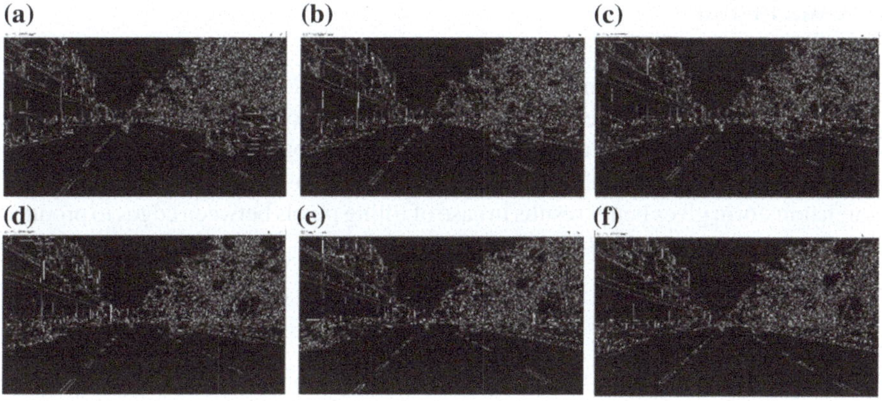

Fig. 6 Edge detection algorithm on input frames

Output images produced by applying edge detection on input samples at 30 frames per second are shown below.

5 Merits Over Existing System

1. Lane detection using a camera installed in the vehicle reduces amount of stress on driver while driving on various road conditions. So, lane detection can be used in lane departure warning in advanced driver assistance systems.
2. Lane detection improves the performance and safety of driver while continuously monitoring lanes on-road and thus helps to keep vehicle in certain defined lane.

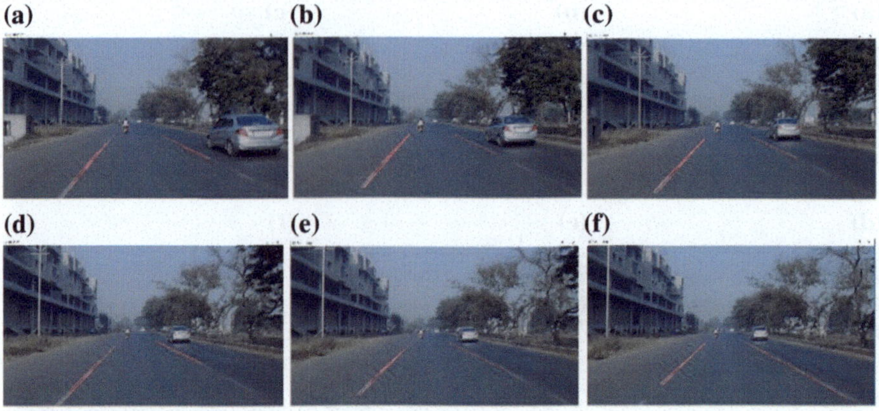

Fig. 7 Output frames after edge detection algorithm

6 Conclusion

This lane detection algorithm works in real time on a video of 30 frames per second. The edge detection technique used is Canny edge detection which gives better results for parameters like signal to noise ratio as compared to other operators. It has higher probability of detecting real edges, thus avoiding non-edge pixels in the output. Hough transform gives better results in case of filling pixels between edges to produce smoothened perfect shape of detected object.

References

1. Chung-Bin W, Wang L-H, Wang K-C (2019) Ultra-Low Complexity Block-Based Lane Detection and Departure Warning System. IEEE Trans Circuit Syst Video Technol 29(2) 582–593
2. Nguyen Van Q, Seo C (2017) A study on detection method of vehicle based on lane detection for a driver assistance system using a camera on highway. In: 2017 11th Asian Control Conference (ASCC) Gold Coast Convention Centre, Australia December 17–20, 2017
3. Kodeeswari M, Daniel P (2017) Lane line detection in real time based on morphological operations for driver assistance system. In: International conference on signal processing computing and control, pp 316–320
4. He J, Rong H, Gong J, Huang W (2010) A lane detection method for lane departure warning system. In: International conference on optoelectronics and image processing
5. Somasundaram G, Kavitha (2011) Lane change detection and tracking for a safe-lane approach. In: Real time vision based navigation systems , CCSEA 2011, pp 345–361
6. Chee W, Lau PY (2017) Implementing lane departure warning system on raspberry Pi. Int J Eng Res Technol 5(06)
7. li H, Zhang Z, Zhao X, He M (2017) Two-stage hough transform algorithm for lane detection system based on TMS320DM6437. In: IEEE international conference on imaging systems and techniques (IST), Beijing, pp 1–5

8. Jahan R, Suman P, Singh DK (2018) Lane detection using canny edge detection and hough transform on raspberry Pi. Int J Adv Res Comput Sci 9(2)
9. Samuel M, Mohamad M (2018) Development of edge-based lane detection algorithm using image processing. Int J Inf Visual 2(1)
10. Akhtar S, Rajayalakshmi D, Sattar SA, A theoretical survey for edge detection techniques and watershed transformation. Int J Comput Technol Electron Eng (IJCTEE) 2(1)
11. Canny J (1986) A computational approach to edge detection. IEEE Trans Pattern Anal Mach Intell Pami-8(6)
12. Kreucher C, Lakshmanan S (1999) Lane extraction algorithm that uses frequency domain features. IEEE Trans Robot Autom 15(2)
13. Kreucher C, Lakshmanan S, Kluge K (1998) A driver warning system based on the LOIS lane detection algorithm. In: Proceedings of IEEE international conference on intelligent vehicles, pp 17–22. Stuttgart, Germany
14. Lei G, Jianqiang W, Keqiang L (2006) Lane keeping system based on THASV-II platform. In: IEEE international conference on vehicular electronics and safety, pp 305–308
15. Borkar A, Hayes M (2009) Robust lane detection and tracking with Ransac And Kalman filter. In: International conference on image processing, pp 3261–3264
16. Cualain DO, Hughes C, Glavin M, Jones E (2012) Automotive standards-grade lane departure warning system. IET Intell Transp Syst 6(1):44–57
17. Felguera-Martín D, González-Partida J-T, Almorox-González P, Burgos-García M (2012) Vehicular traffic surveillance and road lane detection using radar interferometry. IEEE Trans Veh Technol 61(3)
18. Lee JW, Kee C-D, Yi UK (2003) A new approach for lane departure identification. In: Intelligent vehicles symposium, 2003 proceedings, IEEE, pp 100–105. IEEE
19. Hsiao P-Y, Yeh C-W, Huang S-S, Fu L-C (2009) A portable vision-based real-time lane departure warning system: day and night. IEEE Trans Veh Technol 58(4):2089–2094
20. Kim Z (2006) Realtime lane tracking of curved local road. In: Proceedings of the IEEE ITSC IEEE intelligent transportation systems conference Toronto, Canada, September-20

Next-Generation Communication Networks: Wired or Wireless—A Big Question!

Meenakshi Malhotra and Inderdeep Kaur Aulakh

Abstract Network is called as a good network if the packet/data delivery is on time or delay rate is as much as it can be tolerated but if the delay rate is high then the network resources are being affected. The demand for a good network nowadays has increased which is making the need for proper communication mechanism a great pressure. The challenge is to choose a network that will not only satisfy the customer need but also channel reliability and data security. Depending upon some features like bandwidth and services, the wired and wireless communication does the same work but if we talk about the wired communication it will show us some other factors and talking about wireless communication leads us to think about the security. This paper focused on some key elements which can be useful to choose a proper network for communication in the aspect of security, communication medium, real-time application, etc.

Keywords Communication · Channel · Medium · Networks · Security · Technology · Wired · Wireless · Sender · Receiver

1 Introduction

Digital technology depends upon the demand for network resources for fast sending and receiving of the packets over the network. Wired technology uses a physical medium to start communication between sender and receiver. The technology works well until any type of medium or cable failure does not occur during communication. So to make communication easy, the maintenance of the cable requirement is high. Wireless technology on the other hand does not rely on wires or cables, instead it depends on the waves or signals. With such technology the communication is done on short distance or can also cover large areas. The communication can occur between more than two points. The points can be mobile station, satellite, remote

M. Malhotra (✉) · I. K. Aulakh
University Institute of Engineering and Technology Panjab University, Chandigarh, India
e-mail: meenakshi.malhotra@outlook.com

I. K. Aulakh
e-mail: ikaulakh@yahoo.com

S. Fong et al. (eds.), *ICT Analysis and Applications*, Lecture Notes in Networks and Systems 93, https://doi.org/10.1007/978-981-15-0630-7_22

control, wireless mouse, or keyboards, etc. To establish a communication in wireless environment one (Mobile Station) should communicate to base station that base station should be connected to MSC and that MSC will be communicating to other MSC for communication.

Home network can be established after considering wired and wireless networks. For the wired home network, Ethernet cables are needed for communication between devices and in wireless home network, two or more devices can communicate with each other with the help of the router connectivity like at home you can connect printers, phone, computer, laptop at the same time within the same network.

Wired communication has its own advantages but at some stages, the wireless communication can be used as an alternative to that of wired technology.

The remaining paper is presented as Sect. 2 described the communication medium used by various authors for the research. Section 3 covered the security issues of wireless communication. Section 4 will be focusing on the key benefit of using the wired plus wireless communication both on same time and the Sect. 5 covered the conclusion part of the paper.

2 Communication Medium

Transmission medium is used to transfer information from source to destination in the network. Wired communication uses twisted pair cable, coaxial cable, and fiber optic cable for the communication establishment between the two or more than two entities.

Twisted Pair Cable: It is made through twisting a pair of copper cables. The idea behind making a twist is to reduce cross-talk effect and vulnerability.

In [1] four models were used. Two design models were followed by forward and backward model. The results improved the working cable. Manufacturing industries can use this for electrical cable design. In [2], the purpose of the authors was to eliminate the wiring harnesses. Authors did this by making some changes in features of wire—like height, length, and pitch. As a result by changing the features, more segments were used for purposed work but the time taken to complete the work increased.

Coaxial cable: In Coaxial cable, two conductors are placed in parallel. The copper is there as a central conductor. To differentiate between inner and outer conductor, nonconductive material is placed between them.

In [3], authors' purpose was to identify the position of the person working under Tunnel near the leaky coaxial cable. Position Detection Scheme was used and the achieved outcomes can be applied in Wireless LAN and department of Fire but someone in the range of more than 100 m has the possibility of being ignored in the measurement. In [4], authors used Framework-Data Over Cable Service Interface Specification (DOCSIS) for fine measurement to measure the mismatches occur at the time of communication in coaxial cable. Good measurement can be achieved even if micro-reflector is present but with the increase in micro-reflector, the values may get affected.

Fiber Optics: Light uses fiber optic medium to be transferred from one place to another. The glass stand is used as a center of the cable. The light travels from this glass through reflective cladding which makes the transmission effective and there will be no chance of escaping the light.

In [5], the authors aimed to identify the advantages associated with Optical over RF. And as a result, they stated that wireless optical findings give huge bandwidth, no EMI, and better transceiver in terms of cost. In [6], problems related to high speed and better communication between the channels and power consumption were considered. To do the same authors used multiplexing space division, multiplexing technology and multicore fiber which can be used in real-time application for Next Gen communication.

Wireless communication uses RF and IF for the communication establishment. The signals in wireless communication passes through air. Wireless communication gives the advantage to communicate wirelessly and it also increases the chances of interference, bandwidth, and throughput issues and most importantly the increment in the number of devices increases the channel capacity problem.

Radio Frequency: For wireless communication, RF makes use of antenna and transmitter. RF is the measurement of the oscillations produced at the time of communication. When a radio wave is transmitted, RF is measured in terms of the number of cycles per second. Waves are not visible with the human eye.

In [7], the authors focused on better communication between multiple radios without delay and interference. For this, authors used multiple access scheme (as multiple radios are in communication), Combination of wired–wireless environment with optical orthogonal code were considered. As a result, with no time limitation with same frequency, communication between various channels took place but the increase in data rate, channel may get affected.

In [8], authors improved the performance and computation time of the communication system by considering:

1. For Audio communication—Digital Signal Processor
2. Modulation Technique—Binary Phase Shift Keying or Direct Sequence Spread Spectrum
3. Alternative benefits of using RF with the help of DSP.

And two outcomes were identified 1. Data and signal were transmitted over channel effectively

2. Communication made secure with a modulation technique

Infrared Frequency: As a comparison to the Radio waves, Infrared is shorter but longer than the visible light. These frequencies cannot be seen with the human eyes but can be sensed as heat.

In [9], the authors focused on audio communication over infrared medium. Authors first send the IR emitting LED receiver in photodiode, and intensity modulation and direct detection as modulation technique which stated that intensity of infrared light is dependent upon range and coverage area of modulated signal.

In [10], the authors focused on low cost, lightweight and better data rates between two stations. Microcontroller for signal generation and MATLAB for simulation were considered which in term stated that fast access of data within indoor application but data will get affected if the distance between the stations gets increased.

3 Security of the Network

Network security is that part of the network which not only secures the data but also the communication taking place in the network. In wired technology, the security can be broken physically when one unauthentic person makes some unwanted activity into the communicating systems. In wireless communication as the communication takes place with the help of signal transmitting through the air so there is huge chance for the attacker to attack the network or steal the user's private information. So there is a high demand for the security of the wireless networks.

S. No.	Authors	Year	Purpose	Major finding	Outcome
1	Agan et al. [11]	2005	DDoS Attack Prevention	1. Auction Theory 2. Attack Defense Problem was calculated	1. Less packet drop 2. Packet delivery in more reliable way
2	Ramesh et al. [12]	2012	Prevention from DoS Attack and Passive information gathering	1. Symmetric key algorithm comparison with Neural Network 2. Threshold value for comparison	Less computational cost with less memory need and less time to execute process
3	Roy and Nene [13]	2016	Node Replication Attack	1 Signal strength indicator 2. Quality Indicator 3. Packet Sequence number	High probability of detection
4	Shimpi and Shrivastava [14]	2016	Replication Attack	Adhoc on Demand Distance Vector Protocol for Sensing the atmosphere	Increase throughput, packet delivery ratio and end to end delivery of packets

(continued)

(continued)

S. No.	Authors	Year	Purpose	Major finding	Outcome
5	Bhuvaneswari and Ramachandran [15]	2017	1. Prevention from DoS attack 2. Prevention from Black hole, Worm hole, and Grey hole Attack	1. OLSR as effective routing protocol 2. Elliptical curve cryptography for generating keys	Better throughput and packet delivery ratio but no discussion about packet drop ratio as well as average delay in networks
6	Choi et al. [16]	2017	Spoofing Attack	Dynamic random forest algorithm with wireless intrusion detection system	Better result in the presence of noise
7	Nagar et al. [17]	2017	DDos Attack	Intrusion detection scheme for passing and blocking of nodes	Real knowledge of attackers and nodes to the station

4 Combination of Wired and Wireless Technology

After going through some papers it is noticed that nowadays combination of wired and wireless technology is taking place to improve communication better without any hurdle. The combination of both the technology can be used in real-time applications.

S. No	Authors	Year	Application area	Major finding	Outcome
1	Chen Yu et al. [18]	2008	Improvement in time taken to measure temperature in **Hospital**	1.Sensor DS18B20 were used for signal collection 2. SCM AT89C52 for measurement and alarming process 3. nRF905 makes system central system	It can be applied in greenhouse atmosphere as well as in intelligent monitoring
2	Alippi and Sportiello [19]	2009	Hybrid technology for **better sensing**	1. Sensors were considered as clusters and further divided into sub clusters 2. Microcontroller with embedded Analog-to-Digital Convertor 3. Radio module as gateway	1. Increase in Bandwidth 2. Effective clock synchronization 3. Power Saving or less power consumption

(continued)

(continued)

S. No	Authors	Year	Application area	Major finding	Outcome
3	Kobayashi et al. [20]	2010	**Broadband Designing** Companies	1. MMIC design for front end applications 2. Combination of Technologies like E-Mode PHEMT and GaN HEMT technology	1. Low noise measurement 2. High data rate 3. Better Communication
4	Srovnal and Machacek [21]	2010	Better data transfer between wired and wireless node in **Industry**	1. Mathematical model 2. Parallel Signal Processing 3. WiFi and Ethernet was considered	Less time delay for receiving the packets and better routing data
5	Kang et al. [22]	2011	**Improving service** in 16QAM-OFDM for wireless and wired communication	1. Mach–Zehnder module 2. Full duplex Communication 3. Wavelength reuse technique 4. Modulation-Optical Carrier Suppression (OCS)	Performance Measurement of Upstream and downstream signal over communication observed quite improved
6	Rauch and Weeks [23]	2011	**Food-related** manufacturing company	Open-ended, Semi-structured questions under qualitative study	1.Considering feedback from respondent the various factors were highlighted for the improvement like reliability, system failure 2. Difficulty to choose a network when the number increased in large organization
7	Krishna et al. [24]	2017	Utilization of water and Energy to make campus **Smart Campus**	Learning Automata-based architecture for utility management	1. Control over the natural resources 2. Low cost implementation 3. No wired direction and speed was considered

(continued)

(continued)

S. No	Authors	Year	Application area	Major finding	Outcome
8	Kawanishi et al. [25]	2018	Linking radio & optical signal **for 5G network**	1. Radio over Fiber technique (RoF) 2. Free Space Optical (FSO) 3. Sensor over Fiber (SoF)	1. Comfortable for linking wired and Wireless technology for 5G communication or above it

5 Conclusion

The communication whether it is wired, wireless or a combination of both should be secure and fast enough. In this paper, we discussed many factors related to wired technology and wireless technology. Wired technology is secure as it always relies on the wires and the short range is considered for the communication but if we talk about wireless technology, the security-related issues are high in the network. Nowadays, people are using a combination of wired and wireless technology, which is overcoming the disadvantages associated with both communication. After going through a lot of literature, it can be said that no network is a perfect one until or unless we take precautions to make the network.

Acknowledgements The research work is supported by Technical Education Quality Improvement Project III (TEQIP III) of MHRD, Government of India assisted by World Bank under Grant Number P154523 and sanctioned to UIET, Panjab University, Chandigarh (India).

References

1. Lago A, Penalver CM, Marcos J, Doval-Gandoy J, Melendez AAN, Lopez O, Freijedo FD, Vilas JM, Lopez JCL (2011) Electrical design automation of a twisted pair to optimize the manufacturing process. IEEE Trans Compon Packag Manuf Technol 1(8):1269–12681
2. Yamamura M, Kami Y, Murano K, Xiao F (2015) Analysis of transmission characteristics for twisted pair cables using the RLGC parameters of the cable. In: IEEE Asia-pacific symposium on electromagnetic compatibility (APEMC), pp 720–23
3. Nakamura M, Takagi H, Terashima J, Einaga K, Nishikawa T, Moriyama N, Wasaki K (2010) Development of a simple multiple-position identifying system with a long range multiband leaky coaxial cable for rescue operations in tunnels or passages in underground facilities. In: IEEE Asia-pacific microwave conference, pp 163–66
4. Cazarotto M, da Rocha Lopes R (2017) Characterising micro-reflections in coaxial cable network. IET Sci Measur Technol 11(8):1043–1048
5. Mukherjee M (2016) Wireless communication-moving from RF to optical. In: 2016 IEEE 3rd international conference on computing for sustainable global development, pp 788–95
6. Nakajima K, Matsui T, Saito K, Sakamoto T, Araki N (2017) Multi-core fiber technology: next generation optical communication strategy. IEEE Commun Stand Mag 1(3):38–45

7. Rezaei HF, Kruger A (2013) Wireless multichannel bus communication using CDMA coding for single frequency radios. In: IEEE sensors applications symposium proceedings, pp 1–6
8. Shelke AL, Kolhare NR, Sarvadnya RV, Kangane VA (2016) Wireless RF communication based on DSP. In: IEEE 3rd international conference on devices, circuits and systems, pp 109–114
9. Korian A, Sharma S, Mittal VK (2016) Wireless audio communication over Infra-Red medium. In: 2016 IEEE 3rd international conference on signal processing and integrated networks, pp 240–45
10. Shahriar A, Chakraborty M, Hossain S, Halder D, Chowdhury NB (2014) Wireless infrared communication between two computers by MATLAB. In: 9th international forum on strategic technology, pp 60–64
11. Agah A, Basu K, Das SK, (2005) Enforcing security for prevention of DoS attack in wireless sensor networks using economical modeling. In: IEEE international conference on mobile Adhoc and sensor systems conference, 8-p
12. Ramesh MV, Raj AB, Hemalatha T (2012) Wireless sensor network security: real-time detection and prevention of attacks. In: IEEE fourth international conference on computational intelligence and communication networks, pp 783–87
13. Roy S, Nene MJ (2016) Prevention of node replication in wireless sensor network using received signal strength indicator, link quality indicator and packet sequence number. In: IEEE online international conference on green engineering and technologies, pp 1–8
14. Shimpi B, Shrivastava S (2016) A modified algorithm and protocol for replication attack and prevention for wireless sensor networks. In: IEEE international conference on ICT in business industry & government, pp 1–5
15. Bhuvaneswari R, Ramachandran R (2017) Prevention of Denial of Service (DoS) attack in OLSR protocol using fictitious nodes and ECC algorithm. In: IEEE international conference on algorithms, methodology, models and applications in emerging technologies, pp 1–5
16. Choi SH, Hwang DH, Choi YH (2017) Wireless intrusion prevention system using dynamic random forest against wireless MAC spoofing attack. In: IEEE conference on dependable and secure computing, pp 131–37
17. Nagar S, Rajput SS, Gupta AK, Trivedi MC (2017) Secure routing against DDoS attack in wireless sensor network. In: IEEE 3rd international conference on computational intelligence & communication technology (CICT), pp 1–6
18. Chen Y, Zhang H, Wang N, (2008) Body temperature monitor and alarm system used in hospital based on 1-wire and wireless communication technology. In: IEEE international workshop on ETT & GRS, Vol 1, pp 401–04
19. Alippi C, Sportiello L (2009) Energy-aware wireless-wired communications in sensor networks. In: SENSORS, IEEE, pp 83–88
20. Kobayashi KW, Krishnamurthy K, Vetury R, McCleary Y, Hillermeier R, Shealy JB (2010) Broadband design techniques and technology for future wireless and wire-line applications. In: IEEE Asia-pacific microwave conference , pp 924–31
21. Srovnal V, Machacek Z (2010) Industrial communication network for wire and wireless data transfer. In: IEEE international conference on applied electronics, pp 1–4
22. Kang Z, Zou N, Wang D, Liu J, Gao Y, Li P (2011) Seamless amalgamation of full duplex 2.5 Gbps wireless and 10Gbps wired optical access networks based on QAM-OFDM technology, pp 922–925
23. Rauch M, Weeks R (2011) Utilisation of wireless versus wired access technologies in the South African food-related manufacturing industry. In: IEEE proceedings of PICMET'11: technology management in the energy smart world, pp 1–7
24. Krishna PV, Obaidat MS, Misra S, Yenduri S, Hisao KF (2017) Designing and prototyping utility management using hybrid wireless-wired network technologies. In: IEEE international conference on platform technology and service, pp 1–7
25. Kawanishi T, Kanno A, Freire HS (2018) Wired and wireless links to bridge networks: seamlessly connecting radio and optical technologies for 5G networks. IEEE Microwave Mag 19(3):102–111

SYNC—Short, Yet Novel Concise Natural Language Description: Generating a Short Story Sequence of Album Images Using Multimodal Network

M. S. Karthika Devi, Shahin Fathima and R. Baskaran

Abstract Image captioning, which aims at generating automated descriptions for an image, is the large focus in current research while most of the previous works have dealt with the association between the single image and single sentences. This paper proposes to take one step further to investigate the summarized version of the narrative description for the image stream and in more generalized form for a normal user. The major challenge in the proposed work is to consider the visual variance in an ordered image collection and in preserving coherence relation among multiple sentences. Our proposed work is aimed to retrieve a coherent flow of multiple sentences that use multimodal neural architecture and ranking-based summarization to generate the summarized description of possibly larger image streams. With qualitative evaluation, the proposed work has attained significant performance improvement over traditional state-of-the-art method for text sequence generation and has captured the relevant context with syntactic meaning with respect to summarized version of the detailed descriptions.

Keywords Image captioning · Bidirectional recurrent neural networks · Convolutional neural networks · Coherence model · Summarization deep learning · Computer vision · Artificial intelligence

1 Introduction

A quick glance at an image is sufficient for a human to point out and describe an immense amount of details about the visual scene. However, this remarkable ability has proven to be an elusive task of content analysis for visual recognition models.

M. S. Karthika Devi · S. Fathima · R. Baskaran (✉)
Department of Computer Science and Engineering, College of Engineering, Anna University, Chennai, Tamil Nadu, India
e-mail: baaski@annauniv.edu

M. S. Karthika Devi
e-mail: karthikadevi88@gmail.com

S. Fathima
e-mail: shahinfathi1995@gmail.com

S. Fong et al. (eds.), *ICT Analysis and Applications*, Lecture Notes in Networks and Systems 93, https://doi.org/10.1007/978-981-15-0630-7_23

Automatic description of an image is a very challenging task as the system must capture not only what is contained in an image but also how the objects in an image are related to each other, what actions are involved in and how the changes occurred over one image to other are captured. This task is harder than tradition image or object recognition models, as it requires language model to express the semantic knowledge along with the visual understanding.

Visual contents are generally represented by images or videos which is associated with captions or tags as text sentences. This provides the way to learn the representation of multiple data types. The multimodal neural architecture that is used for implementation consists of convolutional neural networks for image descriptions, bidirectional recurrent neural network to represent the content flow of text sentences.

A key requirement for any machine translation system that produces natural language sentences is the coherence of its output. Coherence relations must able to capture the relatedness between the texts with respect to sentence transitions. Local coherence is necessary for global coherence that automatically abstracts a text into a set of entity transition sequences and records distributional, syntactic and referential information about discourse entities, i.e. entity grid representation of discourse, which captures pattern of entity distribution in a text.

General users usually take a large number of images and description of each of the images would possibly create detailed descriptions like paragraphs which could be too long for users. The summarized description must be short, able to preserve the coherent content flow and should be in accordance with the story sequences. However, the shortened story over machine-translated natural language sentences have not been explored in the earlier works.

This proposed work could have a great impact, for instance, helping novice users better understand the context of an image available in a web and can be used for various applications such as remembering the memories of life, recognition of action [1] over a series of events, digital storytelling and many such sequential or historical events through stream of images.

Organization of this paper is as follows. Section 2 presents a literature survey of the work. Section 3 presents the detailed design. Section 4 deals with experimental setup, and Sect. 5 presents the result analysis.

2 Related Works

In recent years, there has been a growing interest in exploring the relation between images and language. Simultaneous progress in the fields of Computer Vision (CV) and Natural Language Processing (NLP) [2–4] has led to impressive results in learning both image-to-text and text-to-image connections. Tasks such as automatic image captioning, image retrieval or image generation from sentences have shown good results.

Image Captioning: Image captioning method is paying drastic attention in the field of computer vision and machine learning community which aims to generate text sentences that describes an input image. Most popular approach of text generation is to retrieve best sentences by learning from training a system through embedding between images and text [3, 5]. This work involves retrieval of text from the learned model and generates story based on compatibility score between the language model and the coherence model.

Many earlier research attempts have exploited multimodal networks that combine deep Convolution Neural Networks (CNN) [6, 7] for image description and Recurrent Neural Networks (RNN) [6, 8–10] for language sequence modelling. There are many variants of combinations used as multimodal architecture includes CNN with bidirectional RNN [11], long-term recurrent convolution networks [6], long-short term memory networks [7]. This work takes an advantage of existing models with distinctive extension to multiple dimensions of input and output.

Huang et al. [12] introduced the first dataset of sequential images with corresponding description. They first collect storyable photo albums from Flickr and then outsourced to crowd workers using Amazon's Mechanical Turk (AMT) to collect the corresponding stories and descriptions.

Retrieval of Image: Most of the existing work involves retrieval of image by keyword for the structured queries. Few earlier works include image ranking and retrieval based on text sentences, multiple attributes and other data structured objects like graphs [13]. In [14], three different data types such as image, text and sketch are combined as a query for image retrieval. Lin et al. [15] proposed a method for video search using a text sentence as a query. Hu et al. [16] retrieve a natural language object that takes an image and associated texts, for each text query that corresponds to the image contouring. Kong et al. [17] took a scene and a sentence as an input to find the relatedness between its regions and text phrases. Similarly in [18], the correspondence between regions to phrases is computed. This work is distinct as it involves images sequences, instead of single image.

Entity-Based Approaches to Local Coherence: Substance based records of neighbourhood intelligence include a long custom inside the phonetic and subjective science writing. Element-based portrayal [19] of talk permits learning the properties of lucid writings from a corpus, without plan of action to manual explanation or a predefined information base. Entity-based theories capture coherence by characterizing the distribution of entities across discourse utterances, distinguishing between salient entities and the other texts.

Text Summarization: Text summarization is an approach that uses Natural Language Processing principles and algorithms to understand the larger text [20] and generate smaller and efficient summaries. Our approach uses certain linguistic elements [21] to identify the most relevant segments of a text and must be able to capture the syntactic and coherent flow of the generated narrative descriptions while reducing it to precise representation of the text.

3 Proposed Architecture

Image captioning technique applied over sequence of images requires learning coherent meaning and the summarization technique aimed at generating concise description of an image. This provides the way to investigate multimodal architecture which has been shown in Fig. 1.

Figure 1 majorly divided into four different components: Convolutional Neural Network (CNN) used for describing an image, Bidirectional Recurrent Neural Network (BRNN) for language modelling, the local coherence learning to capture the smooth flow of sentences, and rank-based summarization technique to produce the crisp story of an image sequences.

3.1 Text Descriptions

The text sentences associated with an image are represented in two ways: paragraph vector to represent the text features and parse tree to represent the grammatical roles of the text sentences.

3.2 Bidirectional Recurrent Neural Network

The BRNN model is used to represent a content flow of text sequences. This bidirectional model helps to consider the previous and next text sentences while modelling forward and backward processing.

Initialize the weights $W_i^{f_c} \psi W_i^{b_c} \psi W_{f^c} \psi W_{b^c} \psi W_{o\wp\psi}$ and bias $\Leftarrow\leftarrow$ $b_i^{f_c} \psi b_i^{b_c} \psi b_{f^c} \psi b_{b^c} \psi b_o \wp$.

For each paragraph vector p_t, set the activation function f to the Rectified Linear Unit (ReLU)

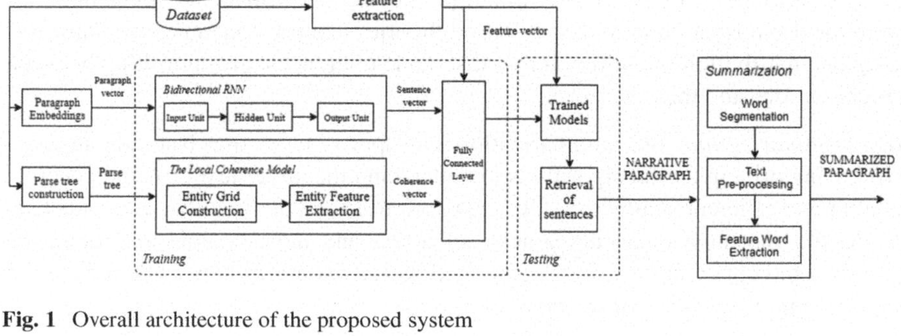

Fig. 1 Overall architecture of the proposed system

$$f(x) = \max(0, x) \tag{1}$$

Compute the activation of input units to the forward units $\left(x_t^f\right)$

$$x_t^f = f\left(W_i^f p_t + b_i^f\right) \tag{2}$$

Compute the activation of input units to the backward units $\left(x_t^b\right)$

$$x_t^b = f\left(W_i^b p_t + b_i^b\right) \tag{3}$$

Compute the activation of forward hidden units $\left(h_t^f\right)$

$$h_t^f = f\left(x_t^f + W_f h_{t-1}^f + b_f\right) \tag{4}$$

Compute the activation of backward hidden units (htb)

$$htb = f\ (xtb + Wb\ ht - 1b + bb) \tag{5}$$

Compute the final activation of BRNN—output unit (ot)

$$o_t = W_o\left(h_t^f + h_t^b\right) + b_o \tag{6}$$

3.3 *The Local Coherence Model*

To learn the coherence among the texts, this work includes a local coherence model. The sequenced parse trees are concatenated, from which an entity grid for the whole sequence is represented. Each text is represented by an entity grid, a two-dimensional array where each row of the grid corresponds to sentences, while the column corresponds to discourse entities. This representation helps to capture the distribution of discourse entities across text sentences.

Each grid column thus corresponds to a string from a set of categories reflecting the entity's presence or absence in a sequence of sentences. Our set consists of four symbols: S (subject), O (object), X (neither subject nor object) and—(gap which signals the entity's absence from a given sentence). It first identifies the entity classes, fills out the grid entries with relevant syntactic information and then determines the constituent structure for each sentence, from which the syntactic roles are identified.

An entity transition is a sequence $\{S, O, X, -\}^n$ that represents entity occurrences and their syntactic roles in n adjacent sentences. Local transitions can be easily obtained from a grid as continuous subsequences of each column. After entity grid construction, entity transition is enumerated and the ratio of the occurrence frequency of each transition is calculated.

Zero padded coherence representation is forwarded as input to Rectifier Linear Unit (ReLU) which would output the vector of the coherence model (q).

3.4 Multimodal Network

The outputs of BRNN $\{o_t\}_{t=1 \text{ to } N}$ and the coherence model (q) are given together as a input to two fully connected layers to decide proper language and coherence match. Dropout rates and the dimensions of the variables are set accordingly.

$$W_{f2}W_{f1}[O|q] = [S|g] \tag{7}$$

where $O = [o_1|\ o_2|\cdots\ o_N];\ S = [s_1|\ s_2|\cdots\ s_N]$.

3.5 Training and Retrieval of Sentences

To train the model, define the compatibility score between an image comprising an album and the corresponding text sequence. The algorithm considers corresponding score between sentence and image of all possible combinations to find out the best matching.

Retrieval of best sentence sequence for a given query image stream is as follows:

1. Select the k-nearest images for each query image from training database using Euclidean distance on the image features.
2. The sentences associated with k-nearest images at location are concatenated as a paragraph sentences. This represents the candidate sentences.
3. Compatibility score between an image stream and a paragraph sequence is computed based on the following method:

   a. The ordered and paired compatibility score between a sentence sequence and an image sequence are defined as:

$$S_t^k * V_t^l. \tag{8}$$

   b. The coherence relevance relation between an image sequence and a textsequence are defined as:

$$G^{k*}V_t^l. \tag{9}$$

c. The score S_{kl} for a sentence sequence k and an image stream l are defined as:

$$S_{kl} = \sum_{t=1,..N} (S_t^k * V_t^l) + (G^k * V_t^l) \tag{10}$$

where V_t^l denotes the 4096-dimensional CNN feature vector for tth image of stream l, and G^k and S_t^k are the output of Eq. (2.7) for a sentence sequence k.

d. The cost function to train the model are defined as follows:

$$C(\theta) = \sum_k [\sum_l \max(0, 1 + S_{kl} - S_{kk}) + \sum_l \max(0, 1 + S_{lk} - S_{kk})] \tag{11}$$

3.6 *Text Summarization*

PageRank algorithm which is used for text summarization is known as Text Rank. It is an unsupervised method for computing the extractive summary of a text. PageRank algorithm is applied over sentence graph, where the graph is symmetrical. The algorithm then built the PageRank transition by building the sentence similarity.

1. Preprocess the text: It includes removing stop words and stemming the remaining words.
2. Create a graph where vertices are sentences.
3. Each sentence is connected by an edge. The weight of the edge is defined by the similarity of the two sentences.
4. Run the PageRank algorithm on the graph.
5. Pick the vertices which represent the sentences with the highest PageRank score.

4 Experiment

Dataset. The Visual Storytelling (VIST) is the first-ever dataset created particularly for sequential image-to-language. The dataset includes 81,743 unique photos in 20,211 sequences, aligned to descriptive and story language. The image streams are extracted from Flickr and the text stories are crowdsourced for written to Amazon Mechanical Turk (AMT).

4.1 Retrieval Task

For experimental evaluation, the dataset is split into 8-1-1 ratio as a training set validation set, and test set, respectively. Each input query image is represented as a query I_q and the corresponding text annotated sentences as groundtruth T_G. The algorithm retrieves the text sequences from training set for each input query album images that should match well with groundtruth sentences.

Given an input album image and the text sequences, an algorithm computes the compatibility score as in Eq. (11). The low-cost text sequence has given more priority and that is the best-matched sequence retrieved. The generation tasks for our approach are evaluated using quantitative measures. The proposed work performs in test set to produce narrative paragraph and the corresponding shortened story. This work exploits two metrics of language similarity (i.e. BLEU [22] and METEOR [23]) which are popularly used in text generation. A better performance is indicated by higher BLEU and METEOR values.

Figure 2 shows various examples of sentence sequences on VIST dataset. Three different stories are generated for each query image stream: *Image description* represents the single image context. *Narrative story* is generated based on other images comprising a query image sequence, and *Summarized story* is generated for the corresponding narrative story. The difference between single image description and the corresponding narrative story is the coherence among the sentences which are indicated by highlighted words.

5 Result and Discussion

The quantitative results of story generation are shown in Table 1. The methods involved in proposed work is partitioned into three groups: (i) image captioning corresponds to the implementation of Recurrent Convolutional Network (RCN), (ii) generation of narrative paragraphs corresponds to RCN and entity-based coherence model, and (iii) generation of summarized results.

Figure 3 clearly demonstrates that executing a coherent model over language modelling has a significant exhibition as for bleu score and has improved execution concerning meteor score, while having summarization capability has improved execution as for both bleu and meteor score.

The sequence of text annotated sentences for each test image sequence is represented as groundtruth T_G, and the generated summarized stories are evaluated with reference to T_G. Since the retrieval method for summarized story is based on the generated narrative story and the evaluations are performed with T_G, the same has captured the coherent meaning and can only generate the similar sentences at best. This can be inferred from Fig. 3 that summarized story is most similar with T_G and from Fig. 2, the coherent meaning which are preserved from the image descriptions and generated narrative stories are shown by highlighted words.

Image description
A castle is underneath a clear blue sky. The calm river runs underneath a suspension bridge. The cobblestone drive way up to the house was fitting. A wooden log shaped like an alligator with a bridge in the background. A river with blue bridge above it and someone is riding a bike across the bridge.

Narrative story
We went to visit a town that is centered around the castle. The river **leads straight to the castle.** Quaint streets lead to gracious neighbourhood and the streets looked like cobblestone. A local wit **carved this log into friendly crocodile into an open field;** it leads the way to **suspension bridge**. There are smaller walking bridges for pedestrians and we enjoyed the views of **blue skyand old town.**

Summarized story
A trip to an old town has river leads to the castle while cobblestone streets drive to the houses. There is a field finds a way to suspension bridge and there are small bridges for pedestrians to enjoy the view.

Image description
A basketball player is saving the play. A basket ball player is shooting the ball. A group of players in blue defending the player in white. A basket ball player wearing the blue shirt is passing the ball. A player steals the ball from the other player.

Narrative story
During the basketball game, the player **attempt to save the ball for the team** from being out of bounds. Then **the blue team player took a shot to score**. After that, he was guarded by the players in opposite team. Then the **blue team player passed the ball to his team**. The blue team tried to get the basketball back.

Summarized story
A blue team and white team **playing a basketball game** with players in **blue team took a shot to score** for the team.

Fig. 2 Examples of generated story on VIST dataset

Table 1 Evaluation of story sentence generation with language similarity metrics (BLEU and METEOR)

	BLEU1	BLEU2	BLEU3	BLEU4	METEOR
Image captioning	0.140	0.027	0.000	0.000	0.026
Narrative story	0.106	0.027	0.000	0.000	0.046
Summarized story	0.162	0.161	0.160	0.159	0.153

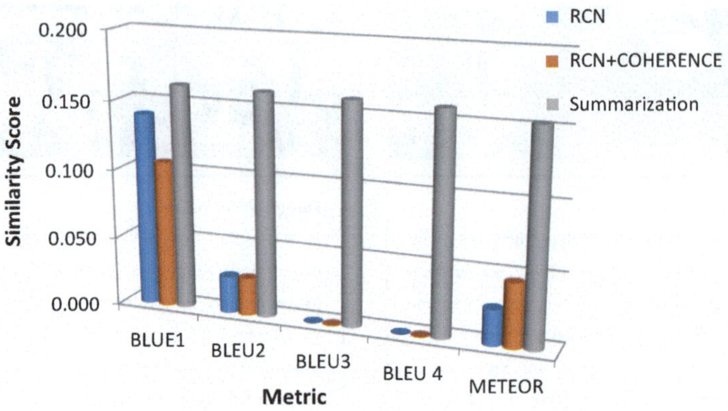

Fig. 3 Comparison of scores for narrative and summarized paragraph

6 Conclusion

Capturing the coherent meaning of a set of images is an important task for generating narrative paragraphs, instead of retrieving a text sentence associated with each image of an image set. Thus, the proposed work implemented a method for generating precise, yet concise story that best describes a sequence of images. With quantitative evaluation, this work demonstrates that generating summarized story from the narrative description has improved performance, and however, it preserves the context of an image set with syntactic and referential information.

Acknowledgements This Publication is an outcome of the R&D work undertaken in the project under the Visvesvaraya Ph.D. Scheme of Ministry of Electronics & Information Technology, Government of India, being implemented by Digital India Corporation (formerly Media Lab Asia).

References

1. Gowsikhaa D, Abirami S, Ramachandran B (2014) Automated human behavior analysis from surveillance videos: a survey. Artif Intell Rev 42(4):747–765 https://doi.org/10.1007/s10462-012-9341-3
2. Park CC, Gunhee K (2015) Expressing an image stream with a sequence of natural sentences. In: Proceedings of the international conference on neural information processing systems, vol 1(NIPS 15), pp 73–81
3. Gowsikhaa D, Abirami S, Baskaran R (2014) Construction of image ontology using low-level features for image retrieval. In: International conference on computer communication and informatics, pp 1–7
4. Richard S, Karpathy A, Le QV, Manning CD, Ng AY (2014) Grounded compositional semantics for finding and describing images with sentences. Trans Assoc Comput Linguist 2 1:207–218 https://doi.org/10.1162/tacl_a_00177

5. Farhadi A, Hejrati M, Amin Sadeghi M, Young P, Rashtchian C, Hockenmaier J, Forsyth D (2010) Every picture tells a story: generating sentences from images. In: European conference on computer vision, pp. 15–29. Springer, Berlin, Heidelberg. https://doi.org/10.1007/978-3-642-15561-1_2
6. Donahue J, Hendricks LA, Guadarrama S, Rohrbach M, Venugopalan S, Saenko K, Darrell T (2015) Long-term recurrent convolutional networks for visual recognition and description. In: Proceedings of the IEEE conference on computer vision and pattern recognition, pp 2625–2634. https://doi.org/10.1109/cvpr.2015.7298878
7. Vinyals O, Toshev A, Bengio S, Erhan D (2015) Show and tell: a neural image caption generator. In: Proceedings of the IEEE conference on computer vision and pattern recognition, pp 3156–3164 https://doi.org/10.1109/cvpr.2015.7298935
8. Ramisa A, Yan F, Moreno-Noguer F, Mikolajczyk K (2018) Breakingnews: article annotation by image and text processing. IEEE Trans Pattern Anal Mach Intell 40(5):1072–1085 https://doi.org/10.1109/tpami.2017.2721945
9. Srivastava Nitish, Salakhutdinov Ruslan R (2012) Multimodal learning with deep boltzmann machines. Adv Neural Inf Process Syst 25:2222–2230
10. Feng Y, Lapata M (2013) Automatic caption generation for news images. IEEE Trans Pattern Anal Mach Intell 35, 4:797–812 https://doi.org/10.1109/tpami.2012.118
11. Karpathy A, Fei-Fei L (2015) Deep visual-semantic alignments for generating image descriptions. In: Proceedings of the IEEE conference on computer vision and pattern recognition, pp 3128–3137 https://doi.org/10.1109/cvpr.2015.7298932
12. Huang TK, Ferraro F, Mostafazadeh N., Misra I, Agrawal A, Devlin J, Girshick R, He X, Kohli P, Batra D, Zitnick CL (2016) 'Visual storytelling', North American Chapter of the association for computational linguistics: human language technology, pp 1233–1239 https://doi.org/10.18653/v1/n16-1147
13. Deborah LJ, Baskaran R, Kannan A (2010) A survey on internal validity measure for cluster validation. Int J Comput Sci Eng Surv 1(2):85–102. https://doi.org/10.5121/ijcses.2010.1207
14. Siddiquie B, White B, Sharma A, Davis LS (2014) Multi-modal image retrieval for complex queries using small codes. In: Proceedings of international conference on multimedia retrieval, pp 321. ACM https://doi.org/10.1145/2578726.2578767
15. Lin D, Fidler S, Kong C, Urtasun R (2014) Visual semantic search: retrieving videos via complex textual queries. In: Proceedings of the IEEE conference on computer vision and pattern recognition, pp 2657–2664 https://doi.org/10.1109/cvpr.2014.340
16. Hu R, Xu H, Rohrbach M, Feng J, Saenko K, Darrell T (2016) Natural language object retrieval. In: Proceedings of the IEEE conference on computer vision and pattern recognition, pp 4555–4564 https://doi.org/10.1109/cvpr.2016.493
17. Kong C, Lin D, Bansal M, Urtasun R, Fidler S (2014) What are you talking about? text-to-image coreference. In: Proceedings of the IEEE conference on computer vision and pattern recognition, pp 3558–3565 https://doi.org/10.1109/cvpr.2014.455
18. Plummer BA, Wang L, Cervantes CM, Caicedo JC, Hockenmaier J, Lazebnik S (2015) Flickr30 k entities: collecting region-to-phrase correspondences for richer image-to-sentence models. In: Proceedings of the IEEE international conference on computer vision, pp 2641–2649 https://doi.org/10.1109/iccv.2015.303
19. Barzilay R, Lapata M (2008) Modeling local coherence: an entity-based approach. Comput Linguist 34 1:1–34 https://doi.org/10.1162/coli.2008.34.1.1
20. Allahyari M, Pouriyeh S, Assefi M, Safaei S, Trippe ED, Gutierrez JB, Kochut K (2017) Text summarization techniques: a brief survey. arXiv preprint arXiv:1707.02268
21. Mihalcea R, Tarau P (2004) TextRank: bringing order into text. In: Proceedings of the 2004 conference on empirical methods in natural language processing, pp 404–411, Association for Computational Linguistics
22. Papineni K, Roukos S, Ward T, Zhu W-J (2002) BLEU: a method for automatic evaluation of machine translation. In: Proceedings of the 40th annual meeting on association for computational linguistics, pp 311–318. Association for Computational Linguistics https://doi.org/10.3115/1073083.1073135
23. Lavie A, Denkowski MJ (2009) The METEOR metric for automatic evaluation of machine translation. Mach Trans 23, 2–3:105–115 https://doi.org/10.1007/s10590-009-9059-4

Dark Data: People to People Recovery

Sangeeta Chakrabarty and Ramprasad S. Joshi

Abstract The Internet of Things (IoT) is exploding disruptively. The IoT is making life easier for ordinary consumers and workers, but it is also generating zettabytes of *dark* data. The real time analytics involving close interaction between humans and instruments on the Internet is the main commercial motivation behind the IoT revolution. This means the data that are consumed instantly and their interpretation that is filed in an indexed and structured form are the main productive outcomes, while dark data and haphazardly stored interpretations add to the tare, bringing down efficiency and increasing costs. Since out of all data stored in the world today, almost all are generated in the recent three years, and the phenomenal growth will soon lead to a crisis, we need to put in place a global framework that never lets dark data clog the information highways but actually harnesses the real time analytics for a better planned future. We propose here a plan to build a "Data Waste Management" or "Data Sewerage" and local "Data Reservoir" system.

1 Introduction

The Internet of Things (IoT) is growing unbridled, in a similar manner as the Internet (of personal computers) itself did around the turn of the century. Reference [7] estimated a year ago that 90% of data stored in the world then was generated in 2016–17. Today we can imagine that it is perhaps 99% or more, and soon almost all the data stored at any given point will be "current" data. It will be like water and sunshine: *data ventilation, data sanitation, data pipeline, data sewerage* will be more apt terms than *information highways*.

In this work, we propose a system of managing ubiquitous dark data such that no accumulation of dark data happens in any centralized form, and productive use of

S. Chakrabarty
Dempo College of Commerce and Economics, Cujira, Bambolim, Goa, India
e-mail: sangeetafromgoa@gmail.com

R. S. Joshi (✉)
BITS Pilani, K K Birla Goa Campus, Pilani, India
e-mail: rsj@goa.bits-pilani.ac.in

S. Fong et al. (eds.), *ICT Analysis and Applications*, Lecture Notes in Networks and Systems 93, https://doi.org/10.1007/978-981-15-0630-7_24

data is possible in real-time while information highways and IoT-empowered smart areas are not clogged and infested by dark data.

In the following sections, we first review the three main sources of our motivation and ideation. We examine the idea of soft sensors in the next section with the purpose of asserting the utility and plausibility of crowdsourcing from digital social networks for data acquisition for greater common good. Following that, we examine a concrete case study of curating data from diverse, including digital and pre-digital, sources of different qualities and periods for model building and social and environmental data acquisition. In the fourth section, we review a long-term wide application framework for actually building a data pipeline and reservoir—*Knowledge Base Construction*. Further we identify what is common among these. We explain our design of a municipal data system and propose a workflow for it.

2 The BAUSPACE Proposal: Soft Sensors

Reference [7] acknowledge that "efficient big data analysis in IoT infrastructure is becoming mandatory to transform [dark data] into meaningful information" but yet point out that it is not enough; "the quantitative information provided by traditional "hard" sensors is not enough to deal with some scenarios where human observations are required". They propose a system of "soft sensors", a term they coined to mean humans acting as context-aware semantic sensors communicating over digital social networks about shared concerns and contingencies and events in their environment. They propose the development of applications able to process and analyze information from social networks for prediction of the event that is about to happen, detection of the event that has already occurred, and management of the event. They cite similar (but limited domain) applications during earthquakes [8–10]. We take their approach further. We propose, like them, oversight of social network soft sensors by public civic administration. But in addition we suggest a modified approach that is more distributed and publicly shared than their centralized system, for organizing public life towards greater environment conservation and greater common good.

3 A Case Study of Data Recovery for Detecting Environmental Trends

Reference [14] note that "many scientific data over the past 50+ years were collected at a time when the technology for curation, storage, and dissemination was primitive or nonexistent and consequently many of these datasets are not available publicly. These so-called "dark data" sets are essential to the understanding of how the ocean has changed chemically and biologically in response to the documented shifts in temperature and salinity (aka climate change)." They report that recovering these data "started with amassing metadata: how, when, and where the zooplankton data were collected... However, the crucial element that makes the effort possible is the input of the scientist who conducted the research for which the samples were collected and who remembers many important details about where to look and what to look for." For rectifying the shortcomings of incomplete information and discrepancies and errors, as many sources as possible were used and corroborated with human expert supervision. We need to also factor in the limitations: for example, the study cited in this section focused on the recovery of just two specific parameters; the dark data that they analyzed contain a lot more, to be recovered. Whether attempting recovery of all that can be recovered will give a better quality of recovery and interpretation of the individual parameters, and what will be the additional costs of this, are open questions.

4 The DeepDive Framework

In a series of publications depicting sustained research, Shin, Zhang, De Sa and their group [4, 11, 15, 16] have proposed a comprehensive approach to handling dark data. They define "The dark data extraction or Knowledge Base Construction (KBC) problem" as "to populate a SQL database with information from unstructured data sources including emails, webpages, and PDF reports" [4, 16]. DeepDive harnesses statistical inference and machine learning in a declarative framework that frees the users from the need to write probabilistic inference programs. Domain knowledge can be encoded into features and rules by domain experts who do not need to know the underlying computational mechanism. The result is such that, as [16] claim, "DeepDive has been successfully applied to domains such as pharmacogenomics, paleobiology, and antihuman trafficking enforcement, achieving human-caliber quality at machine-caliber scale".

5 The Common Thread

A key insight obtained from the DeepDive work [16] is that the KBC building process has to be iterative: performance criteria evolve, data sources and concepts aggregate increasingly as the process builds up to some equilibrium. Thus, an incremental pipeline adaptive to the evolution in data and domain experts' declarative specifications has to be designed from the beginning. This necessitates a higher, meta-level, domain-independent design specification that is not expressed in the same declarative formalism. Further, we observe some common approach features of the systems described in the three sections above:

1. Applications of dark data recovery were identified first.

2. Domain knowledge was crucial in the design and execution throughout, and it came from experts who were as far removed from data science as they were sagacious in their domain.
3. Multiple and growing sources of data were to be incorporated in a complex manner, and the glue was statistical inference.
4. Though the two other groups than the DeepDive developers do not acknowledge so explicitly, the processes are iterative, with the human insight itself getting refined with the development of data pipeline and accumulation of interpretations after data recovery.
5. Dark data sources and even their volumes, vary on a wide range; IoT to social media to pre-digital scientific surveys, and excel sheet kilobytes scale to zettabytes. But the "darkness coefficient" (if we may call it so) depends more on lack of structure and inference or interpretation mechanism embedded at source, within the pipeline, or storage.

In our proposal, we seek to combine the first four insights and envisage the possibility of addressing the problem highlighted in the last one.

6 Our Proposal: Municipal Data System

We propose that for the most common concerns related to the greater public good, namely,

1. Public Health: sanitation, sewerage, water supply, food production and distribution, urban community physical training initiatives such as the spread of Yoga, public awareness of good and hygienic practices, etc.;
2. Common Education: primary to higher secondary education;
3. Psychological Well-Being: urban stress and rural distress causing long-term psychological health degradation;
4. Public-funded resources for all the above (urban water and sewerage systems, garbage disposal, schools and hospitals, and their professional support systems);

we should set in place a data pipeline for creating a curated, library-like source of **local, municipal** data that common citizens can both help build and use in making their daily lives safe, healthy, futuristic, and meaningful. We are emboldened by the works cited in previous sections that we can harness

1. Individual citizens' intuitive knowledge of changing facts and evolving systems in their neighborhood and city;
2. IoT data accumulated by apps via smartphones and other connected devices;
3. Social media and search engine data ("soft-sensors" of BAUSPACE);
4. Ontology-enabled scientific and policy data (from governmental and vetted research sources);
5. Encyclopedic, geographical, and historical aggregated data and facts.

We have a concrete example of "collective intelligence harnessed by data science tools" to learn from: Google Flu Trends (GFT: [6]). GFT could "accurately map the incidence of flu with a lag of roughly one day" using data gathered online "through hundreds of billions of users' queries on the popular search engine to look for small increases in flu-related terms such as symptoms or vaccine availability" [5]. This generated a lot of enthusiasm for Big Data, see [2]. There were some dampeners, though: GFT faltered when it mattered the most, completely missing the onset in April 2009 of the H1N1 pandemic [5]. This was analyzed as "Big Data Hubris", a result of "the often implicit assumption that big data are a substitute for, rather than a supplement to, traditional data collection and analysis" [2]. Citing [2, 6] states that "The initial version of GFT was a particularly problematic marriage of big and small data. Essentially, the methodology was to find the best matches among 50 million search terms to fit 1152 data points. The odds of finding search terms that match the propensity of the flu but are structurally unrelated, and so do not predict the future, were quite high. GFT developers, in fact, report weeding out seasonal search terms unrelated to the flu but strongly correlated to the CDC data, such as those regarding high school basketball. This should have been a warning that the big data were overfitting the small number of cases, a standard concern in data analysis."

Keeping all these inspirations and dampeners in mind, we envisage a crowdsourcing and collective intelligence scheme that equally weighs raw instrumental or hard sensor data, processed soft sensor data (see the BAUSPACE proposal above), people's semantic perceptions and qualitative cognition trends, and statistical and machine learning inference for local trend detection and dissemination, and whets the same with encyclopedic and universal scientific expert knowledge. The complex process that will build such a system will go through the following steps (after conception, garnering public will and funding, and design stages) iteratively:

1. Design of a conceptual framework of ontological "input-output" relationships, the highest level of DeepDive's domain features and rules. For this, we need to perform some bootstrapping process. First, carry out a minor dark data recovery or text information retrieval to glean an ontology of common public concerns that are discussed in an open community uninhibited. The GFT example cited above is a case in point.
2. Then we need to get an expert opinion, going beyond common perceptions that can include half-true urban legends, on what type of inference and what kinds of queries can be handled without public hazard by this system without authentication by continuous centralized and expert supervision. In proposing this design process, we are inspired by the almost four decades of experience of two public health initiatives, namely, the book *Where There is No Doctor* [13] and the Bangladeshi movement for a rational drug policy [1, 12]. Both have depended on crowdsourcing and functional community education, and their success varies in space and time accordingly. This will be the task of creating reinforcement learning mechanisms that incorporate expert-led inference mechanisms and refinement of the inference by continuous expert feedback and *ex post facto* validation by taking into account false positives and false negatives.

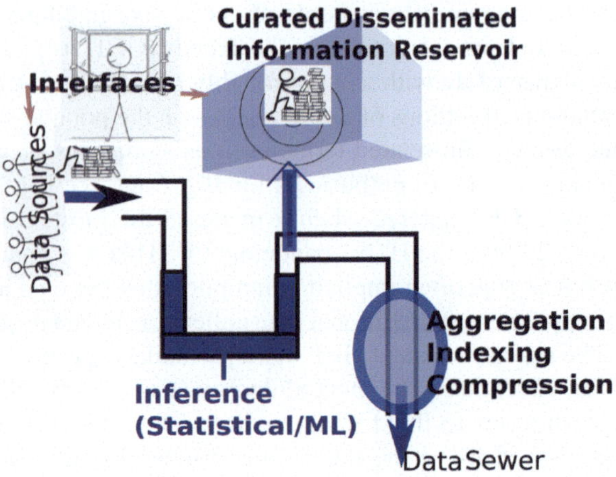

Fig. 1 An Imagination of the proposed Municipal Data System

3. Creating the "flanges, fittings, manholes and pipes": defining, declaratively as in DeepDive and/or procedurally in an algorithmic setup, how different data sources and streams will be merged or corroborated, and how filtration, segregation, convergence, aggregation, and condensation of indexed data and inferences will take place and be stored.
4. Developing
   - the back end database avoiding duplication and data/access expiry;
   - the middle-ware for indexing, query processing, caching and efficient dissemination, interfacing between the database back end and the various front ends via WWW, IoT and periodic processing cycles;
   - various easy access interfaces including natural language question-answer systems and mobile app interfaces such as Google Maps.

6.1 A Graphic Depiction

The Resulting Municipal Data System What we expect as an outcome is depicted pictorially in Fig. 1. The process envisaged in the previous subsection is described in Fig. 2.

6.2 The Wherewithal

Who will do all this and from where will the resources come? As we mentioned above repeatedly, we depend on crowdsourcing of data; but can we also hope for crowdsourcing the design and development/evolution? Of course, for the larger part,

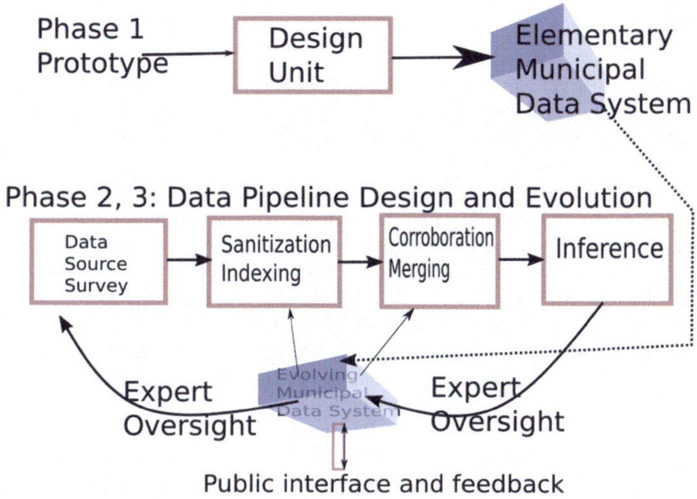

Fig. 2 Envisaged Process of Evolution

yes. But in the beginning, a prototype is to be developed via some R&D project sponsored via corporate or govt. funding (the usual CSR or CSIR route or new initiatives for startups). This prototype will have the pipeline of futuristic capacity but the data reservoir can be of rudimentary structure and capacity. As it takes off, the community resources—time spent, feedback quality, data storage, and computation for inference—can be harnessed gradually. The interfaces can be of the nature of *Wiki* pages building, along with actual Question-Answer systems in natural and semi-formal languages. The recent initiative of the Niti Ayog in getting together industries [3], academic researchers, and consumers on one platform for natural language processing in Indic languages shows that this is possible increasingly with crowdsourcing.

7 Conclusion

Awareness about dark data and development of dark data recovery systems is a recent phenomenon. Three successful dark data recovery systems that we surveyed have certain common features, the most important of which are the role of domain experts, iterative development process, and the significance of application-oriented ontologies. We posit that the role of experts can be augmented and complemented by crowdsourcing from common citizens if the data resource and use are related to their municipal life. Based on this assumption, and inspired by similar initiatives in community health, we propose a Municipal Data System. This can be further taken up for actual project funding.

References

1. Ahmed PSM, Evans TG, Standing H, Mahmud S (2013) Harnessing pluralism for better health in Bangladesh. Lancet 382(9906):1746–1755
2. Cukier KN, Meyer-Schoenberger V (2013) The rise of big data: how it's changing the way we think about the world. Foreign Affairs 92(3)
3. Darshankar P (2018) The unique challenge india presents to natural language processing. Business Today (8 October). https://www.businesstoday.in/opinion/columns/the-unique-challenge-india-presents-to-natural-language-processing/story/283650.html
4. De Sa C, Ratner A, Ré C, Shin J, Wang F, Wu S, Zhang C (2016) DeepDive: declarative knowledge base construction. SIGMOD Rec 45(1):60–67. https://doi.org/10.1145/2949741.2949756, http://doi.acm.org/10.1145/2949741.2949756
5. Eisenstein M (2018) Cloudy with a chance of flu: internet search data, medical records and networks of on-the-ground experts could enable the accurate forecasting and faster control of disease outbreaks. Nature 555:S2–S4
6. Ginsberg J, Mohebbi MH, Patel RS, Brammer L, Smolinski MS, Brilliant L (2009) Detecting influenza epidemics using search engine query data. Nature 457(19 February 2009):1012–1014. https://doi.org/10.1038/nature07634, http://static.googleusercontent.com/media/research.google.com/en/us/archive/papers/detecting-influenza-epidemics.pdf
7. Hernández D, Arcas-Túnez F, Muñoz A, Cecilia JM (2018) Bauspace: a scalable infrastructure for soft sensors development. In: Proceedings of the 47th international conference on parallel processing companion, ACM, New York, NY, USA, ICPP'18, pp 22:1–22:4. https://doi.org/10.1145/3229710.3229744, http://doi.acm.org/10.1145/3229710.3229744
8. Liu SB, Bouchard B, Bowden D, Guy M, Earle P (2012) USGS tweet earthquake dispatch (@ usgsted): using twitter for earthquake detection and characterization. In: AGU fall meeting abstracts
9. Sakaki T, Okazaki M, Matsuo Y (2010) Earthquake shakes twitter users: real-time event detection by social sensors. In: Proceedings of the 19th international conference on World wide web, ACM, p 851–860
10. Sakaki T, Okazaki M, Matsuo Y (2013) Tweet analysis for real-time event detection and earthquake reporting system development. IEEE Trans Knowl Data Eng 25(4):919–931
11. Shin J, Wu S, Wang F, De Sa C, Zhang C, Ré C (2015) Incremental knowledge base construction using DeepDive. Proc VLDB Endow 8(11):1310–1321. https://doi.org/10.14778/2809974.2809991, https://doi.org/10.14778/2809974.2809991
12. Srinivasan S (1996) 'Singing about the dark times': bangladesh drug policy. Economic and Political Weekly 31(21):1252–1255. http://www.jstor.org/stable/4404173
13. Werner D (2010) Where There Is No Doctor. Hesperian
14. Wiebe PH, Allison MD (2015) Bringing dark data into the light: a case study of the recovery of northwestern atlantic zooplankton data collected in the 1970s and 1980s. GeoResJ 6:195–201. https://doi.org/10.1016/j.grj.2015.03.001. http://www.sciencedirect.com/science/article/pii/S2214242815000273, rescuing Legacy Data for Future Science
15. Zhang C, Shin J, Ré C, Cafarella M, Niu F (2016) Extracting databases from dark data with DeepDive. In: Proceedings of the 2016 international conference on management of data, ACM, New York, NY, USA, SIGMOD'16, pp 847–859. https://doi.org/10.1145/2882903.2904442, http://doi.acm.org/10.1145/2882903.2904442
16. Zhang C, Ré C, Cafarella M, De Sa C, Ratner A, Shin J, Wang F, Wu S (2017) DeepDive: Declarative knowledge base construction. Commun ACM 60(5):93–102. https://doi.org/10.1145/3060586, http://doi.acm.org/10.1145/3060586

Deployment of 5G Networks Challenges for Developing Countries

Snahil Indoria

Abstract Network transmission has evolved over a period, beginning from 1G in 1980s, 4G networks have been established and work is done in deploying the standards of 5G technologies so far, 4G is yet to be established across the globe. Despite that, working on the establishment of new technologies has been started as 4G is proving to be insufficient to get along with the extended necessity regarding the highly dense network. So, to surpass these issues implementation of 5G technologies are going on as it provides, high-speed data rate, minimize latency, conserve energy, videos can be transferred without affecting its quality. The main challenges faced by developing countries in implementing 5G technologies are lack of infrastructure which includes poor fiber construction, no proper mechanism for the rapid increase in a number of users, low rate of data speed and the high cost and many more. This paper will represent the challenges in the implementation of 5G in India, discuss its future scope, its applications, comparisons from ongoing networks and the challenges in the implementation of 5th generation in the developing country.

Keywords 5th generation · GPRS · 5G-Infrastructure · Spectrum

1 Introduction

Radio technology has a rapid growth in recent years with the development of the cellular network in 1980. Industries are now on a roll to full fill current demands of the "Growing user around the globe to support the demand of high data rate and high reliability." 5th generation technology is the fourth upcoming network technology which is beyond the expectation of the consumer with ultra-fast rate, 5G has the capacity to change the language of mobile communication in coming years. With a huge capacity of managing devices, high quality video can be transferred to the consumer without compromising the video quality. 5th generation will be implemented by using the available spectrum in the existing LTE frequency ranging from (600 MHz to 6Ghz). The network will also support the implementation of

S. Indoria (✉)
Department of Computer Science, JK Lakshmipat University, Jaipur, Rajasthan, India
e-mail: snahilindoria@jklu.edu.in

S. Fong et al. (eds.), *ICT Analysis and Applications*, Lecture Notes in Networks and Systems 93, https://doi.org/10.1007/978-981-15-0630-7_25

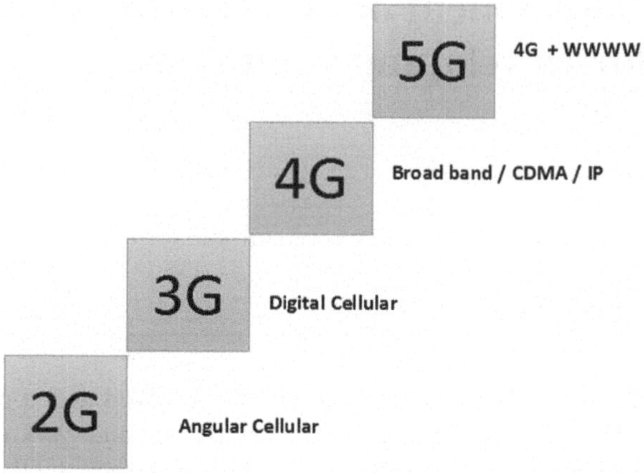

Fig. 1 Cellular generation ladder

the internet of things technology; by 2020 the current 5th generation technology will have all the support of sharing data anytime and anywhere. 5th generation is a combination of current technology which has the capability of working at low power. The objective of this technology is to provide user unhindered service with 24 × 7 high-speed availability. The implementation of 5th generation network includes costing and manpower, but the results will be a more advanced network which will be capable to support the current demand of the user (Fig. 1).

2G—2G or second-generation networks were launched on GSM standards in Finland for a commercial purpose it came into existence when smartphones were introduced. The major benefit of 2G was that phone conversations were digitally encrypted. Its major contributor is TDMA (Time division multiple access) or CDMA (Code division multiple access). **3G**—2G technology anticipated 2.5G technology which overpassed 2G technology to 3G. It not only benefited users with the increased speed of communication but also focused on implementing value-added services

4G—Successor of 3G, 4G is the fourth generation of broadband cellular networks. The main motive behind 4G technology was to provide IP-based voice, data, and multimedia streaming at a speed of minimum 100 Mbit/sec.

5G—The successor of 2G, 3G, and 4G or the latest generation of cellular mobile communication targets high data rates, decreased latency, energy preserving, cost reduction, and immense device connectivity.

2 Literature Review

Arun Kumar & Manish Gupta showcases their work on "major technologies and problem in the deployment of 5G mobile communication system." Their work stated that with the growing demand of mobile network there will be soon urged for more advanced and efficient mobile communication. They have highlighted the key issues for the development of 5G network as the data coverage in current network generation is less efficient in respect to indoor communication & high data rate at a constant speed will be resolved after 5G deployment in the country [1].

Wg.Cdr(Retd) Arif Khan has presented his work on "Deployment Challenges for 5G Technology" 5G is a new concept with limitless possibility which can meet the future demand of the user. However, there are several challenges in 5G deployment, i.e., increase in the number of connection and n number of applications are connected through the network, the current network does not have the capacity to manage current network demand. 5G also provide device to device communication that means the device does not have to rely on the base station for communication or over the network previous technology also had the push to talk technology but it had a limited bandwidth but provide a similar user experience as a device to device communication with lower reliability so the challenge is to have a robust implementation where the network is efficient enough to manage the device to device communication with sufficient bandwidth [2].

Shivani Sharma showcases his work on "Problems in Implementing 5G in India and Solutions for it" his work showcases the comprehensive study of 5G and problems associated with it as India lacks in proper optical fiber infrastructure as it plays a major role in implementation of 5th generation in the county with almost efficiency, India also Lack in governing body which hinders implementation of 5th generation. The author proposed a solution to overcome such problem, i.e., mobile edge computing which is a framework of the cellular network with a feature of cloud computing which provides IT surrounding at the edge of the cellular network. She also proposed the use of smart antennas to boost implementation of 5G [3].

Uvika Kujur & Dr. Ragini Shukla did a detailed and comprehensive study about "Features Analysis and Comparison of 5G Technology." The work provides a detail description of different evolution of wireless technology which promotes the development of the 5G wireless network. Multiple Telecom companies have successfully tested 5G network and at the end of 2020, the first 5G network will be available for public use in simple words it is termed as array of antennas which have multiple antennas which help in projecting the signal in one direction with at max intensity without any signal loss [4].

3 The Need for 5G Technology

5G which is the inheritor of 4G will absolutely change the way how cellular communication is done. This technology will bring huge data capacity, high-speed and astonishingly low latency. 5G also promises to improve efficiency and release the potential of automation. It is anticipated to reinforce notably faster mobile broadband speeds and progressively immense mobile data usage which also ensure the full usage of the Internet of Things (IoT). A present calculation of pinnacle ability is around 10 Gigabits per second (Gbps), which is way rapid then the existing internet services which include mobile and wired networks this will make it easier to download high definition movies in few seconds, video calling can be in full GD, QHD, or even 4 K resolution. Existing networks have up to 50 ms latency on an average connection, while 5G could have latency as low as 1 ms.

4 5G Architecture

The architecture of 5G is complex and dependent on the current network. The architecture of 5G is designed to meet the current demand of the user. 5th generation technology is based on the "Radio technology" which can efficiently manage the growing demand of the network and indoor communication will also be improved with high density network coverage. Multiple repeaters will be used parallelly to maintain signal strength.

5G Terminal—A terminal is an electromechanical hardware device which is used for infiltrating data into and showing it from a computer. The 5G terminals are comprised of software characterized radios and inflection designs along with new error-control designs, easily available on the Internet. The terminals therefore will have access to various cellular technologies at a time.

GPRS/EDGE—5G, absolutely an IP-based model, has various self-reliant and uncontrolled radio access technologies. EDGE (Enhanced Data rates for GSM Evolution) on the other hand, footholds data rate acceptable of 384 kbps and adapts higher inflection schemes like 8-PSK. EDGE footholds MCS1 to MCS9.

LTE—LTE or Long-Term Evolution, a standard for a cellular broadband connection for gadgets such as mobile devices and data terminals depends on GSM/EDGE and UMTS/HSPA technologies.

Data Server—A software plan which is used to afford or produce database services including storing, processing, and securing the data. The database services, therefore, are absorbed by various other software programs.

Servers for Real-Time Communication—RTC (Real-Time Communication) is a medium of communication where information is traded instantly (with no latency). The data is not gathered between transmission and reception.

Policy Server—A safeguarding element of the network which is highly based on policies which produce authorization services and enables tracking of files (Fig. 2).

The model of 5G is IP-based model for all the wireless networks. The model consists of the terminal-user terminal which is the main part of the 5G architecture. The function of the radio technology is to act as an IP link for the internet. The objective of Internet Protocol designee to manage the data routing Master Core Technology is the converging point for other technologies in the wireless network. It allows operating in the parallel multimode including all IP network and 5G mode. The compatibility feature of this technology makes it more efficient and more powerful [5] (Tables 1 and 2).

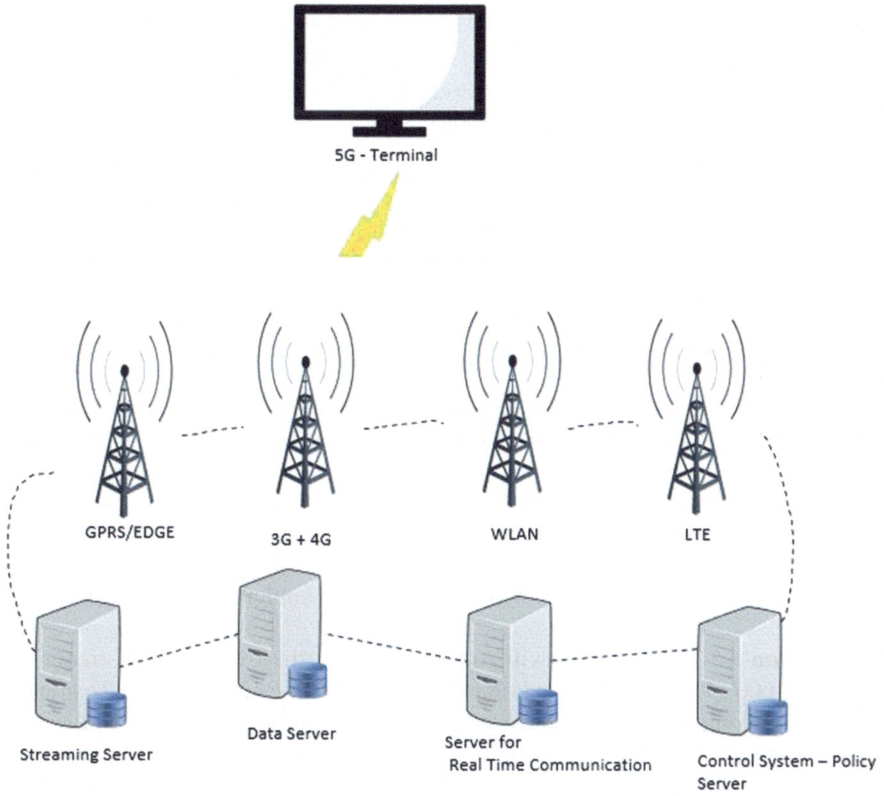

Fig. 2 Architecture of 5G network

Table 1 Comparison of OSI layer & 5G layer

Application layer	Application (Service)
Presentation layer	
Session layer	Open transport protocol
Transport layer	
Network layer	Upper network layer Lower network layer
Datalink layer	
Physical layer	Open wireless architecture
OSI layer	5G layer

5 Implementation 5G

Technologies Used

The 5G technology is the next version of mobile communications which is evolving from 1G to 2G, 3G, 4G, and now the latest 5G. This new 5G technology is very different as it targets a high data rate, reduced latency, energy saving, lesser cost, etc. [6].

Table 2 Comparison 1G–5G

Technology	1G	2G/2.5G	3G	4G	5G
Deployment	1970/1984	1980/1999	1990/2002	2000/2010	2014/2015
Bandwidth	2 kbps	14–64 kbps	2 mbps	200 mbps	>1 gbps
Technology	Angular Cellular	Digital Cellular	Broadband width/CDMA/IP	Unified IP & Seamless combo of LAN/WAN/WLAN/PAN	4G + WWWW
Service	Mobile Telephony	Digital Voice, Short Messaging	Integrated high quality audio, video & data	Dynamic information access, variable devices	Dynamic information access, variable devices with AI capabilities
Multiplexing	FDMA	TDMA/CDMA	CDMA	CDMA	CDMA
Switching	Circuit	Circuit/Circuit for access network &air interface	Packet except for air interface	All Packet	All Packet

The various technologies that will be used in the implementation of 5G are as follows:

Spectrum—5G technology will use spectrum in the existing LTE frequency range, i.e., 600 MHz to 6 GHz and, in mm wave bands, i.e., 24–86 GHz. To attain a higher throughput demand, the licensed carrier spectrum is not enough. Unlicensed spectrum, where the Wi-Fi is operated, will be more useful.

Public Land Mobile Network—It is a wireless communication system considered for use by the earthbound follower in vehicles or on foot. It is interconnected with a fixed system like the public switched telephone network (PSTN).

Sensors: A device which can convert one form of energy into another. It converts some physical process into an electrical instinct that can be translated to determine a reading. For example, a microphone that takes sound waves as an input and converts it to electrical energy.

Actuators: A device that operates in the reverse direction of a sensor, i.e., it takes an electrical input and then converts it into its physical form.

Operating Systems: It is a framework coherence of compound services and end-to-end slices on top of a heterogeneous and multi domain 5G infrastructure.

6 Challenges in 5G Commercial Deployment

Countries like South Korea, Japan, China, and the USA have already implemented 5G technology whereas in India it is still a challenging task to implement it. India, still being a developing country is facing a lot of issues in implementing 5G technology. Some of the major challenges are as follows:

Lack of Fiber Framework: India faces various issues such as call drop, poor quality of services which shows India's low fibers framework.

Low Data Speed and High Rates: low speed of data and huge rates is another issue for the deployment of 5G technology.

4G has Not Achieved a Maturity Stage: Only a rate of 12% penetration is there and it will take time for the penetration rates of 4G to grow which can delay the deployment of 5G.

Huge Data Volume: As the growing number of end-users hence the data consumption rate is increasing so to satisfy the huge data demands the network service provider needs to implement new hardware to generate the data at a high rate.

7 Factors Affecting 5G Implementation

Cost Factor: The Implementation of 5th generation requires optical cable infrastructure which is expensive to deploy with high maintenance costs.

Beamforming: Data loss while transfer is the major concern for the network provider.

Frequency Band: 5G New radio standard is set on working on the 6 Ghz range and millimeter waves of the RF spectrum.

Security and Privacy: security is the major concern for the cellular service provider. 5G network must ensure security and privacy for end-users

8 Conclusion

This paper highlights the challenges and issues faced by developing countries like India, in implementing and deploying 5G technology. Already, most of the industries, research institutes and universities are researching about deployment of 5G by the year 2020. Developed counties like China, Japan, South Korea, and USA are playing front foot to nurture the research in 5G mobile communication. In current scenario, the deployment of 4G in developing counties like India is still in progress, and most of the people are not even aware of 5G. In coming years, India will be able to deploy 5G mobile networks by improving its framework and commanding bodies. So, at this point, the big question to be asked is: Is the whole world ready for 5G?

References

1. Kumar, A, Gupta M (2015) Key technologies and problems in deployment of 5G mobile communication system 1:2394–4714
2. Wg. Cdr. (Retd) Arif Khan (2015) Deployment challenges for 5G technology

3. Sharma S (2018) Problems in implementing 5G IN India and solutions for it
4. Kujurn U, Shukla R (2018) Features analysis and comparison of 5G technology: a review. Int J Adv Res Comput Eng Technol
5. Singh S, Singh P (2012) Key concept and network architecture for 5G mobile technology. IJSRET 1(5):165–170
6. Sen J, Visions of 5G communications Innovation Lab, Tata Consultancy Services Ltd.Bengal Intelligent Park, Kolkata-700091 Email: Jaydip.Sen@tcs.com

A Novel Encryption Using Genetic Algorithms and Quantum Computing with Roulette Wheel Algorithm for Secret Key Generation

Kalavathi Alla, Praneetha and V Ramachandran

Abstract Information security dictates today's digital era of electronic commerce and business applications. DNA cryptography is surfacing as one of the fastest emerging technologies and opens a plethora of optimism in unbreakable algorithms. For transmission and storage of data DNA can be used and its scads of lengthy polymers of linked nucleotides are segregated into Adenine(A), Cytosine(C), Gaunine(G), and Thymine(T) that inherently include nitrogen base. Secret messages are concealed in these DNA sequences and are transformed into RNA sequences. To provide another security layer, quantum cryptography techniques were evolved to process the encoded text. Then the text is processed to an encryption function. To encrypt at this level, a 64 bit secret key is generated using genetic algorithm Roulette wheel selection to encrypt secret messages.

Keywords DNA cryptography · Quantum computing · Genetic algorithms · Roulette wheel selection

1 Introduction

A furor of Cryptographic techniques is available to guard data over the network. Cryptography converts secret messages into human illegible code. Data integrity in the three components of CIA triad was enforced by Cryptography. Security approaches should address the fast changing breaches in a vigilant manner. Studies were made over the braches and Breach Level Index and found that there is a massive increase in past two years, in some specific cases it is even raised over thousand percent. Sectors other than retail and professional services, like finance, academia, public sector, and entertainment experienced a morbid growth in either the number of compromised records or security incidents. The entire above statistical studies over the network data insecurity due to the dynamism in data breaches, create a wide-open space for

K. Alla (✉) · Praneetha · V. Ramachandran
Vasireddy Venkatadri Institute of Technology, JNTUK, Namburu, AP, India
e-mail: Kalavathi_alla@yahoo.com

V. Ramachandran
e-mail: vrc.bhatt@gmail.com

S. Fong et al. (eds.), *ICT Analysis and Applications*, Lecture Notes in Networks and Systems 93, https://doi.org/10.1007/978-981-15-0630-7_26

more effective and efficient security approaches globally. The pivotal point here is the data security and encryption of data with its motility property must be realized by considering the user access control.

2 DNA Cryptography

Orthodox cryptographic techniques embark on cryptic mathematical assertions that are rendered both conceptually and practically. The symmetric key and asymmetric key cryptosystem approaches have typical flaws. The augmented mega keys of new-age cryptography are still vulnerable with a power of billions of distributed grid of computing power with the longest cycles of bursts and can crack the secret key after a trillion years. Data can be hidden in DNA sequences. Every plain text symbol is turned down into a blend of four bases that make up the human deoxyribonucleic acid(DNA) [1, 2]. Using DNA cryptography, myriads of data gets fit into a tiny volume, a gram of DNA consist (2^{10}–3) DNA bases which amounts to data around 108 terabytes [3]. Less than kilograms of DNA can contain the whole data prevailing in the whole world. There were many studies worked on DNA cryptography [4]. Storing DNA sequences as binary data is the current carry forwarded challenge in research. DNA cryptography, being the first ever issue introduced by Leonard Max Adleman in finding the solution for Hamiltonian path problem. In DNA computing information is hidden using a four-character genetic alphabet $\sum = (A,C,G,T)$, rather than the binary alphabet used by traditional computers, $\sum = (0, 1)$. DNA strands by virtue can store huge amount of information than the general storage systems like in the traditional computers [5]. This kind of sequence of these nucleotides in DNA is called DNA Sequencing.

2.1 Genetic Algorithm

Genetic Algorithm (GA) is a search-based optimization method which uses the principles of genetics and selection algorithms. These algorithms play a vital role in finding the optimal or near optimal solutions for complex problems, and also used to solve optimization problems, artificial intelligence, and machine learning. The genetic algorithms are designed based on the Charles Robert Darwin's theory. It is adaptive heuristic exploration algorithm which uses the mechanics of the natural selection principles and genetics. Based on this evolution theory, genetic algorithms belong to the family of evolutionary algorithms.

2.1.1 Population Generation

Population is set of all chromosomes. To generate population it uses two preliminary methods to start a GA population [6]. They are random initialization and heuristic initialization.

Genetic Algorithms are very iterative and they transform the population of generated chromosomes into posterities. Various GA operators are applied, once it identifies the population for individuals selection. These individuals can be selected by using diverse genetic operations and fitness functions.

2.1.2 Crossover

Once generating the population, crossover operation is the most important operation in GA. Crossover operators are used on population and the resulting generation is much more fit. Crossover is the mechanism of taking multiple parent solutions, and generate a child solution. Crossovers are of two types; single point and two point crossover [7].

2.1.3 Mutation

Mutation operator is used to preserve genetic diversity among the generations. This process alters every gene independently centered on the possibility. Search space investigation can also be done using Mutation [8]. There are various kinds of mutation operations: Bit flip, random resetting, swap, scramble, and inversion. These individuals have highest fitness rate.

2.1.4 Selection

Selection is a process of selecting the better individuals. This is applied after generating population and after applying the crossover, mutation operations. Best fit among the present population is prevailed. Roulette wheel algorithm is used for selecting best fit population.

3 Quantum Computing/Cryptography

Quantum cryptography exploits quantum mechanical assets to perform various cryptic tasks. A verity of models in quantum computing, segregated by the basic elements were in existence where the computation is decomposed. One of such model is quantum gate array, where computation is segregated into an order of some qubit quantum

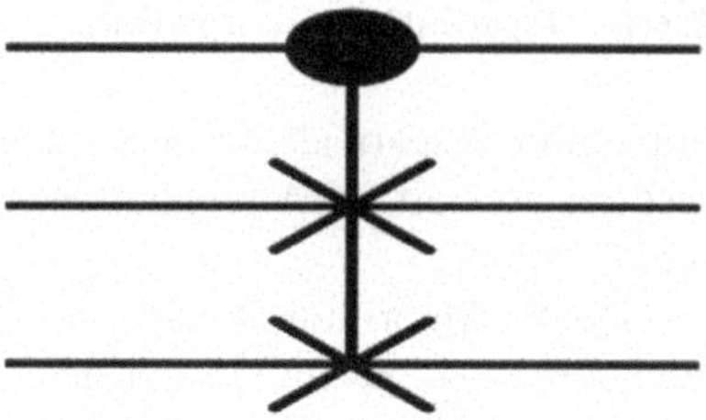

Fig. 1 Fredkin gate

gates. Contrasting many other logic gates, quantum gates use reversible logic. Quantum gates use unitary matrix representation. In these gates, the total number of input and output bits should be equal. These n qubits are represented by 2^n X 2^n. Quantum states can be considered as vectors in 2^n complex dimensions. The base vector/vectors is the possible outcome if measured, and the quantum states are considered as a linear mixture of the above outcomes. Mostly, quantum gates are operated on spaces of one or more qubits, similarly like a logic gate. Single qubit is represented using a vector representation:

$$|a\rangle = v_0|0\rangle + v_1|1\rangle \rightarrow \begin{bmatrix} v_0 \\ v_1 \end{bmatrix}$$

Two qubits in Vector representation is:

$$|ab\rangle = |a\rangle \otimes |b\rangle = v_{00}|00\rangle + v_{01}|01\rangle + v_{10}|10\rangle + v_{11}|11\rangle \rightarrow \begin{bmatrix} v_{00} \\ v_{01} \\ v_{10} \\ v_{11} \end{bmatrix}$$

Computations on a particular quantum state are identified by multiplying the vector |ab > which represents the state by the matrix U: U|ab>

The work done by Griffith & Queensland Universities developed a Quantum Fredkin Gate, is a computational circuit and is suitable for reversible computations, developed by Edward Fredkin as shown in Fig. 1.

All Arithmetic/Logical operations can be realized by Fredkin gates. The simple Fredkin gate is known as a controlled swap gate(CSWAP). A generalized nxnFredkin gate and its functionality depicted below (Table 2).

4 Proposed Work

In this research work, GA is included to create a best fit secret key using Roulette sheel selection [9]. This work also combines the methods of DNA cryptography and quantum computing. The secret key generated by this algorithm has non-repeating

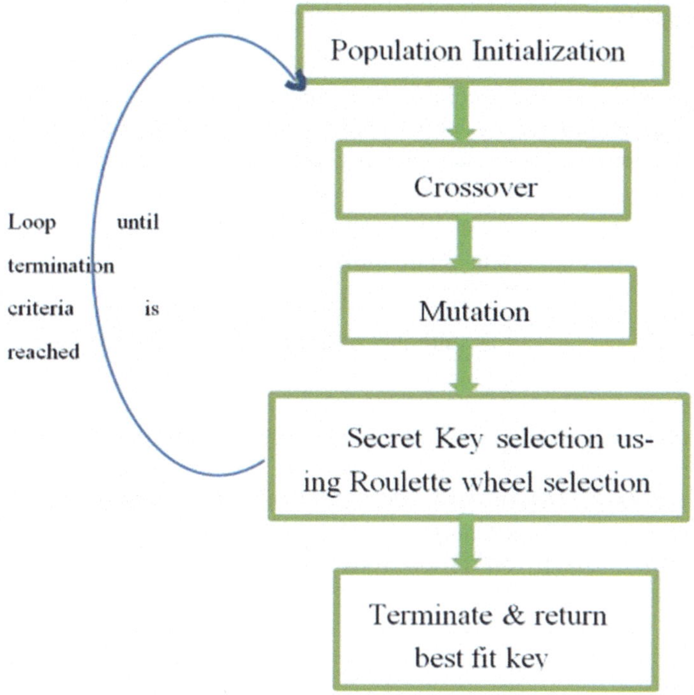

Fig. 2 Secret key generation procedure

nature which makes the ciphertext difficult to decode. The fundamental process in generating a secret key is generating preliminary population, crossover, mutation, and fitness function selection using Roulette wheel GA [10]. The offered key generation solution involves the following steps as illustrated in Fig. 2.

4.1 Population

GA normally starts with the population of random generated secret keys. This algorithm generates a 64 bit key. The population size varies on a large number of probable solutions. The larger the key, gives more computationally secure. Soon after the generation of the initial population, population is exposed to a wide variety of genetic operations which in turn increases the number of chromosomes [11]. These operators are used to select individuals from the population.

4.2 Crossover

Crossover simulates sexual genetic recombination. Crossover is applied on two randomly selected chromosomes [7]. Crossover rate is calculated and the number of crossovers is computed using the following formula:

$$\text{No of crossovers} = \text{CR} * \text{l} * \text{n}$$

Where CR = crossover rate

l is length of the key

n is number of keys

Ex: Parent 1
=1010000010100100101000110001100111100011000110001100000000111010
Parent 2:
0010101100001011010001010110001110000000011111000001110001001010
Crossover point: 37 New Child:
1010000010100100101000110001100111000000011111000001110001001010.

4.3 Mutation

The next step to generate secret key-mutation. Mutation is a procedure where a random chromosome is chosen from the current population, a mutation is selected to invert the corresponding bit. A personalized mutation probability leads to mutation while evolution [12]. Generally, mutation rate must be set to low. If this rate is very high, the search procedure turns out into an original arbitrary look-up. For example, if this is set to 1, which means 1 out of every 100 bits are manipulated. Here in the suggested key search bit inversion mutation operator is used. Mutation rate is calculated using the formula:

$$\text{Number of mutations} = \text{MR} * \text{l} * \text{n}$$

Where MR = mutation rate

l is the length of the key

n is the number of keys

Ex: 1010000010100100101000110001100111000000011111000001110001001
010

Mutation point = 23 1010000010100100101000010001100111000000011111000001110001001010.

Table 1 DNA nucleotides encoding pattern

DNA nucleotide	Binary sequence
A	**00**
C	**01**
G	**10**
T	**11**

4.4 Calculating Fitness Function Using Roulette Wheel

The fundamental process of selection is in line with existence of the fittest. Fittest individuals will have a better probability of survival and weaker individuals can not survive. Table 3 illustrates the fitness calculation for a sample population of 5 individuals.

These individuals consist of 10 bit chromosomes. The input range for x is between 0 and 10. The fitness values are computed using the function of x. The individual number 2 (0100011001) has the highest fitness value. Therefore this will be considered as the best fit secret key. In the proposed work, we took samples of 100 individual chromosomes and applied crossover, mutation operators to calculate fitness values for those. The best value is selected out of the population of 100 with 64 bits. Since it is very difficult to show these findings, the above table illustrates the procedure to calculate fitness value for 10 bits.

4.5 Encryption Algorithm

Step1: Divide the input text into block of 64 bits
Step 2: Encode the text P into DNA nucleotides as shown in Table 1.
Step 3: Translate the DNA strand into mRNA stream
Step 4: Apply Vernam cipher on the resulting bitstream with the pre generated 64 bit secret key
Step 5: Encrypt the resulting bitstream using the reversible quantum gate(Fredkin).

4.6 Decryption Algorithm

Step 1: Decipher the ciphertext using quantum cryptography
Step 2: Apply vernam cipher on the resulting bitstream with the shared secret key which is generated by using GA.
Step 3: Convert the mRNA stream into DNA nucleotides.
Step 4: Translate the DNA sequence into the binary stream using the DNA encoding shown in Table 1.

Table 2 Fredkin gate truth table

Input	Output
0000	**0000**
0001	**0001**
0010	**0010**
0011	**0011**
0100	**0100**
0101	**0101**
0110	**0110**
0111	**0111**
1000	**1000**
1001	**1001**
1010	**1010**
1011	**1011**
1100	**1100**
1101	**1110**
1110	**1101**
1111	**1111**

Table 3 Roulette wheel fitness calculation

S.No	Chromosome	Value_{10}	X	$f(x) = -1/4(x)^2 + 2x + 5$
1	1011001100	716	7.1	6.5
2	0100011001	281	2.05	8.04
3	0000101101	45	0.37	5.71
4	1110001100	908	8.89	3.02
5	0000010011	19	0.12	5.24

Step 5: Retrieve the original secret message using the general ASCII encoding pattern.

5 Results

The proposed work has been implemented and analyzed using Java. Random samples are created by generating the initial population for 100 chromosomes. Numerous number of tests are applied on the samples to find the best fit secret key and also on the robustness of encryption algorithm, the results are satisfactory.

6 Conclusion

The quantum-based DNA cryptography is powerful to generate secret keys and also for encryption and decryption mechanisms. The above cryptographic work using DNA can be extended to other operations like DNA annealing, DNA synthesis. This in turn illustrates that DNA molecules can also be used for non-biological purposes Encryption and decryption mechanisms generally use translation and transcription which are two basic functions of DNA computing. This is observed from the above findings that the generated secret key using GA is very random and is difficult to decode. The genetic algorithms are very complex and when combined with Roulette wheel selection gives the best fit secret key. For any cipher algorithm, security depends on the complexity of secret key. This can be used for any symmetric key algorithm. It makes almost impossible for the cryptanalysts to break the secrecy of message.

References

1. Goldreich O (2001) Foundations of cryptography, Volume 1: Basic Tools, Cambridge University Press, ISBN 0-521-79172-3
2. Limin Q (2008) The study of DNA—based encryption method . Zheng Zhou: Zheng Zhou University of Light Industry
3. Adleman L (1994) Molecular computation of solutions to combinatorial problems. Sci 266:1021–1024
4. NoorulHussain U, Chithralekha T, Naveen Raj A, Sathish G, Dharani A (2012) A hybrid DNA algorithm for DES using central dogma of molecular biology (CDMB). Int J Comput Appl
5. Srinivasa Murthy YV, Satapathy SC, Srinivasu P, Saranya AAS (2001) Key Generation for text encryption in cellular networks using multi-point crossover function. Int J Comput Appl (0975–8887) Volume 32-No. 9
6. Kumar A, Rajpal N (2004) Application of genetic algorithm in the field of steganography. J Inf Technol 2(1):12–15
7. Umbarkar AJ, Sheth PD (2015) Crossover operators in genetic algorithms: a review. Ictact J Soft Comput
8. Delman B (2004) Genetic Algorithms in Cryptography, MS Thesis
9. Koblitz N (1994) A course in number theory and cryptography. Springer, New York, Inc
10. Goyat S (2012) Cryptography using genetic algorithms (GAs). IOSR J Comput Eng (IOSR-JCE), 1(5):0608. Identification of common molecular subsequences. J Mol Biol 147:195–197
11. Bagnall AJ (1996) The applications of genetic algorithms in cryptanalysis. School of Information Systems, University Of East Anglia
12. Liu H, Wang X, Kadir A (2012) Image encryption using DNA complementary rule and chaotic maps. ScienceDirect 2012

Use of ICTs in Financial Engineering Applications in Insurance Business

Venkamaraju Chakravaram, Srinivas Jangirala and Sunitha Ratnakaram

Abstract The research paper evaluates the use and contribution of Information and Communication Technologies (ICTs) in the design and development of financial engineered life insurance policies and financially engineered life insurance business processes in the growth of the life Insurance Business in India. The research study explores the use of ICTs in various product development stages of financially engineered life insurance policies and the use of ICTs in insurance processes. Specifically, this paper reports a theoretical examination that simultaneously considers the effects of these relationships among ICTs, Financial Engineering, Insurance Business and growth of the insurance business in India, and the role and contribution of ICTs in the financial engineering applications in financially engineered life policies and processes. It encapsulates the role of ICTs in FE applications in the design and development of Financially Engineered policies, business processes of insurers including companies' performance management.

Keywords Financial engineering · Insurance business · ICTs · Insurance policies

1 Introduction

We cannot imagine the current era of lifestyle without ICTs contribution and role in each and every field of human life especially the fields like Banking, Insurance sectors. *"Information and Communication Technologies (ICTs) refers to technologies that provide access to information through telecommunications. It is similar to Information Technology (IT) but focuses primarily on communication technologies. This includes the Internet, wireless networks, cell phones, and other communication*

V. Chakravaram (✉) · S. Jangirala · S. Ratnakaram
Jindal Global Business School, O. P. Jindal Global University, Haryana 131001, India
e-mail: vchakravaram@jgu.edu.in

S. Jangirala
e-mail: sjangirala@jgu.edu.in

S. Ratnakaram
e-mail: sratnakaram@jgu.edu.in

S. Fong et al. (eds.), *ICT Analysis and Applications*, Lecture Notes in Networks and Systems 93, https://doi.org/10.1007/978-981-15-0630-7_27

mediums (Tech Terms, 2010)". Without the role and contribution of Information and Communication Technologies, the said two fields are incomplete and they cannot offer their effective and complete services to their customers in time effectively at all stages of their life. ICTs also playing a major role in the day to day operations of the companies performing in these sectors in each and every aspect of their business life cycle.

For example, if an Insurance company tried to introduce a newly financially engineered life insurance policy into the market, ICTs plays a major role in all the aspects of the task. From the beginning of understanding the exact needs and requirements of customer expectations, existing market conditions, competitors and etc., related research studies till to launch of the newly designed products or services into the target market, also till provide effective sales after services to the customers. This is explained in detail in the coming sections.

Financial Engineering is a process of creating an innovative newly designed financial model or life insurance product to provide innovative policies which to satisfies the needs and requirements of customers and which offers innovative solutions to the existing financial problems of the fields. To explain the same in the context of insurance, we can define FE as follows. It is a process of creating innovative life insurance policies by merging the existing life insurance policies to design a new and innovative life insurance policy or model by introducing innovative financial features which satisfy all the core stakeholders of the sector. Here core stakeholders are customers (policyholders), Insurer (Seller or Service provider of the life insurance coverage), Govt. Regulating Agencies (IRDAI).

The below diagram shows the life cycle of Life Insurance Policy. In each and every stage of financially engineered life insurance policies, ICTs plays a major role and the researcher tried to explain in the below sections.

2 Literature Review

Betsy McLeod	2018	The author nicely introduced a list of twelve marketing ideas in the digital world and effectively explained the use of ICTs in the marketing of insurance products
Anshu	2003	The research something about E-Commerce selling of Insurance policies. Explained how to make more business use of internet applications through E-Commerce marketing and how to reduce the costing in another mode of selling channels
O'Brien	2002	Research paper about Information Technology and related applications used in the business management process. He tried to explain various ways use of IT applications in the business management process

(continued)

(continued)

Chaudhury et al.	2002	The paper about electronic insurance and how to sell insurance policies. How this electronic selling of insurance policies will greatly reduce the cost of operations and how to maximize the more insurance business in a span of short time explained
Steven Alter	2002	In this research paper, the author is motivating the insurers to introduce new technologies and ICTs in the business management process. He warned the companies that there is no life to the companies if they neglect the introduction of new technologies in all the phases of business operations
Karahanna et al.	1999	Through this research work, the author encouraging the business people to use ICTs in day to day business operations to achieve the success and good results in all stages of business processes

3 Significance of the Study

The present research is an attempt to examine the contribution and use of ICTs in Financial Engineering Applications with special reference to the life insurance sector in product design, development and insurance business management processes.

Objectives:

1. To study the use of ICTs in Financial Engineered life insurance policies
2. To study the use of ICTs in Financially Engineered life insurance business processes.

Hypotheses:

Ha1: There is a significant usage of ICTs in Financially Engineered Life Insurance Policies

Ha2: There is a significant usage of ICTs in Financially Engineered Life Insurance Business Processes.

4 Reasons for the Emergence of Financial Engineering in the Life Insurance Business

Rapid changes in the life insurance sector and in other financial services of the sector have changed the scenario of financial markets. Due to continuing growing competition in the life insurance business, due to the entrance of more private life insurance companies with abroad based joint venture collaborations with innovative product knowledge and ideas, the traditional life insurance policies are replaced with more complex and multi-benefit oriented life insurance products. The entire situation has given birth to a new discipline in the financial management area known as "Financial Engineering".

The main reasons for Rapid Growth in Financial Engineering: (*Hoda Ahmed Ibraheem, College of Business & Economics, Qassim University, Saudi Arabia*):

- In 1972, sudden floating of exchange rates due to the breakdown of the Bretton Woods agreement has led to major increases in volatility and competition Smith (1990:33).
- Revolution also caused for the Technological improvements dramatically in this period.

Environmental Factors for Financially Engineered Applications: Increase in price volatility, Globalization of the world economy and competition, Deregulation and an increase in competition, Advances in technology and communication, Development of new markets and market linkages, Advances in financial theory, Tax asymmetries, Standardization, Low documentation costs

5 Financially Engineered Products Development—International Perspective

FE insurance products development globally faces many similar and unique issues navigating the insurance product development process. Brazil markets and their regulatory environments are not cooperative and not encouraging the use of Financial engineering applications in the to design and development of new and innovative life insurance products. Italy, the new product development and the use of FE applications are taking place at the reinsurance companies only. Here also the teams of underwriting and claims settlement are hardly involving in this process.

China, Brazil, France and South Africa countries, companies are using Financial Engineering application in the development of products. But in these countries companies are mostly depending on banking institutions to introduce their new FE Insurance products in the market or to the end costumers. Here mostly Financially Engineering products are having saving nature and full of saving features, not on pure mortality based. North America, in this country also administrative systems of life insurance companies are not effective and there is no encouraging atmosphere to use the Financial Engineering applications in the new product development process.

A few international countries, where life insurance companies mostly depend on banking institutions in selling and marketing process of their products are facing difficulty in the use of FE applications in the process of design and development of new and innovative life insurance products. Banking institutions are not encouraging these companies and greatly influencing them to design the products as per their convenience, without giving due importance to the needs and requirements of customers or customization and innovation process.

6 The Big Five ICTs Which Speeded up the Financial Engineering Processes in the Design, Development and Claim Settlement Stages of the Insurance Business Are

Big Data

The main purpose of the big data collection is to assess the risk assessment of prospect customer during the underwriting process and to select the risk also to detect fraudulent or fake information provided by the applicant. To suggest a suitable product and policy plan based on the applicant's behavioural profile.

Block Chain

It is a mutual distributed ledger system developed the financial conduct authority to provide innovative solutions to the service providers. Helping Insurance providers to streamline their paperwork particularly in the claims settlement process.

Internet of things

All most all the electronic and auto gadgets are connected now a day to the internet like mobile phones, computers, laptops, cars and etc., One can easily assess the habits and lifestyle of customers while issuing insurance products, risk coverages, to keep rider coverage, insurance coverage limitations by analyzing the information available in the web world on behalf of prospect applicant.

Drones

Helps to assess and estimate the fire loss and extent of property damage, repairs in settling the claims of the insured property. Also helps to decide the coverage and to count the probability of risk, proportionate premium counting thoroughly the risky points, constructions, processes associated with ensuring the property.

Artificial Intelligence

One of the rapidly growing and frequently using technology in the process of financial engineering of the insurance policies and insurance services is Artificial Intelligence technology. Insurance companies are using AI technology to assess the risk factors, feasibility of the prospect applicant to issue the policy with in a swift time through a quick underwriting process. AI also helping the companies in the process of designing pricing structure of financially engineered life insurance policies. In an article (Daniel Schreiber Jan 1st 2017) about the Lemonade Insurance Company, Mr. Daniel has clearly explained how they are doing insurance business innovatively and settling their claims very quickly by using AI technologies and with the help of their data scientists. Lemonade Insurance Company sets a world record by settling a claim within 3 s after submission of claim application by a customer through his iPhone7 Lemonade app. For this great achievement, Lemonade team cross checked the policy details by running nearly about 18 anti-fraud algorithms during this short period and settled the claim. This is really a miracle and best example of Artificial Intelligence. The company name is Allianz-backed US insurer Lemonade Insurance Company.

7 The Essential Steps in the Design and Development Process of Financially Engineered Life Insurance Products Are

- Identify and list out the life Insurance needs, various protection riders, saving and investment requirements of prospective customers.
- Identify various risks anticipated considering the economic, social, health factors and family profile of an average customer and putting in the place the best way to manage those risks.
- Identify or forecast the challenges of the market, expected threats from competitors, existing and anticipated regulations of IRDAI, SEBI, other regulating agencies of Government.
- Financially Engineered insurance products need to be ensured to manage those risks most effective and economical way to add value to all its stakeholders and to optimize its shareholder value.
- Listen and take appropriate recommendations, suggestions of actuarial professionals, marketing agencies and professionals of the insurance sector.
- Regular watch on opportunities to cover new risks to design a new FE Product or to redesign the existing FE life insurance products. Risks always provide an opportunity for the insurance industry. For Example, the latest identified risks in the sector are cybercrime and cyberterrorism (Table 1).

For example:
The contribution of ICTs there greatly in the following product development stages of Financially Engineered Life Insurance products:

- ***Idea Generation***:
  Generate new and innovative products considering the needs and requirements of prospect customers anticipating new problems or risks associated with the prospect lives.
- ***Product concept/feasibility***:
  Design the concept in order to satisfy all the stakeholders to share a win and win benefits out of a newly developed concept. Feasibility test to be conducted whether the product is finically viable or not.
- ***Underwriting guidelines***:
  Should consider underwriters' guidelines, suggestions, instructions in the process of development of FE products.
- ***Assumption development***:
  Develop hypotheses and test them to go before the product planning stage.
- ***Product planning & design***:
  Plan the product features like minimum and maximum age to cover the policy, riders, multiple coverages, saving and investment options, reinsurance portion of the premium, Administrative expenses part of the premium, risk coverage part of the premium and investment part of the premium and etc.,

Table 1 Contribution of ICTs in financially engineered life insurance processes

Coordinating financially engineered processes in insurance	Aims and culture	Processes	Main forms of coordination	The main Coordinating Information and Communication Technologies (ICTs)
Supply chain management	Efficiency via economies of scale, deep integration	Planning, procurement, production, distribution	Stable network	Supply chain (CPFR), e-commerce systems
Relationship management	Efficiency via economies of scope, customer relationship management (CRM)	Marketing, sales, service policy	Market	CRM- and document-oriented EC-systems
Innovation	Time-to-market, dynamic, high interdependency, promotion of creative stars	Idea generation, concept finding, development	Dynamic networks, virtual teams	E-collaboration platforms
Infrastructure	Efficiency via service culture and standardization; cost reduction, economies of scale	Accounting, human resource management, assets	Internal, stable network	Distributed efficient resource planning (ERP)-systems
Organizational development	"Network-compatible" employees and partnerships	Organizational development	All forms of coordination	All forms of coordination

- ***Traditional rider development***:
  Needs to develop traditional riders by incorporating new or coverage options. Ex: critical illness rider covers earlier six critical diseases only, now try to incorporate a few more newly identified diseases in the covering list.
- ***Living benefits rider development***:
  Design and develop innovative FE riders in order to enjoy the policyholder during his/her policy period or after the policy period with some financial or premium discount benefits.
- ***Product pricing***:
  Fix the competitive price structure of the FE products considering the mutual benefits of all stakeholders.
- ***Reinsurance***:
  Allocate an appropriate portion of coverage, premium towards reinsure of newly developed FE products in order to share safe and secured financial benefits to the main stakeholders.
- ***State/regulatory filings***:
  Identify various regulatory norms and state clearly and file them in time.
- ***Update business procedures***:

Needs to list out the new product launching procedures, target customers, segments, demographics prospect customers, required training sessions, materials, advertisements to make understand the marketing and sales force and prospect customers.

- ***Marketing plans***:
  Needs to plan and formulate a strict and systematized marketing strategy very well in advance after due verification of similar kind of products of competitors.
- ***Product Development Complete***:
  With commercial test trails in selected segments through selected marketing channels, the FE Product development process will reach the completion stage in the product development.

8 The Role of ICTs in the Marketing of Financially Engineered Life Insurance Policies

Use of ICTs in the marketing of Financially Engineered Life Insurance policies in the following forms.

(a) Online Advertising of various financially engineered Life Insurance policies with attractive illustrations, tables.
(b) Marketing of financially engineered life insurance policies through emails.
(c) Marketing of financially engineered life insurance policies through social media websites like Facebook, Twitter, ticktack, What's up, Instagram, LinkedIn, YouTube, Snapchat, Reddit and etc.,
(d) Marketing of Financially Engineered life insurance policies with attractive articles and case studies through blogging.
(e) Target marketing is possible through ICTs applications by sending specific emails, ads to target groups in online (Fig. 1).

8.1 Use of E-Commerce and Online Selling of FE Policies

The following benefits are there to the insurers due to the use of ICTs in marketing.

The insurer can catch easily the number of new customer prospects without any mediators or agents at a deadly cheaper cost.

Increases customer loyalty and goodwill of the company due to timely and effective online services to the customer.

We can make easier the application submission while purchasing insurance policies and while submitting claim forms online. Companies can help customers greatly in saving their valuable time.

Fig. 1 Use of ICTs in marketing. *Source* Syeda Kanwal; 2015 https://www.slideshare.net/hestsky/use-of-ict-in-marketing

Introduction of dissemination of valuable customer information through emails, internet applications, skype interviews, service apps, video conferencing, what's up chats will reduce greatly the cost of marketing in procuring new business (Fig. 2).

9 The Role of ICTs Performance in the Improvement of Financially Engineered Life Insurance Business

In the context of insurance, feasibility test or study happening for the financially engineered insurance policies. Means, a systematic analysis which takes all the relevant factors of the financial engineering project into account. These factors include, legal, financial, economical, technical and etc., to ascertain the likelihood of completing the FE project successfully. Competitive analysis, cost-benefit analysis and multi-year sales projections are items that are frequently included in business cases. However, balance sheets and income sheets are not, implying that business case approval is not dependent on detailed actuarial and financial modelling.

B2C: Policyholders of Insurance Companies or Policyholders of mediating insurance brokers, policyholders of mediating technology groups in the insurance industry.

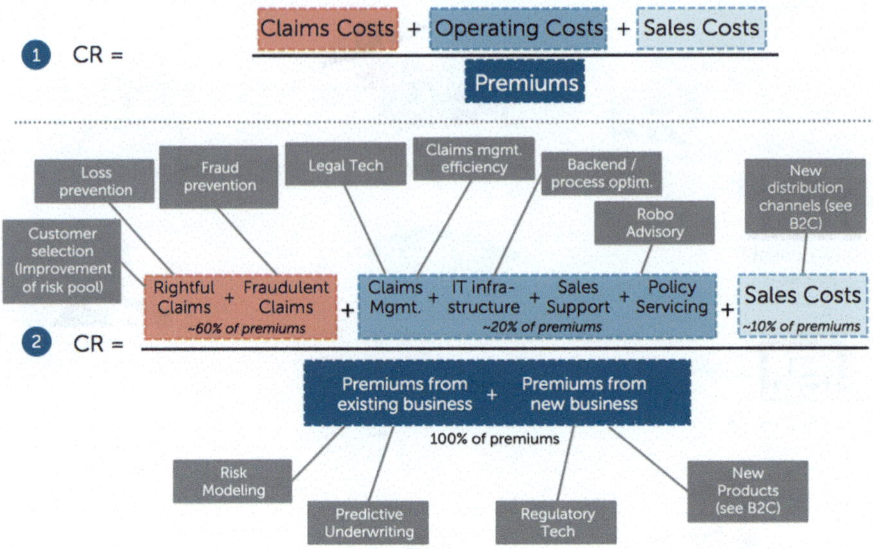

Fig. 2 InsurTech fields in the insurance equations. *Source* Ravi Kurani (2017) https://medium.com/birds-view/dissecting-insurtech-part-1-where-is-the-value-f559d609cce6

ICTs successfully reducing the operating cost of Financially Engineered Insurance Policies (See the model below). One of the most important Key Performance Indicator (KPI) for insurance companies is the Combined Ratio (CR):

Combined Ratio = (Claims Costs + Operating Costs)/Premiums.

10 Conclusion

Based on the above discussions, the researcher concluded that there is a significant role and usage of ICTs in the financially engineered life insurance products and life insurance business processes. Without ICTs, it is not possible to sustain the insurance business practices in the current technological era. More than eighty percentages of the prospects and existing customers of the insurance sector are using mobiles to review the products, its features, feedback and to place an order in the part of final decision making. Hence, ICTs are playing a major role in financial engineering applications in the insurance business and concluded.

References

1. Sapprasert K (2006) The impact of ICT on the growth of the service industry. Proceedings of International Conference "Knowledge, innovation and competitiveness: Dynamics of firms,

networks, regions and institutions. Copenhagen, Denmark, June 18–20
2. The Insurance Business in Albania. Procedia Soc Behav Sci 44:155–162
3. Thuo KJ, Kibera FN, K'Obonyo PO, Wainaina G, Customer relationship management and competitiveness of commercial banks in Kenya
4. Dasgupta P, Sengupta K (2002) E-Commerce In The Indian Insurance Industry: Prospects And Future. J Electron
5. Gupta AK, Govindarajan V (1984) Business unit strategy, managerial characteristics, business unit effectiveness at strategy implementation
6. O'Brien JA (2002) Management information systems: managing information technology in the E. business
7. Karahana E, Straub DW, Chervany NL (1999) Information technology adoption across time, E-Business watch Report
8. Godbole MR, IBMRD's JManag Res, Print ISSN: 2277–7830, Online ISSN: 2348–5922
9. Qarri A, Bazini E, Leskaj B (2011) The impact of ICT use in competitive advantage in SME-s within the service sector in Albania
10. J Inf Technol Econ Dev 2(1):9–19. 11p
11. Inam W (2006) regulatory and institutional challenges of corporate governance in nigeria post banking consolidation. Article Nigeria Econ Summit Group Econ Indicator 12:55–63
12. Julian WF (2004) The role of the regulator in Corporate Governance. Bahamas Director's forum on Corporate Governance in the Bahamas
13. Companies and Allied Matters Acts (2006) Companies proceedings rules and companies winding up rules, case and materials. Princeton, Nigeria
14. https://techterms.com/definition/ict

Sentiment Extraction from Image-Based Memes Using Natural Language Processing and Machine Learning

Devika Verma, Rohit Chandiramani, Pranay Jain, Chinmay Chaudhari, Anmol Khandelwal, Krishnanjan Bhattacharjee, S. ShivaKarthik, Swathi Mithran, Swati Mehta and Ajai Kumar

Abstract The widespread use of image-based memes on socioeconomic or political issues has witnessed a booming effect unparallel to any form of media in the recent years. The ability to go viral on social media in seconds and the popularity of memes on online platforms give a wide scope and pathway for research as it will help in understanding the usage patterns of the public and in turn be used for analyzing their sentiment toward a specific topic/event. In this paper, initially gap analysis on the features used for sentiment extraction on memes is presented. Exploring the correlation of image based and textual features, this paper gives a novel approach (correlating the facial features along with the text in the meme itself) for the extraction of sentiment from image-based memes. This paper also addresses the challenges

---

D. Verma (✉) · R. Chandiramani · P. Jain · C. Chaudhari · A. Khandelwal
Vishwakarma Institute of Information Technology, Pune, India
e-mail: devika.verma@viit.ac.in

R. Chandiramani
e-mail: rohit.chandiramani@viit.ac.in

P. Jain
e-mail: pranay.jain@viit.ac.in

C. Chaudhari
e-mail: chinmay.chaudhari@viit.ac.in

A. Khandelwal
e-mail: anmol.khandelwal@viit.ac.in

K. Bhattacharjee · S. ShivaKarthik · S. Mithran · S. Mehta · A. Kumar
Center for Development of Advanced Computing (C-DAC), Pune, India
e-mail: krishnanjanb@cdac.in

S. ShivaKarthik
e-mail: shivakarthiks@cdac.in

S. Mithran
e-mail: swathimithran@cdac.in

S. Mehta
e-mail: swatim@cdac.in

A. Kumar
e-mail: ajai@cdac.in

S. Fong et al. (eds.), *ICT Analysis and Applications*, Lecture Notes in Networks and Systems 93, https://doi.org/10.1007/978-981-15-0630-7_28

faced in this relatively new area of sentiment extraction on memes. Finally, this paper concludes with insightful results.

Keywords Memes · Sentiment extraction · Machine learning

1 Introduction

Memes as coined and described by ethologist Richard Dawkins, are an "element" of evolution that unlike natural Darwinian selection which is propagated from species to species, is transferred from "brain to brain". This philosophical idea has been drastically modified for its current twenty-first century interpretation. The "Internet Meme" today is widely used in social media for expressing different sentiments or opinions of netizens [1]. They are in the form of short media which is transferred and spread along the social media platforms like Twitter, Facebook, and Snapchat in a second. This media is majorly represented as images, which also contain short, crisp, and mostly satirical text used to express emotions in more detail. These characteristics make the memes more holistic and give researchers the impetus to look into them in great depth.

In a recent study done by the Forbes [8], millennials spend more than 3 h of their day on social media surfing and interacting with Internet memes! Any major or minor event that happens in the world today is being converted to a meme. The fact that there are websites and tools that are used to create memes [1] is a clear evidence of this. Owing to the massive scale on which memes are generated every second, there is a need to understand and devise a system which automatically derives the overall sentiments from them thereby giving us insights as to what the meme is conveying along with its major significances in political campaigns, brand placements and marketing, advertising, alteration in administrative reforms, etc.

In this paper, our goal is to devise a system to automatically infer the sentiments from Image-based memes by proposing an approach by coalescing the textual as well as visual features, unlike conventional sentiment analysis approaches which take only into consideration the individual attributes (only text/image). Also, it has been observed that the relatively unexplored area of memes provides great potential as agents to understand the dynamic and vast opinions of people over the social media today.

2 Literature Survey

According to our survey as elaborated in [3], the work that has been done in the area of memes and memetic images have been niche and only concentrated to few certain approaches like clustering the memes for detection and retrieving social impact [7], event-specific analysis [5, 11, 12], escalation and growth analysis [2, 14]. This is maybe due to the challenges that are faced while dealing with memes due to its wider context, diversity, and types (text, image, and video). French et al. [9] proposed a system for using image-based memes as agents for predicting sentiments using the metadata like comments and caption on image. This survey depicts the trends in which the work on memes is progressing. Work on memes has been done on text-based content with two broad spectrums left relatively unexplored, viz., image- and video-based memes. This stands for our identified gap area to explore further.

3 Methodology

3.1 Dataset

Due to the sheer volume of memes, there exists numerous variety and forms of memes. The visual features of the image have various technical characteristics but the content/domain is a vital one for sentiment. The text is present in various languages, font styles, font sizes, and colors making it difficult for extraction. Hence, the scope is narrowed down to memes having impact font. The domain is majorly restricted to political memes as it has a wide spectrum of opinions and emotions to work with. The dataset is stored as a collection of image files consisting of around 5000 images.

3.2 Input Attributes

3.2.1 Face Recognition

A big share of memes is generated on famous celebrities and trending person(s) on the social media. Hence, by recognizing the face and thereby deriving its personality status helps to get an idea of the sentiment the meme is trying to convey.

Detection of faces in the meme can easily be done using the Haar feature-based cascade classifiers. A Haar cascade is nothing but a classifier used for detecting objects for which it is trained on. The K-Nearest Neighbors (KNN) algorithm is used for recognizing the detected faces. The training data consisted of 6 high resolution face images for each person with varying orientation and expressions. For every image, the face encodings or facial features were extracted. These face encodings are 128 dimensional vector which represents that particular face and consists features like location of the eyes, nose, lips, etc. These facial encodings are saved along with their corresponding output label (Name of the person). Similar operation is then performed on the test image and the closest neighbors are then identified using the Euclidean distance metric for quantifying similarity with a fixed threshold. At last, majority voting of the outputs from the K-neighbors is performed and the face is identified.

3.2.2 Face Emotion

The visual features mostly consist of the face of the person(s) for whom or by whom the meme was generated. The facial expression of the image gives a further depiction of the nature of the meme. For example, if the text in the meme is satirical along with a playful emotion of a face enhances the sarcastic nature of the meme. On the contrary, if the text is a serious negative jibe and the facial features also depict negative tones like sorrow or anger, it enhances the gravity of the situation or the event being depicted. Hence, getting face emotions from meme is imperative to get an all-embracing comprehension of the meme.

For the same, training of classifier is done on the FER2013 [4] dataset. This dataset consists of about 35 k grayscale, 48 × 48 sized face images with 7 various emotions namely, fear, happy, sad, surprise, neutral, angry, and disgust. The 48 × 48 image size is extrapolated to a higher dimension by duplicating the layers to three dimensions and is used as input. This is done as the classifier is trained using the pretrained weights of ResNet50 [15]. ResNet is a convolution neural network architecture that had participated in the ImageNet challenge [13] and was one of the prime algorithms getting near human performance. This concept of using pretrained weights (to learn high-level features) is called Transfer Learning. ResNet50, as our base convolution layer, involves the addition of multiple dense layers with 0.5 dropout to prevent overfitting, and softmax is the classification layer. The entire architecture is compiled using the categorical cross-entropy cost function.

3.2.3 Text Extraction

Preprocessing: The images obtained are generally raw and need to be preprocessed in order to derive and extract useful and valuable features from them.

- Gray Scaling: Gray scaling the image is to remove the multiple color channels leaving behind only one color dimension.
- Binary Image Conversion: Binary image conversion process is used to convert the image into a binary image. It is the process of converting the gray scale image into a digital image in which pixels can have only two possible values.
- Gamma Correction: The operation used for encoding and decoding luminance nonlinearly and based on the value of the gamma exponent (γ).

$$V_{out} = AV_{in}^{\gamma} \tag{1}$$

If $\gamma <$;1: image toward the darker end

If $\gamma >$;1: image appears lighter which is used for background threshing along with binarization operation.

Segmentation is carried out for separating individual characters in an image. This is usually done by the help of line and character recognition algorithms. The main purpose as the name suggests is to find lines that divide the image in specific number of segments. Once these segments are well defined, character detection algorithms are used which, essentially, make tighter bounds on the segments by taking binarized image values to create the boundaries. The matrix representation of the binarized character image (obtained after preprocessing and segmentation) is our main feature. The matrix image is rasterized as a single-dimension tensor and fed to a neural network. Weights are updated using backpropagation. The final output layer contains 36 nodes (26 letters + 10 digits) with a sigmoid activation. The entire model is compiled with the Adam optimizer and categorical cross-entropy loss function.

3.2.4 Text Emotion

Work on sentiment analysis on text from tweets or some form of metadata or manual transcriptions of some written data has thoroughly been explored with accuracies reaching up to about 96%. Such data being generally descriptive highly qualifies them for sentiment extraction with corresponding polarity to quantify the same. Whereas, the text that occurs in memes is very short, about 10–12 words in length, making it very difficult to automatically attain sentiments with certainty as was in the prior. Among the limited annotated datasets that go beyond positive, negative, and neutral as sentiment classes, the emotion in text dataset by CrowdFlower [6] provides shades of emotions like fear, anger, happiness, surprise, etc. as output. Training of model is done using LSTMs on this dataset using GloVE embeddings [10] of the input sentences as features. On validation set, accuracy of about 74% is obtained. But, on the test set of meme text, the accuracy was meager, and hence proving the difficulty on meme text as stated above.

Hence, we observed a conventional NLP approach of a bag of words model, by identifying trigger words for the various output emotions like happy, angry, sad, neutral, surprise, and disgust. These corpuses are then expanded by appending their corresponding synsets. Lastly, preprocessing is done on the text to remove stop words, emoticons, special characters, and generate our own test bag of word and calculate a weighted average gradient score (Table 1).

Table 1 Evaluation results for input attributes

	Training data	Test data	Accuracy (%)
Face recognition	5–6 images of each personality	5000 memes	82.56
Face emotion	FER 2013 Dataset	5000 memes	74.38
Text extraction	On labeled character images in impact font	5000 memes	Character level: 77.89 Word level: 73.12
Text emotion	Emotion trigger words	5000 memes	76.88

3.3 Overall Sentiment

This module is used to derive the overall sentiment of the meme. It is clear from the literature survey that until now sentiment extraction on memes has been done either using only the facial features or only the textual ones including the metadata of the meme. But, in our algorithm, use of the above mentioned textual as well as facial features is done in amalgamation to create a rule-based system. The input features to the correlation module consist of faces identified along with their facial emotion and textual sentiment. Along with it, some additional parameters are also considered such personality status of the person(s) which have been dynamically obtained from hashtags of tweets about the person identified in the meme. The tweets pass through the text sentiment extraction module to get an average polarity and hence personality of the feature in consideration.

These input features are then characterized as rules which can be lucidly expressed as a First Order Logic (FOL). For example, rule for sentiment:

$$F_{1,k} \wedge F_{2,k} \ldots . \wedge F_{n,k} \rightarrow \text{Positive/Negative} \tag{2}$$

where $F_{i,k} \rightarrow$ kth unique value for ith feature

These rules that are formed are then given as input to various machine learning models for training. Since the nature of data and intention of generating memes are to majorly portray some subtle humor with a touch of satire, which in totality is interpreted as negative, the data of the output classes are widely imbalanced. Hence, up-sampling of data is done using Random over-sampling and Synthetic Minority Over-Sampling Technique (SMOTE) [16]. This data is run on acclaimed algorithms for imbalanced data such as Decision Tree, Random Forest, Ensemble Boosting, Penalized Scalable Vector Classifier (SVC). The results of the experiments carried out are summarized below.

4 Results and Experimentation

The following measures are obtained using 5000 memes (Table 2),

Based on the above performance measures obtained it can be inferred that the Random Forest model gives the highest accuracy with adequate F-score. Even Though, the SVC model promises a higher F-score, on testing it shows signs of overfitting hence owing to a lower accuracy. The model can be enhanced by also improving the individual attributes accuracy and training it on a larger dataset with lesser class imbalance (Fig. 1).

Table 2 Performance metrics for overall sentiment

	Random over-sampling				SMOTE			
Model name	Accuracy (%)	Precision	Recall	F-score	Accuracy (%)	Precision	Recall	F-Score
Balanced random forest classifier	78.78	0.52	0.83	0.64	76.51	0.48	0.73	0.58
Penalized scalable vector classifier	69.35	0.77	0.48	0.59	79.54	0.54	0.63	0.58
Boosting classifier	67.64	0.73	0.66	0.66	79.51	0.54	0.66	0.59

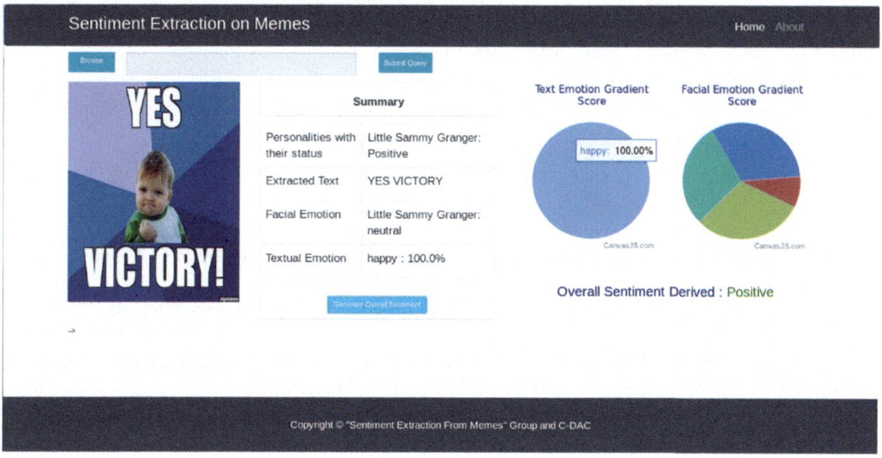

Fig. 1 Integrated dashboard depicting results

5 Conclusion and Future Work

The proposed system for performing sentiment analysis on different memes heavily used in social media platforms, explores various features to predict the sentiment better. It also explores the correlation of image-based features and textual features to predict the sentiment of the meme. A meme being a very complex form of data, needs a high-level understanding of the language and image-based features along with the global knowledge of the domain. These features show a positive pathway into this new field of meme sentiment analysis. The proposed system mainly targets an approach to the better understanding of the image and text using various modern day techniques to extract features that will help to decide a sentiment. Results show success on simple and straightforward memes with little to guess. But, memes due to their indirect and complex nature still remain difficult to extract sentiment from them, also as it may be perceived as different by different people. It is inferred that major improvements need be made into the text analysis of the memes, due to short, indirect and complex nature to better understand the directed message. The lack of gold standard dataset is also an important point to be looked into. Also, an improved mechanism for global knowledge needs to be integrated to access the problem, as most memes are based on current topics or some common references based on real lives. This being a relatively new area, more experimentation needs to be carried out to find better features and architecture to find overall sentiment. We believe that our novel approach is the way forward. With improvements made into modules being more capable, and a stronger and active research community can turn things around and better predict a clean and unambiguous sentiment from images.

Acknowledgements We are grateful to Dr. Krishnanjan Bhattacharjee, Mr. Shiva Karthik, Ms. Swati Mehta and associated guides in Applied AI Group of Centre for Development of Advanced Computing (C-DAC) for the project supervision and suggestions. We also thank our internal guide from College, Mrs. Devika Verma for her support.

References

1. Asur S, Huberman BA (2010) Predicting the future with social media. In: WI-IAT, vol 1. IEEE, pp 492–499
2. Bai J, Li L, Lu L, Yang Y, Zeng D (2017) Real-time prediction of meme burst. IEEE. 978-1-5090-6727-5/17
3. Chandiramani R, Jain P, Chaudhari C, Khandelwal A, Verma D, Bhattacharjee K, Karthik S, Mithran S, Mehta S, Kumar A (2019) Survey and gap analysis on sentiment extraction from image based memes, presented In: International conference on science, technology, engineering and mathematics. IEEE
4. Challenges in representation learning: facial expression recognition challenge. https://www.kaggle.com/c/challenges-in-representation-learning-facial-expression-recognition-challenge/data. Accessed 06 Apr 2019
5. Colbaugh R, Glass K (2011) Detecting emerging topics and trends via predictive analysis of ‘meme’ dynamics. IEEE. 978-1-61284-4577-0085-9/11
6. Crowdflower text data. https://www.crowdflower.com/wp-content/uploads/2016/07/text_emotion.csv. Accessed 06 Apr 2019
7. Emilio F, Jafari M, Varol O, Qazvinian V, Menczer F, Flammini A (2013) Clustering memes in social media. arXiv:1310.2665v1 [cs.S1]
8. Forbes study on memes. https://www.forbes.com/sites/forbesagencycouncil/2018/07/19/more-than-a-trend-meme-marketing-is-here-to-stay/#4bbc96e1248. Accessed 06 Apr 2019
9. French J (2017) Image-based memes as sentiment predictors In: International conference on information society
10. GloVe: global vectors for word representation. https://nlp.stanford.edu/projects/glove/. Accessed 06 Apr 2019
11. He S, Zheng X, Bao X, Ma H, Zeng D, Xu B, Tian G, Hao H (2014) Ranking online memes in emergency events based on transfer entropy. IEEE. 978-1-4799-6364-5/14
12. He S, Zheng X, Wang J, Chang Z, Luo Y, Zeng D (2016) Meme extraction and tracing in crisis events. IEEE. 978-1-5090-3865-7/16

13. ImageNet Object localization challenge. https://www.kaggle.com/c/imagenet-object-localization-challenge. Accessed 06 Apr 2019
14. Luo C, Zheng X, Zeng D (2015) Inferring social influence and meme interaction with Hawkes process. IEEE. 978-1-4799-9889-0/15
15. Resnet-50, https://www.kaggle.com/keras/resnet50. Accessed 06 Apr 2019
16. SMOTE: Synthetic Minority Over-Sampling Technique. https://jair.org/index.php/jair/article/view/10302. Accessed 06 Apr 2019

Prediction of Sedimentation in an Arid Watershed Using BPNN and ANFIS

Sandeep Samantaray, Abinash Sahoo and Dillip K. Ghose

Abstract Suspended sediment model and predicting its concentration in a natural stream are important fundamentals in managing water recourses policy worldwide. Present investigation considers adaptive neuro-fuzzy inference system (ANFIS) and backpropagation neural network (BPNN) to model suspended sediment load (SSL). Rainfall, temperature and SSL data are used to train and validate the model from Mahanadi river in Odisha, India. The estimation results obtained by using the neuro-fuzzy technique are tested and contrasted to those of artificial neural networks (ANNs). Root mean squared errors (RMSE) and coefficient of determination (R^2) are utilized as assessing criterion to evaluate the model performances. Based on research finding ANFIS provides superlative value of R^2 is 0.9625 and 0.9814, but for BPNN it delivers 0.9376 and 0.9592, respectively. Assessment outcomes show that ANFIS is better suited to apply for estimating suspended sediment daily.

Keywords Rainfall · BPNN · ANFIS · Suspended sediment · Watershed

1 Introduction

To understand sediment transport phenomenon in rivers huge steps have been considered in past years. Suspended sediment (SS) movement relies on flow velocity in perspective to flow variability at different segments of waterway. Capacity of basin, waterway routing, existence of production of power energy from water, and water habitants are influenced to a great deal because of shipping of SS. Hence, SSL and its transportation have fascinated the interest of engineers from different facets and various techniques have been employed to estimate SSC.

S. Samantaray (✉) · A. Sahoo · D. K. Ghose
Department of Civil Engineering, NIT Silchar, Silchar, Assam, India
e-mail: samantaraysandeep963@gmail.com

A. Sahoo
e-mail: bablusahoo1992@gmail.com

D. K. Ghose
e-mail: dillipghose_2002@yahoo.co.in

S. Fong et al. (eds.), *ICT Analysis and Applications*, Lecture Notes in Networks and Systems 93, https://doi.org/10.1007/978-981-15-0630-7_29

Kisi [1] scrutinized capabilities of neuro-fuzzy (NF) and neural network (NN) methods for modeling connection amid streamflow and suspended sediment (SS) at Quebrada Blanca and Rio Valenciano station. Rajaee et al. [2] examined suspended sediment load in USA having NF, conjunction of wavelet analysis and sediment rating curve (SRC). Nourani et al. [3] employed ANFIS for modeling Suspended sediment Ajichay watershed, located in the northwest of Iran. Khan et al. [4] predicted variation in the suspended sediment concentration with water discharge using ANNs for the Ramganga River, Ganges Basin. Tayfur and Guldal [5] urbanized ANN model for predicting total suspended sediment on a daily basis. Nivesh and Kumar [6] developed models namely multiple linear regression (MLR), ANFIS and SRC for estimating suspended sediment concentration (SSC) from Vamsadhara River. Kisi and Zounemat-Kermani [7] proposed an ANFIS c-means clustering technique to estimate SSC. Kaveh et al. [8] investigated potential and efficacy of ANFIS trained by means of Levenberg-Marquardt algorithm to estimate SSC in a river. Bisoyi et al. [9] correlated sediment load unpredictability with rainfall and runoff by applying BPNN algorithm to a huge tropical stream. Kisi et al. [10] explored efficacy of ANFIS in estimating suspended sediment in Kizilirmak Basin, Turkey. Nagy et al. [11] used ANN model using BP algorithm for estimating sediment discharge in natural streams in SSC terms. Buyukyildiz and Kumcu [12] investigated abilities of support vector machine (SVM), ANNs, and ANFIS models among the artificial intelligence methods to estimate the SSC in Ispir Bridge gauging station on Coruh River. Kisi and Shiri [13] compares the accurateness of various soft computing techniques, ANNs, ANFIS, and gene expression programming (GEP), in estimation of SSC on natural streams taking hydro-meteorological records. Rajaee et al. [2] investigated potential and efficacy of standard artificial intelligence (AI) techniques to predict SSC of Little Black and Salt River. The objectives of this research are to predict suspended sediment load using BPNN and ANFIS.

2 Study Area and Data

Angul (Fig. 1) lies at the center of Odisha state, India and is located amid 84° 15 E and 85° 23 E longitude and 20° 31 N and 21° 40 N latitude. This region spreads over 6232 km^2 area. Angul is abundant with natural resources. The maximum temperature is 50.9 °C which is usually in the month of May and lowest temperature being 5.2 °C in the month of January with average of 26 °C. Maximum rainfall in this region is 325.3 mm with wind speed at 37 km/h and 85% humidity. Various machine learning algorithms are used for predicting climatic parameters in a catchment [14–16].

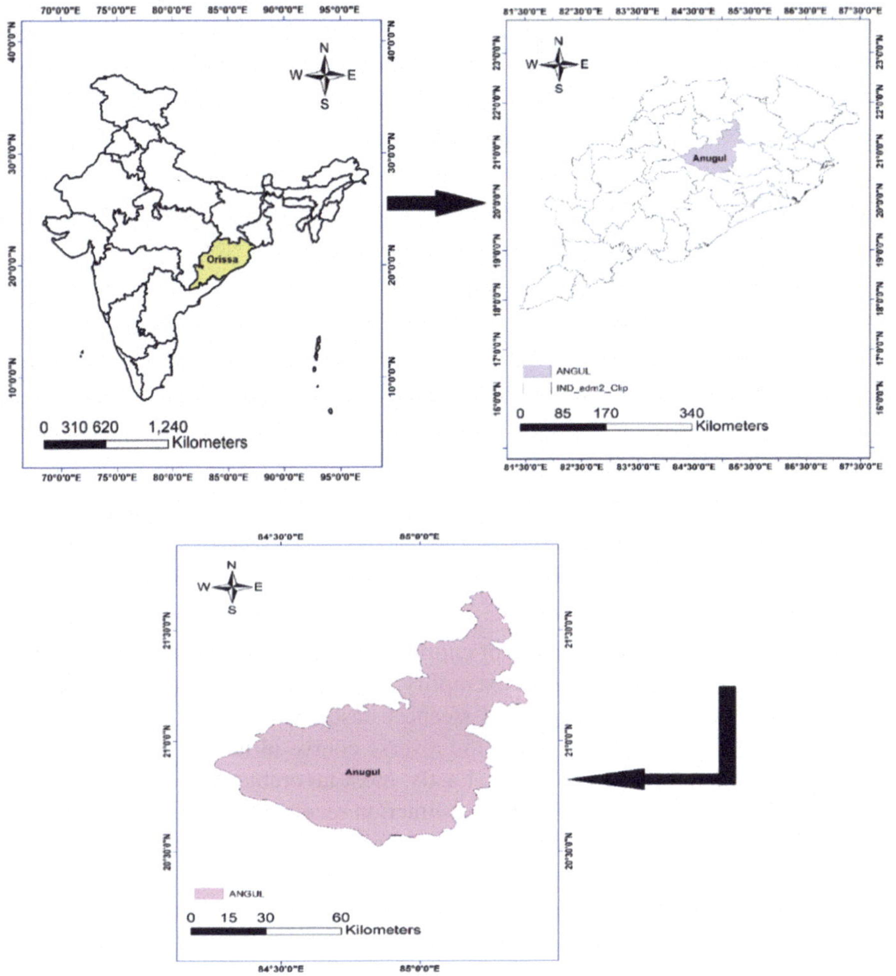

Fig. 1 Proposed watershed

3 Methodology

3.1 BPNN

BPNN is an emblematic multilayer ANN which utilizes backpropagation for training networks. Structure commonly utilized in hydrology for mapping every incessant nonlinear function comprises of three layers: input, hidden, and output layer as

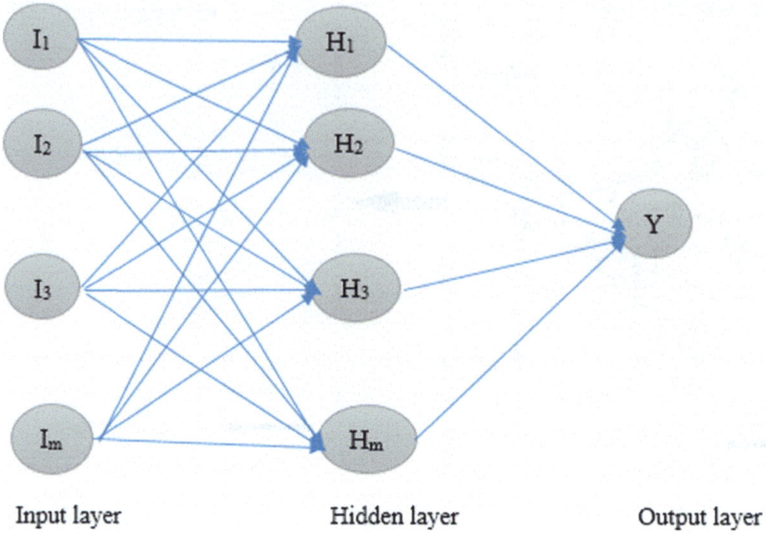

Fig. 2 Architecture of BPNN

depicted in Fig. 2. NN is a collection of enormous nodes having thresholds, creation functions and correlation weights to exemplify network structure. The BP algorithm is an organized technique on basis of steepest descent technique for minimizing overall faults. Output faults are fed in the reverse course of network for modifying threshold data and correlation weights. Lastly, most favorable data can be found by means of iterative modification. Objective function receives root mean square error.

3.2 ANFIS

ANFIS is an amalgamation of fuzzy inference system (FIS) and ANNs that is a thriving innovation by Jang [17]. This method is competent to approximate any genuine incessant function on a compressed set to any extent of accurateness. ANFIS is a feed-forward network having multiple layers that utilizes NN learning algorithms and fuzzy logic for mapping input to output space. Fuzzy assessment policies are applied as membership functions (MFs) and model finds out the best fitting parameters of MFs. MF is a curve which illustrates how every point in input space is mapped to a membership value amid 0–1. ANFIS has prospective for capturing advantages of both techniques in a solitary structure. Therefore, ANFIS is regarded as a universal

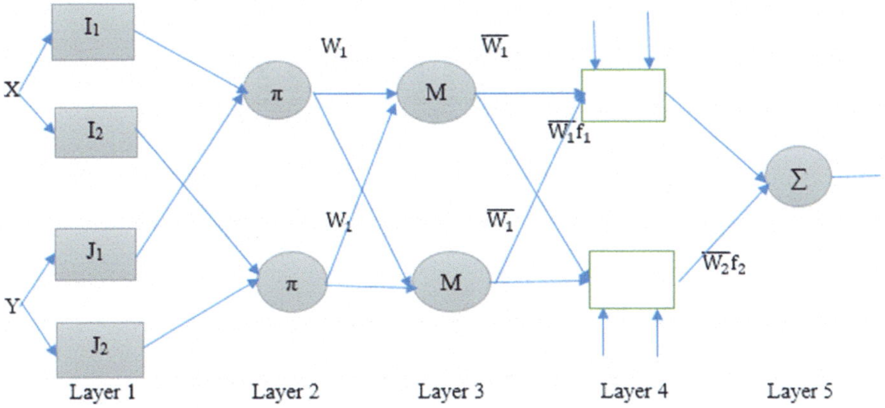

Fig. 3 Architecture of ANFIS

estimator. Its presumption system refers to fuzzy IF-THEN rule which has capability to learn approximating nonlinear functions. Structure of ANFIS is shown in Fig. 3.

4 Results and Discussion

All transfer function performance measured applying ANFIS and BPNN are shown in Tables 1 and 2 correspondingly. Three diverse transfer functions for BPNN and six diverse membership functions for ANFIS are taken into consideration for finding the superlative model which helps in predicting sediment load efficiently. It can be noticed from the table that all models presented suitable outcomes for predicting sediment load. In case of BPNN Tan-sig transfer function with 4-4-1 architecture gives best value of coefficient 0.9592 and 0.9376 in testing and training phase, respectively. But in case ANFIS, Gbell transfer function gives prominent value of R^2 is 0.9625 and 0.9814 for training and testing phase, respectively. Performance value of all transfer functions are given below. Actual versus predicted value for R^2 are presented in Fig. 4.

5 Conclusion

The present study validated efficacy of two different soft computing techniques namely ANN, ANFIS to estimate daily SSL on natural streams. ANFIS results were contrasted with that of BPNN. On basis of results after assessment ANFIS is observed to be performing superior than that of BPNN. Numerous input amalgamations consisting of rainfall, temperature and evapotranspiration of current as well as previous

Table 1 Performance value of BPNN

Model input	Sigmoid function	Architecture (L-M-N)	MSE		RMSE		R^2	
			Training	Testing	Training	Testing	Training	Testing
Precipitation Maximum temperature Minimum temperature Evapotrnspiration	Tan-sig	4-2-1	0.000821	0.004309	0.028392	0.064165	0.9139	0.9358
		4-3-1	0.000789	0.005823	0.027456	0.074887	0.9027	0.9221
		4-5-1	0.000735	0.005009	0.043267	0.058356	0.8878	0.9011
		4-9-1	**0.000622**	**0.003916**	**0.031869**	**0.049967**	**0.9376**	**0.9592**
	Log-sig	4-2-1	0.000498	0.003007	0.019934	0.036856	0.8562	0.8699
		4-3-1	0.000547	0.004034	0.034555	0.063949	0.8725	0.8995
		4-5-1	**0.000669**	**0.004881**	**0.024271**	**0.063460**	**0.9234**	**0.9409**
		4-9-1	0.000558	0.003879	0.034379	0.061133	0.9003	0.9248
	Purelin	4-2-1	0.019289	0.001298	0.138902	0.036011	0.8355	0.8678
		4-3-1	0.000751	0.003339	0.027378	0.057999	0.8549	0.8825
		4-5-1	**0.000483**	**0.001025**	**0.023163**	**0.030858**	**0.9109**	**0.9345**
		4-9-1	0.000394	0.000958	0.019809	0.030932	0.8997	0.9117

Table 2 Performance value of ANFIS

Function	MAE		RMSE		R^2	
	Training	Testing	Training	Testing	Training	Testing
Tri	0.062595	0.174878	0.082746	0.195276	0.8694	0.8978
Trap	0.062998	0.169945	0.084829	0.262856	0.8865	0.9017
Gbell	**0.050034**	**0.048834**	**0.072854**	**0.087219**	**0.9625**	**0.9814**
Gauss	0.062648	0.092176	0.083951	0.129674	0.9569	0.9701
Gauss2	0.058859	0.091654	0.082158	0.133287	0.9458	0.9628
Pi	0.068435	0.014834	0.087956	0.217654	0.9232	0.9437

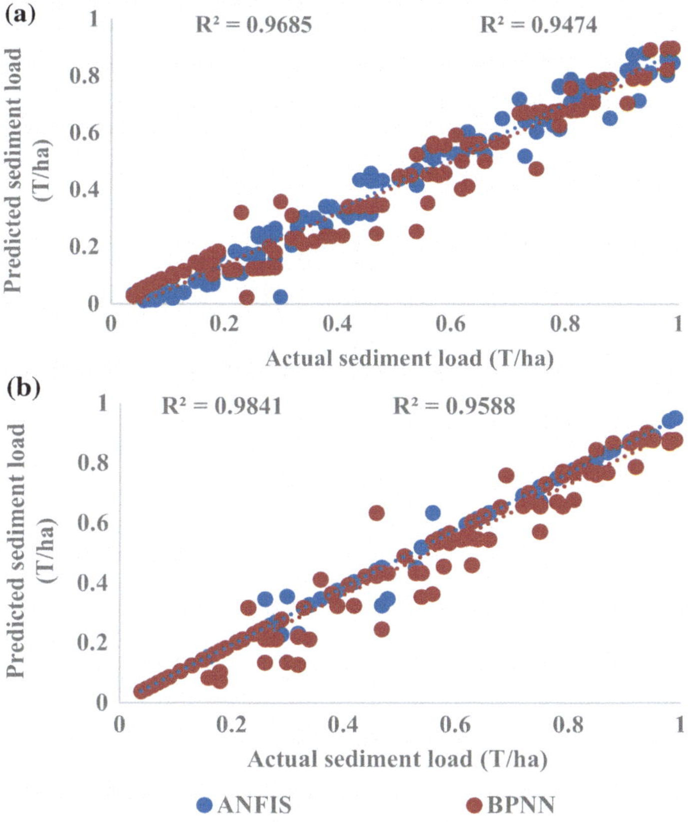

Fig. 4 Actual versus predicted sediment load using BPNN and ANFIS for **a** training and **b** testing phase

days were assembled. Finding SSL is extremely significant for designing and operating different hydrological systems. It will also assist the water resources engineers in a lot of facets if at all it is rightly projected. Present findings can supposedly guide researchers to use AI techniques in assessment of SSL in natural streams.

References

1. Kisi O (2005) Suspended sediment estimation using neuro-fuzzy and neural network approaches/Estimation des matières en suspension par des approches neurofloueset à base de réseau de neurones. Hydrol Sci J 50(4)
2. Rajaee T, Mirbagheri SA, Zounemat-Kermani M, Nourani V (2009) Daily suspended sediment concentration simulation using ANN and neuro-fuzzy models. Sci Tot Environ 407(17):4916–4927
3. Nourani V, Alizadeh F, Roshangar K (2015) Capability of adaptive neuro-fuzzy inference system to model suspended sediment transport (A Case Study: Ajichay River)
4. Khan MYA, Tian F, Hasan F, Chakrapani GJ (2019) Artificial neural network simulation for prediction of suspended sediment concentration in the River Ramganga, Ganges Basin India. Int J Sediment Res 34(2):95–107
5. Tayfur G, Guldal V (2006) Artificial neural networks for estimating daily total suspended sediment in natural streams. Hydrol Res 37(1):69–79
6. Nivesh S, Kumar P (2018) River suspended sediment load prediction using neuro-fuzzy and statistical models: Vamsadhara river basin India. World 2:1
7. Kisi O, Zounemat-Kermani M (2016) Suspended sediment modeling using neuro-fuzzy embedded fuzzy c-means clustering technique. Water Res Manag 30(11):3979–3994
8. Kaveh K, Bui MD, Rutschmann P (2017) A comparative study of three different learning algorithms applied to ANFIS for predicting daily suspended sediment concentration. Int J Sediment Res 32(3):340–350
9. Bisoyi N, Gupta H, Padhy NP, Chakrapani GJ (2019) Prediction of daily sediment discharge using a back propagation neural network training algorithm: a case study of the Narmada River India. Int J Sediment Res 34(2):125–135
10. Kisi O, Haktanir T, Ardiclioglu M, Ozturk O, Yalcin E, Uludag S (2009) Adaptive neuro-fuzzy computing technique for suspended sediment estimation. Adv Eng Softw 40(6):438–444
11. Nagy HM, Watanabe KAND, Hirano M (2002) Prediction of sediment load concentration in rivers using artificial neural network model. J Hydraul Eng 128(6):588–595
12. Buyukyildiz M, Kumcu SY (2017) An estimation of the suspended sediment load using adaptive network based fuzzy inference system, support vector machine and artificial neural network models. Water Res Manag 31(4):1343–1359
13. Kisi O, Shiri J (2012) River suspended sediment estimation by climatic variables implication: comparative study among soft computing techniques. Comput Geosci 43:73–82
14. Samantaray S, Sahoo A (2020) Assessment of sediment concentration through RBNN and SVM-FFA in arid watershed, India. In: Satapathy S, Bhateja V, Mohanty J, Udgata S (eds) Smart intelligent computing and applications. Smart innovation, systems and technologies, vol 159. Springer, Singapore
15. Samantaray S, Sahoo A, Ghose DK (2019) Assessment of runoff via precipitation using neural networks: watershed modelling for developing environment in arid region. Pertanika J Sci Technol 27(4):2245–2263
16. Samantaray S, Ghose DK (2019) Sediment assessment for a watershed in arid region via neural networks. Sādhanā 44(10):219
17. Jang JSR (1993) ANFIS: adaptive-network-based fuzzy inference system. IEEE Trans Syst Man Cybern 23(3):665–685

Design of Microservices Architecture for Home Automation

R. Pushpalatha, Siddhant Verma, Vineeta Tiwari and S. Lakshmi

Abstract Cognitive content, that a lot of IoT application on home automation have implemented, the methodical to build an IoT device is still an obscure. So building an architecture based on microservices for the home automation makes things a little better. The primary objective of this paper is to design microservices architecture for home automation. The service for each sensor is done through microservices based on microservices architecture. The designed system follows rather than a Monolithic Architecture. Microservices architecture is purely a concept when we have so many sensors, where service is created for each sensor which is not dependent on other sensor services. The prototype has been designed to prove that architecture is workable.

Keywords Internet of Things (IoT) · Microservices architecture · Home automation · Prototype

1 Introduction

Nowadays, Internet of things makes wonders in human's day to day life. From a specialized design aspect, IoT can be divided into many layers where it contains more number of devices to connect and analyze them. So the architecture is built using SOA for a database. The paper presents the review on SOA based on microservices. Microservices are the services which are decoupled, that build an application as it is a software elaboration technique, i.e., a variant of SOA. Its architecture has

R. Pushpalatha (✉) · S. Lakshmi
Sathyabama Institute of Science and Technology, Chennai, India
e-mail: pushparajendran887@gmail.com

S. Verma
Pune, India
e-mail: sid0308v@gmail.com

V. Tiwari
Cdac Acts, Pune, India
e-mail: vineetat@cdac.in

S. Fong et al. (eds.), *ICT Analysis and Applications*, Lecture Notes in Networks and Systems 93, https://doi.org/10.1007/978-981-15-0630-7_31

fine-grained services and lightweight protocols. Advantages of decaying an application into acceptable services will improve its standards. Microservices are the software-oriented entity which have the following criteria: execution environment will be based on the virtualization container, microservices can be moved, duplicated independently without any codes or commands, it's an open standard and efficiency and available communication model, and finallyeach microservice should handle its own task.

2 Related Work

Using microservices in home automation has become more challenging, as there are so many related works relevant to the architecture.

Zhiliang et al. [1] proposed an approach in which IoT contains numerous kinds of things, in which it is hard to build a communication layer which is typically a heterogeneous network. So to overthrow this problem a solution to build a conventional SOA-based IoT communication middleware has been found. As a result of this they got time-consuming cure. To make unpackaging, packaging, and network communication process once more, they use gateway transmission mode communication. Throughput gradually increases with the growing prompt time with low number of threads. The test reveals the work of the middleware that it can service about 60 requests per second.

Leu et al. [2] proposed an approach to deduce the service scale of the entire system as there are numerous number of devices engaged in the IoT system. To overcome this he proposed the SOA technique. The actualized planning plan supported by a priority queue model can effectively adjust the reaction messages from the scattered IoT sensors as per each customer's demand. Development of an IoT situation proposes down to earth skeleton dependent on HTTP strategies, it results in transmission wastefulness. The OSGi structure has been institutionalized as an entryway in Service-Oriented Architecture.

Zhang et al. [3] proposed an approach, event-driven SOA, for IoT service to integrate Event-driven Architecture(EDA) and Service-oriented Architecture for IoT services, and step by step instructions to acknowledge versatile EDSOA, so as to fill the gap and keep consistency with event-driven methodology. The theory of Hybrid Process Algebra (HyPA) is described to depict IoT organizations, where IoT organizations' discrete lead can be straightforwardly installed into the greater crossover framework model. A couple of uses and examinations are given to exhibit the thought proof for such event-driven SOA for IoT administration.

Chen et al. [4] proposed an approach since IoT gadgets are physically associated by means of communication networks. Interoperability among heterogeneous IoT gadgets can be given by the SOA. To defeat this they utilized SOA procedure. They build up a method dependent on dispersed collective separating to choose input utilizing closeness. To consolidate coordinate trust and aberrant trust progressively and furthermore to limit combination time and trust estimation predisposition within the sight of noxious hub performing pioneering administration and agreement assaults they build up a novel versatile sifting method. They demonstrated via simulation the superiority of our adaptive IoT trust protocol over EigenTrust and PeerTrust in trust convergence, accuracy, and resiliency against pernicious hubs performing self-advancing, sassing, vote stuffing, and pioneering administration assaults. They additionally considered on persevering assailants.

Cheng et al. [5] proposed an approach to transfer procedure of coal mine well-being checking and control mechanization situations, and furthermore report the measurement and analysis of the platform's performance. It supports the effective dispatching sensory data that provide both provider and consumer using the reliable real-time information distribution model, to arrange those merchants into lattice overlay organize dependent on the network majority to give the offbeat correspondence in an extensive scale, appropriated and approximately coupled IoT application condition. The publish/subscribe communication model can distribute the messages in an asynchronous fashion on demand, however when the scale of the brokers in publish/subscribe communication model is growing up in IoT environment, and the packet delivery rate of data packet will decrease dramatically. An expansion of existing lightweight service mash up middleware, visualization technology, such as 3D technology, can further improve the visibility of sensor objects.

Cheng et al. [6] proposed an approach to upgrade the adaptability and readiness to react rapidly to the dynamic changes in the physical world in an opportune way. Occasion-driven and administration-situated spry IoT administrations correspondence and organization stage is proposed. The lithe IoT administration correspondence and organization is portrayed, which receives occasion-driven SOA component, and comprises of uniform tangible gadgets to get to the board, asset the executives, basic message correspondence foundation, and administration's coordination motor. The brought-together message space is a dispersed overlain arrangement, which depends on distributed/buy-in instrument. It demonstrates the region warming situation with a wide range of boilers, heat exchange stations, and remote room heat metering in brilliant city, it can understand continuous observing of physical gadgets and remote control of physical hardware, and give coordinated IoT administration to warming, vitality sparing, the executives and activity in the region warming industry. The region warming IoT administration situations. It permits the IoT administration correspondence and arrangement react rapidly to the dynamic changes in the physical world in an auspicious way.

Lan et al. [7] suggested to examine IoT sensing service characteristics and proposes future services architecture. It is centered around middleware engineering and interface introduction innovation. To improve that another detecting administration framework dependent on EDSOA (Event-Driven SOA) design is proposed. To help ongoing, occasion driven, and dynamic administration execution. It introduces new IoT program highlights, including utilizing enlarged reality innovation for information and yield and understanding the superposition of the physical world and conceptual data. Here SOA assumes a job where it can sort out and utilize all figuring capacity conveyed in various possession spaces to satisfy client prerequisites. The administrations in EDSOA can be executed simultaneously on the off chance that they get the occasion in the meantime. So the exhibition of EDSOA is superior to SOA in which the administrations execute in an arrangement called by another administration.

Porrmann et al. [8] suggested to tackle the significant attributes and exercises learned by the advancement and model execution of an occasion situated and cloud-based SCADA framework that is built utilizing a microservice design. The microservice structure has executed with two isolated however coordinated territories, an interior and an outer zone. The outer zone incorporates the portrayal of an advantage, similar to the item or a work station, for example a drill station. The portrayal depends on the RAMI4.0 and contains the administration shell. Also, another methodology is acquainted with broadening the RAMI4.0 digitalisation dimension.

Djogic et al. [9] suggested to redesign SOA (service-oriented architecture) integration platform by following principles of microservices design. Recognized group of microservices. Implementation of DB microservice with REST interfaces for main database utilization. Component decomposition into small independent microservices. Removing service contract dependencies. Switching DB consumption to be by using created microservice for that purpose. Creation of local microservice configuration where it was needed. Switching ALL configurations to be local. Setting up queue monitors and queues per microservice. With microservices platform redesign they have made ability for resolving issues, they are maintenance, deployment scalability, improvement of resource management, platform hosting, and production deployment.

Butzin et al. [10] proposed an approach to investigate patterns and best practices that are used in the microservices approach and how they can be used in the internet of things. So that he briefly the overview on some new patterns and best practices that have emerged from the microservice approach or have made it possible. The container technology innovation was explored just as if orchestration or movement ought to be utilized to assemble services together. As a result of this paper we can see that the architectural goals of both, microservices and the internet of things, are quite similar.

3 Proposed System

By the above-related works, the proposed architecture is made on microservices based on services created on both sides. As it is with most of our strategies. The most obvious reason for this architecture is that when something happens in one sensor or service created for the sensor it will not affect the other services created for other sensors, it will be working as usual without any disturbance. The update also can be done with individually. In general the services which are created are independent of each other. So if your job is to change anything in your software, the only thing you will have to work on is the service and not the routine.

Table 1 Comparison between the architecture

	Monolithic architecture	Service-oriented architecture	Microservice architecture
Components	Libraries of reusable codes	Independently deployed services	Services of reusable codes
Inter-component communication	In Process calls	Tightly coupled messages	Publish- subscriber events
Orchestration	Application logic	Service bus	Event queue
Interface definition	Code documentation	Discoverable, provider contracts	Discoverable consumer- driven contracts
Technology stack	Single application platform	Disparate application platforms	Independent application
Persistence	Centralized persistence	Centralized persistence	Ploygot persistence
Data management	Single database	Single database	Multiple database
Consistency	Transactional consistency	Transactional consistency	Eventual consistency

Comparison between thee Architectures for a general architecture is given below (Table 1).

4 General Block Diagram

The General block diagram for home automation based on the analysis made above and the architecture shown below (Fig. 1).

In general, the working of this architecture is based on the services on both the ends (i.e., server side and client side). The sensor layer will consist of sensors which are included in home automation. The data will be gathered by data acquisition protocol from the sensor layer. The data will be sent in the form of request to the proxy layer (The proxy layer is used in the user side like when the user is often searching for one url or site in the webpage, again in case if that user search for the same side then, next time it will be loaded soon on the webpage). Then it goes to the load balancer (The load balancer will balance the data send to the server in case large amount of data send to the server).

The data then sent to the API gateway (The function of API gateway is to know which service layer is suitable for the current sensor service in the service layer). The data will be sent to the service layer. All the above actions are done bidirectionally. The database, service maintenance, analytics, and reporting layers are the function which is monitored from outside. The logger service is the layer which will denote the current status of the service which is created.

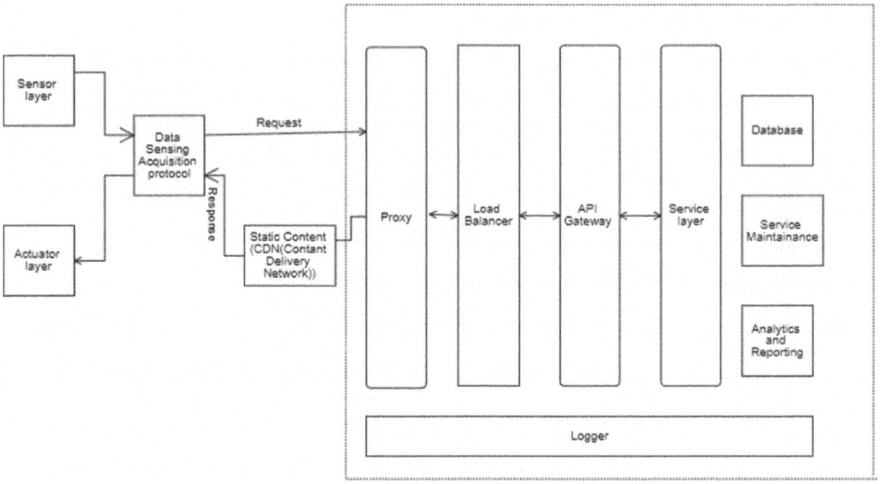

Fig. 1 General architecture for home automation

Once the process is over, the final output will be displayed in the form of notification message or alarm or mobile application or through webpage with url or with the IP address along with the port number.

The generic block diagram is also done, in which the possible sensors for home automation along with the related service layer which is elaborated. The generic architecture is shown below (Fig. 2).

The above generic architecture is done based on the possibility of the sensors that can be applicable for home automation but it will be an elaboration of general architecture. The actuator layer will be a layer in which the final output will be displayed through either notification messages or alarm or mobile application. The service layers have an individual service for each sensor along with the inbuilt database.

The Actuator layer will also have inbuilt database in the service layer. This will be applicable when the entire home is automated. The services which are created at the sensor end and the service layer will be independent of each other.

5 Prototype Implementation

Prototype is done to show that the above architecture is workable. The prototype implementation part is done with three sensors (i.e., Temperature sensor, Moisture Sensor, LDR Sensor). Three sensors are created with services at the sensor end. The data will be collected through raspberry pi3 B + model which will act as a data acquisition protocol sensor (as shown in Fig. 1) from the sensors. Different services are created for the sensors using the docker cluster formation done on both server side service and sensor side service.

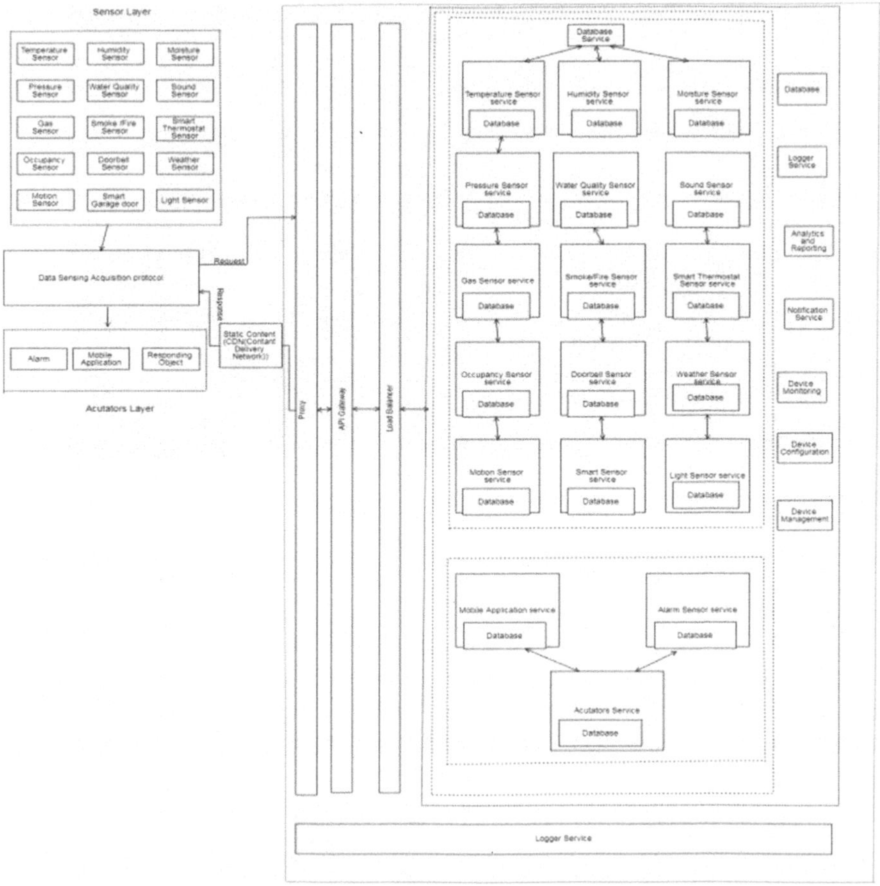

Fig. 2 Generic architecture for home automation with possible sensors

The setup for the prototype implementation is shown below (Fig. 3).

5.1 Service Created at Sensor Side

The three services which are created at the sensor side for the three sensors are independent of each other. The codes are done with python, since raspberry pi will be more adapted to python code, in which this process will be implemented on mqtt protocol (Publish subscribe communication). The below are the three pictures which show the services created for the three sensors on sensor end.

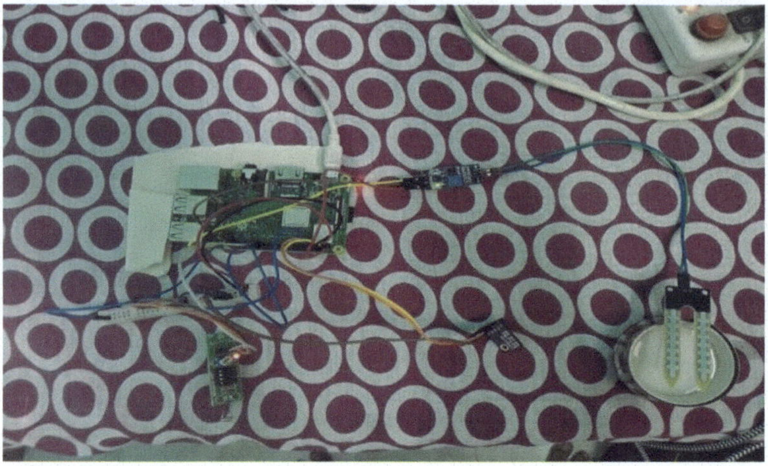

Fig. 3 Setup for the prototype implementation

The code for the moisture sensor and the LDR sensor is given below (Figs. 4, 5 and 6).

The above are the three services at the sensor end; the docker cluster formation is done to join those services using docker containers. Here the docker is used as the API Gateway to identify the appropriate service for the sensor which goes to the API gateway. The docker cluster formation based on container is shown below (Fig. 7).

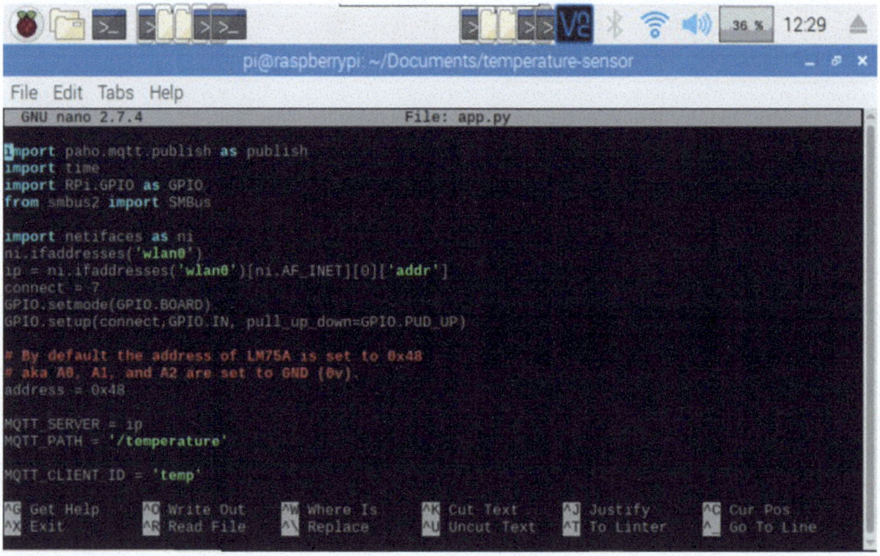

Fig. 4 Code for temperature sensor service

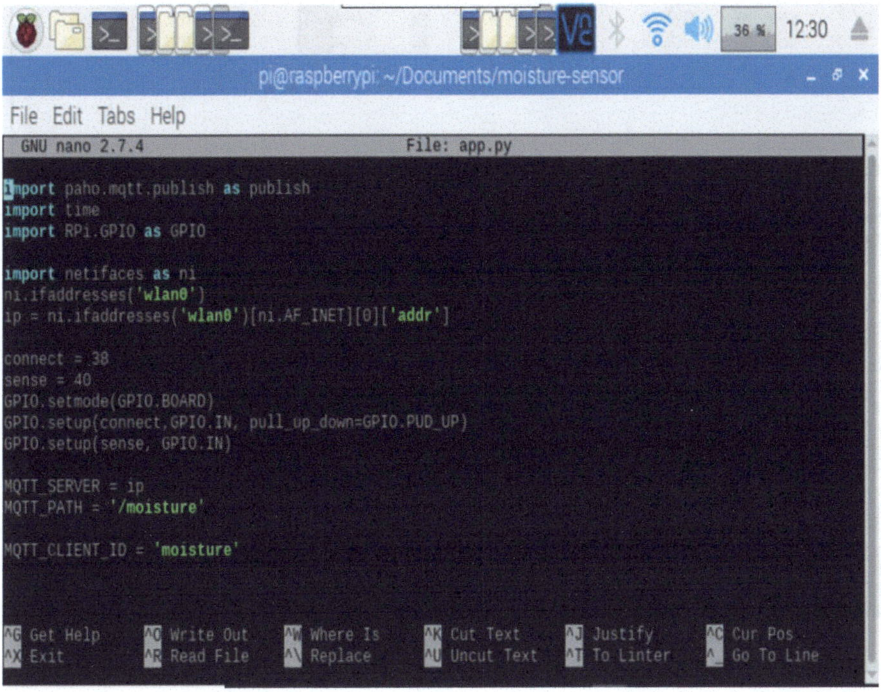

Fig. 5 Code for moisture sensor service

5.2 Services Created at Server Side (Service Layer)

The services which are created for the server side (i.e., service layer) will be independent of each other. All the three sensors will create with the service individually in the service layer (Figs. 8, 9 and 10).

The three codes on the service layer will be initiating that in case if some error or if the sensor is unplugged from the place it got connected will also be displayed. The display will be in the form of "The sensor got Unplugged" The logger service also will be activated through this process either connection is successful or not.

The above are the three services at the service layer; the docker cluster formation is done to join those services using docker containers. Here also the docker is used as the API Gateway as same as the sensor service to identify the appropriate service for the sensor which goes through the API gateway. The docker cluster formation based on container is shown below (Fig. 11).

The Mongodb is used for the database over the service layer which joins all the service at the service layer. The main objective of Mongodb is to increase the performance, availability and manages the application more efficiently scalable.

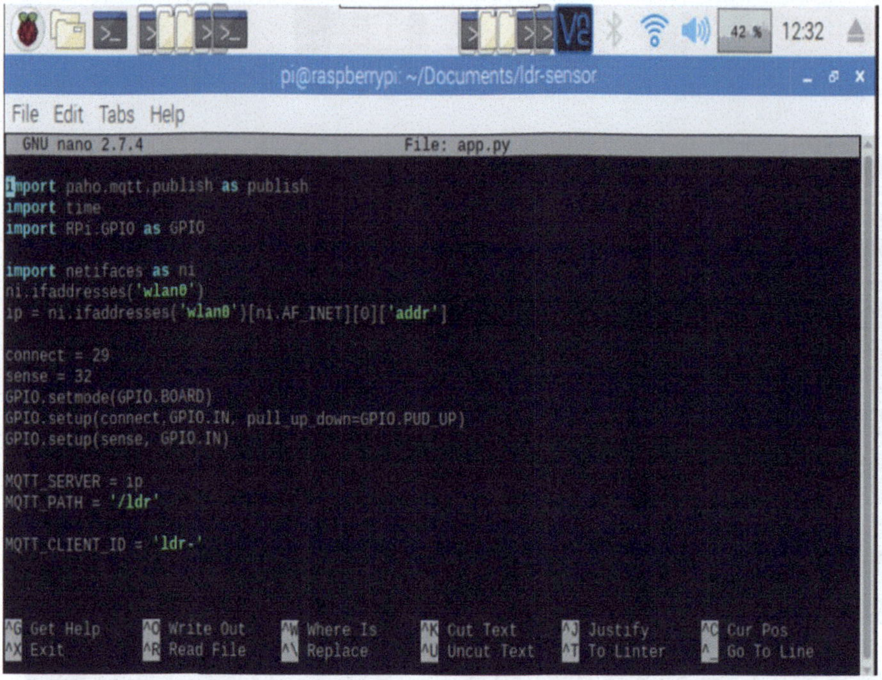

Fig. 6 Code for LDR sensor service

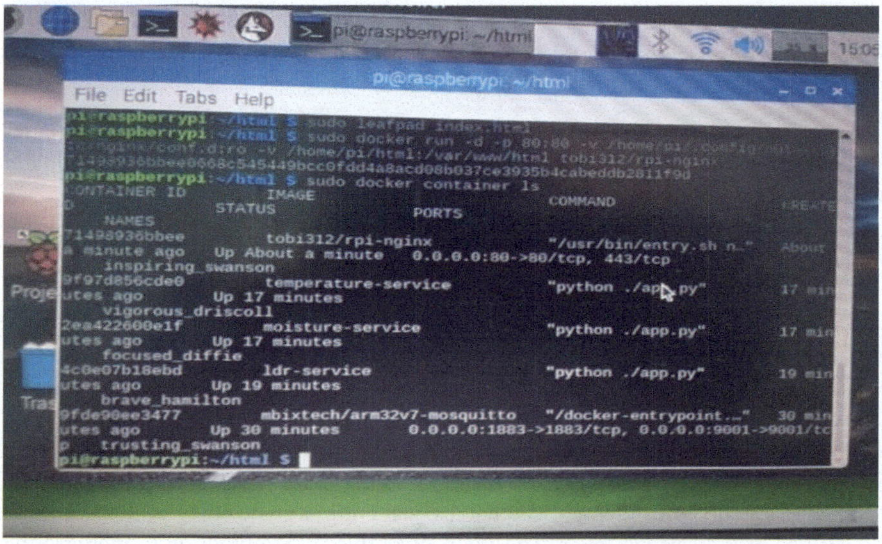

Fig. 7 Formation of docker container for service at sensor side

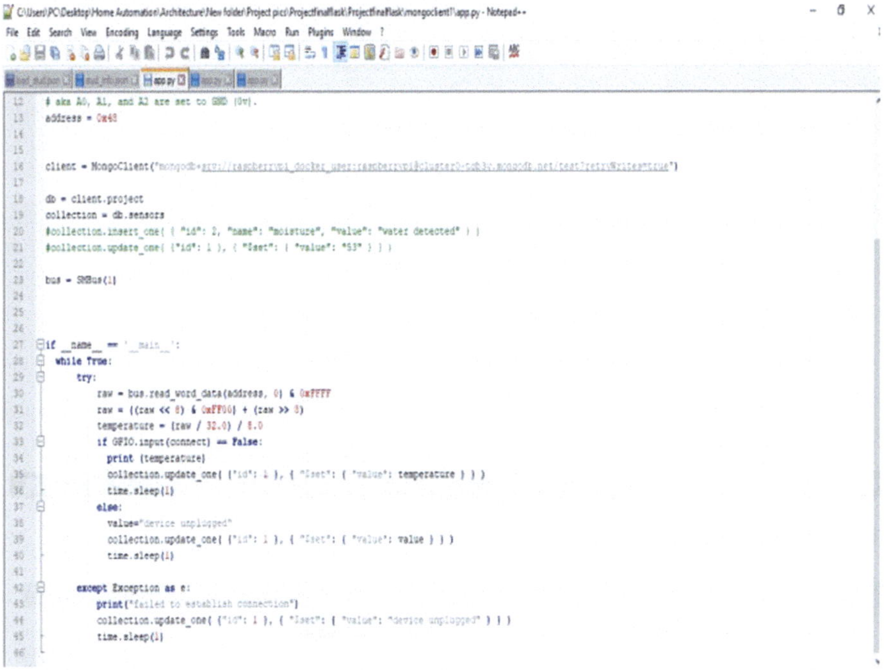

Fig. 8 Code for temperature sensor service for service layer

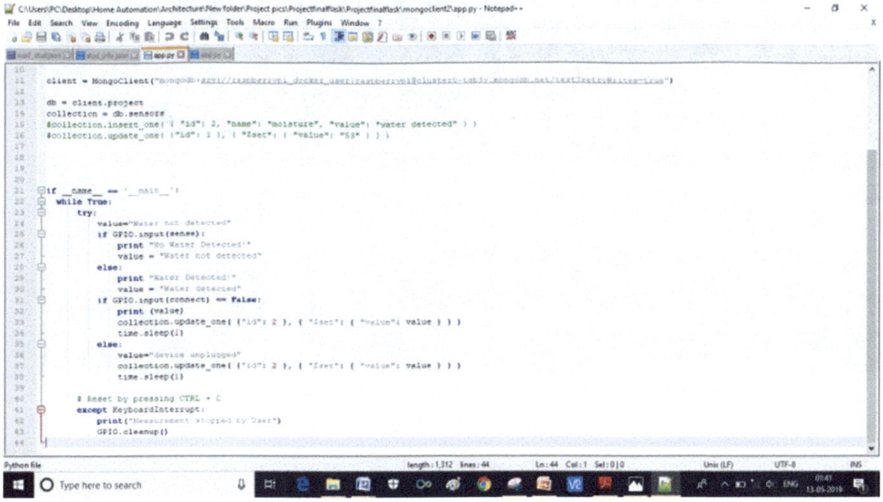

Fig. 9 Code for moisture sensor service for service layer

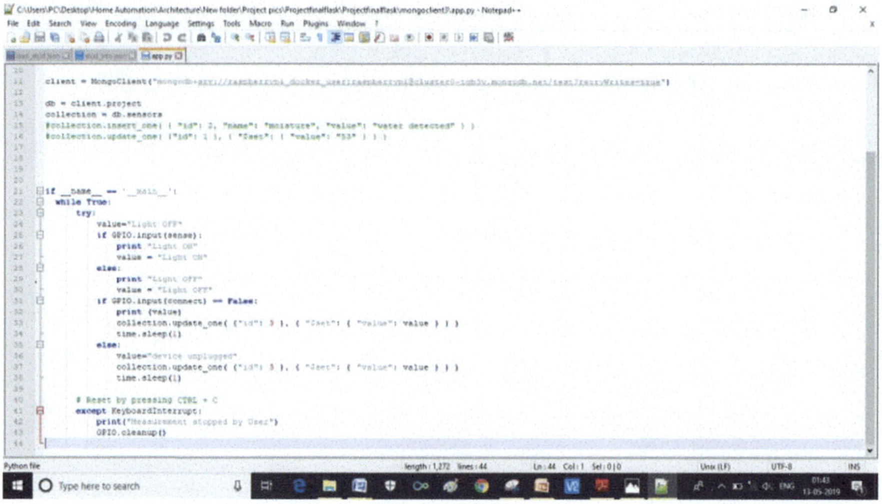

Fig. 10 Code for LDR sensor service for service layer

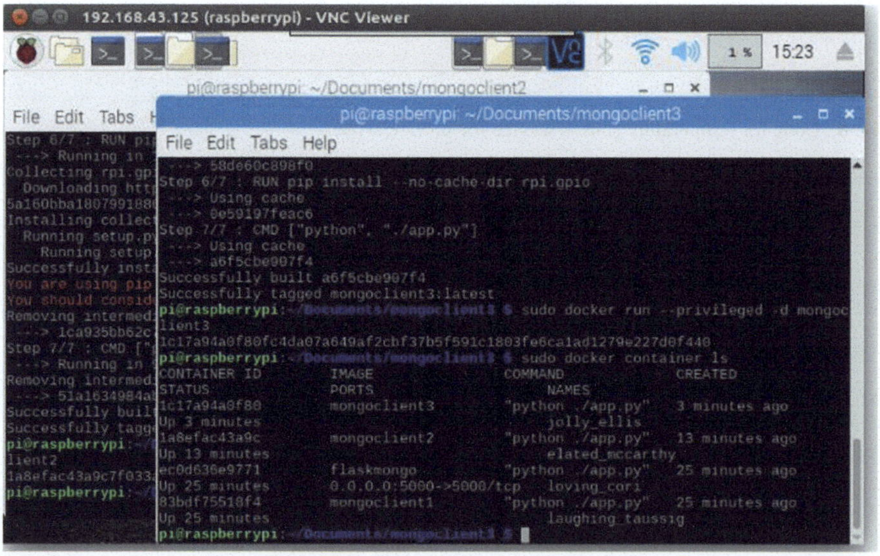

Fig. 11 Formation of docker container for service at the service layer

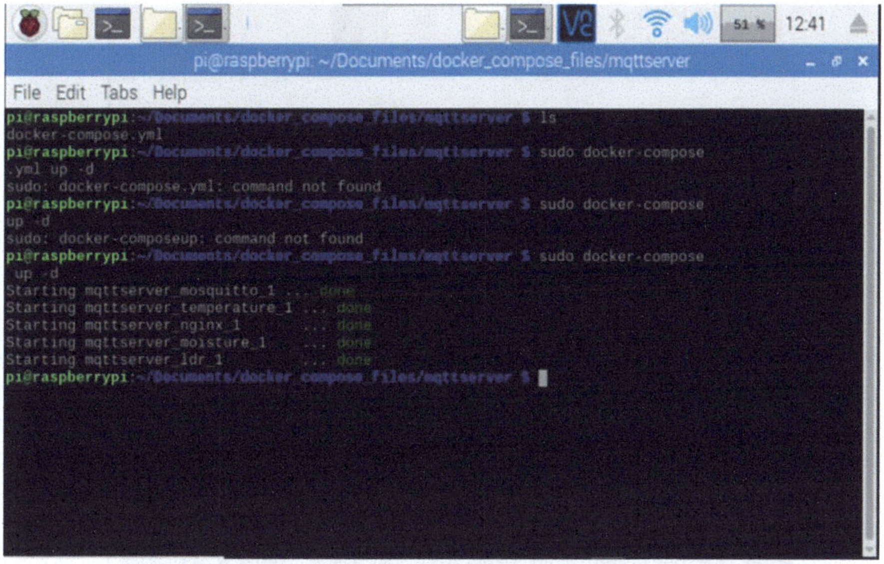

Fig. 12 Running of docker container at MQTT on the sensor layer

5.3 Result and Analysis

The output will be in the form of digital values for temperature whereas for LDR sensor it will detect the presence of light whether it's ON or OFF and for moisture sensor it will show whether detected or not. The running of docker containers at both ends (Sensor Layers and the Service Layer) will be shown in Fig. 12.

The Sensor layer is done with MQTT protocol which uses Publish/Subscribe Communication. Where on the other side for service layer they are done with MQTT client and got together with Mongodb on the database server where Flask mongo and also reverse proxy is implemented on the docker container. The running of docker container on service layer is shown in Fig. 13.

5.4 Output

The output will be in the form of digital values for temperature sensor, and detecting the presence of light whether it's ON or OFF and also detecting the presence of moisture content in water as we discussed before. The output will be shown below in which all three sensors are working.

The running docker with both the sensor and service layer that are with docker container cluster formation is shown below (Fig. 14).

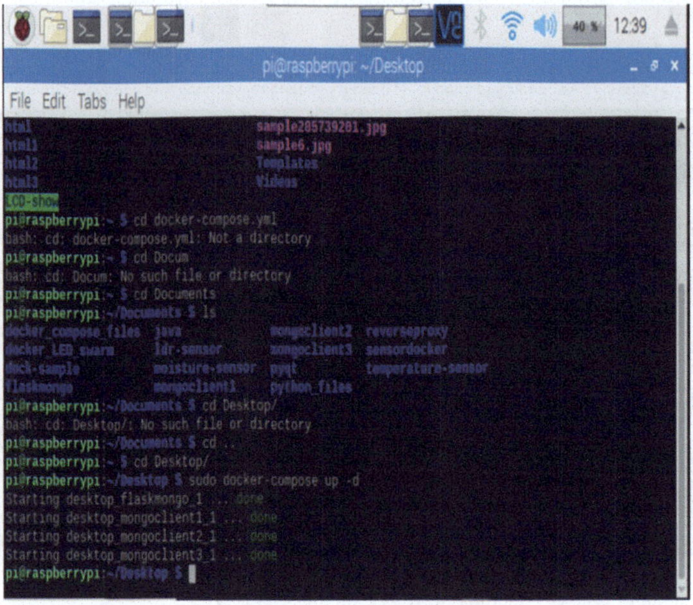

Fig. 13 Running of docker container at MQTT on Service layer

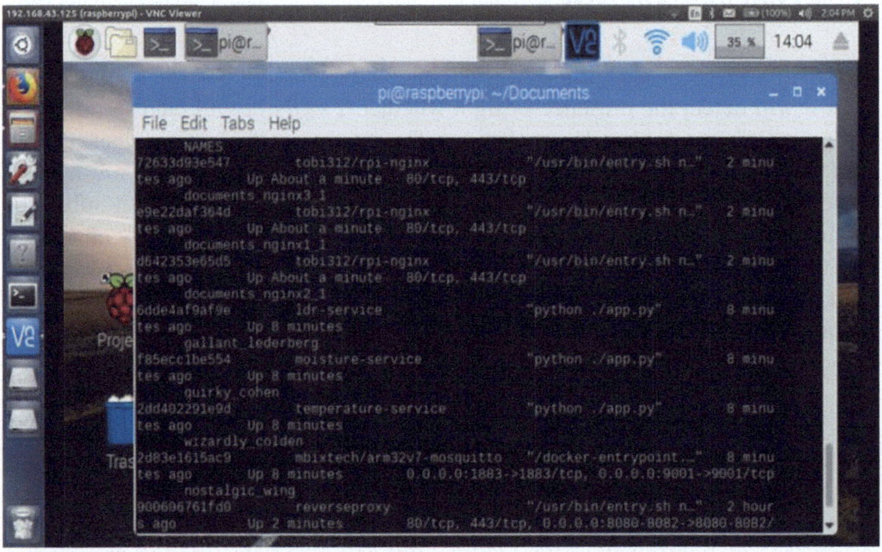

Fig. 14 Running of docker container at MQTT on service layer and sensor layer

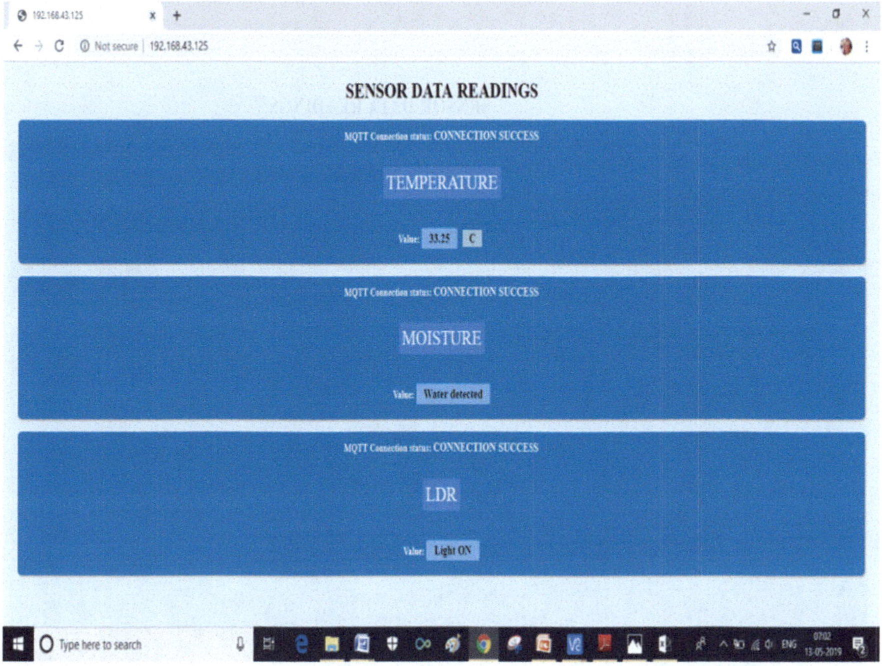

Fig. 15 The working of three sensors

In case sometime if the sensor is removed or unplugged from its place where it got connected will also be displayed as "Device unplugged", but it won't affect the working of the other sensors which are working (Figs. 15, 16, and 17).

6 Conclusion

This paper describes the various analysis of three sensors by using microservice architecture. In this architectural style, the services given by software arrangements are separated into littler parts and concentrated on the particular administration of some usefulness. The methodology of creating microservices with the development of littler programming segments has various favorable circumstances over the conventional solid engineering, for example, expanding the flexibility of the product actualized as a microservice and the simplicity of scaling the arrangement executed through the microservices. The advantage of MQTT is when it is displayed in the

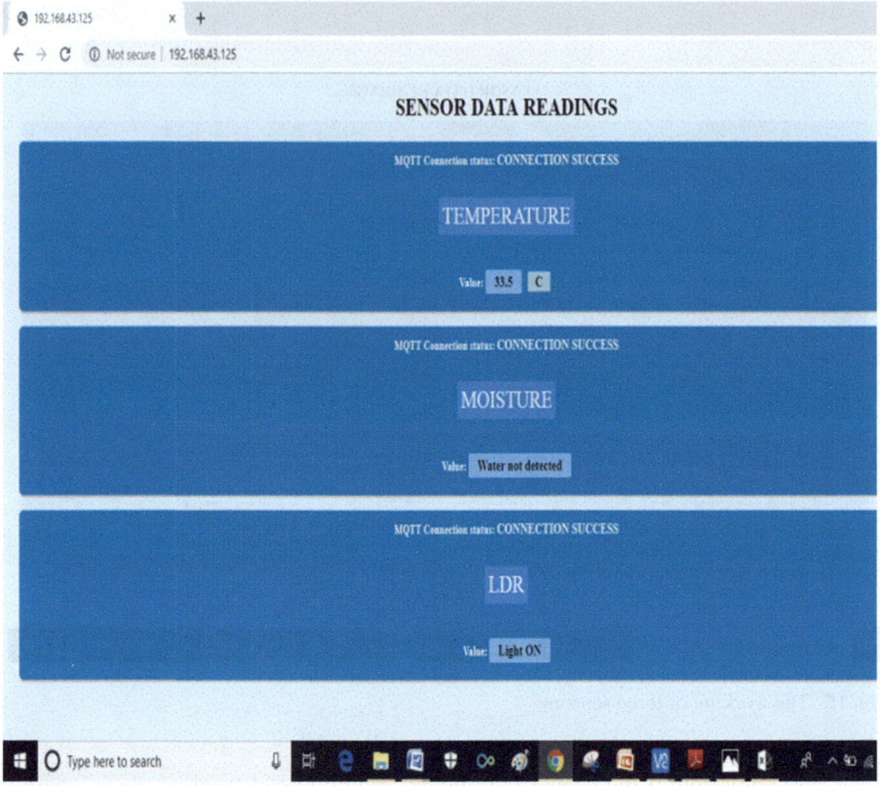

Fig. 16 The working of two sensors when one is not working

webpage there is no need of refreshing the pages, the changes are done automatically in the real-time applications. If we use REST architecture the only disadvantage it has is when it is displayed in the webpage there is a need to refresh the pages every time there are updates or the changes are done.

7 Future Work

The future work of this can be done with so many sensors which will be applicable for home automation.

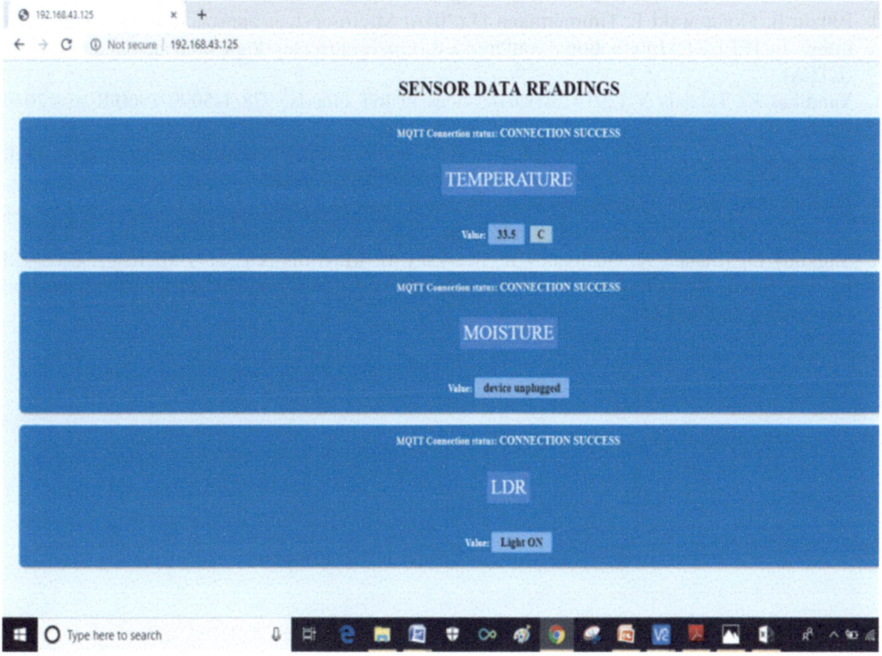

Fig. 17 The working of two sensors when one is removed/unplugged

References

1. Zhiliang W, Yi Y, Lu W, Wei W (2011) A SOA based IoT communication middleware. In: 2011 International conference on mechatronic science, electric engineering and computer, 19–22 Aug 2011. Jilin, China
2. Leu JS, Chen CF, Hsu KC (2014) Improving heterogeneous SOA-based IoT message stability by shortest processing time scheduling. IEEE Trans Serv Comput 7(4)
3. Zhang Y, Duan L, Chen JL (2014) Event-driven SOA for IoT services. In: 2014 IEEE international conference on services computing
4. Chen R, Guo J, Bao F (2016) Trust management for SOA-based IoT and its application to service composition. IEEE Trans Serv Comput 9(3)
5. Cheng B, Zhu D, Zhao S, Chen J (2016) Situation-aware IoT service coordination using the event-driven SOA paradigm. IEEE Trans Netw Serv Manag 13(2)
6. Cheng B, Zhao S, Niu M, Chen J (2018) Agile IoT service communication and orchestration platform using event driven SOA paradigm. In 2018 IEEE conference on computer communications poster and demo (INFOCOM'18 Poster/Demo)
7. Lan L, Wang B, Zhang L, Shi R, Li F (2018) An event-driven service-oriented architecture for the internet of things service execution. In: 2018 IEEE conference on computer communications poster and demo
8. Porrmann T, Essmann R, Colombo AW (2017) Development of an event-oriented, cloud-based SCADA system using a microservice architecture under the RAMI4.0 specification: lessons learned. In: 2017 IEEE conference
9. Djogic E, Ribic S, Donko D (2018) Monolithic to Microservices redesign of event driven integration platform. In: IEEE 2018 41st international convention on information and communication technology, electronics and microelectronics (MIPRO)

10. Butzin B, Golatowski F, Timmermann D (2016) Microservices approach for the internet of things. In: IEEE 21st International conference on emerging technologies and factory automation (ETFA)
11. Vandikas K, Tsiatsis V (2016) Microservices in IoT clouds. 978-1-5090-4960-8/16@2016 IEEE
12. Power A, Kotonya G (2018) A microservices architecture for reactive and proactive fault tolerance in IoT systems. 978-1-5386-4725-7/18 @2018 Crown
13. Babaei Z, Rahmani AM, Rezaei A (2016) Real-time reusable event-driven architecture for context aware systems. In: 2016 24th Iranian conference on electrical engineering (ICEE)
14. Salikhov D, Khanda K, Gusmanov K, Mazzara M, Mavridis N (2017) Microservice-based IoT for Smart Buildings. In: 2017 31st International conference on advanced information networking and applications workshops

Internet of Things Driven Gesture Mimicking SMART Robotic Palm

Jignesh Patoliya, Dhruvang Shah, Uzma Shaikh, Kandarp Rastey and Mohammad Tausif Shaikh

Abstract SMART Robotic Palm is developed to mimic the motion of a human thumb, fingers, and palm. This paper proposes embedded technology-driven electromechanical setup capable of imitating human palm gestures and movements with the use of electronics. This setup is capable of generating voltage corresponding to motion of fingers and then feeding it to the controller in the transmitting glove, which translates the voltage values to angles (in degrees). The angle values (or the value change) are transmitted to the Cloud-based Internet of Things platform using transmitter controller's Wi-Fi capability via the Internet in form of data packets. The receiver end or the artificial robotic palm's controller receives these angle values in real time to duplicate and replicate this motion or imitation. The SMART Robotic Palm's dexterity and its finesse in imitation of motion is achieved using linear potentiometers, NodeMCU DEVKIT 1.0 as the Internet of Things platform which consists of ESP8266EX as its microcontroller and the servo motors in the robotic palm's fingers actuating motion. More precise control and motion shall be useful in numerous applications especially in the health sector for performing surgeries remotely in critical and/or emergency situations.

Keywords Internet of Things · Imitation · Embedded technology · SMART Robotic Palm · Linear potentiometers · Servomotors

J. Patoliya (✉) · D. Shah · U. Shaikh · K. Rastey
Department of Electronics and Communication Engineering, Chandubhai S Patel Institute of Technology, CHARUSAT, Anand 388421, India
e-mail: jigneshpatoliya@charusat.ac.in

D. Shah
e-mail: dhruvang.shah.ds@gmail.com

U. Shaikh
e-mail: uzma.shaikh630@gmail.com

K. Rastey
e-mail: kprastey@live.com

M. T. Shaikh
Department of Electrical and Computer Engineering, Herbert Wertheim College of Engineering, University of Florida, Gainesville, FL 32601, USA
e-mail: mshaikh786@ufl.edu

S. Fong et al. (eds.), *ICT Analysis and Applications*, Lecture Notes in Networks and Systems 93, https://doi.org/10.1007/978-981-15-0630-7_32

1 Introduction

Palm is an interesting part of the human body that has inspired the biomechatronic development of the robotic palm similar to other instances wherein many species and nature's biological phenomenon have inspired numerous engineering feats. The challenge for researchers and engineers is to replicate the basic palm motions as well as capture the dexterity of the human palm. There have been many instances where robotic palms have been designed and developed, however the main challenge with many such replications is the high engineering costs (both recurring and nonrecurring costs) involved in mass-producing the dexterity of the palm. Such robotic palms, if available at reasonable costs have the potential to be used in wide applications based on the robotic palm's finesse in replicating the human palm's motion.

The cost challenge has motivated the development of low-cost, effective, and dexterous SMART Robotic Palm capable of imitating human palm's degrees of freedom that can be used in research projects over other high-cost designs and implementations. Also, the ability to remotely control the robotic palm wirelessly using the Internet of Things (IoT) based on the Cloud technology is a step over the primitive robotic palm prototypes which are controlled using wired external peripherals. The only prerequisite for Internet of Things (IoT) based robotic palm technology to work is the Internet availability to remotely imitate actions based on the human-controlled transmitting glove.

Moreover, other advanced technologies such as gesture control, use of machine learning technology to train the palm perform tasks and actions based on learning, and the use of mind waves to control palm's motion have been researched and demonstrated for their potential [1]. But with the cost trade-off, the Internet of Things (IoT) based gesture mimicking SMART Robotic Palm seems to be feasibly reliable for research and for commercial use in current time-space as compared to the other technologies.

2 Theory and Principle of Operation

2.1 Internet of Things

Internet of Things (IoT) is an emerging technology that helps extend the network by interconnecting numerous devices at the same time. The foundation is laid on the principle of anytime and anywhere connection primarily based on high-speed data transfer and connectivity. Several sensors, actuators, and systems formed by clusters of such electronics in a network are connected with cloud as an interacting medium. The connected devices communicate with each other and form a larger system which transfers with cloud as the central storage. The information sent in form of data packets [2, 3].

There are mainly three layers responsible for this communication to the Cloud: perception layer, network layer, and the application layer [4, 5]. The perception layer also known as recognition layer, forms the lowest layer of conventional IoT architecture. It identifies the embedded systems connected and aggregates the important information from the embedded systems. It identifies each system uniquely with the RFID tag assigned to each system. The next layer is the network layer, acting as the brain and controls all the communication between the devices interfaced with IoT. It is responsible for processing and transferring data as well as network management. The third layer is the application layer, which helps in authentication and key agreement between various networks. This technology is popular in various areas such as biomedical and healthcare, military, industrial automation, consumer electronics, agriculture, and for smart system development [6]. It is efficient, fast, accurate, reliable, (stores data in a Cloud platform) and provides ease of access.

2.2 Biomechatronics and Biomedical Robotics

Biomechatronics is a vast discipline which encompasses the domains of medicine, electronics, mechanics, neurosciences as its subsets. To get the movements of the artificial palm we need to calibrate artificial robotic palm against the human palm movements with the help of electronic system [7].

The concept of Internet of Things is used to create biomechatronic robotics (inspired by human physiology) and is capable of remote control and operation without any physical connections (or intervention). One such attempt is the SMART Robotic Palm which uses this concept to remotely control the palm movement. With the help of such technology one can think of remote surgeries in emergency conditions. This will open new doors in the field of biomedical technology and possess the potential to provide surgical assistance to the remote and poorest parts of the world.

3 Logic Development and Constructions

The construction of the IoT-driven SMART Robotic Palm comprises of two physical parts and an Internet of Things (IoT) based Cloud Platform connecting them.

3.1 Main Parts and Their Operations

Transmitting Glove: The first physical part, which is a wearable glove with five linear potentiometers (acting as sensors) attached, translates motion to voltage. The NodeMCU DEVKIT 1.0 [8] with an ESP8266EX SOC helps map these voltage changes (values) to angle values (in degrees) based on the amount of degree the

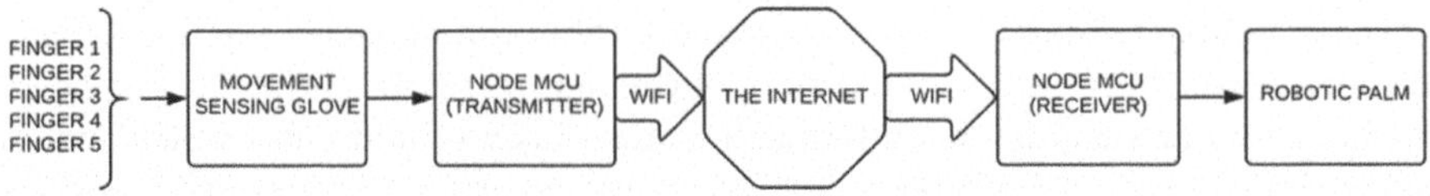

Fig. 1 Working flow of the IoT gesture mimicking driven SMART Robotic Palm

fingers have moved using a mapping function in Arduino IDE and transmit it to the online IoT Cloud, using the Wi-Fi module available on ESP8266EX SOC.

Internet of Things (IoT) Based Cloud Platform: The Adafruit IO is a Cloud-based IoT platform. It receives the angle values from the transmitting glove as signals, stores and relays them in real time to the receiver artificial palm.

Receiving Palm: The second physical part is an actual electromechanical robotic palm inspired from the biomechanics of the human palm. The receiver comprises of NodeMCU DEVKIT 1.0 with an ESP8266EX SOC to receive the Wi-Fi signals from the IoT Cloud. The ESP8266EX SOC receives the angle values (in degrees) and coverts these values back to voltage values based on the mapping function logic. These voltages are fed to the five servo motors present—one on each finger of the SMART Robotic Palm aligned to the actuators. This is the receiver module acting as the artificial palm. The motion of human palm on the transmitting glove is replicated, based on actuators corresponding to the voltage change. These actuators are connected to the fingers using strings (Fig. 1).

3.2 *Transmission End Logic Based Construction*

The transmitting side verifies the WLAN-SSID and WLAN-PASSWORD of the router to which the NodeMCU will connect in order to create a secure communication channel between the transmitter and Adafruit cloud. Once the controller is connected to the Internet it will search for the Adafruit cloud. The cloud requests username and the unique AIO key from the NodeMCUs connected at both transmitting and receiving side. On verification of both credentials, only the cloud accepts the data and allows the communication channel between the two NodeMCUs (Fig. 2).

The NodeMCU uses UART protocol for communication. Once the connection is established the voltage values that are mapped to corresponding angles values (based on map function in Arduino IDE) are transmitted to the cloud. These values are then received by the receiving NodeMCU. All the five connected linear potentiometers are transmitted independently and are received by the corresponding servo motor. Potentiometer 1 is mapped to Servo motor 1, Potentiometer 2 is mapped to Servo motor 2, so on and so forth. The values are transferred in the presence of an internet

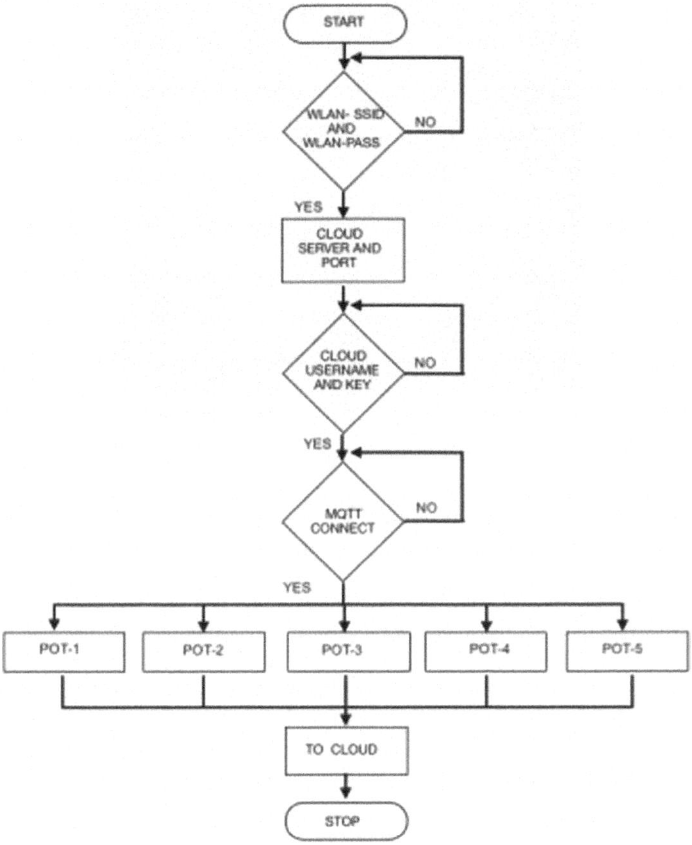

Fig. 2 Transmission logic flow

connection. In case of no internet connectivity, the data sent on cloud will not be transferred and the packets received will be discarded. This leads to inactive robotic palm.

The faster the internet connectivity, more effectively the data will be transferred within a short time span. The internet connectivity's, data speed's and stability's direct relation to the performance of the SMART Robotic Palm is particularly critical for robot-assisted surgeries (Fig. 3).

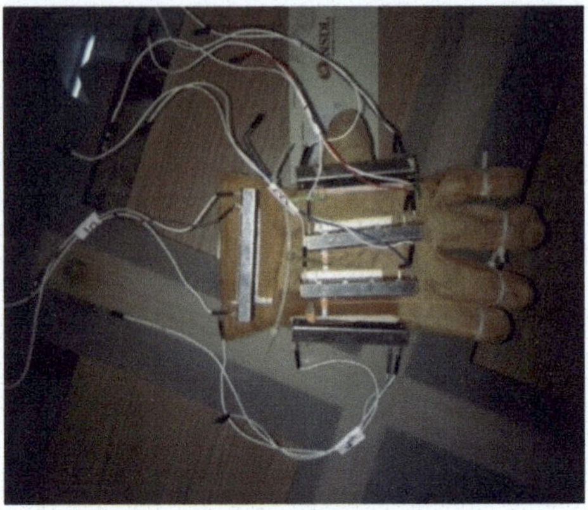

Fig. 3 Transmitting glove

3.3 Receiver End Logic Based Construction

At the receiving end, ESP8266EX controller built on NodeMCU DEVKIT 1.0 receives the transmitted data from Adafruit IO cloud and drives the servo motors to move according to the received angle values. The five servo motors independently work according to the potentiometer they correspond. Once the secure communication channel between the transmitting and receiving NodeMCUs is established and the values are received, these angle values (in degrees) are feed to the servo motors to move accordingly. Each servo motor's actuator is connected to the finger with the help of a string. As the servo rotates the essential amount of degrees, the strings move the fingers accordingly (Table 1).

Hardboard is used as a cost-effective material for the artificial robotic palm. However, plastics and medical-grade polymers can be used for industrial and biomedical implementations. Each finger has two hinges that are connected to the strings and move accordingly. The material selected is lightweight so that the movements are

Table 1 Bill of materials for hardware components

Sr. No.	Hardware	Part number	Make	Quantity	Cost per unit
1	Linear potentiometer	PTA45432015DPB103	Bourns	5	100 INR
2	Servo motor	Tower Pro SG90	Tower Pro	5	200 INR
3	NodeMCU	ESP8266EX	Espressif	2	150 INR
4	5 V power supply	Raspberry Pi AC 100-240 V DC Plug USB power supply adapter	Arduino	2	250 INR

quick and accurate. The servo motors which are having step size of:

$$5\,\text{V}/180\,\text{degrees} = 27.77\ \text{mV}/\text{degree} \tag{1}$$

This value of 27.77 mV/degree represents the minimum amount of change in voltage that can be replicated by the servo motors. These five servo motors each operating at 5 V can provide a torque of up to 0.1765197 Nm which is sufficient to drive the fingers at the receiver end (Figs. 4 and 5).

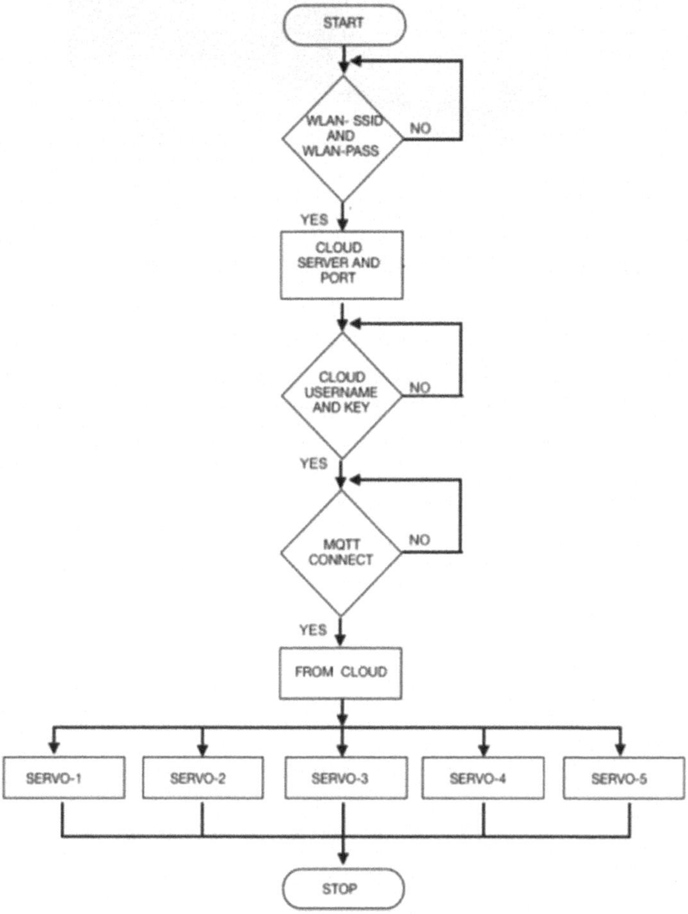

Fig. 4 Receiver logic flow

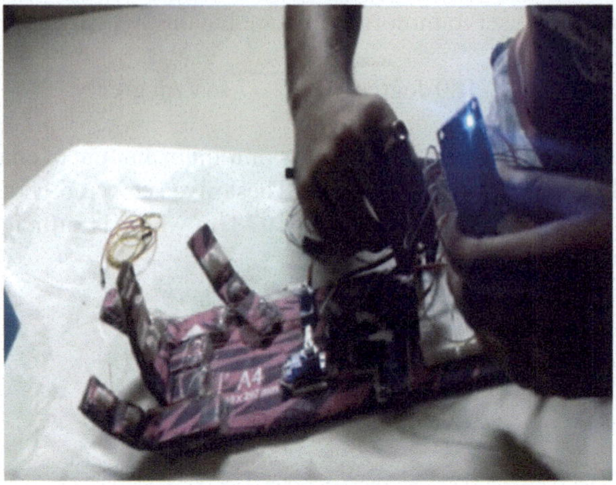

Fig. 5 Receiving SMART Robotic Palm

4 Hardware Schematic Design and Architecture

The complete IoT-driven gesture mimicking SMART Robotic Palm Architecture setup with hardware at both ends have been synchronized to transmit data using Adafruit IO Cloud. Altium version 10 has been used to create the block level (Fig. 6).

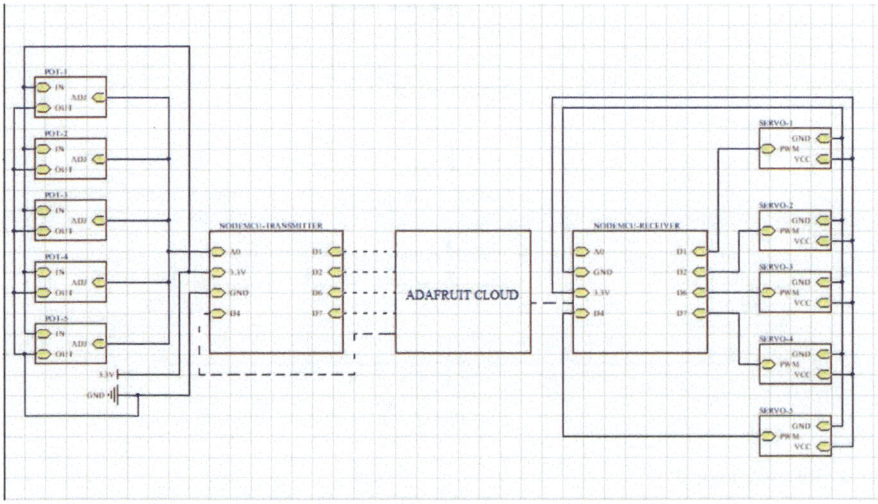

Fig. 6 IoT-driven gesture mimicking SMART Robotic Palm schematic

5 Results

The following was observed based on the operation of the IoT-driven gesture mimicking SMART Robotic Palm:

- The receiver SMART Robotic Palm is capable of receiving data packets and mimicking the human palm movements at the receiver end, and was tested at 10 Mbps internet upload and 7 Mbps download speed to transmit data in 2 s and to receive data in 3 s.
- The linear potentiometers accurately translate the corresponding finger movement to the voltage based on the calculated servo motor's step value of 27.77 mV/degree.

Few values that were transmitted from Adafruit IO are captured (in .csv format) and have been depicted and plotted in the form of graph representing the values sent versus time.

The total cost of implementation based on hardware is 2300 INR (33 USD) and other miscellaneous expenses comes out to be 200 INR (3 USD) bringing the total cost of implementation to 36 USD (Fig. 7).

Time	Finger 1	Finger 2	Finger 3	Finger 4	Finger 5
10:30:20	0	0	0	0	0
10:30:25	20	0	0	0	0
10:30:35	20	32	10	0	0
10:30:42	28	28	28	0	12
10:31:03	94	32	0	0	0
10:31:07	120	116	123	118	92
10:31:15	12	87	91	7	143

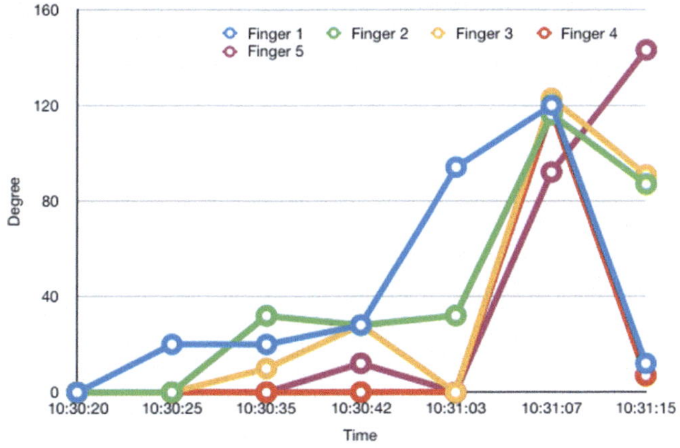

Fig. 7 Finger motion depicted and plotted using angle (in degrees) versus time instant

6 Challenges

The main challenges in the implementation of the IoT-driven gesture mimicking SMART Robotic Palm are:

- Decreasing the latency between transmission and reception.
- Mimicking actual human fingers movements using artificial robotic palm made up of hardboard.

7 Future Scopes and Applications

To improve the mechanical strength of the structure so that it can be more durable, one can opt for 3D printed palm constructed out of high strength plastic polymers. In case of remote operation, the receiver (robotic palm) can also have continuous video streamed back to the transmitter end, such that a real-time feedback is received by the human performing gestures using the glove palm. This helps one monitor, analyze, and correct (if needed) the actions performed by the artificial palm. This work can also be extended to perform surgeries and procedures in emergency conditions and in remote locations after achieving improved dexterity, precise motion control in presence of high-speed and stable internet connectivity, and with the availability of continuous electric power. The automation industry can also benefit with improvements such as instructing the robotic palm to place objects at a distance without the need for a human to touch them physically in industries, especially while dealing with hazardous chemical and radioactive substances and performing experiments involving corrosive and highly reactive chemicals. The improved artificial palms can be used in space missions to provide maintenance in space stations and shuttles. They can be used in current form to create art remotely and perform other activities.

References

1. Gupta A, Eppner C, Levine S, Abbeel P (2016) Learning dexterous manipulation for a soft robotic hand from human demonstrations. In: 2016 IEEE/RSJ international conference on intelligent robots and systems, pp. 529–551, 9–14 October 2016
2. Chen H, Jia X, Li H (2011) A brief introduction to IoT gateway. In: IET international confernece on communication technology and applications (ICCAT 2011). ISBN 978-1-84919-470-9
3. Shah SH, Yaqoob I (2016) A survey: Internet of Things (IoT) technologies, applications and challenges. In: 2016 IEEE smart energy grid engineering. ISBN: 978-1-5090-5111-3
4. Lohan V, Singh RP (2017) Research challenges for internet of things: a review". In: 2017 International conference on computing and communication technologies for smart nation (IC3TSN). 978-1-5386-0627-8/17
5. Guoqiang S, Yanming C, Chao Z, Yanxu Z (2013) Design and implementation of a smart IoT gateway. In: IEEE international conference on green computing and communication, IEEE

internet of things and IEEE cyber, physical and social computing, 20–23 August 2013. ISDN: 978-0-7695-5046-6
6. Fernandez F, Pallis GC (2014) Opportunities and challenges of the Internet of Things for healthcare: Systems engineering perspective. In: 2014 4th International conference on wireless mobile communication and healthcare—transforming healthcare through innovations in mobile and wireless technologies (MOBIHEALTH), 3–5 Nov 2014. ISBN: 978-1-63190-014-3
7. Ali AMM, Jamil MMA (2013) Biomechatronics design of a novel artificial arm. In: The 6th 2013 biomedical engineering international conference. ISBN: 958-1-4799-1467-8
8. ESP8266EX version 4.3 datasheet. https://www.espressif.com/sites/default/files/documentation/0a-esp8266ex_datasheet_en.pdf

A Texture-Based Analysis and Classification of Fruits Using Digital and Thermal Images

Varsha Bhole, Arun Kumar and Divya Bhatnagar

Abstract A business in the fruit market is totally dependent on the quality of the fruit, its physical appearance, color, shape and size. The fruit market needs fast and effective methods in identifying the worth of the fruits. The present research work employs outer texture of the fruits captured through a digital and a thermal camera. In this present work, eleven different varieties of fruits have been used as main elements for creating a texture-based image database using digital and thermal image capturing devices. The fruit is kept on the revolving tray which can be rotated from 0° to 360° and the entire system is controlled through an interface of stepper motor and the Arduino system and the images are captured from both the image capturing devices. The videos thus captured have been converted to frames for the final creation of the images database having normal RGB and thermal images for further application of image preprocessing techniques. The two image databases upon conversion to gray scale are subjected to texture-based feature extraction technique for studying the outer texture of the fruits. The eleven texture features extracted are subjected to further classification and analysis process for final demarcation of fruits being classified into infected and noninfected grades.

Keywords Thermal images · Feature extraction · Classification

1 Introduction

Fresh fruits are eaten and used every day as snacks, meals, and decorations. It is needed in every individual's diet to stay healthy and function well. The accurate grading of fruits plays an important role in agricultural and food industry to increase

V. Bhole (✉) · A. Kumar · D. Bhatnagar
Sir Padampat Singhania University, Udaipur, Rajasthan, India
e-mail: vybhole@acpce.ac.in

A. Kumar
e-mail: arun.kumar@spsu.ac.in

D. Bhatnagar
e-mail: divya.bhatnagar@spsu.ac.in

S. Fong et al. (eds.), *ICT Analysis and Applications*, Lecture Notes in Networks and Systems 93, https://doi.org/10.1007/978-981-15-0630-7_33

the profitability and to enhance competitiveness. Thus the quality of fruits plays a vital role as it is used in varieties of applications like export, producing food item like fruit juice, jam, jellies, etc. India ranks second in the world in the production of fruit [1]. So it offers India tremendous opportunities for export. This is the era of ICT technologies. As the world is getting automated, manual work is getting machine-oriented. Therefore, there is a need for effective and fast methodology to automatically identify and classify infected fruits. The quality assessment of fruits has proved to be a foremost research area in computer vision to get nearer to human levels of recognition. The growth of computer vision is due to the advancement in hardware and software for image processing.

In recent years, researchers tried to use different imaging techniques to deal with such problems. One of the latest techniques is the use of thermal imaging. Thermal imaging is simply the process of converting infrared (invisible) radiation into visible images that illustrate the spatial distribution of temperature differences in objects viewed by a thermal camera. The application of thermal images into machine learning based systems is very popular as by using this technique, the object of interest can be examined internally and externally without rupturing or dissecting it, without touching it, in a minimum span of time.

A computer vision based technique using GLCM is proposed for training ANN and system proves beneficial to sort banana with accuracy 98.8% [2, 3]. In [4], features like geometrical, statistical, and textures are extracted from the infected area and comparing the performances between SVM, MLP, and KNN classifier, it is observed that SVM works best for this system to grade apples for accuracy 92.5% and 89.2%—healthy and defected, respectively. For enhancement of fruit detection, LPT and fuzzy logic techniques have been applied for fusing orange canopy normal and thermal images and system gives improved fruit detection in comparison to thermal images exclusively by using both methods [5]. In [6], effect of thermal temporal variation on citrus fruit thermal images had been analyzed and its potential in fruit detection was evaluated. In this histogram tail based segmentation observed effectual in discerning the fruit canopy for the periods of large temperature differences with true positive and false positive rate of 0.70 and 0.06, respectively. The technique of active thermographic inspection has been applied to bruised apples and they are heated with halogen lamps to produce perfect visualization of bruises in amplitude and phase thermal images and give varying depths of phase differences between them [7, 8].

The objective of the present work is to classify infected or healthy fruit by observing not only from the outer side (skin) but also from the inside. The detection of infected inner part of fruit with thermal images as well as outer part of fruit with digital images using texture features takes place. In the present work, we are concerned with automatic classification between infected fruit and healthy fruit with whatever type of infection it has. In this work, eleven GLCM texture features were extracted for predicative classification of both digital and thermal fruit image dataset. The flow of paper is in following order: Sect. 2 details techniques adopted for data collection and preparation, feature extraction, and classification. The results of the classifier

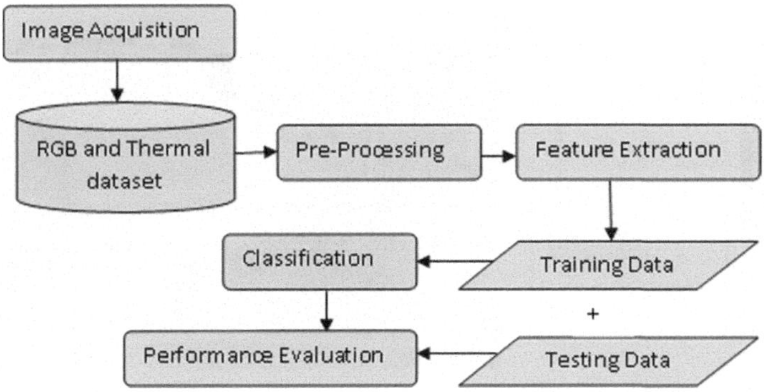

Fig. 1 Overall system

are presented in Sect. 3 for both RGB and thermal data. The conclusion is given in Sect. 4 and deliberates future work.

2 Methodology Adopted

2.1 Steps of Proposed System

All the steps given below are applied to both the RGB and thermal images and shown in Fig. 1.

1. Image acquisition: Videos are acquired and converted into frames to create the dataset of RGB and thermal images.
2. Preprocessing: Resizing and converting images into grayscale images.
3. Feature extraction: Texture-based eleven GLCM features are extracted and stored into a.csv file and used for classification.
4. Classification: Classification is done with the k-nearest neighbor and random forest classifiers and performance matrices, i.e., accuracy and kappa values are evaluated.

2.2 Data Collection

Two main devices are used to capture the images for this study: digital camera and thermal camera (SEEK). The thermal camera-SEEK has a 36° of field view, with a high range thermal sensor of the order of 206 × 156. It has a temperature-based detection range of the order of −40°–626° Fahrenheit. The wide temperature detection range helps in obtaining the temperature of every pixel. The digital camera

(a) (b)

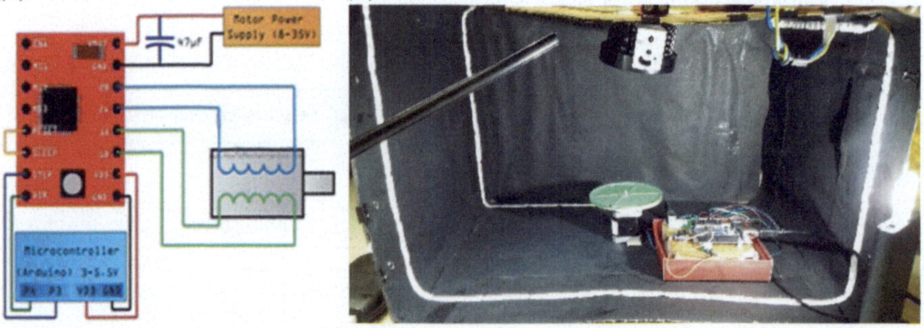

Fig. 2 Data collection system **a** Functional block diagram of stepper motor **b** Prototype designed for hardware system

can acquire only outer texture of fruit and the thermal camera can acquire inner as well as outer texture of fruits.

But in the present work, we have used both cameras to study the outer texture of fruits. The hardware part designed to create the image database consists of fruit carrying a revolving tray, stepper motor, Arduino Uno, ICA4988, digital camera, and thermal camera. Arduino Uno is connected with ICA4988. This ICA4988 is connected to stepper motor. A 12 V power supply is given to stepper motor via adapter. VDD and GND pin of Arduino Uno is connected to IC A4988 of VDD and GND. Also +5 V power is given by Arduino Uno. The functional diagram of stepper motor and prototype designed for hardware system are shown in Fig. 2. Both the cameras are placed in front of the stepper motor to capture the images of same angle. The revolving tray is fit at tip of the stepper motor. The fruit is kept on the revolving tray which can be rotated from 0° to 360° and the entire system is controlled through an interface of these hardware devices. Then the videos are captured and converted to frames for the final creation of the images database having normal RGB and thermal images for further application of image-based preprocessing techniques.

2.3 Data Description and Preparation

To demonstrate the objective of present work, we have used dataset of eleven different categories of fruits which comprises both RGB and thermal images: (i) Apple_Golden-Delicious (ii) Apple_Red (iii) Apple_Washington (iv) Guava(v) Orange_Malta (vi) Orange_Nagpur (vii) Pear (viii) Pomegranate (ix) Papaya (x) Kiwi (xi) Sharda. The different classes of dataset with sample normal RGB images and sample thermal images are presented in Figs. 3 and 4, respectively.

For classification purpose, 170 images per class have been considered, i.e., a total of (i) 1870 images for RGB dataset and a total of (ii) 1870 images for thermal dataset.

Fig. 3 Examples of each class in our normal image dataset

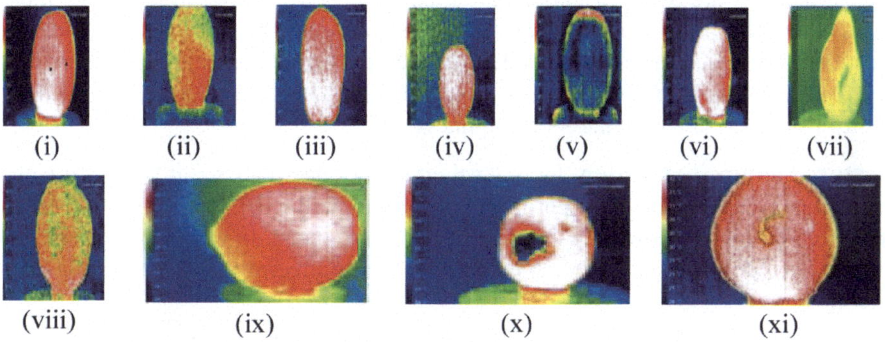

Fig. 4 Examples of each class in our thermal image dataset

Out of 1870 images of both the datasets, 1408 and 462 images are considered for training and testing, respectively.

2.4 Preprocessing

The captured (both RGB and thermal) images are resized into 256 × 256 and further converted into gray scale images. The sample gray scale images of RGB and thermal images have been presented in Fig. 5.

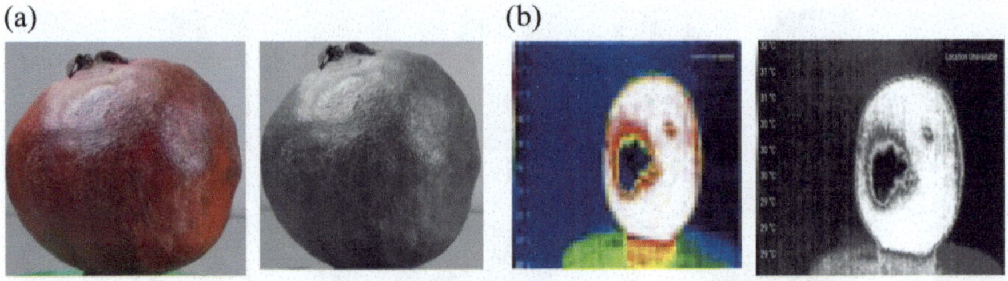

Fig. 5 Sample gray scaling for **a** RGB image and **b** Thermal image

2.5 *Feature Extraction*

Image features are used for classification and decision-making process. The most significant thing is to determine the correct features and choose appropriate classification algorithms. In this study, GLCM-based texture features are used to extract the features of both RGB and thermal images using MATLAB R2015a [9]. Texture indicates degree of smoothness or roughness.

Texture Feature Extraction. Gray level co-occurrence matrix (GLCM) is a well-liked popular texture-based feature extraction method. In [10], co-occurrence matrix and second order statistical texture features are introduced. In which, co-occurrence matrix is calculated first and then texture values are measured from co-occurrence matrix.

GLCM texture basically studies the relation between two neighboring pixels in one offset [11]. The gray values are transformed into the co-occurrence matrix over the four possible orientations ($\Phi = 0°, 45°, 90°$, and $135°$) which gives positions of the pixels with similar gray level values for displacement d. Every element (i, j) in GLCM states frequency of value of pixel i horizontally adjacent to value of pixel j. The gray level co-occurrence matrix $P_{ij}(d, \Phi)$ is as follows:

$$P_{i,j}(d, \Phi) = P(i, j/d, \Phi), 0 < i, j \leq Z \tag{1}$$

where, Z is the maximum gray level. In this work, GLCM matrix is obtained for $d = 1$ and $\Phi = 0°$.

The desired GLCM features used in this study are as follows: Angular Second Moment, Inverse Difference Moment, Contrast, Energy, Entropy, Homogeneity, Variance, Correlation, Shade, Prominence, Inertia. All these GLCM features are calculated for 1870 images of both RGB and thermal images. These features are stored in comma-separated values file (.csv) and then further used for classification.

2.6 *Classification Algorithm*

Classification is the most important step in any recognition system [12]. This depends on the quality and effective extracted features that successively define the efficient and suitable classifier for the system. We tend to used k-nearest neighbor and random forest (RF) classifiers with statistical features. KNN classifier is usually used for its simple interpretation and fast execution time whereas RF has more predictive power as compared to KNN.

KNN Algorithm. KNN finds the minimum distance (using Euclidean, Minkowski, Manhattan, Hamming distance) between test dataset and training dataset to determine k-nearest neighbors. KNN worked as follows: **1.** Set value k. **2.** To get the predicted class, repeat from 1 to total number of training data **2.a.** Find the minimum distance using Euclidean distance between test dataset and every record of training dataset. **2.b.** Arrange the computed distances in ascending order. **2.c.** Find first k rows from the sorted set. **2.d.** Find the most repeated class of those rows. **2.d. i.** If two or more classes represent majority within the set of k-nearest neighbors. **2.d. ii.** Then, calculate the total distances of neighbors in every majority class. **2.e.** Result the predicted class.

Random Forests (RF) Algorithm. Random forest builds multiple decision trees and merges them together to get a more accurate and stable prediction. Random forest classifier is mostly used when there is large set of samples [13]. The important parameters for random forest are: no. of decision trees used (ntree) and in each tree no. of random variables used (mtry). RF classifier works in 2 phases:

1. Random forest creation: It generates number of decision trees from the selected training set. Here first we select x features out of m features, i.e., $x < m$.
2. Random forest prediction: After creation of RF, aggregate the majority votes of different decision trees to predict the class of unknown sample.

3 Results and Analysis

In this work, we have used RGB and thermal dataset of fruits for classification. Classification was done for KNN and random forests for each test data and then performance measures are evaluated. In R using Caret package [14], both RGB and thermal datasets divided into training and testing datasets considering 75:25 proportions. Tenfold cross-validation technique repeated with 3 times is employed to access the statistical relevance of the classifiers. Then the performance measures are evaluated.

Performance Measure. Performance measures are evaluated herein using accuracy and kappa values.

Cohen's Kappa (Kappa). Kappa considers the accuracy that could be expected due to chance alone. Kappa picks the values from −1.0 to +1.0; +1.0 and −1.0 shows perfect agreement and disagreement respectively between prediction model and the actual labels. The form of the measure is shown as follows:

$$\text{Kappa} = \frac{\text{Observed Accuracy} - \text{Predicted Accuracy}}{1 - \text{Predicted Accuracy}} \tag{2}$$

where,

$$\text{Observed Accuracy} = \frac{\text{TP} + \text{TN}}{\text{TP} + \text{TN} + \text{FP} + \text{FN}} \tag{3}$$

$$\text{Predicted Accuracy} = \frac{(\text{TP} + \text{FP})(\text{TN} + \text{FN}) + (\text{FN} + \text{TP})(\text{FP} + \text{TP})}{(\text{TP} + \text{TN} + \text{FP} + \text{FN})^2} \tag{4}$$

where, TP = True Positive, FP = False Positive, TN = True Negative, and FN = False Negative

The result of Accuracy and Kappa values for each of the datasets are matriculated in Table 1 and shown in Fig. 6. It shows that accuracy and kappa value are highest for both the dataset using RF algorithm. The resampling results for both KNN and RF

Table 1 Predicted values for RGB and thermal dataset

Algorithm	RGB dataset		Thermal dataset	
	Accuracy (%)	Kappa (%)	Accuracy (%)	Kappa (%)
KNN	95.97	95.57	82.53	80.78
RF	97.37	97.5	84.97	82.69

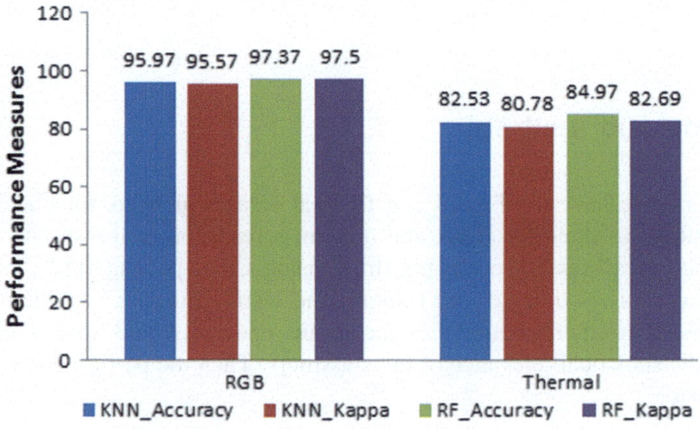

Fig. 6 Performance measure of KNN and RF for RGB and Thermal data

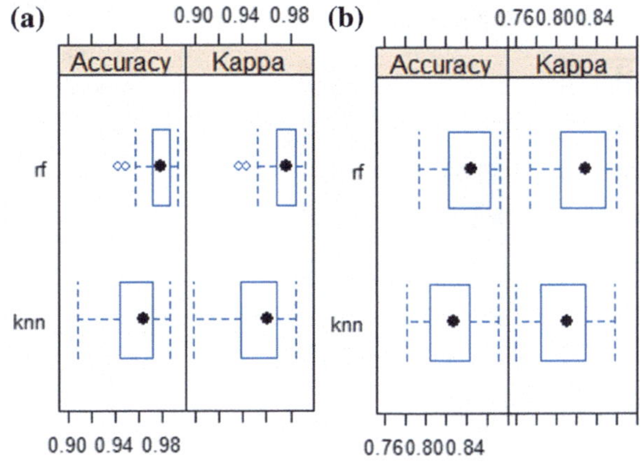

Fig. 7 Resampling results of **a** RGB and **b** Thermal data

across accuracy and kappa values are shown in Fig. 7 and the *difference* resampling results for both KNN and RF across accuracy and kappa values are shown in Fig. 8.

The accuracy of this work has been compared with [15, 16]. In [16], technique to identify and classify the fruit diseases with GLCM and KNN classifier was proposed. The result shows an accuracy of 87.46% which is lesser than an accuracy of 95.97% which has been obtained in this work by using RGB dataset with KNN classifier.

In [15], pulsed thermographic imaging system was developed for blueberry fruit and classification of these, LDA, SVM, RF, logistic regression, and KNN classifiers are considered. The average accuracy is 83.5% which is closer to accuracy of 84.97% which we obtained in this work of thermal dataset with RF classifier. The comparison of present work with [15, 16] is shown in Fig. 9.

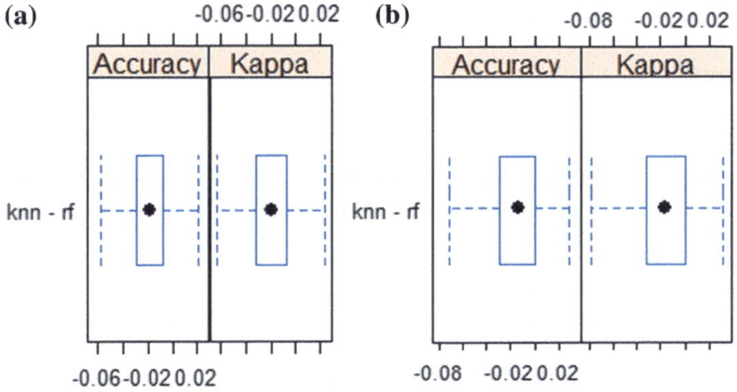

Fig. 8 Difference resampling results for **a** RGB and **b** Thermal data

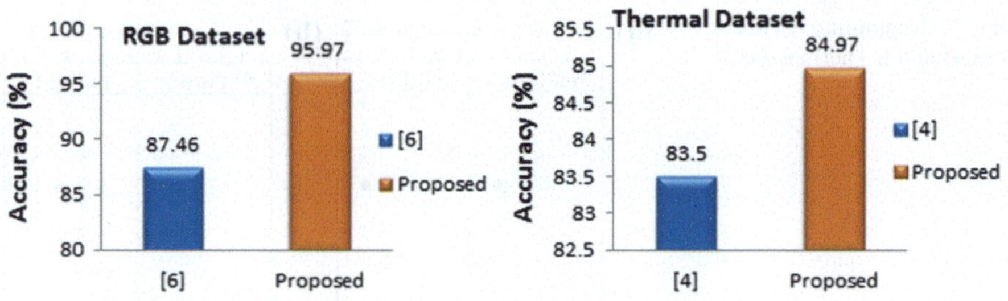

Fig. 9 Comparison of present work with [15, 16]

4 Conclusion

The present research work has been carried out to assess the performance of GLCM-based texture features extracted from the RGB and thermal image data set prepared for fruits over the maximum classification accuracy results achievable from such a data set. The Kappa and Accuracy measures were calculated for every dataset. A comparison between the accuracy results obtained using KNN and RF have been compared. Because of the several salient features of random forest (RF) and k-nearest neighbor (KNN), both the algorithms have become a considerable alternative in the classification tasks and therefore, both RF and KNN are separately used in this study for classification of fruits. The results show that random forest classifiers significantly outperform than KNN for GLCM-based features for both RGB and thermal data sets. When compared to thermal images, RGB dataset gives good results with slightly high accuracy. In the future, this work will be extended for color-based image analysis with Gabor features.

References

1. http://www.apeda.gov.in/apedawebsite/six_head_product/FFV.htm
2. Olaniyi EO, Adekunle AA, Odekuoye T, Khashman A (2017) Automatic system for grading banana using GLCM texture feature extraction and neural network arbitrations. J Food Process Eng 40:1–10
3. Olaniyi EO, Oyedotun OK, Adnan K (2017) Intelligent grading system for banana fruit using neural network arbitration. J Food Process Eng 40:1–9
4. Moallem P, Serajoddin A, Pourghassem H (2016) Computer vision-based apple grading for golden delicious apples based on surface features. Inf Process Agricul 4:33–40. Elsevier. https://doi.org/10.1016/j.inpa.2016.10.003
5. Bulanon DM, Burks TF, Alchanatis V (2009) Image fusion of visible and thermal images for fruit detection. Biosyst Eng 103:12–22. Elsevier
6. Bulanon DM, Burks TF, Alchanatis V (2008) Study on temporal variation in citrus canopy using thermal imaging for citrus fruit detection. Biosyst Eng 101:161–171. Elsevier
7. Baranowski P, Lipecki J, Mazurek W, Walczak RT (2008) Detection of watercore in gloster apples using thermography. Postharvest Biol Technol 47. Elsevier

8. Baranowski P, Mazurek W, Wozniak J, Majewska U (2009) Detection of early apple bruises using pulsed-phase thermography. Postharvest Biol Technol 53:91–100. Elsevier
9. MATLAB and Statistics Toolbox Release (2015) The Math Works, Inc., Natick, Massachusetts, United States
10. Kumar A, Patidar V, Khazanchi D, Saini P (2016) Optimizing feature selection using particle swarm optimization and utilizing ventral sides of leaves for plant leaf classification. Proc Comput Sci 89:324–332. Elsevier
11. Haralick RM, Shanmungam K, Dinstein I (1973) Textural features of image classification. IEEE Trans Syst Man Cybern SMC-3, 6:610–621
12. Agarwal A, Sarkar A, Dubey AK (2019) Computer vision-based fruit disease detection and classification. In: Tiwari S, Trivedi M, Mishra K, Misra A, Kumar K (eds) Smart innovations in communication and computational sciences. Advances in intelligent systems and computing, vol 851. Springer, Singapore
13. Cutler A, Cutler DR, Stevens JR (2012) Ensemble machine learning: methods and applications. Springer, New York
14. Kuhn M, Wing J, Weston S, Williams A, Keefer C, Engelhardt A (2010) Caret: classification and regression training. R package version 4.58. http://CRAN.R-project.org/package=caret
15. Kuzy J, Jiang Y, Li C (2018) Blueberry bruise detection by pulsed thermographic imaging. Postharvest Biol Technol 136:66–177. Elsevier
16. Sharath Kumar YH, Suhas G (2017) Identification and classification of fruit diseases. In: Santosh K, Hangarge M, Bevilacqua V, Negi A (eds) Communications in computer and information science RTIP2R 2016, vol 709. Springer, Singapore

A Comparative Analysis of Text Classification Algorithms for Ambiguity Detection in Requirement Engineering Document Using WEKA

Shilpi Singh and L. P. Saikia

Abstract The volume of digital documents is increasing day by day and thus the task of automatic categorization of document is very important for information and knowledge discovery. Classification is the most common method for finding the mine rule from the large databases. Ambiguity is the major problem in Requirement Engineering (RE) documents. Our proposed work uses WEKA text classification technique to identify and classify ambiguity in the RE document. The present study uses different algorithms on the ambiguity detection dataset and on the basis of different statistical measures like accuracy, time, and error rate we find suitable algorithms for this purpose. The main aim of this paper is to do a comparative study of various classification techniques and methodologies and a detailed analysis of different statistical parameters that are used in classification algorithms in order to analyze the quality of classification.

Keywords Text classification · Ambiguity detection · Decision tree algorithms

1 Introduction

The most important phase of Software Development Life Cycle (SDLC) is Requirement Engineering, where it is very important to understand the accurate requirements or the need of the system before implementing it. This task of requirement extracting and gathering can be unmanageable and clumsy, since the requirements are specified by the clients and the users, who may not have the adequate knowledge to specify the requirements unambiguously and thus, may lead to erroneous analysis. The main goal of requirement analysis and specification phase is to precisely understand the client's requirement and to chronologically organize the requirement into a specification document. Requirement Engineering is mainly concerned with extracting,

S. Singh · L. P. Saikia (✉)
Department of Computer Science and Engineering, Assam down town University,
Guwahati 781026, India
e-mail: lp_saikia@yahoo.co.in

S. Singh
e-mail: shilpi24_singh@yahoo.co.in

S. Fong et al. (eds.), *ICT Analysis and Applications*, Lecture Notes
in Networks and Systems 93, https://doi.org/10.1007/978-981-15-0630-7_34

arranging, analyzing, specifying, and validating user's requirement that is mostly documented using natural language and thus is prone to ambiguity. The quality of any RE document depends on how accurately and concisely the document is summarizing or documenting the systems or the user's requirement. Since the RE document is always written in informal English language and thus introduces variety of ambiguities. The requirements are specified by the stakeholders who are completely from different backgrounds and have inadequate knowledge to specify the requirement unambiguously. The main objective of requirement analysis and specification phase is to clearly understand the user's requirement and to chronologically and systematically organize the requirements into a specification document. On the basis of the literature available we can follow different methods to detect ambiguity in the RE document at both manual and automatic approaches.

1.1 Manual Approaches in Detecting Ambiguities

In this approach, the RE documents are given to the people having different backgrounds of software engineering and they analyze a text in order to identify the ambiguity in any RE document. But this process is considered to be very time-consuming and costly. There are mainly two methods for detecting ambiguity manually and they are inspection and review techniques.

1.2 Inspection Techniques

It is one of the most common approaches to detect and resolve ambiguities from the RE documents. The study in this area was initially conducted by Meyer [1] by identifying the features of natural language text that can induce ambiguity to any RE text. The seven sins identified by Meyer targets only the conceptual level understanding of any RE document. Another study that is worth mentioning here is done by some researchers who introduced the different classes of ambiguity like lexical, syntactic, semantic, and pragmatic [2]. The inspection techniques have good potential to detect ambiguities but are considered to be expensive and time-consuming.

1.3 Semiautomatic Approach Using Machine Learning Techniques

In this approach, the software engineers use both the natural language tools and the advice of human experts in writing a RE document. The different machine learning algorithms are used in order to analyze the quality of RE document and also the datasets used in the study by means of different statistical parameters. The common semiautomatic approaches that use machine learning techniques are decision tree, Naïve Bayes, Support Vector Machine, etc.

1.4 Decision Tree Text Classification Technique

In classification problem we can classify a text into predefined categories based on the features selected to perform classification. In this process, each feature is identified during the search process signified by a node and each node represents a decision point between varieties of unlike possible values of a feature. A classification model used to predict the categories of some unknown records. The decision trees are very efficient in supervised learning. It divides the dataset into homogeneous groups in terms of the variable to be partitioned. It takes as input a set of classified data and generates a tree in which each leaf node is a decision and non-leaf node is a test condition to be performed. The tree is simple and easy to use. Ross Quinlan originally developed ID3 [3] at the University of Sydney. It follows a supervised learning algorithm [4] that generates a decision tree from the fixed set of samples. And the output is used to classify future samples. It generates tree from the information gain obtained from the training instances and then uses the same to classify the training data. The demerit of ID3 is that it is sensitive to the features with large number of values. And thus we consider C4.5 algorithm that uses "Information Gain" with low conditional entropy values. The attribute with highest normalized information gain is chosen to make the decision. The Decision tree J48 is the implementation of algorithm ID3 developed by the WEKA project team. We analyzed the sample dataset for ambiguity detection using J48 algorithm and the result generated is compared with the performance of other algorithms to find the best suitable algorithm for analyzing the quality of classification model generated.

Naïve Bayes Text Classification

The Naïve Bayes classifier classifies a text on the basis of count and word probability. It classifies a text on the basis of class and also by dealing with the probabilities of ambiguous words that appeared on the text. It classifies a text by estimating the class labels "Ambiguous" or "Unambiguous" based on the classification model. It is a probalistic classifier based on Bayes theorem. It generates the class conditional probability by assuming that the attributes are conditionally independent. It is a rule-based classifier that classifies records using a collection of "if …then…." rules. It is one of the popular machine learning algorithms because of its simplicity and reasonable performance. The [5] explained that its performance is extremely good for classification tasks where the probability calculated by itself is not so important. The Naïve Bayes shows surprisingly very good performance for spam detection and thus can also be considered for our task of text classification.

Table 1 Comparative summary of different machine learning techniques

Techniques	Advantages	Disadvantages
Decision tree	It is simple and can be easily interpreted by nonexpert users. It can be reproduced using simple mathematical algorithms and provides a compact view of classification logic	If the dataset contains large features then it can lead to poor performance The learning and modification of decision rules need human intervention
Naïve Bayes	It works well in numeric and textual data. It is simple to implement and works well with real world situations	Its performance is unsatisfactory when features are correlated and do not consider frequency of occurrences of words
Support vector machine	It is considered to be the best method for text classification. It can handle high volume documents	It is relatively complex in training and categorizing algorithms. It takes more memory and time during training and classification stage

Support Vector Machine

In this classification method, the stakeholders use Parts of Speech (POS) tagging to tag every element in the given corpus to represent a word from the text. It then assigns a weight to the corpus and is then compared with the threshold value [6]. And if the weight is larger or difficult to identify by SVM, then the errors are detected. It is based on the computational learning theory. It is one of the discriminative classification methods that is comparatively more accurate. In our proposed work on ambiguity detection we prepared the dataset using the methodology followed by support vector method. The SVM considers both the positive and negative training sets which are not very common for other classification methods. The document representatives closed to the decision surface are called support vector [7]. The SVM is considered to be the best method for classification problems. The Hussain [8] concluded that it is able to build pattern recognition system to detect ambiguity patterns in any text document by analyzing the different word appearances. By suing SVM we can detect textual ambiguity (Table 1).

2 Classification Model

The ambiguity in the RE document is a serious issue that needs to be resolved and detected in the early stage of software development process. The classification technique can be used in Machine Learning to classify the data samples into some known classes. In our study we prepared sample training datasets to learn models that can classify a text as “Ambiguous” or “Unambiguous”. The experimental process uses WEKA to train the classifier and generating models for classification.

The major steps involved in this process are:

(a) Create sample training datasets.
(b) Identify attributes and its classes that can induce ambiguity to any text data.
(c) Identify the important features for classification.
(d) Learn the model using training samples.
(e) Use the model to predict the unseen dataset.

3 Experimental Study

All experiments are conducted with Windows 7 professional operating system with Intel core 3.3 GHz processor with 2 GB RAM. And for the implementation of the algorithm the proposed study uses WEKA tool. The dataset prepared follows decision rules to identify ambiguity in any textual document.

3.1 Ambiguity Detection Process

The ambiguity detection in a Requirement Engineering document is a classification problem that consists of two stages: (1) Training phase (2) Testing phase.

In the training phase, the dataset is provided with classification rules that can identify whether a sentence is "Ambiguous" or "Unambiguous". These classification rules are used to prepare the training data file in .arff format which is fed to the classifier. And then we apply different classification algorithms on the provided dataset in order to identify the best classifier that is able to classify a text. After finishing the training the classifier is able to analyze the quality of classification with the help of different statistical parameters. Finally, the performance of the classifier is analyzed in the background of different statistical measures like kappa index, confusion matrix, true positive and false positive, etc. and after that the best classifier is selected based on its performance in the background of accuracy, time, and error rate. The result generated shows that the quality of dataset used is substantial and thus can be used for the experiment with some minor changes.

3.2 Datasets

The dataset used in this study is created by Hussain [9]. The original dataset consists of twelve features but we included only ten features for our classification problem. And thus in order to identify the features that have the potential to induce ambiguity an RE document is selected for feature selection and extraction process [10]. The study started by analyzing the ambiguity detection process manually in order to identify the features that introduce ambiguity. The study concluded that proper annotation guidelines are needed in order to rate a sentence as "Ambiguous" or "Unambiguous". And thus we used Stanford Parser by Klein and Manning [10]. The frequency of occurrences of any feature can be very easily counted since the Stanford Parser automatically tags parts of speech in a sentence. The frequency of occurrence of different parts of speech in sentence is stored in Bag of Words. And after that the likelihood ratio is calculated that can be used to identify important features [10]. The annotation guidelines were prepared after a thorough study on the literature available for writing a good RE document [1]. The feature extractor tool [9] [Ishraar Hussain] is used to extract feature values from an RE document. We embedded Stanford Syntax Parser and morphological stemmer with Java code (JDK 1.6) by using Machine Learning Packages (JML) to create the training datasets at both sentence level and passage level by comparing likelihood ratios of the keywords. The resulting instances are stored in the MS Excel and then are converted to CSV format.

WEKA

It is a data mining tool designed by University of Waikato in New Zealand. The tool was developed in Java language. And in the tool different data mining algorithms are implemented. These algorithms are grouped into different groups according to classification rules generated by the algorithms. And we apply the datasets on different classification algorithms to generate results using different statistical parameters. The tool is simple to use and available free on the Internet. From the table shown above we can conclude that the dataset is best suitable for J48 when compared to other algorithms. The test mode for all algorithms used is "10-fold cross-validation" and classifier model is full training set. Each fold is used as test dataset while other nine folds are used as training dataset. The tool is simple to use and available free on internet. The WEKA tool provides GUI to the user with many facilities.

Explorer—It is the main interface in WEKA. It has different panels to perform different tasks. After loading the dataset the panel can be used to perform experimental analysis.

Experimenter—It's an environment to perform experiments and statistical tests between learning models.

Knowledge Flow—It has the same function as an explorer but with a drag and drop facility.

Simple CLI—It has command-line interface for direct execution of WEKA commands.

Execution in WEKA—It is a step by step process. The first step is to load the dataset from the different sources like files, URLs, and databases, etc. The dataset must be in CSV format and then we can store it directly to the .arff format which is compatible with WEKA classifier.

4 Experimental Result

The ambiguity detection datasets are analyzed in the WEKA classifier. The classifier is fed with the classification model in order to use the training datasets prepared and is then analyzed by using different statistical measures. The kappa index ratio shows the substantial quality of our dataset in J48 is substantial but for Naïve Bayes it is almost perfect. The percentage of correctly classified instances is highest in J48 when compared to other algorithms. The following tables show simulation of different classification algorithms on ambiguity detection dataset using different parameters like kappa index, mean absolute error, root mean squared error, relative absolute error, true positive, recall, and precision. The training dataset can be used to train the WEKA classifier and can be used to predict the unseen dataset. The run information generated using the WEKA classifier is summarized in the Tables 2 and 3.

Prediction of Unseen Dataset Using WEKA

After learning the model we can classify new unseen data using the classification rules. The following section describes the process of devising a model using graphical method. Firstly we need to create a file with different cases to predict that have the same structure that the file used to learn the model. The difference is that the question mark (?) indicates the unknown class whose value is to be predicted. After learning a decision tree using the ambiguity detection datasets the following sample dataset can be used to predict the six cases included in the .arff file.

The above dataset is given as input to the WEKA explorer and the training dataset is fed to the classifier that predicts the unseen dataset shown above. The run information generated using the WEKA classifier is summarized below. From the result we can observe that the training dataset used as a model that trains the classifier shows similarity between the model and the output of the unseen dataset (Fig. 1).

Table 2 Simulation of quality of classification on the basis of TP, FP, precision and recall

Algorithm	TP	FP	Precision	Recall
J48	0.913	0.086	0.914	0.913
Naïve Bayes	0.891	0.109	0.891	0.891
Random tree	0.778	0.222	0.778	0.778

Table 3 Simulation of time and kappa index

Algorithm	Percentage of correctly classified instances (%)	Percentage of incorrectly classified instances (%)	Kappa ratio	Time
J48	91.3043	8.6957	0.8262	0.09
Naïve Bayes	89.1304	10.8696	0.7825	0.05
Random tree	77.7778	22.2222	0.5554	0.01

Fig. 1 Unseen dataset used to predict the class

```
@RELATION sentence_corpus
@ATTRIBUTE bad_DT REAL
@ATTRIBUTE bad_RB REAL
@ATTRIBUTE bad_MD REAL
@ATTRIBUTE bad_JJ REAL
@ATTRIBUTE vb_in_p REAL
@ATTRIBUTE tokn_in_p REAL
@ATTRIBUTE parentheses REAL
@ATTRIBUTE fragment {TRUE, FALSE}
@ATTRIBUTE adverbs REAL
@ATTRIBUTE passives REAL
@ATTRIBUTE adjectives REAL
@ATTRIBUTE class {Ambiguous, Unambiguous}
@DATA
1,1,0,0,0,0,0,FALSE,2,0,1, ?
1,1,0,2,0,0,0,TRUE,1,0,2, ?
0,0,0,0,0,0,0,FALSE,0,0,0, ?
0,1,0,0,1,0,1,TRUE,1,1,0, ?
1,1,0,0,0,0,0,FALSE,2,0,1, ?
```

The run information after the prediction of unseen dataset is summarized below. The predicted class of unseen data and the error rate and kappa index is shown in Fig. 2.

5 Observation

The classification algorithms for ambiguity detection are analyzed in the background of different statistical parameters. From Table 1 it is clearly observed that J48 and Naïve Bayes algorithms can be considered for classifications, since in the other algorithms the number of correctly classified instances is very poor. The reason for this is

```
=== Re-evaluation on test set ===

User supplied test set

Relation:   sentence_corpus

Instances:   unknown (yet). Reading incrementally

Attributes:  12

=== Predictions on user test set ===

  inst#   actual  predicted error prediction

     1 1:Ambiguous 1:Ambiguous      1

     2 1:Ambiguous 1:Ambiguous      1

     3 1:Ambiguous 1:Ambiguous      1

     4 2:Unambiguous 2:Unambiguous     0.958

     5 2:Unambiguous 2:Unambiguous     0.958

     6 2:Unambiguous 2:Unambiguous     1

=== Summary ===

Correctly Classified Instances        6         100     %

Incorrectly Classified Instances      0           0     %

Kappa statistic                  1

Mean absolute error                  0.0139

Root mean squared error                0.0241
```

Fig. 2 Run information generated using WEKA

that the classification model designed for ambiguity detection problem is not properly working with these rules. And thus we started focusing on J48 and Naïve Bayes. The kappa index generated for J48 and Naïve Bayes is 0.82 and 0.78, respectively, which indicates that the quality and the validity of the dataset are substantial. But when it comes to execution time and accuracy J48 takes 0.09 s and Naïve Bayes takes 0.05 s, and this is because the number of correctly classified instances is more in J48 as compare to naïve Bayes. The proposed study also shows a very commonly used indicator like mean of absolute error and the root mean squared error. The average error is considered in order to derive conclusions. The result of the experiment is summarized in Table 2. The result indicates high error rate of J48 (0.988) as compared to Naïve Bayes (0.1602). But the root mean squared error shows almost similar error rate for J48 and Naïve Bayes. And thus we can continue considering J48 algorithm for our classification problem.

The proposed work also considered the other statistical measures like true positive, false positive, precision, and recall in order to validate the hypothesis that a particular text is considered as "Ambiguous" or "Unambiguous". The values for precision and recall in J48 were 0.914 and 0.913, respectively and thus indicates that rules developed for classification are working perfectly with the J48 algorithm.

6 Conclusion

The ambiguity detection in an RE document is one of the major issues in today's world. To identify ambiguity in a text or an RE document will definitely improve the quality of an RE document and thus speeds up the process of software development and saves both time and money. Our present study analyzes the ambiguity detection dataset using WEKA classifier. And we concluded that the majority of parameters like kappa index, precision, recall, true positive, and false positive rates show that J48 algorithm is best suitable for our classification problem. The error rates are showing good results since the number of correctly classified instances is more in J48 algorithm as compared to Naïve Bayes. The present study shows that J48 algorithm has better accuracy than other algorithms.

References

1. Meyer B (1985) On formalism in specifications. IEEE Softw 2(1):6–20
2. Kamsties E, Berry BM (2001) et al Detecting ambiguities in requirements documents using inspections. In: Proceedings of the first workshop on inspection in software engineering (WISE), Paris, France, pp 68–80
3. Quinlan JR (1993) Programs on machine learning. Morgan Kaufman, San Francisco (CA)
4. Shrivastava A, Choudhary V (2012) Comparison between ID3 and C4.5 in contrast to IDS Surbhi Hardikar, VSRD–IJCSIT 2(7):659–667
5. Domingos P, Pazzani MJ (1997) On the optimality on the simple Bayesian classifier under Zero—one—Loss. Mach Learn 29(2/3):103–130
6. Singh S, Saikia LP (2018) Feature extraction and performance measure of RE document using text classification technique. In 4th International conference on recent advances in information technology. 978-01-5386-3019-6. IEEE
7. Lin Y (1999) Support vector machines and the Bayes rule in classification. Technical Report No. 1014, Department of Statistics, University of Wiscousin, Madison
8. Hussain I (2007) Using text classification system to automate ambiguity detection in SRS document. Master's Thesis, Computer Science and Software Engineering Department, Concordia University, Montreal Canada, August 2007
9. Hussain I et al (2007) Automatic quality assessment of SRS text by means of decision—tree based text classification. In: Seventh international conference on quality software
10. Klein D, Manning CD (2003) Accurate lexicalized parsing. In: Proceedings of the 41st meeting of the association for computational linguistics

Clustering Based Algorithmic Design for Cab Recommender System (CRS)

Supreet Kaur Mann and Sonal Chawla

Abstract An efficient Cab Recommender System (CRS) assists the cab drivers with the shortest distance for the next passenger location. For this, it becomes imperative for a CRS to generate clusters for Geolocations. Clustering of Geolocations faces major challenges like noise, identification of meaningful clusters, semantic locations, etc. Therefore, the objectives of this research paper are fourfolds. Firstly, to extensively review the literature for Geolocations and identify the existent clustering techniques. Secondly, to propose an algorithm for generating clusters for Geolocations. Thirdly, to implement and test the proposed algorithm on standard dataset pertaining to different clustering techniques and finally, to analyze and compare the results of the proposed algorithm for effective clustering of Geolocations.

Keywords Cab recommender system · Clustering techniques · Geolocations

1 Introduction

The recommender system is an intelligent application to assist the user to make the best decision to choose one item amongst several alternative choices. Recommender systems are used to predict a user's behavior on the basis of historical information. They have been used to improve the decision-making process and quality. Reference [1] The recommendation system has shown high practical values for cab services too. The cab recommendation system can be for both the cab drivers as well as passengers. In order to recommend cab drivers with the next high passenger potential pickup location, it is essential to generate clusters for Geolocations. Clustering of Geolocations faces major challenges like noise, identification of meaningful clusters, semantic locations, etc. which can be overcome using an efficient clustering technique. Clustering techniques can be broadly classified into three categories as Partition-Based

---

S. K. Mann (✉) · S. Chawla
Panjab University, Chandigarh, India
e-mail: supreet@pu.ac.in

S. Chawla
e-mail: sonal_chawla@pu.ac.in

S. Fong et al. (eds.), *ICT Analysis and Applications*, Lecture Notes in Networks and Systems 93, https://doi.org/10.1007/978-981-15-0630-7_35

Clustering, Density-Based Clustering, and Hierarchical Clustering. Partition-Based Clustering algorithms split the data points into k groups or clusters. A set of n data points are partitioned into k clusters, where $k \leq n$. The cluster center represents the complete cluster [2]. Each cluster should contain at least one data point and every data point is a part of at least one cluster. The main drawback of partition-based clustering algorithms is it does not give result for data points which lie close to the center of another cluster. Density-Based Clustering algorithms [3] defines clusters based on density. It separates densely populated clusters from sparsely populated. If data points in a specified diameter are equal to greater than the minimum points defined, then they form a cluster else they are marked as outliers. Hierarchical Clustering [4] involves cluster formation on the basis of predetermined ordering from top to bottom. Either it follows the divisive method (top-down) or agglomerative method (bottom-up). Many researchers have used different clustering techniques for Geolocations. Most prominent clustering techniques for Geolocations includes K-Means(Partition-Based) [5, 6], DBSCAN(Density-Based) [7–9] and BIRCH(Hierarchical) [10].

2 Research Objectives

After an extensive literature survey on CRS, the research identified few gaps that need to be addressed. The research identified the lack of standard clustering techniques for cab GPS trajectory data. Though the clustering of GPS trajectory datasets has been implemented using various clustering techniques like K-Means, DBSCAN, etc. yet there is no comparative analysis of clustering techniques for Geolocations. Therefore, the objectives of the research were:

1. To conduct an extensive literature review of existing CRS and to identify their clustering techniques.
2. To propose an algorithm for a system to generate clusters for GeoLocations.
3. To implement and test the proposed algorithm on standard dataset pertaining to different clustering techniques.
4. To analyze and compare the output of different clustering techniques which shall be of significant use for CRS recommending high passenger potential pickup locations.

3 Research Methodology

In order to propose an algorithm for Geolocation Clustering, the research methodology involves the following three phases:

Phase I: Identification of clustering techniques for Geolocations.

Phase II: Design and Development of an algorithm to generate clusters for geolocations.

Phase III: Testing of the proposed algorithm using standard dataset and analyze the output of different clustering techniques.

3.1 Phase I: Clustering Technique Identification

Based on an extensive literature survey, it was observed that many researchers have used different clustering techniques for Geolocations. Most prominent clustering techniques used by researcher for Geolocations includes K-Means(Partition-Based) [5, 6], DBSCAN(Density-Based) [7–9], and BIRCH(Hierarchical) [10].

K-Means

K-Means is a well-known partition-based clustering model [11] which represents its cluster by its centroid, which is the mean of all the data points of that cluster. The number of clusters in the case of K-means is fixed to k. Firstly, The data is divided into a predefined number of clusters. Later, A data point is assigned to a cluster with a minimum distance between the centroid and the data point. The complexity of K-Means is $O(knt)$ [12] where k is the number of clusters, n is the number of data points, and t is the number of iterations. K-Means partitions N observations, i.e., s_1, s_2,... s_n into k clusters i.e. S_1, S_2, ..., S_k so as to minimize the intra cluster sum of squares:

$$argmin \sum_{i=1}^{k} \sum_{s_j \in S_i} \parallel s_j - \mu_i \parallel \tag{1}$$

where $\mu_i = \frac{1}{k_i} \sum_{S_j \in S_i} s_j$ and k_i is the number of observations in the i-th cluster [13].

DBSCAN

DBSCAN(Density-Based Spatial Clustering of Applications with Noise) [3] is a data clustering algorithm that groups together the nearby points into one cluster and marks far off points as outliers. It is a density-based clustering algorithm as it clusters the data into high-density points in a cluster and marks low-density points as outliers.

BIRCH

BIRCH(Balanced Iterative Reducing and Clustering using Hierarchies) [14] is used to perform hierarchical clustering over a particularly large dataset. It takes a set of N data points as input and cluster these data points into K clusters. The number of clusters depends on the diameter of all entries in the leaf node and number of entries in each cluster.

3.2 Phase II: Proposed Algorithm

Based on the clustering techniques identified, the research proposed an algorithm for using these techniques for Geolocation identification having a high potential of the passenger.

Input: The cab data includes GPS trajectory data stored in a comma-separated values(.csv) file which act as a input to the proposed system. The pickup locations(latitude and longitude) are extracted from csv file. The dataset of the city is in the form of .osm file which is an XML file.

Steps : It involves the steps followed by the proposed algorithm so as to generate clusters using GPS trajectory Dataset.

Output : The proposed algorithm generates output in the form of two plots: Folium-Based plots and Graph Plots. Folium-based plots are plots on the city map that represents the clusters according to the latitude and longitude on the World map. Graph plots depict the clusters on an X-Axis and Y-Axis where Latitude is plotted on X-Axis and Longitude is plotted on Y-Axis.

Figure 1 represents the overall structure of the proposed algorithm for clustering Geolocation in CRS and Algorithm 1 defines the steps followed in the proposed algorithm.

Algorithm 1: Steps for Proposed Algorithm

Input: Cab Data(.csv) and City Map(.osm)
Steps:

1. *Import Libraries and Data Acquisition*: Import the required libraries and acquire the dataset containing Cab GPS Trajectory data and the map file of City (.osm file).
2. *Data PreProcessing*: Since the data collected from the GPS signal is noisy and contains numbers of outliers, the data needs to preprocessed. This involves two steps: GPS data cleaning and GPS point matching. The data which is out of range and repeated is removed. Due to the GPS measurement errors, the GPS location of the cab might not match with the road network. In order to match the cab location with the road network GPS point matching is done.
3. *Transform Data*: Data is transformed using aggregation of features such as longitude and latitude, to get a GPS location on a map.
4. *Clustering*: Three clustering techniques were implemented on the data:
   - K-Means(Partition-Based)
   - DBSCAN(Density-Based)
   - BIRCH(Hierarchical-Based)

Result: Folium-Based Plots and Graph Plots

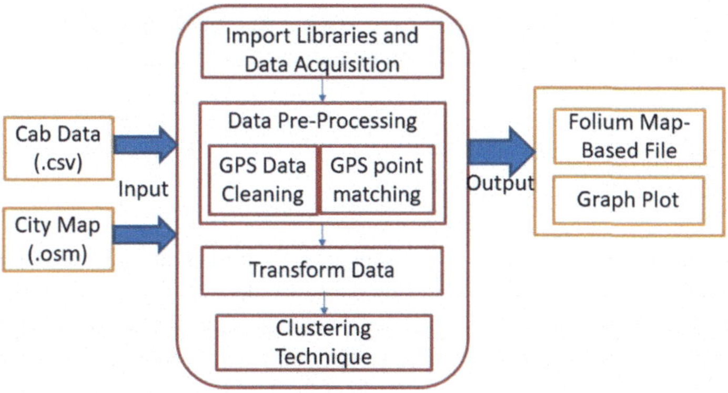

Fig. 1 Proposed system for clustering geolocations in CRS

3.3 *Phase III: Testing and Analysis of Proposed Algorithm*

The testing involved the following steps:

Step 1: Acquiring the datasets.
Step 2: Testing the proposed algorithm for the acquired dataset.
Step 3: Analysis of the outcomes.

3.3.1 Acquiring Dataset

The experiment is performed on GPS trajectory data of Uber Cabs in New York city that contains data of cabs in April 2014. It contains 564516 records. The dataset is downloaded from https://github.com [15]. Randomly 100 unique records are selected and clusters are formed. The graphs are plotted for clusters. The parameters of the dataset include the following fields (Table 1).

The city map is downloaded from OpenStreetMap.org which allows free access to the full city map. The city map consists of four elements (Table 2).

Nodes represent a specific point defined by its latitude and longitude on the earth's surface. Ways is a path joining nodes in a city. Way can be: open way, close way or

Table 1 Parameters of .csv file

Date/Time	Lat	Lon	Bas

Table 2 Parameters of .osm file

Node	Ways	Relation	Tags

an area. Relation is the relationship between two different elements like nodes, ways, etc. Tags are the map features of the other parameters that describe its geographic attribute.

3.3.2 Testing of Proposed Algorithm

The algorithms are implemented in Python(2.7). Python is an open source programming language. The cab dataset is preprocessed by data cleaning where the missing values are replaced by the mean values as the mean value of latitude and longitude represents the central location of the city. The data points that are out of range and repeated are removed. The data points are matched on the city map. The nodes from OSM file are extracted and plotted on the map. Data is transformed by extracting Latitude and Longitude fields from .csv file and creating a data point with their combination which represents the pickup points on the city map. The clustering techniques under investigation were K-Means(Partition-Based), DBSCAN(Density-Based) and BIRCH(Hierarchical-Based).

3.3.3 Analysis of Outcomes

K-Means algorithm combines n data points into k clusters. Each data point is assigned to the cluster with the nearest mean. In this experiment, K-Means is implemented with n = 10. The black points mark the cluster heads. The data points in the same color belong to same cluster. Figure 2a, b depicts the graphs for K-Means.

DBSCAN is a density-based clustering algorithm that groups together the data points that are close to each other. The low density data points are marked as outliers. It accepts $minPts$ and ϵ as input. If the number of points is more than $minPts$ in a radius of ϵ, it forms a cluster else it is marked as outlier. In this experiment $minPts$ is set to 4 and ϵ is 0.5. The number of cluster formed are 2. Figure 3a, b depicts

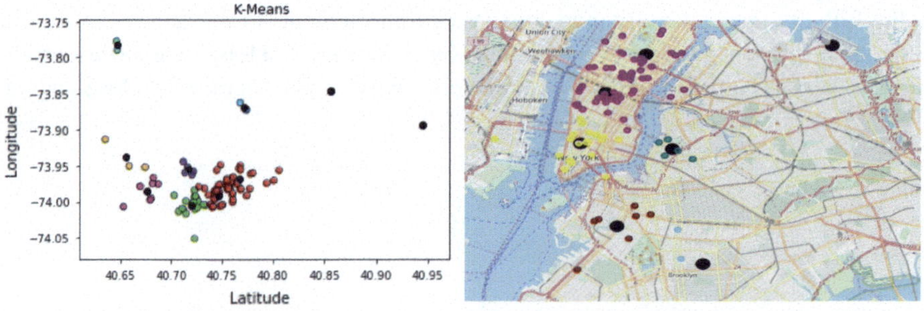

(a) K-Means Graph Plot (b) K-Means Folium Plot

Fig. 2 K-Means clustering

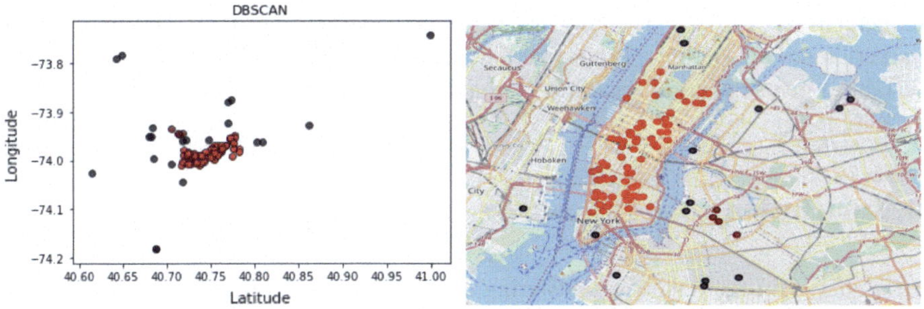

(a) DBSCAN Graph Plot (b) DBSCAN Folium Plot

Fig. 3 DBSCAN clustering

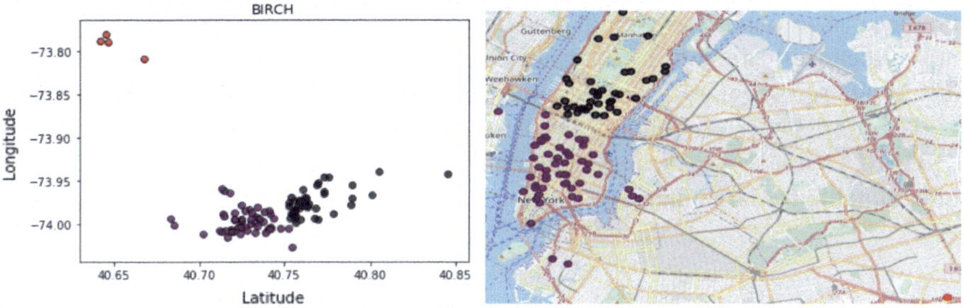

(a) BIRCH Graph Plot (b) BIRCH Folium Plot

Fig. 4 BIRCH clustering

Table 3 Comparision of clustering algorithms

Algorithm	Parameters	Outliers	Time taken to form clusters (in sec)	Drawbacks
K-Means	Number of cluster	No	1.71020	Difficult to predict number of cluster, doesnot work well with density-based data
DBSCAN	Minpts and ϵ	Yes	1.98206	Difficult to estimate ϵ value
BIRCH	Branching factor and diameter of cluster	Yes	2.02251	Handles only numeric data

the graphs for DBSCAN. Here data points marked in black are outliers and data points marked in same color belong to same cluster. BIRCH performs hierarchical clustering over the large data set. It is agglomerative hierachical clustering. It accepts two inputs: number of entries in each cluster and the diameter of all entries in a leaf node. In this experiment, the number of entries in each cluster is set to five and diameter is set to 0.5. The number of cluster generated are two. The data points marked in red are outliers. Figure 4a, b depicts the graphs for BIRCH.

Table 3 gives the comparision between these algorithms.

4 Conclusion

After analyzing the results of clustering algorithms, the following conclusions can be drawn. K-Means generates a better result for compact clusters, sensitive for outliers and noise and the number of clusters is to be detected. DBSCAN algorithm clusters the data according to the low and high-density of data points hence it does not require the number of clusters. The single link effect is reduced due to MinPts parameter. The parameter minPts handle the outliers effectively. DBSCAN is not entirely deterministic since it doesnot clearly handles the border points. It cannot cluster data sets having large differences in densities. The clusters are generated on the basis of distance ϵ, which is a difficult task to choose a correct threshold point. BIRCH does not require scanning all data points in order to generate clusters. It is better suitable for large datasets.

References

1. Pathak B, Garfinkel R, Gopal R, Venkatesan R, Yin F (2010) Empirical analysis of the impact of recommender systems on sales. J Manag Inform Syst 27(2):159–188
2. Chitra K, Maheswari D (2017) A comparative study of various clustering algorithms in data mining. Int J Comput Sci Mob Comput 6(8):109–115
3. Ester M, Kriegel H-P, Sander J, Xu X (1996) A density-based algorithm for discovering clusters in large spatial databases with noise. In: Proceedings of the 2nd international conference on knowledge discovery and data mining, Portland, OR, USA, 2–4 Aug 1996
4. Manning CD, Raghavan P, Schütze H (2009) Hierarchical clustering. Cambridge University Press, pp 346–368. https://doi.org/10.1017/CBO9780511809071.018
5. Ma S, Zheng Y, Wolfson O (2015) Real-time city-scale taxi ridesharing. IEEE Trans Knowl Data Eng 27(7):1782–1795
6. Zhang D, He T, Liu Y, Stankovic JA (2003) CallCab: a unified recommendation system for carpooling and regular taxicab services. In: Proceeding IEEE international conference on big data, pp 439–447
7. Zhang D, He T, Liu Y, Stankovic JA (2013) CallCab: a unified recommendation system for carpooling and regular taxicab services. In: proceedings of the IEEE international conference on big data, pp 439–447
8. Yuan J, Zheng Y, Zhang L, Xie X, Sun G (2011) Where to find my next passenger? In: Proceedings of the 12th ACM international conference on ubiquitous computing, pp 109–118
9. Zhang J, Meng W, Liu Q, jiang H, Feng Y, Wang G, Efficient vehicles path planning algorithm based on taxi GPS big data. Optik 127(5):2579–2585
10. Li Q, Zeng Z, Yang B, Zhang T (2009) Hierarchical route planning based on taxi GPS-trajectories. In: Proceedings of the 17th international conference on geoinformatics, pp 1–5
11. Hartigan JA, Wong, MA (1979) Algorithm AS 136: a k-means clustering algorithm. Appl Stat, 100–108

12. Xu D, Tian Y (2015) A comprehensive survey of clustering algorithms. Ann Data Sci 2:165. https://doi.org/10.1007/s40745-015-0040-1
13. Wang R, Chow C, Lyu Y, Victor C, Kwong S, Li Y, Zeng J (2017) TaxiRec: recommending road clusters to taxi drivers using ranking-based extreme learning machines. IEEE Trans Knowl Data Eng, 1–1. https://doi.org/10.1109/TKDE.2017.2772907
14. Zhang T, Ramakrishnan R, Livny M (1996) BIRCH: an efficient data clustering method for very large databases. In: SIGMOD '96: proceedings of the 1996 ACM SIGMOD international conference on management of data. ACM, pp 103–114. https://doi.org/10.1145/235968.233324
15. Uber Data. https://github.com/ginacordova/Portfolio/blob/master

Tweet-Based Sentiment Analyzer

Gresha Bhatia, Chinmay Patil, Pranit Naik and Aman Pingle

Abstract People, these days, express their opinions regarding any particular topic or issue widely on social media. One such popular social media platform among masses is twitter with over 320 million monthly users. Users also express their thoughts on any political announcements or decisions taken by a particular party. Analyzing these tweets on a specific topic can help in determining what people think about measures undertaken by the government. It will give an idea on how many percent of people are in favor of any announcement, and how many of them stand against it. This will in turn provide areas of improvement for the ruling or opposition party. This paper thus aims on finding sentiments of tweets on a political leader, some party or announcements like a union budget. This can further be generalized to any particular measure undertaken by any organization.

Keywords Sentiment · Twitter API · Scraping · Live graphing · Multinomial Naive Bayes

1 Introduction

The number of users on Internet is increasing day by day with over 3.9 billion people in the world accessing Internet for their day to day tasks. This also leads to people utilizing various social media sites and connecting to their acquaintances. Thus, individuals tend to posit their views on various issues or advancements on social media networks. Such arguments or thoughts of people are found to be useful for

G. Bhatia · C. Patil (✉) · P. Naik · A. Pingle
Vivekanand Education Society's Institute of Technology (V.E.S.I.T), Chembur, Mumbai, India
e-mail: 2016.chinmay.patil@ves.ac.in

G. Bhatia
e-mail: gresha.bhatia@ves.ac.in

P. Naik
e-mail: 2016.pranit.naik@ves.ac.in

A. Pingle
e-mail: 2016.aman.pingle@ves.ac.in

S. Fong et al. (eds.), *ICT Analysis and Applications*, Lecture Notes in Networks and Systems 93, https://doi.org/10.1007/978-981-15-0630-7_36

analysis of propensity of people toward a particular side through their contention on social networks. Advantage of such analysis can be used by various political parties to recognize the mindset of the population. This forms the base of a Sentiment Analyzer.

Twitter is a popular microblogging site which lets people put forth their thoughts on to a worldwide platform [1]. The popularity of twitter is bolstered by the fact that 500 million tweets are posted everyday. This in turn provides large amount of data available for any topic with people casting their views on it. Twitter allows users to put forth their thoughts in a short text with a limit of 280 characters per tweet. Also, being in the category of blogging, twitter has more users using textual format to contend their views on a pool of issues. Twitter allows us to view various trends and lets us determine the inclination of users on it, i.e., either positive or negative. The above factors make analysis on the data faster, easier, and more accurate. As a result, we narrowed down our approach to a Tweet-Based Sentiment Analyzer from a Sentiment Analyzer [2].

Despite having an enormous amount of data, the text tweeted by users often contains discrepancies. The tweets are often multilingual, have acronyms used, contain various emojis, misspelt words, incorrect grammar, etc.

Such inconsistencies act as noise in data [3]. However, we can still classify the data into positive and negative tweets with the help of significant words or terms stated in the text. The paper thus depicts the approach of the project to classify such data by tackling these difficulties [2, 4].

2 Related Work

Twitter being a humongous platform for people to express their opinions, its data was analyzed by different researchers for varying purposes. Azam et al. [2] built a system to cluster tweets based on similarity using Markov Clustering technique with each cluster depicting an event. Various scenarios considered were for Israel—Gaza conflicts, Delhi assembly election and union Budget 2015. This was carried out by considering tweets as nodes in a social graph and weighted edge between them representing the similarity between the tweets. Norman et al. [5] performed sentiment analysis by gathering English tweets on demonetization and Indian Budget 2017. They used a Naive Bayes Classifier to predict sentiment of tweets fetched in real time to classify them into either positive, negative or neutral. The results helped to determine the feeling and estimation of the general population about the government's call to demonetization and its outcome on the proposed Budget in 2017. Naiknaware [6] worked upon estimating the inclination of people by scrutinizing the tweets of Union Budget of India from 2016 to 2018 in order to classify them into three classes, viz., positive, negative, or neutral. Kaur [7] worked upon improving the accuracy of a sentiment analyzer by proposing a system design that combined Lexicon-based and machine learning approaches. This also brought into light an approach of hybrid model which may consist of multiple machine learning methods like SVM, Naive Bayes, etc. to determine the polarity of a tweet. Sarlan et al. [8] in their research, developed a model to obtain opinion of customers on an organization or company which will turnout to be beneficial for the company by measuring the perceptions of their customers. The model gave output in the form of a pie chart on an HTML page after classifying the tweets into two classes, i.e., positive and negative. The accuracy was improved by incorporating Natural Language Processing before actually classifying the data. Verma et al. [9] worked upon opinion mining for movies to be released in India in real time. The tweets were streamed in real time with help of a Twitter Streaming API. This helps in determining the mood of viewers and how the movie will perform in box office upon its release. Rahman et al. [10] proposed an approach to sentiment analysis on various topics by categorizing tweets into sentiments by a trained model on Machine Learning algorithm, i.e., Naive Bayes Method. Guha et al. [4] proposed a system for analyzing twitter data of SemEval 2015 by training a linear SVM.

2.1 Lacuna of Existing Systems

The existing systems classify data for English-based tweets only, i.e., there is no multilingual support. Also, these systems do not fetch data in a dynamic manner or produce tweet sentiments in real time along with no functionality to get and classify historical tweets for analysis purpose.

3 Proposed System

- The user first logs into the system.
- Then, he can choose between two options:
  - Get the sentiments of old tweets.
  - Get the sentiments of live tweets.
- The old tweets are fetched using web scraping and the live tweets are fetched using Twitter API.
- These tweets are processed and then fed to the trained model.
- Training the model:
  - First, the dataset is preprocessed.
  - Then it is fed to the algorithm which outputs the model.
- The results obtained from the model are plotted as graphs and presented to the user (Fig. 1).

Fig. 1 Block diagram

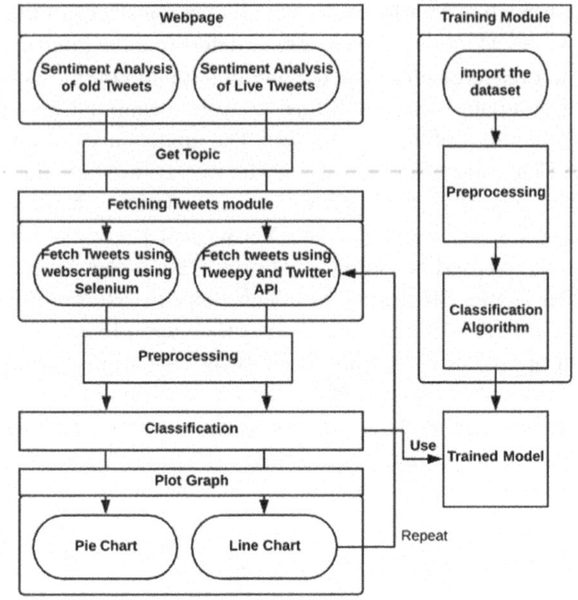

4 Methodology

The project was developed in various phases. First, dataset was collected for training the classifier after which an appropriate algorithm was selected to categorize tweets into different classes. Tweets were captured based on users' input text using two main approaches, i.e., through Twitter API and Web Scraping. The input tweets were fed to the trained model to predict the sentiment. The output of the algorithm was then displayed to the end user.

4.1 Dataset Used

A predefined dataset of tweets by Indians on a Union Budget or government decision is not readily available. As a result, a movie review dataset for short text is used for training purpose. The dataset comprises of two different types of tweets in two files—positive and negative—with each having over 5000 movie reviews.

These files are then combined and shuffled randomly to obtain a mixed dataset with both positive as well as negative tweets. From a line of text, according to the study, an adjective plays the most vital role in determining the polarity of the sentence [11]. A ratio of positive to negative occurrence of an adjective is calculated to find in which of the two cases positive or negative the word is more associated. If the word "excellent" occurs 30 times in positive classified data and only 3 times in negative data, then it is more closer to the positive side. Thus, adjectives are extracted as features of a sentence and mapped to the respective polarity. All such words form feature sets which are classified according to frequency of occurrence in the dataset. Among them, top 5000 or most frequent 5000 feature sets are picked and the corresponding model is trained based on these selected frequent feature sets along with picking 3000 sentences from the shuffled dataset. The trained model is then tested on next 1000 sentences to test the accuracy of the classifier.

4.2 Classification Algorithm

The classification algorithm used for predicting sentiment of tweets is Multinomial Naive Bayes Classifier as it is suitable for classification with discrete features such as text based classification [12]. Under Naive Bayes assumption we have:

$$p(f_1, f_2, \ldots, f_i|c) = \prod_{i=1}^{n} p(f_i|c)$$

$$p(f_i|c) = p(c|f_i) * p(f_i)/p(c)$$

The term Multinomial Naive Bayes lets us know that each *p(fi|c)* is a multinomial distribution, rather than some other distribution. *p(fi|c)* denotes probability that *fi* lies in class *c* [13].

This works well for data which can easily be turned into counts, such as word counts in text. Consider following training dataset (Table 1).

Let us determine whether the statement "overall budget is good" results in a positive statement or negative statement [14] (Fig. 2).

Since positive probability is greater as compared to negative, the text "Overall budget is good" is classified as "Positive".

Table 1 Training dataset

Text	Sentiment
"Education has become easier"	Positive
"It is affordable. Overall, a good move"	Positive
"A truly horrible decision by XYZ"	Negative
"It is a very good decision"	Positive
"They will suffer"	Negative

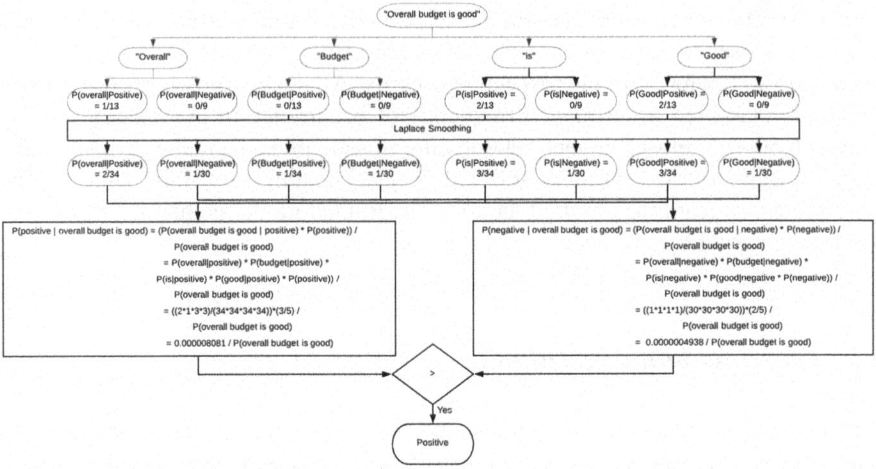

Fig. 2 Calculation of probability

4.3 Gathering Tweets

The project includes analysis of live tweets as well as old tweets or historical tweets. Thus, two different approaches have been used. The former uses Tweepy while the latter applies concept of web scraping using Selenium. Scraping is used, as twitter's privacy policy provides truncated old tweets using its API [15].

Live Tweets Using Tweepy. In order to fetch live tweets, Tweepy and Twitter API are used [16]. Steps to get Twitter keys:

- Apply for twitter developer account.
- Click on "Create an Application".
- Fill the details of the Application.
- The access tokens will then be available.

Tweepy handles the authentication, connection, creation, and destruction of the session.

Web Scraping using Selenium Webdriver. For fetching the old tweets, scraping of Twitter pages is done using Selenium Webdriver. Basically, Selenium is an automation testing tool. It can be used to perform various browser actions by writing a program [17]. Now, twitter is a dynamic website which loads more content upon scrolling with changing HTML as compared to other static web pages which have a fixed HTML code. As a result, we require a dynamic web scraping tool. Selenium is thus apt for the requirement. It simulates a human browsing the twitter pages loading more tweets by pressing page the down button. More the webdriver scrapes, more tweets are acquired from the HTML for sentiment prediction.

Based on the topic to be analyzed, a URL of search query is generated. Then, the corresponding page is visited on a browser. Followed by, all the content having "body" as the tag name being fetched. From the body, the "div" tag which has tweet text is reached and the tweet is captured.

4.4 Processing Tweets

A captured tweet contains variety of languages, emoticons, and noise. All of this has to be processed first to obtain a generalized format before predicting its polarity. So, it is passed through three different phases after which the sentiment is determined.

Convert Emojis to Literal Meanings. Tweets comprise of emoticons which play a vital role in expressing the sentiment of the user [18]. As a result, the system converts them into their meaning in textual format. This functionality is achieved using python module "Emoji" and the function being called as "demojize" in its documentation. E.g., the emoji ":)" is converted to the text "smile".

Cleaning of Tweets. Tweets contain URLs if it has an image associated to it. In this case, the URL is also fetched with the text and it acts as noise in data and should be eliminated [19]. This is removed using python module called "preprocessor" [20]. It also removes hashes from the text in case there is any hashtag present. Input statement: "The decision is good. https://xyz.com/image.png" is processed to give "The decision is good."

Translation. Tweets captured can be in various languages and not just English. However, the model recognizes and is capable of processing only English language. Hence, "GoogleTrans" Python library is used to detect the language of tweets and translate them into English wherever required. The API is also capable to translate a tweet in some other language typed in English to English language.

4.5 Output Representation

Both the methods incorporated for sentiment analysis depict the results in different fashion. They are as follows:

Line Graph. Live capturing of tweets generates a live graph which updates continuously based on time. The X-axis represents time and the Y-axis shows sentiment value. The graph initially starts with the value of Y-axis being 0. When a tweet is classified as positive, the Y value increments by 1 else it is decremented by 1. This shows current mood of people on a particular topic. An upward moving graph denotes that people are happier about a decision or there is a positive feedback and a downward moving graph suggests a negative feedback (Fig. 3).

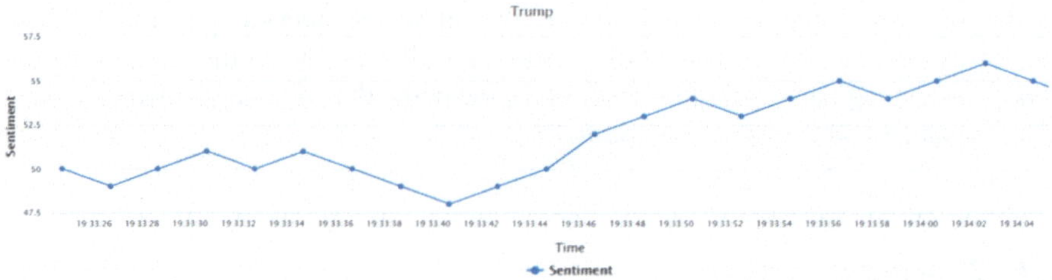

Fig. 3 Live sentiment analysis

Fig. 4 Scraped tweets analysis

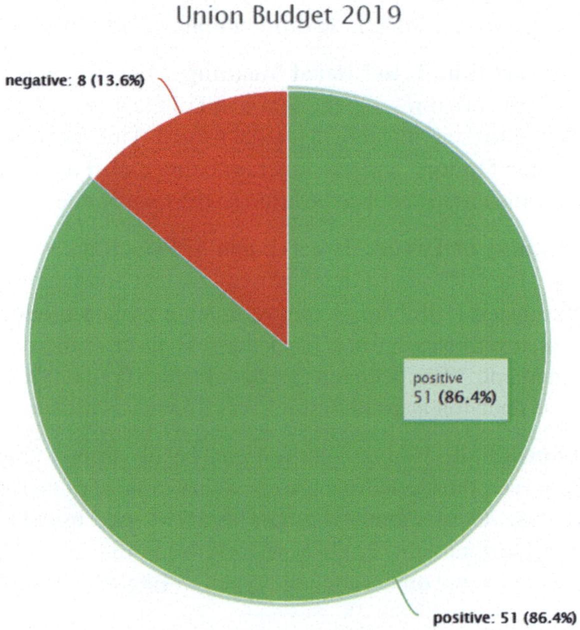

Circular Statistics. Scraping of twitter gives its analysis in the form of a pie chart with a view of the number of tweets analyzed based on amount of scrolling done. The pie chart shows the percentage of tweets classified as positive and negative (Fig. 4).

4.6 Conclusion

The developed Tweet-Based Sentiment Analyzer can be used for analyzing various decisions or policies undertaken by the government to get an overall view of the public reaction. The sentiment analyzer provides an accuracy of about 76% upon taking multilingual input and an accuracy about 85% for input tweets with English language only. The multilingual inputs' text is converted to English using GoogleTrans Python library each time for the classifier to recognize the input text. The accuracy of the classifier upon considering multiple languages decrease as the python library may at times incorrectly translate the tweet thus resulting in wrong predictions at a few instances. This problem, however, does not occur when considering English tweets, thus resulting in a better accuracy. This is achieved using Multinomial Naive Bayes algorithm to classify tweets and output is represented in the form of line graph when graphing the tweets live and pie chart when using web scraping to predict sentiment of old tweets.

References

1. Phand SA, Phand JA (2017) Twitter sentiment classification using stanford NLP. In: 2017 1st International conference on intelligent systems and information management (ICISIM)
2. Azam N, Jahiruddin, Abulaish M, SMIEEE, Haldar NAH (2015) Twitter data mining for events classification and analysis. In: Proceedings of the 2nd international conference on soft computing and machine intelligence (ISCMI'15). IEEE CPS, Hong Kong, Nov 23–24
3. Wang H, Can D, Kazemzadeh A, Bar F, Narayanan S (2012) A system for real-time twitter sentiment analysis of 2012 U S presidential election cycle. In: Proceedings of the 50th annual meeting of the association for computational linguistics. pp 115–120, Jeju, Republic of Korea, July 8–14
4. Guha S, Joshi A, Varma V (2015) Sentibase: sentiment analysis in twitter on a budget. In: SEM 4th joint conference on lexical and computational semantics denver, Colorado, USA. Report No: IIIT/TR/2015/-1
5. Norman J, Mangayarkarasi R, Vanitha M, Praveen Kumar T, UmaMaheswari G (2017) A Naive-Bayes strategy sentiment for sentiment analysis on demonetization and Indian budget 2017-case-study. Int J Pure Appl Math 117(17):23–31. ISSN: 1311-8080
6. Naiknaware B (2018) Peoples opinion on Indian budget using sentiment analysis techniques. Int J Res Eng Appl Manag (IJREAM). ISSN: 2454-9150, Special Issue-NCCT (2018)
7. Kaur J (2016) A review paper on twitter sentiment analysis techniques. Int J Res Appl Sci Eng Tech (IJRASET) 4(X). Guru Nanak Dev Engineering College, Ludhiana, Oct 2016, IC Value: 13.98, ISSN: 2321-9653
8. Sarlan A, Nadam C, Basri S (2014) Twitter sentiment analysis. In: 2014 International conference on information technology and multimedia (ICIMU). Putrajaya, Malaysia, Nov 18–20
9. Verma A, Singh KPA, Kanjilal K (2015) Knowledge discovery and twitter sentiment analysis: mining public opinion and studying its correlation with popularity of Indian movies. Int J Manag (IJM) 6(1):697–705. ISSN 0976–6502
10. Rahman E-U, Sarma R, Sinha R, Sinha P, Pradhan P (2018) A survey on twitter sentiment analysis. Int J Comput Sci Eng 6(11). Open Access Survey Paper, India, e-ISSN: 2347-2693
11. Kouloumpis E, Wilson T, Moore J, Twitter sentiment analysis: the good the bad and the OMG! In: Proceedings of fifth international AAAI conference on weblogs and social media (ICWSM)
12. Xu S, Li Y, Wang Z, Bayesian multinomial naive bayes classifier to text classification. In: International conference on multimedia and ubiquitous engineering international conference on future information technology
13. Heba M, Ismail, Harous S, Belkhouche B (2016) A comparative analysis of machine learning classifiers for twitter sentiment analysis. In: 17th International Conference on Intelligent Text Processing and Computational Linguistics-CICLing
14. Shinde PD, Rathod S (2018) A comparative study of sentiment analysis techniques. Int J Innov Adv Comput Sci 7(3). ISSN 2347–8616

15. Tugores A, Colet P (2013) Mining online social networks with Python to study urban mobility. In: Proceedings of the 6th European conference on python in science
16. Dhanush M, Ijaz Nizami S, Patra A, Biswas P, Immadi G (2018) Sentiment analysis of a topic on twitter using tweepy. Int Res J Eng Tech 5(5):2881. e-ISSN: 2395-0056
17. Jagannatha S, Niranjanamurthy M, Manushree SP, Chaitra GS (2014) Comparative study on automation testing using selenium testing framework and QTP. IJCSMC 3(10):258–267. ISSN 2320–088X
18. Anand N, Kumar T (2017) Text and emotion analysis of twitter data. Int J Comput Sci Eng 5(6). Open Access Research Paper, e-ISSN: 2347-2693
19. Chirawichitchai N (2013) Sentiment classification by a hybrid method of greedy search and multinomial naïve bayes algorithm. In: 2013 Eleventh international conference on ICT and knowledge engineering
20. Gupta B, Negi M, Vishwakarma K, Rawat G, Badhani P (2017) Study of twitter sentiment analysis using machine learning algorithms on python. Int J Comput Appl 165(9):(0975–8887)

Smart Portable Neonatal Intensive Care for Rural Regions

Samden Lepcha, Suraj S. Jain, Sonal Kumar, C Puneeth, Naman Singhal, Anurag Kumar Saw, Abhijeet Singh Batra, Sameeksha Shukla and Aynur Unal

Abstract Every year, an increasingly large number of neonatal deaths occur in India. Premature birth and asphyxia are being two of the leading causes of these neonatal deaths. A well-regulated thermal environment is critical for neonatal survival. In the current scenario, it is impossible for the health centers in the rural areas of India to afford a neonatal incubator for every newborn due to its price and transportability. The successful delivery of neonates is hampered in India due to its increasing population along with limited technology and resources. Thus, a prototype of an incubator has been designed that is affordable, transportable, and energy saving for the health centers in the rural regions, with an AI-based decision support system.

S. Lepcha · S. S. Jain · S. Kumar · C. Puneeth · N. Singhal · A. K. Saw · A. S. Batra · S. Shukla (✉)
Computer Science and Engineering, Christ (Deemed to Be University), Bangalore, India
e-mail: sameeksha.shukla@christuniversity.in

S. Lepcha
e-mail: sam.lepcha@outlook.com

S. S. Jain
e-mail: surajsjain@hotmail.com

S. Kumar
e-mail: skbrahee@gmail.com

C. Puneeth
e-mail: pnthraj37@gmail.com

N. Singhal
e-mail: prince.naman1999@gmail.com

A. K. Saw
e-mail: kanurg101@gmail.com

A. S. Batra
e-mail: abhijeetbatra29@gmail.com

A. Unal
Digital Monozukuri, Stanford, USA
e-mail: aynurunal@alumini.stanford.edu

S. Fong et al. (eds.), *ICT Analysis and Applications*, Lecture Notes in Networks and Systems 93, https://doi.org/10.1007/978-981-15-0630-7_37

Keywords Healthcare · Neonatal healthcare · Rural healthcare · Medical electronics · Smart medical electronics · Neonatal incubator · Decision support system

1 Introduction

The most vulnerable time for an infant's survival is the first twenty-eight days in the life of a newborn—the neonatal period. Newborns face the highest death risk in their first month of lives. In 2017, eighteen deaths for every one thousand lives born was the average global newborn death rate. Globally in 2017, 2.5 million children died in their first month—approximately seven thousand neonatal deaths every day—most of which occurred in the first week, with about a million dying on their first day and close to a million in the next few days.

Globally, every year almost 2.6 million babies do not survive the first month of birth. That means an average of 7,000 neonates every day. As per UNICEF, the country's income level is closely linked to newborn survival rate. Generally, in low-income countries, the average newborn mortality rate (NMR) is twenty-seven, while in high-income countries the figure is only three. This significant gap has been noted that if every country brought its newborn mortality rate to the high-income average, or below, by 2030, 16 million newborn lives could be saved. A study reveals that the causes of 80% of such deaths are preventable and treatable. Also, newborn deaths in India was one of the highest in the world in the year 2017.

Therefore, in this paper, we will discuss our approach to solve this problem with not only a product but also with a system design that incorporates the various underlying processes as a whole.

2 Thermoregulation of Neonates

There must be a balance that needs to be ensured between the production and loss of heat by the body of the neonates. Heat loss might affect the functions of various important organs as well as the metabolic activity in the body of the neonates. The loss of heat from the body of the neonates happens mostly through the skin as the skin will be in direct contact with the external environment. Also, if the air around the infant is cold, it circulates within the body of the neonates as they breathe in and this will increase the rate loss from the bodies of the neonates.

The optimum temperature of the neonates' body is between 36.5 and 37.5 °C. This optimum temperature must be maintained in the body of the neonates in order to ensure proper metabolism in the body. If the body temperature falls below or exceeds the optimum temperature range, it leads to life-threatening complications such as hypothermia and hyperthermia that occurs when the temperature of the neonates' body falls below and exceeds the optimum range respectively. These conditions are very dangerous due to the complications that they cause in the body of the neonates.

The humidity of the surroundings of the body of the neonates must be maintained above seventy percent in order to ensure the proper functioning of the body. Therefore, it becomes critical to maintain and continuously monitor an environment that is required to maintain the body temperature as well as ensure the comfort for the neonates.

3 Proposed Design of the Incubator

The incubator works on the principle of automating the setting of thermal and humidity conditions of the incubator by using sensors to continuously monitor the current conditions in the incubator; use computer fans to regulate and maintain the temperature inside the incubator, and water heater to produce water vapor to maintain the humidity inside the incubator. Four temperature and humidity sensors would be used to monitor the temperature and the humidity conditions inside the incubator. A thermal bed would be used for the baby in the incubator as it would help in maintaining the baby's body temperature by keeping the baby warmer.

The incubator would be built using cheap, recyclable, and readily available wood such as plywood. A cheap processor such as a Raspberry Pi would be used for running the Artificial Neural Networks (ANN) and other programs that would automate the incubator.

The incubator can be made portable for use in an ambulance or outside the hospital by using a battery to power it. This portability of the incubator allows the transport of the baby from the ambulance to the hospital uninterruptedly by just carrying the incubator from the ambulance into the hospital by hand.

When the incubator is used in countries with poor transport infrastructure in the rural regions where the ambulances reach the place of birth late, each village can be provided an incubator which is battery-powered so that it can be powered up until the ambulance reaches the village or it can also be carried easily to the nearest point that the ambulance can reach and moved easily from outside into the ambulance by just carrying it (Figs. 1, 2, and 3).

3.1 Raspberry Pi 3 Model B+

The Raspberry Pi 3 model b+ would be used as a processor in the incubator. This can be replaced by any other cheap processor that can run a Linux distribution operating system on it. The Raspberry Pi used here will be running Raspbian OS.

The processor will be used to interface with the sensors, fans, cameras, and a relay module and for interfacing with the water heater and the incandescent bulb.

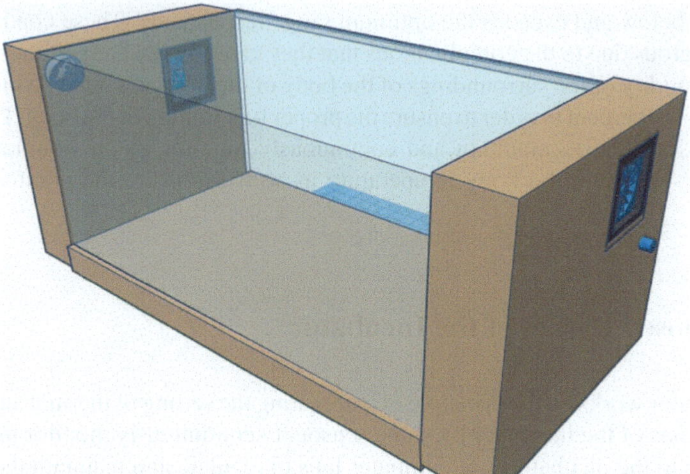

Fig. 1 Outside view of the incubator

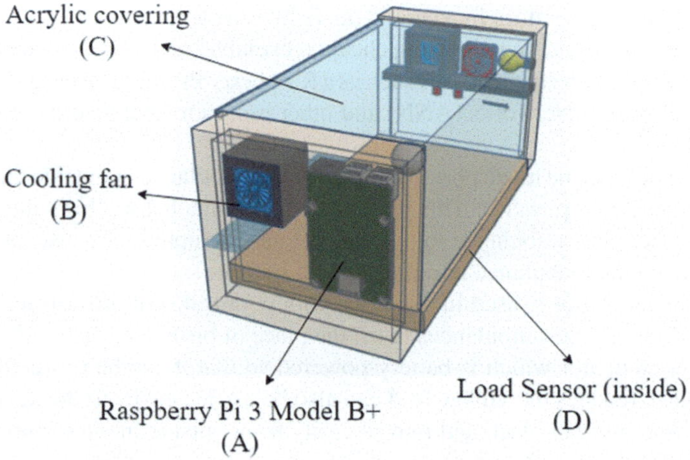

Fig. 2 Left–side diagonal view of the incubator and the parts inside

The computer will continuously fetch the readings from the sensors and the footage from the camera for determining the conditions of the baby as well as the conditions inside the incubator. Then, it will run various algorithms to set and maintain the conditions inside the incubator that suits the best for the neonatal baby by controlling the heating, cooling, and humidity systems.

The Raspberry Pi would also send data continuously to the servers, which would allow the doctors and nurses to monitor the conditions by using a mobile app to log into the dashboard.

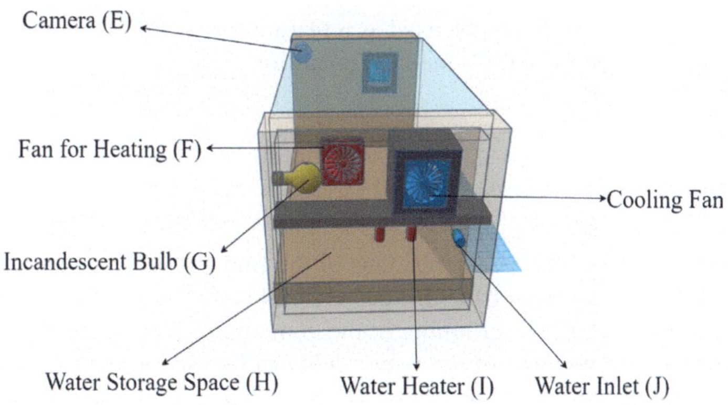

Fig. 3 Right–side view of the incubator and the parts inside

3.2 Cooling System

The cooling system would be used to decrease the temperature inside the incubator when the baby's body temperature is rising up so as to maintain the baby's body temperature between 36.5 and 37.5 °C.

The cooling system would use two computer fans controlled by the processor. Whenever the computer detects a rise in body temperature of the baby, it will turn the cooling fans on. The cooling fans would then take the hot air out of the incubator, which would decrease the temperature inside the incubator and cool the baby body down so that the optimum temperature is maintained.

3.3 Acrylic Covering

The covering of the incubator will be made of a thin and light acrylic sheet. It is easy to open and close.

Acrylic is highly durable and a good heat insulator. This allows the incubator to be resistant to damage when used in ambulances and also to rough usage. The temperature inside the incubator would not be affected easily by the outside temperature due to the acrylic covering, which makes it easy to maintain the temperature and the humidity inside the incubator efficiently.

3.4 Load Sensor

The load sensor would be placed on the bottom of the inner surface of the incubator. It would be connected to the computer and gives the body weight of the baby to the computer. The body weight can be used as a feature for an ANN running inside the computer to determine the baby's medical conditions.

3.5 Camera

The incubator would use two cameras. One of the cameras will be a cheap webcam with a microphone. The input from this camera will be fed as an input to a Convolutional Neural Network (CNN) running in the computer to determine the skin color and other behavioral patterns of the baby. This can be used to detect jaundice and other conditions of the baby. The cry of the baby can be monitored by the microphone and can be fed into another ANN to determine other behavioral patterns and conditions of the baby.

The other camera will be an infrared camera. This can be used to measure the body temperature, heartbeat, and breathing patterns of the baby. This camera's input will be fed into different Artificial Neural Networks (ANN) running in the computer to determine the body temperature, heartbeat, and breathing patterns of the baby, and the conditions inside the incubator will be set accordingly to best suit the baby.

3.6 Heating System

The heating system comprises a computer fan and an incandescent bulb. Whenever a fall in the body temperature of the baby is detected by the processor, it activates the heating system to raise the temperature inside the incubator which would, in turn, increase the body temperature of the baby in order to maintain it in the optimum range.

The heating system is activated automatically by the processor by turning on the incandescent bulb to heat the air and the computer fan to push the hot air into the incubator to raise the temperature in the incubator.

3.7 Incandescent Bulb

A 100 W incandescent bulb will be used in the incubator which when turned on, heats the surrounding air. The automatic turning on and off of the incandescent bulb would be controlled by the computer by means of a relay module.

3.8 Water Storage

The water storage space is a part of the humidity control system in the incubator. It can store water that can be heated to produce vapor. The vapor will further be directed into the incubator in order to maintain the humidity level in the incubator above 70%.

3.9 Water Heater

The water heater is a part of the humidity control system in the incubator. It can be turned on and off by the computer by means of a relay module.

The computer turns the water heater on automatically when it detects a fall in the humidity level below 70%. The water heater will use two coils, which when heated by electricity, heats the water in the water storage to produce water vapor.

3.10 Water Inlet

The water inlet to the incubator is a part of the humidity control system. It allows the water storage to be refilled easily by using a pipe and a tap. A funnel can also be attached to it in order to fill water into it from other sources.

4 The Back End

The Incubator sends data continuously to the back end. The data contains information such as the sensor readings of the conditions inside the incubator, the conditions of the baby as determined by the cameras, and also the automation activities being performed by the Artificial Intelligence running in the computer in the incubator. The Artificial Intelligence will be performing activities such as regulating the temperature and humidity conditions in the incubator to best suit the baby in the incubator based on the conditions of the baby in the incubator.

Mobile and web applications will be built for the incubator. These applications will display a dashboard with the data received from the incubator at the back end. The dashboard can also be displayed on the incubators that have an LCD screen attached to them. The doctors and nurses will have access to this dashboard where they can monitor the temperature and humidity conditions in the incubator as well as the conditions of the baby such as the body temperature, breathing, and heartbeat conditions.

The mobile and web applications will also provide an option for the doctors and the nurses to set their own values for temperature and humidity conditions in the incubator as the Artificial Intelligence in the incubator just acts as a decision support system and there might be situations where the doctors and the nurses would want to set their own conditions in the incubator based on the conditions of the baby in the incubator. When the doctors or the nurses set the temperature and the humidity values in the mobile or the web application that value will be sent to the back end. The back end will then send those values to the incubator and the incubator will ignore the regulations suggested by the Artificial Intelligence and will maintain the regulations set by the doctors or the nurses.

The dashboard in the mobile and web applications can display data from multiple incubators at one place in an organized way. This makes it easier for the doctors and the nurses to handle multiple incubators as the status of all the incubators would be available at one place (Fig. 4).

The back end runs various ANN and other algorithms that would monitor the data received from the incubators and alerts the nurses and doctors by sending a notification to their phone in case if emergency care is required.

The Artificial Intelligence running in the back end will also act as a decision support system for nurses by providing necessary instructions to the nurses. These instructions help nurses to take the final call. This substitutes for the need of doctors as the nurses can get the instructions from the back end running an Artificial Intelligence that monitors the incubator and generates necessary instructions for the nurses. This allows the incubators to be used more easily and efficiently by the nurses in the regions where the availability of doctors is less.

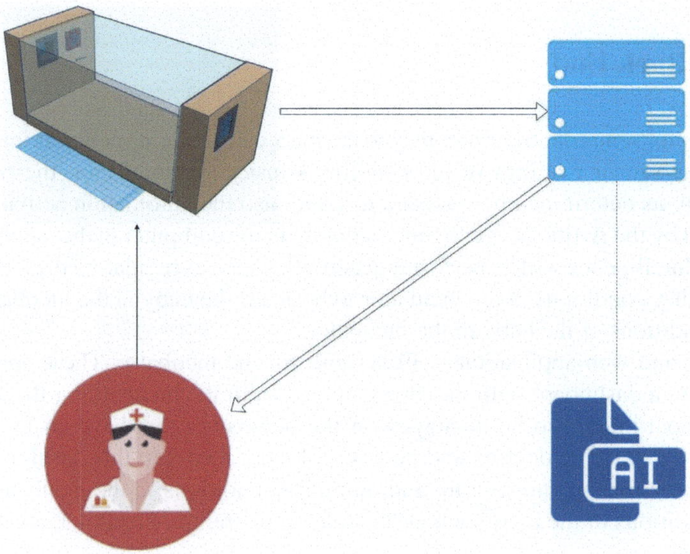

Fig. 4 Overview of the role of the back end in the incubator system

Table 1 Various components used to build the incubator and their cost

Product	Cost ($)	Quantity	Total ($)
Raspberry Pi 3 B+	40.00	1	40.00
Acrylic sheet	15.00	1	15.00
Plywood	10.00	1	10.00
Humidity sensor—DHT 11	2.00	4	8.00
Temperature sensor—LM35	0.20	4	0.80
IR camera AMG8833	40.00	1	40.00
USB camera—logitech c310 webcam	27.00	1	27.00
Relay module—4 channel	3.00	1	3.00
100 W incandescent bulb	1.50	2	3.00
12 V DC fan	3.00	4	12.00
Bed (thermal bed)	22.00	1	22.00
Water heater	1.50	2	3.00
Backup battery	50.00	1	50.00
500 W 12 V DC to 220 V AC inverter	15.00	1	15.00
Battery charger	5.50	1	5.50
Connections and accessories	8.00	1	8.00
3.5' LCD touch display	26.00	1	26.00
USB microphone	4.50	1	4.50
Weighing sensor	1.50	1	1.50
Total cost			294.30

5 Cost Analysis of the Incubator

The incubator we have designed uses cheap and readily available components used for Do It Yourself (DIY) projects. This allows the incubator to be fixed easily and in a faster way by just replacing the necessary components in case of breakage.

The rough cost of the incubator turns out to be $249.30, which is cheaper when compared to the other incubators in the market. This makes it to be affordable and usable by most of the government run hospitals, health centers, and private hospitals of the third world countries (Table 1).

6 Comparison with Prevailing Incubators

The incubator proposed has major advantages over the other incubators available in the market. It is significantly cost-effective and intelligent unlike the other incubators available in the market as it is driven by Artificial Intelligence. Acrylic sheets make the whole incubator build stronger, lighter and easier to clean and handle.

Maintenance costs are lesser when compared to general hospital grade incubators. As the incubators are designed in a simple and robust way, they are easy to be carried around in the rural areas and are even suitable to be carried in a moving vehicle like an ambulance. The need for special attention could be decreased as most of the calls are made by the AI system built into the incubator unlike other incubators available in hospitals today.

7 Conclusion and Future Scope

This work proposes a prototype of a neonatal incubator that can be used at the health centers in the rural regions, emergency vehicles, and also in hospitals. This neonatal incubator could provide a proper and conducive environment to the newborn. Primary care can be provided within a short period during the critical condition, thus reducing the mortality case among neonatal infants. This will also be monitored by an AI-based system to substitute the constant supervision of doctors and nurses and assist them in decision-making. Hence, the cost and resources for the operation will be reduced.

The proposed system has few limitations which could possibly be overcome by continuous technology advancement in the future. In the recent development stage, as there could be some critical situations which are dangerous for the life of the infant, Artificial Intelligence may not be the best option that people can trust upon. The incubator is connected to the remote back-end servers, the network connection between the servers and the incubator could be a problem in rural areas. Although the incubator is powered by several sophisticated batteries that will give a power backup, there can be a few cases of the incubator running low on power due to excessive long use on battery power that happens especially in places where there is no electricity or has long power cuts due to which the incubator cannot be powered from an external power supply through the power chord of the incubator.

Acknowledgements This project was started as a research initiative at Faculty of Engineering, CHRIST (Deemed to the University), Bengaluru, India. We thank all the members concerned for the same.

References

1. Neonatal mortality—UNICEF Data. UNICEF Data. https://data.unicef.org/topic/child-survival/neonatal-mortality/
2. Asmidar N, Binti N, Fudzi M, Binta N (2016) Development of infant incubator for rural area of Malaysia. In: IECBES. IEEE
3. Fei J, Pavlidis I (2006) Analysis of breathing air flow patterns in thermal imaging. In: Conference proceeding of IEEE engineering in medicine and biology society. IEEE

4. Kim Y, Park Y, Kim J, Lee EU (2018) Remote heart rate monitoring method using infrared thermal camera. Int J Eng Res Tech
5. Frischer R, Penhaker M, Krejcar O, Kacerovsky M, Selamat A (2014) Precise temperature measurement for increasing the survival of newborn babies in incubator environments. Sensors (Basel)
6. Elahi F, Nisha NA (2014) Low cost neonatal incubator with smart control system. In: 8th international conference on software, knowledge, information management & applications

Women Empowerment Through Social Media: Insights from India

Rajesh R. Pai and Sreejith Alathur

Abstract In recent times, social media has been used by people to participate in a particular event and has resulted in the generation of a large amount of data online. These data can be helpful for the decision-maker in promoting and devising necessary policies at the right time. The purpose of this paper is to understand the peoples' sentiments and emotions about a recent social movement. Based on the result and analysis, the possible inferences have been presented.

Keywords India · Mental health · Mobile health · Sentiment analysis · Social media analytics

1 Introduction

Social media has transformed the traditional system of service delivery online and has benefitted many business and organizations to reach a large number of people worldwide [1–3]. Sequentially, [4] indicated that the Internet and social media are also used for good or for evil purposes [4]. It has been used in education for improving instruction and to promote active learning among students [5]. Realizing these benefits, healthcare professionals, government officials, and policymakers started using social media for better governance [6]. For example, national health portal of India basically helps an individual to get updated about various health and disease conditions through their social media webpages. In some cases, social media is used as a medium to target any social or political issues and to seek public support for a cause which is either done by making its meme or creating awareness about the problem faced by an individual because of its existence and practice [7–9]. For example, during social media movement in India and abroad many individuals have used this platform to share their difficulties and experiences faced by them from society [7, 10]. With the availability of an enormous amount of these data online and

R. R. Pai (✉)
Manipal Institute of Technology, Manipal Academy of Higher Education, Manipal, India
e-mail: pairajesh.m@gmail.com

S. Alathur
National Institute of Technology Karnataka, Surathkal, India

S. Fong et al. (eds.), *ICT Analysis and Applications*, Lecture Notes in Networks and Systems 93, https://doi.org/10.1007/978-981-15-0630-7_38

on social media, its analytics has largely gained importance among researchers and professionals in decision-making.

Social media analytics plays an important role in understanding the peoples' opinion, which is expressed as real-time messages or as tweets in social media. Among which the most commonly used social media analytics for understanding the opinion is the sentiment analysis. In fact, the sentiment of the people has an influence on measuring the acceptance and rejection level of certain products, services, technologies, or any ordinances [11, 10]. Twitter is frequently used to study the people's sentiment because of its popularity based on maximum number of users and billions of page views [12]. In Twitter, a user can post messages or tweets followed by the establishment of hashtags thereby forming a base for data extractions and analysis [13].

This study involves social media analytics for a recent social movement, which is considered to be an independent outgrowth which is inclined toward the international campaign against women being subjected to sexual harassment at the workplace [14]. This movement has been analyzed by extracting 71,769 tweets collected from November 1, 2018 to December 7, 2018. Each tweet extracted has been cleaned before performing the sentiment analysis. The purpose of this paper is to carry out a sentiment analysis on the tweets dataset and to understand the peoples' sentiments and emotions about the movement.

2 Literature Review

This section reviews the prior literature related to this study.

2.1 Social Movements

A social movement is a kind of action group of individuals or organizations which are sometimes informal and focused on implementing or preventing changes (either radical or minor) to societies' political or social issues [15, 16]. It is defined as "organizational groups and policies that may permit an oppressed individual to provide effective challenges and fight against their peers or leaders" [17]. These movements are basically dependent on whom the movement is attempting the change and how much change a movement is being advocated [18]. Literatures suggests that "freedom of expression, edification, urbanization, industrialization, and other economic liberations are responsible for establishing various social movements" [19–21]. However, "the origination of this movement is still not clear and is found to be increasing and becoming part of the general and global expression of dissent" [22]. Some of these movements include anti-vaccination movement, civil rights movement, cultural movement, etc. [23], which are found to be predominant across various disciplines and sectors. These movements often used various technologies and Internet to facilitate people by engaging themselves in decision-making. Social media platforms such as Facebook, Twitter, etc., and video sharing platforms such as YouTube are the commonly used medium for participating online [24]. These online movements have not only saved time and cost of an individual but also provided an opportunity for them to create awareness about a particular disease condition, etc., by making it widespread across different regions and countries. For example, the Ice Bucket Challenge, was initiated to promote awareness about a disease called amyotrophic lateral sclerosis, thereby encouraging for donations which are later used for the purpose of research [25, 26].

Across the world, the literature indicates that women are often the victims of sexual harassment [27, 28]. Online social movements started against problems related to sexual assault and sexual harassment faced by an individual at their workplace and got the momentum over a period of time [27, 29, 28]. Shortly, these movements have taken all forms of social media, and people across the world started posting their experiences or endorsing their support through social media in the form of tweets which had created a large amount of data [28]. This paper uses tweets extracted from the Twitter website to understand peoples' sentiments and emotions associated with social media movement.

3 Methodology

The methodology followed in this study is qualitative, wherein the data are collected from the social media site "Twitter". The process used in the extraction of tweets and for performing sentiment analysis has been adapted from the research paper of Bharti et al. [30]. With the help of RStudio tool and by using Rcodes, 71,769 tweets using hashtags were collected from November 1, 2018 to December 7, 2018. The data extracted were cleaned and processed for further analysis.

4 Results

This section discusses the findings from the sentiment analysis of the tweets for identifying the specific themes surrounding a social media movement.

4.1 Sentiment Analysis

The data extracted were analyzed using word cloud which was drawn according to polarity and emotion values classified according to Syuzhet package function

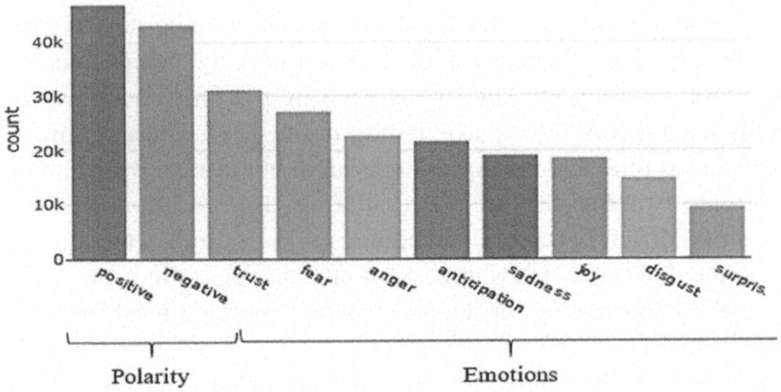

Fig. 1 Sentiment analysis based on polarity and emotions

available in RStudio library and correspondingly boxplot, and its related sentiment score has been calculated.

From the text analysis, it was identified that the word "women" had been found frequently in the tweets of the users, which is acting either in the direction of support or found criticizing this movement. Similarly, when calculated sentiment values based on polarity and emotions, it was identified that the total number of positive sentiment values is more than that of negative (Fig. 1). This can be because of the fact that the people who once remained silent and experienced or experiencing sexual harassment and violation at their workplaces or homes are now coming forward to reach out to the public to make others aware and to get support or seek justice against it. The emotion values indicate that individuals have repeatedly mentioned about trust (31345), fear (27207), anger (22776), and anticipation (21720) over others.

4.2 Sentiment Score Values

From Figs. 2 and 3, we can identify that the mass of the data distribution is concentrated in its center and its tail on the left side is longer than that of the right. This indicates that the distribution is negatively skewed as mean value is less than that of the median value in the negative region. The graphs signify that people may be supportive for women empowerment movement and are likely to hide the larger

Fig. 2 Boxplot

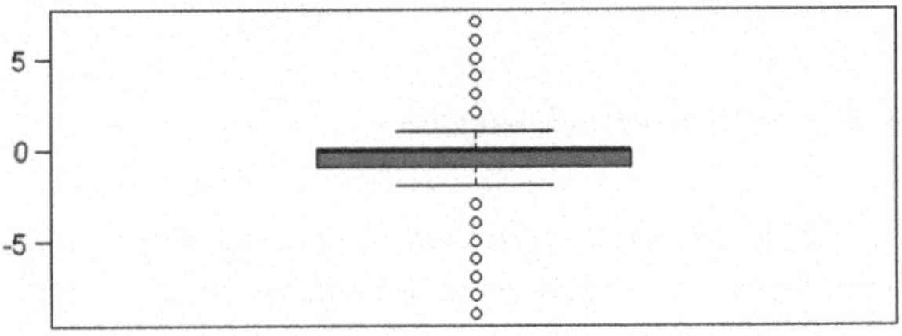

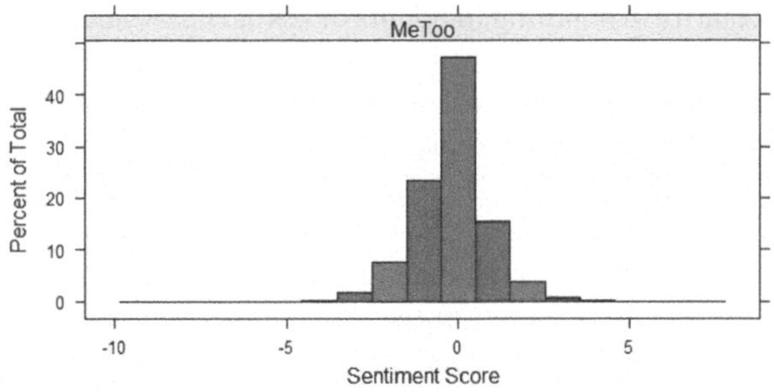

Fig. 3 Sentiment scores

amount of people who are supportive toward the elimination of all forms of sexual harassment and other violations expanding the average value.

In fact, people have a bad experience, and opinions about sexual harassment are circulated across various sectors and are often trying to bring it to a large audience for creating awareness and support which all together help this country in developing effective policies and actions against such offenses.

5 Inferences

The present study investigates the sentiment and emotions of the social media users through sentiment analysis and also discusses specific themes that the individuals are concerned about. The results showed that women had been largely associated with harassments and violence at their workplaces and/or at homes. The polarity value also indicates that people are feeling positive about the social media movement as it may make others aware and can seek support. However, the emotion values that show trust, fear, anger, and anticipation have been repeatedly used over others. This signifies that the individuals and victims feared harassments as they suffered from a problem relating to mental health when eventually losing trust with the employers and others. The analysis of the tweets identified major themes such as harassment, gender, sexual penetration, women, sectors, reform movements, culture, political, health risk, story, and right time are found to be commonly repeated keywords in the tweets.

Sexual harassment and abuse have captured the attention of a large population using social media. This is mainly because of the accessibility and availability of social media and rapid developments in information and communication technologies like mobile phones, tablets, etc. Though the government has implemented certain policies and actions against harassment and other illegal activities happening across the social media and other online forms, its involvement in all the forms of harassment and violation are yet to covered comprehensively in various policies. For example, the Sexual Harassment of Women at Workplace (Prevention, Prohibition and Redressal) Act, 2013, i.e., the PoSH Act has been implemented in India to eradicate harassment of women at their workplace [31]. However, deadlines for filing the complaints, lack of clarity in handling incidents of sexual harassment, etc. are often discussed in media [31]. It is also criticized that the social media movement has created an influence on dislodging scores of an individual from their high-power positions [32]. Some of which are explained in the sections below.

Political Sector: The use of social media to a large extent, especially in political communication often starts during the election. Social media was used in most of the cases for campaigning and communicating with the electorate, which subsequently may cause an increase in the poll results and create a winning moment for that candidate [33]. This led to various other parties to use social media platforms for campaigning and other political communications. Thus, the use of social media is considered to be a significant development in Indian politics and women empowerment.

Film and Entertainment Sector: In this sector, social media movements have created an influence on the stakeholders if involved in a case either as a plaintiff or a defendant. For example, in the case of celebrities, if an individual files a complaint against any stakeholder, it is likely that the number of offers for acting in movies might be reduced maybe because of disagreement from the other stakeholders or fear in involving them in their movie projects. This has created a negative impact on their carrier, which might influence other individuals from refraining themselves away from filing complaints against sexual harassment and abuse in the upcoming years. The same may be the case in any art industry.

Newspaper Industry: It was reported that the newspaper industry had also been affected by the storm of social media movements and many of the stakeholders had also received complaints regarding their involvement in sexual misconduct with an individual. Stakeholders of the newspapers may have stepped down from their positions.

For relevant keyword search apart from normal tweets about sexual harassment and abuse allegations against individuals irrespective of different sectors, it was also observed that some of the tweets had been indicated regarding their impact on individual health. Thus, it can be concluded that sexual harassment and abuse has an influence on the mental and physical health of an individual and may require a proper governing structure toward early prevention of individuals from becoming the victim of such activities thereby saving people from improper mental and physical health. A tweet has been found in support of early prevention from violence for supporting healthy and empowered individuals. The text analysis also revealed that users are promoting awareness about mobile applications so that the victims can report about sexual harassment instantly.

When it comes to sexual harassment and violence, even the organization plays an important role in providing justice to an individual or to protect them from such complaints. For example, in financial institutions such as banks, mergers, and acquisitions, various advisory members or representatives have been appointed to protect themselves during negotiations and contract agreements [34]. Morris et al. (2018) reported that healthcare organization is particularly at risk when it comes to sexual harassment claims, as healthcare professionals' main task is to work closely with the victims to seek evidence and support without blurring any of the body or body parts

[34]. In such cases, the healthcare organization should act vigilantly, quickly address complaints, document misconduct, and comply with healthcare-specific obligations to address problems of social media [34]. These processes can be coupled with a smartphone application, as many of them use it for day-to-day activities. The smartphone application helps during potential harassment situations and provide early preventive guidance for the individual.

Due to complexities and fear in reporting sexual harassment about losing the position, career, social status, etc., many individuals may try to remain silent, thereby making them get involved in problems related to mental health. In such circumstances, these individuals can be connected to their respective healthcare professionals through smartphone applications and receive consultations for the same, which addresses the problem of mental health over time. However, doctors also indicated that in the present generation both the young and adult population search health material from online source rather than speaking to their parents or peers so that their privacy and anonymity is protected [35]. Thus, social media applications can be a possible solution for tackling the problems of harassment and mental health.

6 Conclusion

The current study analyzes the use of social media understanding the sentiment and emotions of people and also tries to identify specific themes surrounding the social media movement. Upon analysis, it was identified that women have largely been associated with sexual harassment and violation at their workplaces or homes. The sentiment analysis of social media movement indicates positive sentiment as the harassment which was happening within the doors are now being brought into public for creating awareness, requesting support and seeking justice toward sexual harassment and other violations. In addition, the tweets indicated trust, fear, anger, and anticipation as the frequently repeated emotion components of the sentiment analysis. It means that people have feared and lost trust in other individuals because of its impact on their physical and mental health. However, the analysis of the tweets identified various themes which are grouped under the major themes of harassment, gender, sexual penetration, women, sectors, reform movements, culture, political, health risk, story, and right time classified based on frequently repeated keywords. Finally, to conclude, at this moment it is difficult to analyze the case of social media movements as it is not possible to generalize women empowerment potential of social media using secondary data in a particular time period. But the overall results from the sentiment analysis of social media movement have been found positive, and people encourage more individuals to report such crimes and offenses such that right action can be taken against the harasser.

References

1. Chan NL, Guillet BD (2011) Investigation of social media marketing: how does the hotel industry in Hong Kong perform in marketing on social media websites? J Travel Tour Mark 28(4):345–368
2. Stephen AT, Galak J (2010) The complementary roles of traditional and social media publicity in driving marketing performance. In: Fontainebleau: INSEAD working paper collection
3. Trainor KJ, Andzulis JM, Rapp A, Agnihotri R (2014) Social media technology usage and customer relationship performance: a capabilities-based examination of social CRM. J Bus Res 67(6):1201–1208
4. Barak A, King SA (2000) The two faces of the Internet: introduction to the special issue on the internet and sexuality. Cyberpsychology Behav 3:517–520
5. Tess PA (2013) The role of social media in higher education classes (real and virtual)–a literature review. Comput Hum Behav 29(5):A60–A68
6. Lee G, Kwak YH (2012) An open government maturity model for social media-based public engagement. Gov Inf Q 29(4):492–503
7. Joseph N, Grover P, Rao PK, Ilavarasan PV (2017) Deep analyzing public conversations: insights from twitter analytics for policy makers. In: Conference on e-Business, e-Services and e-Society. Springer, Cham, pp 239–250
8. Yeung D (2018) Social media as a catalyst for policy action and social change for health and well-being. J Med Internet Res 20(3):e94
9. Aswani R, Kar AK, Aggarwal S, Ilavarsan PV (2017) Exploring content virality in facebook: a semantic based approach. In: Conference on e-Business, e-Services and e-Society, Springer, Cham, pp 209-220
10. Singh P, Sawhney RS, Kahlon KS (2018) Sentiment analysis of demonetization of 500 & 1000 rupee banknotes by Indian government. ICT Express 4(3):124–129
11. Pai RR, Alathur S (2018) Assessing mobile health applications with twitter analytics. Int J Med Inform 113:72–84
12. Gautam A (2014) Top microblogging sites list with high pr and best. http://mywptips.com/top-microblogging-sites-list/. Accessed 7 Nov 2018
13. Lakhiwal A, Kar AK (2016) Insights from twitter analytics: modeling social media personality dimensions and impact of breakthrough events. In: Conference on e-Business, e-Services and e-Society. Springer, Cham, pp 533–544
14. Goel V, Venkataraman A, Schultz K (2018) After a long wait, India's #MeToo movement suddenly takes off. https://www.nytimes.com/2018/10/09/world/asia/india-sexual-harassment-me-too-bollywood.html. Accessed 7 Nov 2018
15. Abbott A (2016) Processual sociology. University of Chicago Press, Chicago, IL
16. Simandan D (2018) Being surprised and surprising ourselves: a geography of personal and social change. Prog Hum Geogr. 0309132518810431
17. Glasberg DS, ShannonD (2010) Political sociology: oppression, resistance, and the state. SAGE Publications
18. Aberle D (1966) The peyote religion among the Navaho. Aldine, Chicago. ISBN 0-8061-2382-6
19. Castells M (1983) The city and the grassroots: a cross-cultural theory of urban social movements, no 7. University of California Press
20. Little K (2013) Urbanization as a social process: an essay on movement and change in contemporary Africa. Routledge
21. Safa HI (1990) Women's social movements in Latin America. Gend Soc 4(3):354–369
22. Tilly C, Wood LJ (2015) Social movements 1768–2012. Routledge
23. Wilson J (1973) Introduction to social movements. Basic Books, New York, pp 233
24. Rotman D, Vieweg S, Yardi S, Chi E, Preece J, Shneiderman B, Pirolli P, Glaisyer T (2011) From slacktivism to activism: participatory culture in the age of social media. In: CHI'11 extended abstracts on human factors in computing systems. ACM, pp 819–822
25. Koohy H, Koohy B (2014) A lesson from the ice bucket challenge: using social networks to publicize science. Front Genet 5:430

26. Fawzy F (2016) Ice Bucket Challenge's 2nd anniversary celebrates its gene discovery. https://edition.cnn.com/2016/07/27/health/als-ice-bucket-challenge-funds-breakthrough/index.html. Accessed 13 Nov 2018
27. Frye J (2018) From politics to policy: turning the corner on sexual harassment. https://www.americanprogress.org/issues/women/news/2018/01/31/445669/politics-policy-turning-corner-sexual-harassment/. Accessed 7 Nov 2018
28. Manikonda L, Beigi G, Liu H, Kambhampati S (2018) Twitter for sparking a movement, reddit for sharing the moment:# metoo through the lens of social media. arXiv:1803.08022
29. Hebert LC (2018) Is 'MeToo' only a social movement or a legal movement too? Emp Rights Employ Policy J 22(2). Ohio State Public Law Working Paper No. 453
30. Bharti SK, Vachha B, Pradhan RK, Babu KS, Jena SK (2016) Sarcastic sentiment detection in tweets streamed in real time: a big data approach. Digit Commun Netw 2(3):108–121
31. KelpHR (2018) Advantages and disadvantages of PoSH Act 2013. https://www.kelphr.com/blogs/advantages-and-disadvantages-of-posh-act/. Accessed 21 Nov 2018
32. Brown D (2018) 19 million tweets later: a look at #MeToo a year after the hashtag went viral. https://www.usatoday.com/story/news/2018/10/13/metoo-impact-hashtag-made-online/1633570002/. Accessed 21 Nov 2018
33. Tiwari R (2018). What to expect in 2019: the new language of politics. https://indianexpress.com/article/explained/narendra-modi-rahul-gandhi-lok-sabha-elections-sabarimala-entry-gender-rights-metoo-ram-temple-5512545/. Accessed 21 Nov 2018
34. Morris FC, Hoerner J, Smith K (2018) 4 ways to address the #MeToo era in health care. https://www.healio.com/gastroenterology/practice-management/news/online/%7B4d01cda5-0d75-471e-80d2-ade0e7a94d82%7D/4-ways-to-address-the-metoo-era-in-health-care. Accessed 21 Nov 2018
35. Young L (2018) What to look for in mental well-being apps. https://medicalxpress.com/news/2018-10-mental-well-being-apps.html/. Accessed 21 Nov 2018

Automation System Using Otsu's Method for Production Line Quality Assessment of Welded Structure

D. D. Bhilegaonkar and D. S. Deshpande

Abstract This paper presents an automation system that mechanizes the post quality review of the weld which was performed by human welders already. A foundation is established to effectively extract information of the welded structure surface to facilitate quality check and inspection of the weld joint using different algorithms. The idea is to provide cost effective solution to the small scale manufacturing industries for robotizing there post welding activities, decrease the time required for these tasks, and increment the speed eventually. The proposed framework built up a GUI which can be utilized by nonspecialized laborers to make their work simple. The framework utilizes various algorithms like Otsu and BHT for estimation and examination of different geometric parameters of the weld which aided in post welding quality assessment of the welded structure. Results obtained from both the algorithms are compared and analyzed. Geometrical parameters thus calculated were then compared with the values of these parameters as specified by industry standards for accepting or declining the welded structure under inspection. Lot of research has been done in measuring characteristics of the welded structure. However, a limited research is done in the quality inspection and post welding operations efficiency performance check. The proposed system satisfies this necessity. In this system three things are remembered while building up the framework, i.e., cost proficiency, broadly utilized welding procedure, and industry interest for mechanization of post welding quality assessment. For actualizing a similar thought, Common CCD (Commercial charge coupled device) cameras are utilized to catch images, GTAW which is a broadly used welding procedure is used and simple to use however proficient GUI is built.

Keywords Balanced histogram thresholding (BHT) · GTAW (gas tungsten arc welding) · GUI (graphical user interface) · CCD (commercial charge coupled device) · Geometrical parameters · Robotization framework · Analysis · Welded structure · Quality assessment · Automation

D. D. Bhilegaonkar (✉) · D. S. Deshpande
MGM's Jawaharlal Nehru Engineering College, Aurangabad, Maharashtra, India
e-mail: dbdipty@gmail.com

D. S. Deshpande
e-mail: deepadsd@yahoo.com

S. Fong et al. (eds.), *ICT Analysis and Applications*, Lecture Notes in Networks and Systems 93, https://doi.org/10.1007/978-981-15-0630-7_39

1 Introduction

Welded structure is the essential input which encourages productive investigation for a talented welder in a complicated welding process. By simply observing the welded structure, he is able to monitor and inspect the quality of the welded structure. The proposed system has designed a GUI in Matlab R2013a which automates the post welding quality inspection. Otsu and BHT techniques are used which facilitate the geometric parameter calculations of welded structure by both these techniques and results are then compared and analyzed. These values are then compared by the skilled inspectors with the industry specified standard values in order to accept or decline the weld under inspection. Images of the welded structure are captured during GTAW process by using CCD camera and these are sent to the computer for further processing the flow of industry process explained in Sect. 3. After acquiring images, image binarization and geometric parameter calculations are performed by using the two techniques Otsu and BHT and the results of both these methods are compared and analyzed.

2 Related Work

In the modern era, there has been quick development in remote sensor systems, large data, Iot, and mobile Internet are entering the modern segment, which establishes the framework for new mechanical upheaval, i.e., industry 4.0. To be specific, German specialized foundation and different organizations have mutually proposed the fourth era of industry—which is intended to guarantee the future intensity of German assembling industry. Welding is one of the key pieces of assembling industry, which is a significant pointer of industrialization level, so the smart welding innovation has turned into an intriguing issue. Estimation of the welded structure geometry can be executed by processing images of the welding procedure from a vision framework. A few researchers have effectively settled vision frameworks and relating scientific strategies to quantify welded structure geometry. Different strategies are proposed to remove the qualities of welded structure like image processing, reversing, local threshold determination, fast Fourier transform, and advanced morphological operations which obtain welded structure width [1]. A minimal cost framework is created utilizing common commercial charge coupled gadget (CCD) camera and joined with a composite light filter. Welded structure image is caught obviously amid steady current gas tungsten arc welding (GTAW). Distinctive image processing methods are utilized to get a quality upgraded image. [2]. A versatile powerful linear model is proposed in paper [3] to demonstrate the dynamic welded structure geometry in GTAW process progressively. Huge numbers of the scientists concentrated on estimating welded structure geometry by evacuating impedances covered with the welded structure area in images in which the estimation is accomplished through PC vision [4]. A neurofuzzy system is developed in paper [5] which estimates the back side width of weld from its geometry.

This paper presents an automation system for the post welding quality inspection of the welded structure, thus allowing for calculation of various geometrical parameters of weld like area, mean, standard deviation, entropy, contrast, kurtosis, skewness, IDM (Inverse difference moment), variance, RMS (Root mean square), energy, correlation, homogeneity for analysis purpose. The proposed system works on images captured by CCD camera for GTAW process. These images are analyzed for post welding process of quality inspection of the welded structure. Efficiency of the welding operations can affect the quality of the welded structure which can help an experienced welder to assess the quality inspection post welding operations. Proposed configuration utilizes an imaginative way to deal with build up a propelled control framework to analyze the welded structure by imitating the estimation and basic leadership procedure of the human welder utilizing GUI in Matlab for nonspecialized laborers. Otsu and BHT techniques are chosen for threshold selection and the binarization of welded structure image. Geometrical parameters of the welded structure are determined, which are then compared with company norms for tolerating or declining the welded structure by quality review. The outcomes of both the techniques are compared and analyzed and the quality of welded structure is examined.

3 Proposed Method

Figure 1 gives the proposed framework flow. Images are captured by CCD camera and then sent to the computer for further processing. The image is acquired first and displayed on GUI of Matlab. Matlab R 2013a was used. To investigate the given image the most significant part is gray scaling. A histogram is taken after gray scaling. Subsequently global thresholding is performed using Otsu algorithm explained in [4] which is the best compared to available thresholding algorithms which also does binarization of image to discover the quantity of pixels in a given image except if and until we get the binary information for mapping of the image. The proposed framework utilizes two strategies Otsu and BHT to get the required results, a brief explanation of which is given below. Then geometrical calculation and area estimation is done. If the welded pool matches or stays between the desired parameters as per industry standards, it is passed forward or discarded. The same flow will be there when applying BHT method. Here only the flow of Otsu is shown.

The same flow will be there for BHT method. Just the difference is that BHT method is used for threshold selection and binarization step. The outcomes of both the techniques are compared and analyzed at the end.

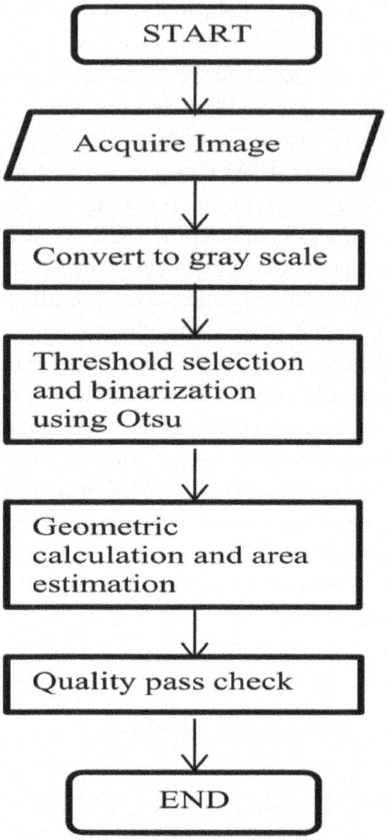

Fig. 1 Proposed framework

3.1 *Otsu's Method*

Otsu's strategy was founded by Nobuyuki Otsu. It was utilized to naturally perform clustering-based image thresholding, to convert a gray level image to a binary image. Image is gained and changed over to gray scale image. After which Otsu technique explained in [4] is applied which computes the ideal threshold isolating the two classes with the goal that their (intra-class variance) is minimum. Finally the geometrical parameters are calculated. Otsu's method is optimum, in the sense that it chooses the threshold value that minimizes the within-class variance.

To find the threshold that minimizes the weighted within-class variance.

Initialisation

$$q_1(1) = P(1); \;\; \mu_1(0) = 0. \tag{1}$$

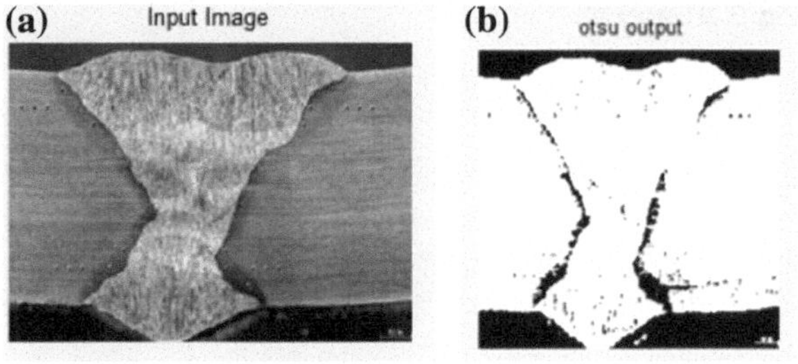

Fig. 2 **a** Original image **b** Otsu output image

Recursion

$$q_1(t+1) = q_1(t) + P(t+1) \tag{2}$$

$$\mu_1(t+1) = \frac{q_1(t)\mu_1(t) + (t+1)P(t+1)}{q_1(t+1)} \tag{3}$$

$$\mu_2(t+1) = \frac{\mu - q_1(t+1)\mu_1(t+1)}{1 - q_1(t+1)} \tag{4}$$

where $q_1(1)$ is probability of class 1 and $P(1)$ is probability function and μ_1, μ_2 are means for two classes [6].

Figure 2a and b show the input and output images after applying Otsu.

3.2 Balanced Histogram Thresholding

The balanced histogram thresholding method [7] is a very simple method used for automatic image thresholding. Like Otsu's method, this is a histogram-based thresholding method. Assuming that the image is divided into two main classes: the background and the foreground, this method tries to find the optimum threshold level that divides the histogram into two classes. This method weighs the histogram, checks which of the two sides is heavier, and removes weight from the heavier side until it becomes lighter. It repeats the same operation until the edges of the weighing scale meet.

Histogram of the image is shown in Fig. 3 below.

Figure 4a and b show the input and output images after applying BHT.

3.3 Experimental Setup

Setup at industry is as shown in Fig. 5. Gas tungsten arc is used for welding process which is fixed in a jig of robotic arm. High-speed CCD camera is placed parallel to

Fig. 3 Histogram

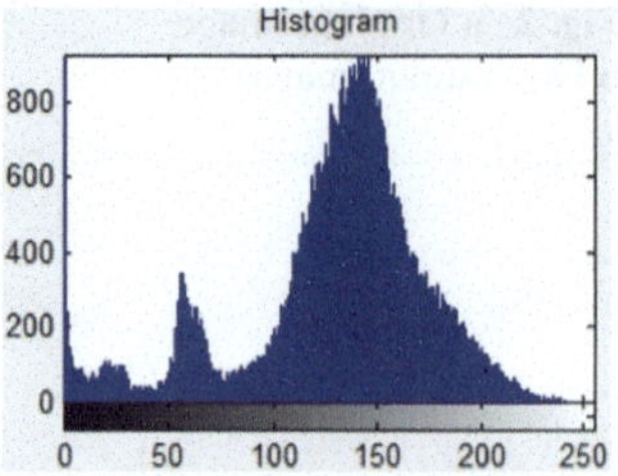

Fig. 4 a Original image b BHT output image

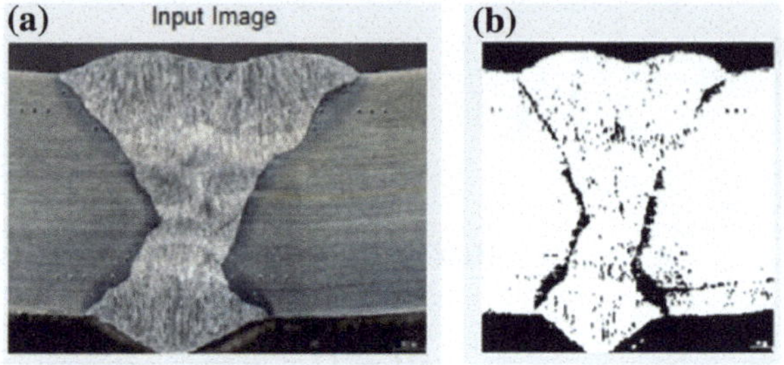

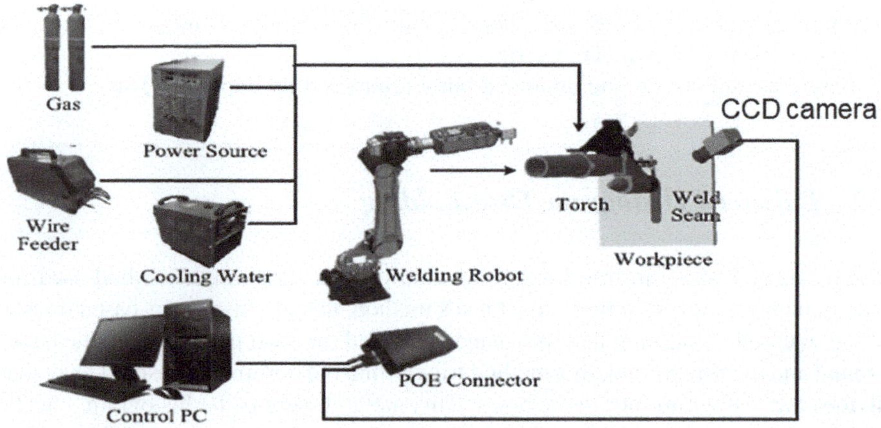

Fig. 5 Exp. setup

the workpiece to capture images. These images are sent in a segment of batches of 10 images to the POE (power over ethernet) controller which acts as the interface between the GUI and the camera. Once these images are stored in the computer, it can be utilized as per our need. The setup shown in Fig. 5 was established by the industry at their end where shielding gases which were inert gases were used for shielding purposes which were supplied through gas cylinders. Wire feeder was used that provides the welded structure material in the form of bundles. High voltage power source was used. Cooling water is supplied through pipes from the holding tank for exhaust purpose. This setup was used in industry when the GTAW process was carried out. The proposed methodology was applied post welding operations and after the welded structure images from the CCD camera were acquired on computer.

4 Experimental Results

Design of the proposed GUI is shown in Fig. 6.

4.1 Features Analyzed

The following features are analyzed.

1. **Mean**: Average material required for the weld.
2. **Standard deviation**: It describes deflection from the original value.
3. **Entropy**: When different materials are used as alloy entropy determines the presence of the material.
4. **RMS** (Root mean square): It is used to find out the central frequency of the image.
5. **Variance**: It finds out enlarged or bubbled faults in welding.
6. **Smoothness**: Evenness with the surface.
7. **Kurtosis**: It tells us how frequency varies from the mean value.
8. **Skewness**: It measures the angle formed between the T joints.
9. **IDM** (In process depth meter): It finds out the penetration of the arc in the given object.

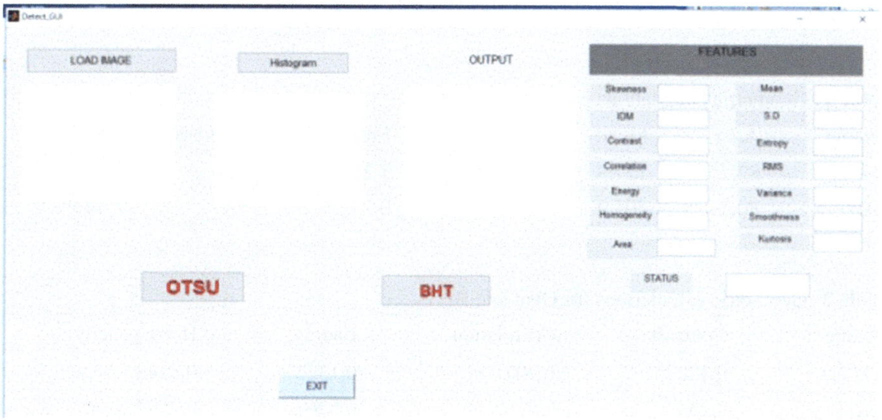

Fig. 6 GUI

10. **Contrast**: Value of the original material needed to be welded is considered in contrast to an image.
11. **Correlation**: When the pixels are closely packed, correlation tells us nearness to this value.
12. **Energy**: It tells us the saturation of black spots.
13. **Homogeneity**: It tells us the state of having a uniform structure.
14. **Area**: Total surface formed during welding is called area.

4.2 Comparative Results of Both Algorithms

Figure 7 shows graph for the comparative study of both the techniques for important features.

Table 1 mentions the calculated values of important features for both the techniques.

Tables 2 and 3 show the values of remaining geometric parameters calculated with Otsu and BHT.

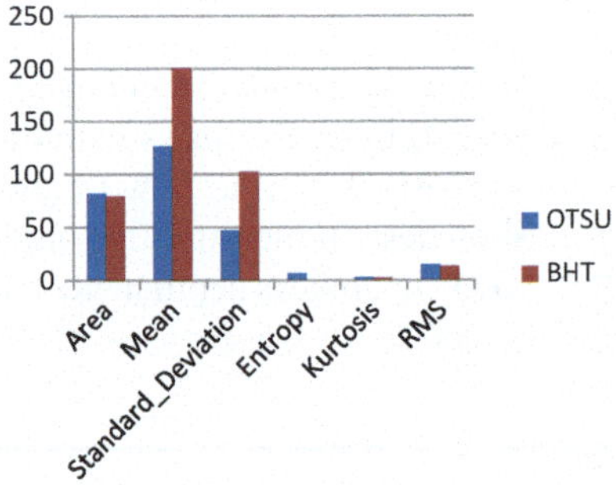

Fig. 7 Geometric calculations for Otsu and BHT

Table 1 Geometric calculations for Otsu and BHT

Name	Area	Mean	SD	Entropy	Kurtosis	RMS
Otsu	82.7	127.3	48.27	7.28	3.63	15.62
BHT	80.7	201.9	103.4	0.73	3.07	14.2

Table 2 Geometric calculations for Otsu and BHT

Name	Contrast	Correlation	Energy	Homogeneity
OTSU	0.4227	0.907	0.1322	0.8448
BHT	2.2541	0.8603	0.6267	0.9597

Table 3 Geometric calculations for Otsu and BHT

Name	Variance	Smoothness	Kurtosis	Skewness	IDM
OTSU	2.12E+03	1	3.6302	−0.9013	255
BHT	1.05E+04	1	3.0721	−1.4395	255

5 Conclusion

In this paper, we proposed a robotization framework that mechanizes the post quality review of the welded structure. We used two techniques, Otsu and BHT. After the examination of both the techniques we found that Otsu and BHT both are not dependable to be utilized as a solitary calculation for welded structure investigation. Though when Otsu and BHT are thought about, the deviation of Otsu is less when contrasted with BHT which shows that it is somewhat superior to BHT. Anyway both the calculations experience the ill effects of less noise immunity. Otsu has an additional bit of leeway of global thresholding, though histogram thresholding has restrictions for deciding the threshold value. Thinking about each one of these focuses, Otsu has performed superior to BHT in welded structure geometry investigation.

Future work

Other methods could be used in the future in order to get better accuracy. The system is working after the welding process is completed. This system can be further enhanced to make a dynamic system which works in real time analysis of welded structure.

References

1. Jiang C, Zhang F, Wang Z (2017) Image processing of aluminum alloy welded structure for robotic VPPAW based on visual sensing. In: IEEE access volume 5
2. SWu C, Gao JQ, Liu XF, Zhao YH (2003) Vision-based measurement of welded structure geometry in constant-current gas tungsten arc welding. In: SC04302 # IMechE Proceedings of the Institution of Mechanical Engineers Vol. 217 Part B: J. Engineering Manufacture
3. Liu Y, Zhang MY (2013) Adaptive modeling of the welded structure geometry in gas tungsten arc welding. In: 978-1-4673-5200-0/13/IEEE
4. Liu J, Hu Y, Wu B, Frakes DH, Wang Y (2016) A specific structuring element-based opening method for rapid geometry measurement of welded structure. Received: 22 March 2016/Accepted: 15 September 2016 © Springer London
5. Kovacevic R, Zhang MY (1996) On-line measurement of weld fusion state using welded structure image and neurofuzzy model. In: Proceedings of the 1996 international symposium on intelligent control dearbom. IEEE, MI, pp 15–18
6. Ratha NK, Bolle R (2011) Automatic fingerprint recognition systems. Springer, New York
7. Anjos A, Shahbazkia H (2008) Bi-level image thresholding—a fast method. Biosignals 2:70–76

Shape Memory Alloy Actuated Cantilever Structure for Sensing in Intelligent Engineering

G. Thenmozhi, M. BanuSundareswari and K. Dhanalakshmi

Abstract Shape memory alloys (SMAs) possess a certain extraordinary ability that is very attractive for several applications. Selecting a profile, configuration, size, cross-section, and material in order to meet the functions a structure has to perform is structural design. This paper describes the design considerations for actuating cantilever beam structures using shape memory alloy wire. An experimental arrangement for detecting the behavior of the cantilever beam and actuation properties of shape memory alloy (SMA) wire is presented in this work. The setup is incorporated with displacement and current sensors and the design is validated by computing the tip displacement of the cantilever beam for various dimensions of beam and shape memory alloy wire. Shape memory alloys can be effectively used to actuate the cantilever beam by introducing variable stiffness and providing push–pull effects due to the bias added with the structure. Some geometrical parameters such as thickness and length of the beam and, diameter and length of SMA wire are analyzed in this work. These investigations on the effect of the geometrical parameters are carried out experimentally. The information of beam displacement and the current required to excite SMA wire are communicated through suitable DAQ to the computer. The values of the response of the structure are presented and optimum structural configurations are to be considered to build a smart sensor in Intelligent Engineering.

Keywords Shape memory alloy wire · Joule heating · Actuation · Cantilever beam · Smart sensing

---

G. Thenmozhi · M. BanuSundareswari · K. Dhanalakshmi (✉)
Department of Instrumentation and Control Engineering, National Institute of Technology, Tiruchirappalli 620015, Tamil Nadu, India
e-mail: dhanlak@nitt.edu

G. Thenmozhi
e-mail: thenmozhi.gp@gmail.com

M. BanuSundareswari
e-mail: jjbanu78@gmail.com

S. Fong et al. (eds.), *ICT Analysis and Applications*, Lecture Notes in Networks and Systems 93, https://doi.org/10.1007/978-981-15-0630-7_40

1 Introduction

The practice of shape memory alloys (SMAs) for actuation represents a technological opportunity for the development of smart structures. The characteristic of SMA wire to undergo mechanical actions with temperature changes or with electrical current allows the development of simple, more compact and reliable smart structures. The shape memory alloys are quite attractive materials that are characterized by shape memory effect and pseudo-elasticity which other ordinary materials do not possess. This unique behavior of shape memory alloy makes this material a promising candidate for the fabrication of smart structures. The shape memory effect (SME) of the alloy is utilized in this work for the actuation of the beam. This effect of the alloy can be used to produce force, while pseudo-elasticity allows energy storage. Shape memory defines the effect of regaining the original shape of a deformed material by increasing its temperature on joule heating. At normal temperatures, shape memory alloys are in the martensitic state. In this condition, their structure is considered as self-accommodating twins. At this state the alloys are brittle and at this state, the alloys can be distorted by de-twinning. This material can be converted to a high strength austenitic state by heating the material above the transformation temperature.

The benefit of integrating these special types of materials into a structure is that the sensing mechanism becomes a part of the structure and can directly sense and excite the beam accordingly. In order to build smart sensors, the shape memory alloy excited cantilever beam is constructed and analyzed. Cantilever structures are extensively employed for sensing physical, chemical, and biological parameters. The exploitation of cantilever structures as sensing devices has led to the development of various sensors that are extremely sensitive to quantities intended to be measured. These cantilever-based sensors possess numerous benefits over the conventional procedures in terms of sensitivity, technique, and cost. The protocol used is nonhazardous and the response is quick. Moreover, the cantilever-based sensing technology has been developed and in the last few years, it has led to the fabrication of Nano-cantilevers for smart sensing applications, thereby giving rise to nanoelectromechanical systems (NEMS). This technology leads to the development of miniaturized and extremely sensitive sensors. Cantilever acts as a sensor by sensing variations in vibrational frequency. Cantilever structure is vibrated at a regular rate on proper excitation. This rate of vibration varies when the mass is changed and makes the system behave like a mass sensor. The design of a smart sensor system requires accurate structural modeling. To design shape memory alloy based smart structures for sensing application, both structural design and geometrical design need to be considered. Unlike the normal applications of SMA as control actuator, vibration damper, and in self-sensing actuation, this work attempts to vibrate a structure by using SMA as an excitation actuator appropriately, which can lead to sensing application.

This paper is structured as follows: Sect. 2 conveys the hardware setup used to generate and measure the vibration of a cantilever beam. Section 3 deals with the interpretation of experimental results on various design considerations and finally Sect. 4 concludes the paper.

2 Experimental Setup

2.1 Hardware Description

The experimental arrangement consists of an aluminum—6063 grade cantilever beam attached to the SMA wire of 0.15 mm diameter with a steel balance weight. The wire is connected at the free end of the beam with an offset of 5 mm. When the SMA is supplied with a pulsed current by joule heating it undergoes shape change which pulls the beam in one direction thereby the beam vibrates. There arises a necessity to provide a biasing force for the SMA to return to its original neutral position, this is achieved by connecting a deadweight (100 g) at the free end of the SMA wire [1]. The vibration of the beam is measured by the laser displacement sensor (Acuity AR200) and the current through the SMA wire is measured by placing current transducer (LEM LTS6-NP) in series with the SMA wire as shown in Fig. 1. To actuate the beam, an electric current of a suitable level is applied to the SMA wire via a driving circuit. The duration and the amplitude of the current applied to the wire are set with the Arduino processor. The current sensor connected in series with the SMA wire measures the current supplied by the self-designed driver circuit. The Data acquisition card (MCC DAQ) is used to acquire data from the sensors and PC records the data. The regulated power supply is used to excite all the sensors used for instrumentation in the experiment. The stiffness of the beam is calculated from the force and tip displacement of the beam and is recorded [2].

(a) (b)

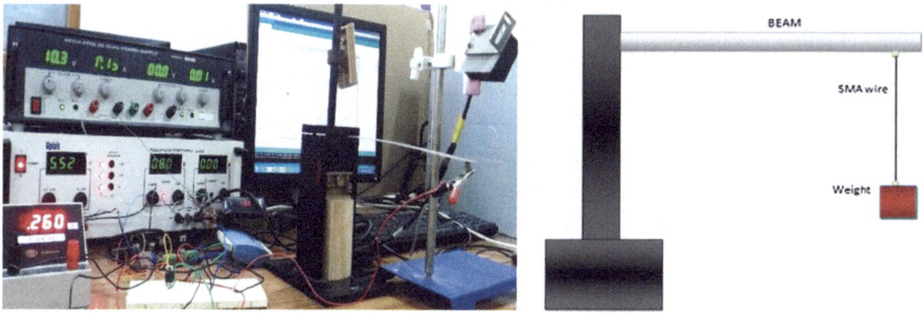

Fig. 1 **a** Experimental setup for the excitation of Cantilever Beam **b** Schematic of the experimental arrangement

Design of Driving Electronics

The driving circuit is designed to actuate SMA wire. SMA wire is excited by providing high temperature which is done by joule heating. To actuate SMA, pulse width modulated signal is generated with Arduino Uno processor and this pulse is provided to the gate terminal of the MOSFET. The drain terminal of the MOSFET is connected to OPA547. The OPA547 is a cheap, voltage to current converter and it can be used to excite various range of loads. It detects the load and issues current signal from 0 to 750 mA based on the input and load values. In addition, it can also be designed to provide intermediate values using current limiting resistor. Arduino controlled MOSFET gate inputs the current to the SMA to produce force for actuating the structure during the turn on time of the pulse. The excitation waveforms are depicted in Fig. 2.

Material properties

The properties of shape memory alloy are enlisted in Table 1. The cantilever beam made of aluminum has Young's modulus of 69 GPa and its density is 2700 kg/m^3.

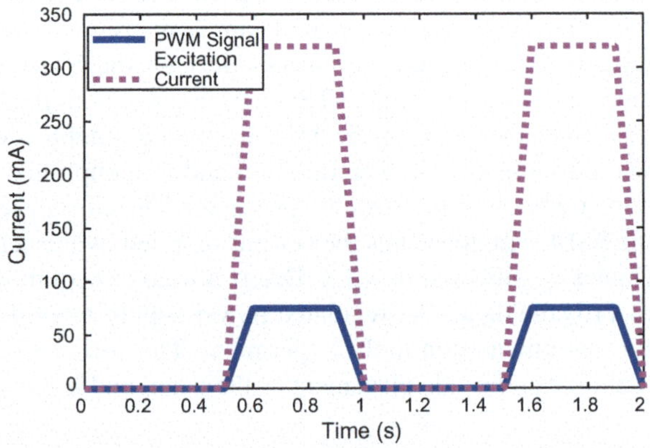

Fig. 2 Excitation signals with 1 Hz frequency

Table 1 Properties of shape memory alloy

Parameter	Description	Value
Young's modulus	Austenite (E_A)	75 GPa
	Martensite (E_M)	28 GPa
Phase transformation temperature	Austenite start (A_s)	68 °C
	Austenite finish (A_f)	78 °C
	Martensite start (M_s)	52 °C
	Martensite finish (M_f)	42 °C

3 Design Considerations and Analysis

3.1 Design Considerations

A preliminary study of the effect of the geometric parameters of each element (length and thickness of the beam, length and diameter of the SMA wire) of the flexible vibrating structure and also the effect of the various excitation frequencies on the beam for the vibration considered as design parameters is necessary for the optimal design of smart sensor.

The vibrating flexible structure has its stiffness indirectly as the influencing parameter that characterizes the nature of the configuration for sensor design. The natural frequency of the structure is affected by its stiffness. So the stiffness is calculated from the natural frequency for the analysis of the structural design.

Natural frequency of the Beam

The natural frequency of the beam is given by

$$\omega = \sqrt{\frac{K}{M}} \text{ where } K = \frac{K_1 K_2}{K_1 + K_2} \text{ and } K_1 = \frac{3EI}{L^3} \tag{1}$$

where, K is the stiffness of the structure (N/m), M is the mass of the beam (kg), K_1 is the stiffness of the beam (N/m), K_2 is the stiffness of the SMAwire (N/m), E is the Young's modulus of the Beam (GPa), I is the Moment of Inertia of the beam (m^4), L is the length of the beam (m) [3].

Effect of Length of the Beam

See Fig. 3.

Effect of the Thickness of Beam

See Fig. 4.

Effect of the Length of SMA Wire

See Fig. 5.

Effect of the Dimension of SMA Wire

See Fig. 6.

Effect of Excitation Frequency

See Fig. 7.

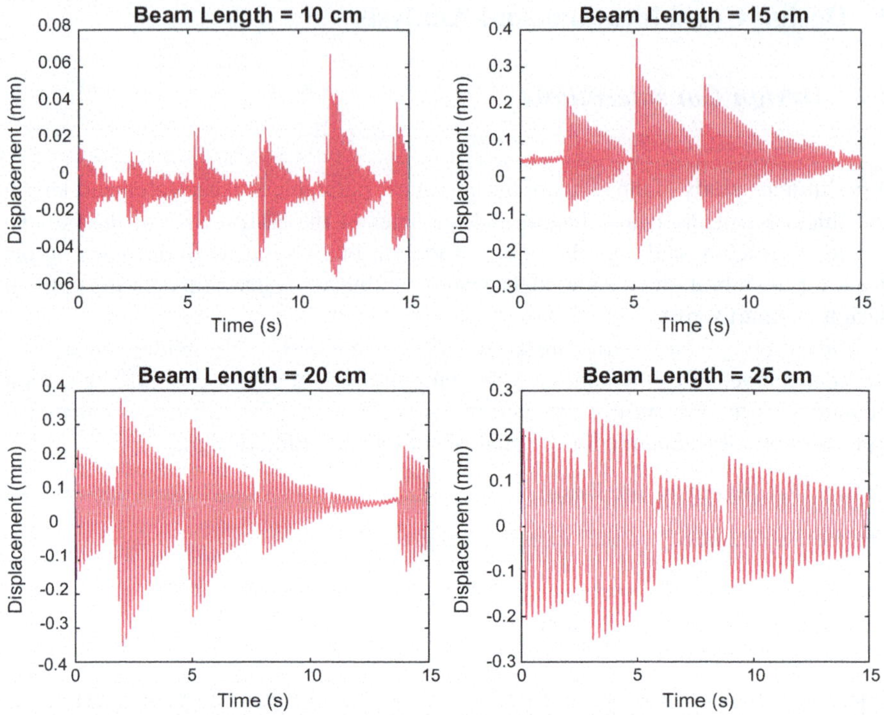

Fig. 3 Effect of length of the Beam at 0.33 Hz excitation frequency

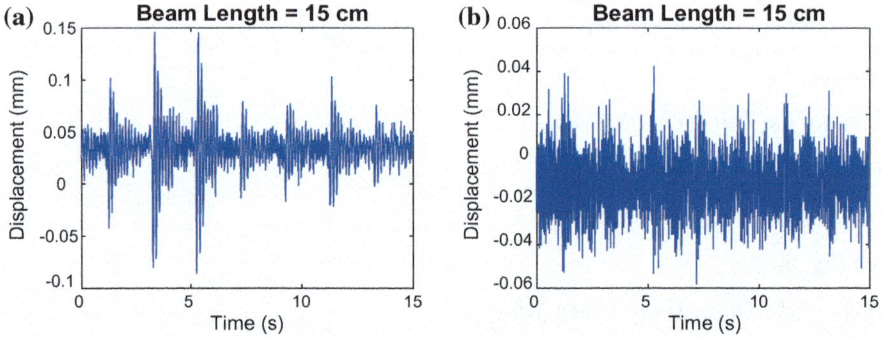

Fig. 4 Effect of the thickness of the beam at excitation frequency of 0.5 Hz **a** 1 mm **b** 2 mm

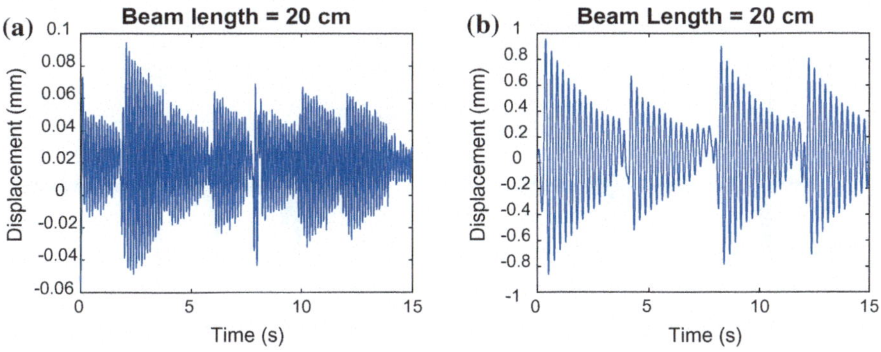

Fig. 5 Effect of the length of SMA wire at the excitation frequency of 0.5 Hz **a** 25 cm **b** 30 cm

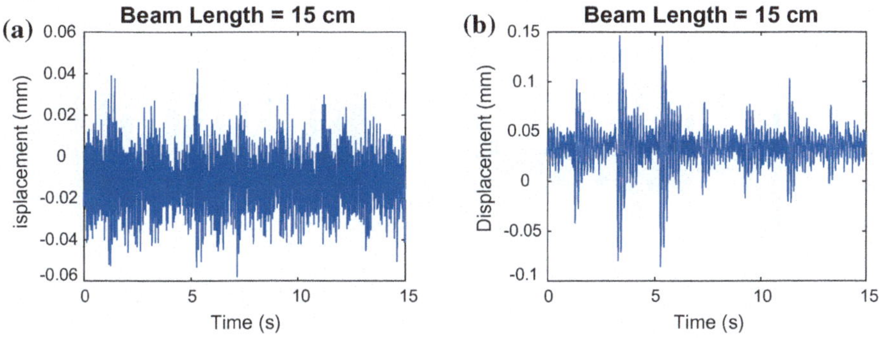

Fig. 6 Effect of the dimension of SMA wire at excitation frequency of 0.5 Hz **a** 0.1 mm **b** 0.13 mm

4 Conclusion

The study is about analyzing the scope of a structural (cantilever-based) design actuated by SMA wire, for mechanical sensing. This paper discusses the effect of the geometrical properties of the elements which include the cantilever beam and the SMA wire for excitation actuation, on the vibration of the structure.

The stiffness is inversely proportional to the length of the beam as shown in Table 2. The same is verified from experimentation that the magnitude of vibration increases with the increase in the length of the beam as shown in Fig. 3. The increase in thickness of the cantilever beam decreases the magnitude of vibration as shown in Fig. 4, which is due to the increase of Moment of Inertia of the beam. The amplitude of vibration is enhanced with increase in the length of SMA wire and diameter as displayed in Figs. 5 and 6, since the force offered by the SMA is more. With the decrease in excitation frequency, continuous, repetitive, and sustained vibration

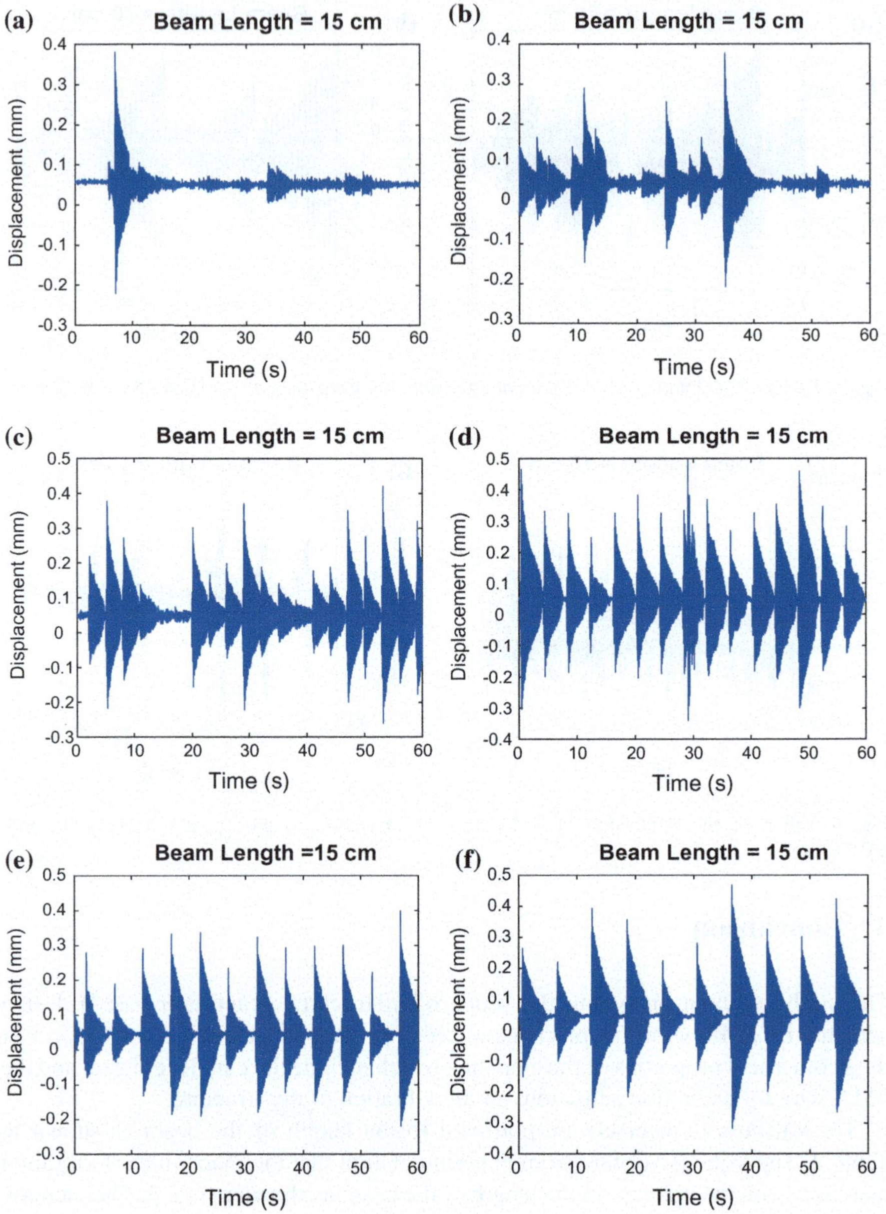

Fig. 7 Effect of excitation frequency **a** 0.75 Hz **b** 0.5 Hz **c** 0.33 Hz **d** 0.25 Hz **e** 0.2 Hz **f** 0.16 Hz

Table 2 Stiffness of the beam for various beam dimensions

Length of the Beam (cm)	Stiffness of the Beam for various thickness (N/m)			
	0.5 mm	1 mm	2 mm	3 mm
25	2.75	22.08	176.63	596.16
20	5.38	43.21	344.99	1164.37
15	12.75	102.43	817.76	2760
10	43.05	345.69	2759.9	9315

is observed due to the decrease in stiffness of the structure, as shown in Fig. 7. By changing any of these parameters by mechanical means, the vibrating structure's characteristic can be altered which can be used as a principle of mechanical sensing in Intelligent Engineering.

References

1. Srinivasan AV, Cutts DG, Schetky LM (1991) thermal and mechanical considerations in using shape memory alloys tocontrol vibrations in flexible structures. Metall Trans 22A:623–627
2. Daudpoto J, Dehghani-Sanij A, Richardson R (2015) A measurement system for shape memory alloy wire actuators. Meas Control 48 9):285–288
3. Skoblar A, Zigulic R, Braut S, Blazevic S (2017) Dynamic response to harmonic transverse excitation of cantilever Euler-Bernoulli beam carrying a point mass. FME Trans 45(3), 367–373

Cloud-Based E-Learning Service: Insight from India

P. S. Vanitha and Sreejith Alathur

Abstract The factors influencing the adoption of a cloud platform in the development of e-learning service is identified in this study. The e-learning service based on a cloud platform is analyzed from single/multiple data center dimensions. In spite of developing e-learning infrastructure, the service of the cloud platform is often adopted. The cloud-based e-learning simulation environment is created using CloudAnalyst tool. The efficiency of the cloud is analyzed based on e-learning hosted on a single data center and multiple data centers. The service time and overall response time of the datacenter are analyzed through CloudAnalyst. The infrastructure cost estimation for both models is also calculated. The study identifies that these factors influence the development of e-learning services in a cloud platform. Earlier studies less analyze the influencing factors of the online courses in the cloud environment. In future research, the cost factor can also be considered in the development of e-learning services. Fewer studies are reported on e-learning service based on a cloud platform in the Indian context. The current study demonstrates how cloud Infrastructure as a Service (IaaS) improves the performance of the e-learning system.

Keywords Cloud-based e-learning · E-learning challenges · Cloud benefits · CloudAnalyst · India

1 Introduction

Many universities in developing countries focus on cloud technologies for the improvement of e-learning service. The cloud provides asynchronous and collaborative online learning environment anywhere, anytime [2, 3]. The benefit of cloud over online course improves the performance of e-learning service. It also provides scalable and reliable service to the end user [20]. Compare to traditional e-learning,

P. S. Vanitha (✉) · S. Alathur
National Institute of Technology, Karnataka, India
e-mail: sm16f10.vanitha@nitk.edu.in

S. Alathur
e-mail: sreejith.nitk@gmail.com

S. Fong et al. (eds.), *ICT Analysis and Applications*, Lecture Notes in Networks and Systems 93, https://doi.org/10.1007/978-981-15-0630-7_41

the cloud provides a better infrastructure service at low cost. It also improves the effectiveness of online learning in the collaborative environment [6].

The drawback of the traditional online environment is limited storage space. Due to this problem, the response time of the learning management system also increased. It automatically slows down the performance of the e-learning system. Whereas the implementation of cloud-based e-learning system improves the scalability feature. The cloud-based e-learning system enables flexible IaaS to the educational institutions [14]. It also reduces the implementation cost of expensive devices and increases the network transmission rate. In addition, it enables dynamic storage device according to the participation in the collaborative environment [18]. Due to the cloud benefit, many developing countries adopt cloud-based e-learning system.

The paper is structured into five sections. Section 2 describes the theoretical background of e-learning challenges and benefits of e-learning in cloud platform across different countries. In Sect. 3, the simulation environment is demonstrated using CloudAnalyst and tested. In Sect. 4, result analysis is discussed. In Sect. 5 the paper concludes with findings.

2 Theoretical Background

"E-learning is a knowledge-intensive industry" [21]. E-learning is referred to as the use of electronic media to develop a virtual learning environment. The growth of internet facilities and other network utilities improves the e-learning service to the next level. Even though the e-learning has many advantages, it faces some problems during implementation. The infrastructure service and lack of finance are considered as major challenges in the institution level [6, 16].

2.1 E-Learning Challenges

This study identifies the adoption of e-learning system based on a cloud platform in higher education. The challenges are classified into technical, personal, organizational challenges as often reported in the literature. In this study, the Contextualized Barrier Framework (CBF) is considered as an e-learning barrier framework. Based on CBF, technical challenges are analyzed [17, 19]. To overcome the technical challenges, the organizational policy framework should be considered [22]. This study concentrates only on the technical challenges faced during the e-learning implementation. The list of e-learning challenges is shown in Table 1.

The technological barrier is classified into two categories: an organizational and social perspective. In an organizational perspective, inadequate infrastructure facilities and improper maintenance of the system will decrease e-learning adoption. The social perspective includes unequal access and use of Internet bandwidth, insufficient speed, and internet connectivity. The Internet is considered an important factor

Table 1 E-learning challenges identified in the study

Authors'	Categories	Factors	Remarks
Chawla and Joshi [5], Mell and Grant [10], Iqbal and Ahmad [7]	Organizational	Inadequate Infrastructure	Hardware, software, and network device. Network connectivity speed and quality
		Lack of maintenance	Installation and maintenance of equipment
		Interoperability issue	Heterogeneous platform
Rao [13]	Social	Low internet facilities	Unequal internet bandwidth, low speed, and connectivity

for e-learning because the availability of the bandwidth deserves the effectiveness of the online course (SEDTA 2012). The poor utilization of online resources and the Internet increase the financial crisis of the education institutions [15]. Therefore, proper technical strategy should be used to develop an e-learning environment.

2.2 *Cloud-Based E-Learning: Selected Country Studies*

Cloud computing provides online resources based on-demand to the organization [4, 10]. The benefits of cloud computing is to help to reduce the implementation cost of the online courses in the universities. It also reduces the financial crisis of the education institution. The Internet is considered as a main source for the service provider to enable the cloud service. The cloud benefits are identified and solutions suggested by different countries' case studies. Based on the cloud QoS parameter, the benefits of the cloud-based online courses are measured. Through literature support, the countries that already adopt the e-learning system based on a cloud platform are found. Based on that the benefits of the cloud-based e-learning are analyzed. The selected country study is shown in Table 2.

South Africa in the cloud service is used to bridge the digital divide gap in higher education. With the help of a high-speed broadband connection, the cloud service is offered to the educational institutions. The cloud-based e-learning system provides a reliable and robust service to the learners' [8]. In Saudi Arabia, the e-learning system faces a problem due to the dynamic growth of learners'. It does not have enough infrastructure service to adopt the learners' in the e-learning environment [1]. The cloud-based e-learning system offers a solution for this problem through reliability, scalability service. It also improves the performance of the e-learning system.

Table 2 Cloud-based e-learning system—selected country studies

Authors'	Countries	Cloud QoS parameter			
		Availability	Scalability	Reliability	Performance
Le Roux and Evans [8]	South Africa	Supported	–		–
Alharthi et al. [1]	Saudi Arabia	–	Supported	Supported	Supported
Riaz and Samson [14]	Pakistan	Supported	–	–	Supported
Stoffregen et al. [18]	Taiwan	Supported	–	–	Supported
Radenkovic et al. [12]	Belgrade	Supported	Supported	–	Supported

In the case of the Pakistan education system, the adoption of cloud-based e-learning system increases the adoption of an online course. It automatically increases the performance of online learners' [14]. The cloud-based e-learning system improves the reliability service in Taiwan universities [18]. The University of Belgrade also adopts e-learning service based on a cloud platform to provide scalable and reliable service. With the help of Open Nebula in the learning management system (LMS), the effectiveness of the e-learning system in a cloud environment increases. Many countries implement a cloud-based learning service to improve the efficiency of online courses.

3 Simulation of an E-Learning Course in CloudAnalyst

The good quality online course requires sufficient Internet connectivity and best quality video/graphics cards. To solve the cloud adoption problem, synchronous tools are used. In the simulation environment, the different zonal time internals is clearly defined. Most nations including India are also facing inadequate infrastructure and Internet bandwidth as an e-learning barrier [5, 13]. When the e-learning course provides a synchronous mode of online courses, the zonal timing is also considered as a critical factor. Therefore, the online synchronous tools used to create an active online environment. In this study, the cloud-based online environment is created using CloudAnalyst simulation tool. There are predefined options are available to create a live online environment in the CloudAnalyst tool.

The CloudAnalyst uses different region id to represent six different continents. The region id for "North America-0, South America-1, Europe-2, Asia-3, Africa-4, and Australia & Oceania-5" [9, 11]. In this study, the users' who registered for a popular e-learning service in Asia is simulated. This is because, India consists of a number of e-learners' next to US. This study focuses on the e-learners' who register for an online course from India. Therefore, the online platform is simulated according to the Asian zonal time. It is assumed that online users are active 7 h per day between the time interval of 13.00 and 22.00. By default, it is expected that 1/10 of the users' are active during peak hours and off-hours. It is assumed that online users are activated 7 h per day between the time interval of 13.00 and 22.00. By default, it is expected that 1/10 of the users' are active during peak hours and off-hours. In this study, half of the learners' population is considered. Only 15,00,000 online learners' are reflected in the cloud-based e-learning simulation environment.

4 Result Analysis

The configuration setup for the CloudAnalyst is subdivided into three steps. In step-1, the main configuration user base is configured. In step-2, the datacenter configurations are performed. In step-3, an advanced configuration like grouping factors is configured. Both models use the same simulation environment except the data center configuration.

In step-1, the main configuration first the simulation time is set as 60 min. The user base name UB4 for region 3 is configured with 15,000 average peak users. The 1/10 of the average users is assumed as average off-peak users. The data size request is 1000 bytes with peak hours 15.00–22.00. The application deployment is configured with 50 virtual machines (VM) with 1000 MB of available bandwidth and memory.

In step-2, the data center is configured with Xen-VMM with LINUX-OS, ×86 architecture. The cost for hosting coursera in the e-learning environment is: "cost per VM$/hr is 0.1, memory cost$ is 0.05, storage cost$ is 0.1 ad data transfer cost$/Gb is 0.1" with one physical hardware unit [9]. The physical hardware details of the datacenter consist of 2 GB memory with 10 × 100 GB dual channel SAS disks of storage. The number of processors is 4 with 10000 processing speed with VM time-sharing policy.

4.1 Case 1—Single Datacenter

The online course coursera is hosted on a single datacenter and analyzed in Case 1. It is deployed in region 3 (Asia). The data center consists of 50 VM with each of 1024 MB of memory. The processing speed of VM is 37006 million instructions per seconds (MIPS). Based on this setup, the simulation is run and results are obtained. The overall response time summary is shown in Table 3.

Table 3 Summary of response time and processing time – Single data center

	Aver (ms)	Mini (ms)	Maxi (ms)
Overall response time	239.89	108.39	329.17
Data center processing time	181.86	56.36	268.11

Table 4 Response time of UB based on region 3 – Single data center

User base	Aver (ms)	Mini (ms)	Maxi (ms)
UB4	239.89	108.39	329.17

Table 5 Request servicing times of single data center

	Aver (ms)	Mini (ms)	Maxi (ms)
Data center 1	181.86	56.36	268.11

Table 6 Cost estimation – Single data center

Data center	VM cost$	Data transfer cost$	Total$
Data center 1	5.00	2129.31	2134.31

The userbase hourly average response time is measured for UB4. It analyzes how the peak load of the userbase (UB4) is distributed for the time period of 7 h. The peak time of the userbase varies according to the zonal time. The simulation result shows the hourly average response time of the userbase for region.

The user base response by region is analyzed and the results are displayed in Table 4. The "average time, minimum and maximum time" of the user base is calculated based on region-3. The data center requesting service time is measured based on the user request. The data center request servicing time is displayed in Table 5.

The VM cost and data transfer cost is combined together to estimate the cost. The total cost estimation of single datacenter is shown in Table 6.

4.2 *Case 2—Multiple Datacenter*

When online users are more, then the service provider uses multiple data centers. The main purpose of the multiple data centers is to increase the response time and quality of e-learning service. All the cloud setups are the same as Case 1, only one more datacenter is added in the simulation environment. The 50 VM is split into two half and given to two centers. The "overall response time and the processing time of the data centers" are shown in Table 7.

The overall response time of the multiple data center is significantly less compared to the single datacenter. This is because of the additional infrastructure facility is provided for the online course. The closest data center service broker policy is the main reason for the improvement. The data center with two different locations makes

Table 7 Summary of response time and processing time – Multiple data centers

	Aver (ms)	Mini (ms)	Maxi (ms)
Overall response time	149.37	68.03	257.24
Data center processing time	92.08	15.77	198.58

Table 8 Response time of UB based on region 3 – Multiple data centers

User base	Aver (ms)	Mini (ms)	Maxi (ms)
UB4	149.37	68.03	257.24

Table 9 Request servicing times of multiple data center

Data center	Aver (ms)	Mini (ms)	Maxi (ms)
Data center 1	92.50	15.77	198.58
Data center 2	91.67	15.78	197.60

Table 10 Cost estimation – Multiple data centers

Data center	VM cost$	Data transfer cost$	Total$
DC1	2.50	1070.30	1072.81
DC2	2.50	1059.00	1061.50

easy access facility to the learners. This, the efficiency of the e-learning course is improved in multiple data center model. The response time by region is in Table 8.

The requesting service time of the multiple data centers is calculated based on the user request. The data center average, minimum and maximum time is shown in Table 9. The overall cost estimation of the infrastructure service is calculated based on VM cost and data transfer cost in multiple data center. The total cost estimation is shown in Table 10.

According to the simulation result, the average processing time of the data center is increased due to multiple VM. On the other hand, the distribution of VM to two different data centers increases the peak loading time in both centers. Thus, traffic time is less in model-2 compared to model-1. In multiple data center, the data loading traffic value is less.

5 Conclusion

The e-learning adoption in the cloud environment is established through CloudAnalyst. The online infrastructure is deployed using a simulator tool. The performance of two different infrastructure models is compared. The loading, processing, and response time of the two models is analyzed. Even though the infrastructure implementation cost is the same for the two cases (single and multiple), the multiple datacenter model has more benefits. The data center requesting, service time and overall response time of Case 2 is a significant high compared to Case 1. Thus, it is confirmed that the deployment of more datacenter in different places will increase the efficiency of an online course in the cloud platform. Although the time factor is varied for the two different models, the cost estimation value for both models is the same.

The two different data center cases are implemented using cloud infrastructure service. To test the e-learning system, the online course named "coursera" is deployed on the cloud platform. Thus, the cloud-based e-learning system is simulated using the CloudAnalyst simulator. Through simulation results, this study confirms that the adoption of cloud service will improve e-learning efficiency through cloud infrastructure service. The response time of the data center models is estimated through CloudAnalyst. It is also shown that multiple data centers improve load balancing at the application level. Thus, online course time is improved in the multiple center models. In addition, the two models are compared together and multiple datacenter models are recommended for the real-time implementation. Future research may focus on other features like cost estimation in the e-learning service based on a cloud platform.

References

1. Alharthi A, Alassafi MO, Walters RJ, Wills GB (2017) An exploratory study for investigating the critical success factors for cloud migration in the Saudi Arabian higher education context. Telemat Inf 34(2):664–678
2. A-Samarraie H, Saeed N (2018) A systematic review of cloud computing tools for collaborative learning: opportunities and challenges to the blended-learning environment. Comput Educ 124: 77–91
3. Attaran M, Attaran S, Celik BG (2017) Promises and challenges of cloud computing in higher education: a practical guide for implementation". J High Educ Theory Pract 17(6):20–38
4. Bouhnik D, Marcus T (2006) Interaction in distance-learning courses. J Am Soc Inf Sci Technol 57(3): 299–305
5. Chawla D, Joshi H (2012) Management education through e-learning in India: an empirical study. Campus-Wide Inf Syst 29(5): 380–393
6. Hew T-S, Kadir SLSA (2016) Predicting instructional effectiveness of cloud-based virtual learning environment. Ind Manag Data Syst 116(8): 1557–1584
7. Iqbal M, Ahmad M (2010) Enhancing quality of education through e-learning: the case study of Allama Iqbal Open University. Turk Online J Distance Educ 11(1):84–97
8. Le Roux CJB, Evans N (2017) Can cloud computing bridge the digital divide in South African secondary education? Inf Dev 27(2): 109–116
9. Meftah A, Youssef AE, Zakariah M (2018) Effect of service broker policies and load balancing algorithms on the performance of large scale Internet applications in cloud datacenters. Int J Adv Comput Sci Appl 9(5):219–227
10. Mell P, Grant T (2011) The NIST definition of cloud computing. http://csrc.nist.gov/publications/nistpubs/800-145/SP800-145.pdf
11. Mezcal YS, Madhi HH, Abd T, Khaleel SK (2018) Cloud computing investigation for cloud computer networks using cloud analyst. J Theor Appl Inf Technol 96(20):6937–6947
12. Radenkovic B, Despotovic-Zrakic M, Bogdanovic Z, Vujin V, Barac D (2014) Harnessing cloud computing infrastructure for e-learning services. Electron Energ 27(3): 339–357
13. Rao P (2011) E-learning in India: the role of national culture and strategic implications. Multicult Educ Technol J 5(2):129–150
14. Riaz S, Samson D (2017) An evaluation of public cloud adoption for higher education: a case study from Pakistan. In: International symposium on mathematical sciences and computing research (MSC). 978-1-4799-7896-0/15
15. Rodríguez G, Perez J, Cueva S, Torres R (2017) A framework for improving web accessibility and usability of open course ware sites. Comput Educ 109:197–215

16. Santiuste E, Gallego-Arrufat M (2016) Barriers in computer-mediated communication: typology and evolution over time. J e-Learning Knowl Soc 12(1):107–119
17. Stoffregen J, Pawlowski JM, Pirkkalainen H (2015) A barrier framework for open e-learning in public administrations. Comput Hum Behav 51: 674–684
18. Stoffregen JD, Pawlowski JM, Ras E, Tobias E, Šćepanovic S, Fitzpatrick D, Mehigan T, Steffens P, Przygoda C, Schilling P, Friedrich H, Moebs S (2016) Barriers to open e-learning in public administrations: a comparative case study of the European countries Luxembourg, Germany, Montenegro, and Ireland. Technol Forecast Soc Chang, 0040–1625
19. VenkateshNai N, Madhav K (2017) Cloud computing architecture for collaborative e-learning system-IEEE Xplore document, vol 23. https://ieeexplore.ieee.org
20. Wang J, Jou M (2016) Qualitative investigation on the views of inquiry teaching based upon the cloud-learning environment of high school physics teachers from Beijing, Taipei, and Chicago. Comput Hum Behav 60:212–222
21. Wang Q, Zhu Z, Chen L, Yan H (2009) E-learning in China. Campus-Wide Inf Syst 26(2):77–81
22. Yang T-M, Maxwell TA (2011) Information-sharing in public organizations: a literature review of interpersonal, intra-organizational and inter-organizational success factors. Gov Inf Q 28(2): 164–175

Sanskrit Stopword Analysis Through Morphological Analyzer and Its Gujarati Equivalent for MT System

Jaideepsinh Raulji and Jatinderkumar R. Saini

Abstract The identification and removal of a stopword is a common preprocessing task in many natural language processing implementations. The morphologically parsed information of stopword is also relevant in analysis of various NLP tasks. The list of most common seventy-five Sanskrit stopwords are evaluated using rule-based morphological analyzer. Most stopwords were classified as indeclinables and pronouns. The Gujarati equivalent of stopwords is retrieved using bilingual dictionary so as to cache the data for faster retrieval during MT process.

Keywords Stopword · Information retrieval (IR) · Natural language processing (NLP) · Morphological analyzer

1 Introduction

The natural language processing (NLP) entails detailed study and analysis of text so as to generate meaningful information. Every elemental unit of text information is equally important in different scenarios of NLP implementation. Stopwords are words which appear frequently, acting as fillers in sentences and convey very little semantic value. Such words with respect to information retrieval and other related applications play a minimal role. But use-cases like information generation, defining entity role, anaphora resolution, machine translation require judicious use of stopwords. Such words exist in most natural languages. In Sanskrit, the most common seventy-five stopwords are identified based on frequency from digitally available text [1]. Most Indian languages are morphologically rich languages and inflectionally complex. Due to such characteristics they have free word order, not adhering to

J. Raulji (✉)
Ahmedabad University, Ahmedabad, India
e-mail: jaideepraulji@gmail.com

J. Raulji · J. R. Saini
Babasaheb Ambedkar Open University, Ahmedabad, India
e-mail: saini_expert@yahoo.com

J. R. Saini
SICSR, Pune, India

S. Fong et al. (eds.), *ICT Analysis and Applications*, Lecture Notes in Networks and Systems 93, https://doi.org/10.1007/978-981-15-0630-7_42

particular arrangement of words. Hence morphological analyzer plays a suitable role in identifying grammatical aspects of stopwords. Such words normally are pronouns, articles, conjunctions with respect to the English language. Similarly in Sanskrit such words are majorly categorized under pronouns and indeclinables. The part-of-speech information can be used for various NLP tasks like machine translation. In a rule-based MT system, searching for an equivalent word from a bilingual dictionary is an intensive task. Hence frequently used stopwords which exist in a sentence can be cached in primary memory so as to improve search time complexity issue rather than searching from whole bilingual dictionary.

2 Related Work

The extent of research work found on generation and removal of stopwords for various natural language processing task specifically—information retrieval are ample. But meager work is found in analysis of stopwords morphologically, especially for Indian languages like Sanskrit.

The effects of removal of stopwords for information retrieval on the English language and Indian languages were experimented and analyzed by Ghosh and Bhattacharya [2]. Their findings stated of removal of stopwords in highly inflected Indian languages has less effect than languages with less inflectional features. The reason is, the meta-information is clubbed with the main word itself in the form of inflection than having different constituent words around the main word in the English language. An information retrieval efficiency is improved by removal of stopwords from the Gujarati language statements as evaluated by Joshi et al. [3] It improved Mean Average Precision (MAP) value—A metric for measuring information retrieval task. Jha, et al. developed deterministic finite automata and pattern search based stopword removal algorithm for the Hindi language [4]. Ghat et al., measured efficiency of sentiment analysis using classification algorithms with and without removal of stopwords from the dataset. The traditional classifier was affected by no removal of stopwords, while results from ARTFSC, sentiTFIDF, RTFSC were not affected [5]. Lazarinis F measured effect of noun lemmatization and stopword removal for greek language web searching showing better results [6]. The extraction of root words from devanagari scripts (Marathi) was carried out using morphological analyzer by Govilkar et al. [7]. The set of Marathi inflections was identified and used to remove to extract root words. Raulji et al., developed morphological analyzer for the Sanskrit language using grammatical features of the language and dictionary containing approximately thirty thousand words [8]. An inflectional morphology analyzer for the Sanskrit language is built by Girish Nath Jha et al. using avyaya and verb database and rules for subanta forms [9]. Namrat et al., built rule-based POS tagger using database and suffix stripping techniques [10]. Akshar et al. built Sanskrit morphological analyzer including modules for Sandhi—*Samasa* analyzer and formation, *Subanta*, *Tinanta*, and *Kridanta* Analyzer [11].

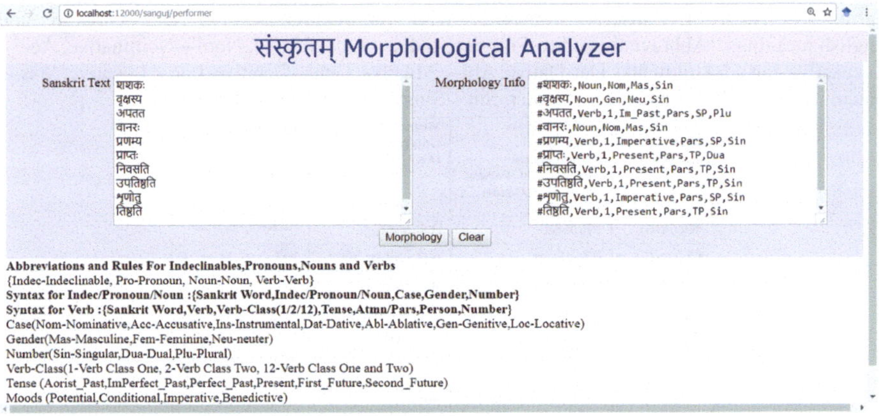

Fig. 1 Screenshot of Sanskrit morphological analyzer

3 Experimental Setup

The standard list of Sanskrit stopwords is not available as compared to English and other languages due to reasons as digitally scarce availability of corpora and other factors too. Raulji and Saini [1] generated a list of most common seventy-five Sanskrit stopwords from digital corpora containing approximately seventy-six thousand words and manually edited by referring non-digital Sanskrit text data like students school and college Sanskrit textbooks.

The morphological analysis of Sanskrit stopwords was carried out using Sanskrit morphological analyzer implemented by Raulji and Saini [8]. The architecture of morphological analyzer is built upon lemmatization, inflectional removal using Sanskrit grammatical rules for different grammatical constituents augmented with a dictionary of approximately thirty thousand words.

Using bilingual Sanskrit–Gujarati dictionary the Gujarati equivalent word for each stopword is retrieved for machine translation process from Sanskrit to Gujarati. The stopwords are most common words frequently appearing in text hence in order to improve time complexity of searching process the Gujarati equivalent of Sanskrit stopword is data-cached (Fig. 1).

4 Results

The seventy-five stopwords were analyzed in morphological analyzer. The results are enlisted in Table 1. The first column contains stopword with ITRANS transliteration scheme, the second column contains output from morphological analyzer, the third column contains output from Sanskrit to Gujarati translator and the fourth column shows English equivalent of the corresponding word for the sake of understanding.

Table 1 Listing of Sanskrit Stopwords with morphological POS tag, Gujarati and English meanings. (Abbreviations: Pro-Pronoun, Indec—Indeclinable, Nom—Nominative, Acc—Accusative, Ins—Instrumental, Dat-Dative, Abl—Ablative, Gen—Genitive, Loc—Locative, Mas—Maculine, Fem—Feminine, Neu—Neuter, Sin—Singular, Dua—Dual, Plu—Plural)

Sanskrit Stopword with ITRANS	POS TAG from "SanGuj" Morphological Analyzer tool (Format : POS, Case,Gender, Number)	Meaning in Gujarati with ITRANS	Meaning in English
अतः (ataH)	अतः,Indec	તેથી (tethI)	therefore, so, hence
अत्र (atra)	अत्र,Indec	અહીં (ahIM)	here
अथ (atha)	अथ,Indec	પરંતુ (paraMtu), બીજું (bIjuM)	but, else
अपि (api)	अपि,Indec	પણ (paNa)	but, also, even, too , and
अयं (ayaM)	अयं,Pro,Nom,Mas,Sin	આ (A)	this
अयम् (ayam)	अयम्,Pro,Nom,Mas,Sin	આ (A)	this
अस्ति (asti)	अस्ति,Indec	હાજર(hAjara)	existent, present
अस्मि (asmi)	अस्मि,Indec	છું (ChuM)	am
अस्य (asya)	अस्य,Indec	મા (mA)	is, in
अहं (ahaM)	अहं,Pro,Nom,Neu,Sin	હું (huM)	I
अहम् (aham)	अहम्,Pro,Nom,Neu,Sin	હું (huM)	I
आम् (Am)	आम्,Indec	હા (hA)	yes
इति (iti)	इति,Indec	માટે (mATe)	to, thus
इदं (idaM)	इदं,Indec	હવે (have)	this, here, now
इदम् (idam)	इदम्,Indec	હવે (have)	this, here, now
इमे (ime)	इमे,Pro,Nom,Mas,Plu	આ (A)	these
इयं (iyaM)	इयं,Pro,Nom,Fem,Sin	આ (A)	this, that
इयम् (iyam)	इयम्,Pro,Nom,Fem,Sin	આ (A)	this, that
एतत् (etat)	एतत्,Pro,Nom,Neu,Sin	આ (A)	this
एतद् (etad)	एतद्,Pro,Nom,Neu,Sin	અહીં (ahIM)	this, here, now, thus, so
एते (ete)	एते,Pro,Nom,Mas,Plu	આ (A)	these
एव (eva)	एव,Indec	માત્ર (mAtra)	only
एवं (evaM)	एवं,Indec	આમ (Ama)	thus
एष (eSha)	एष,Pro,Nom,Fem,Sin	આ (A)	this
एषा (eShA)	एषा,Pro,Nom,Fem,Sin	આ (A)	this
कं (kaM)	कं,Indec	હા (hA)	yes, well, bliss
कः (kaH)	कः,Pro,Nom,Mas,Sin	કોણ (koNa)	who
कथं (kathaM)	कथं,Indec	કેવી રીતે (kevI rIte)	how
का (kA)	का,Pro,Nom,Fem,Sin	કોણ (koNa)	who
कानि (kAni)	कानि,Pro,Nom,Neu,Plu	શું (shuM)	what
किं (kiM)	किं,Pro,Nom,Neu,Sin	શું (shuM)	what
किम् (kim)	किम्,Pro,Nom,Neu,Sin	શું (shuM)	what
कुत्र (kutra)	कुत्र,Indec	ક્યાં (kyAM)	where
के (ke)	के,Pro,Nom,Mas,Plu	કોણ (koNa)	who
क्वचित् (kvachit)	क्वचित्,Indec	ક્યાંક (kyAMka)	somewhere, somewhat
खलु (khalu)	खलु,Indec	હવે (have)	now
च (cha)	च,Indec	અને (ane)	and, also
तं (taM)	तं,Pro	તેને (tene)	him, they
ततः (tataH)	ततः,Indec	પછી (paChI)	therefore, later, then
तत् (tat)	तत्,Pro,Nom,Neu,Sin	તે (te)	that
तत्र (tatra)	तत्र,Indec	ત્યાં (tyAM)	there
तदा (tadA)	तदा,Indec	પછી (paChI)	then, always
तनि (tani)	तनि,Pro	તેઓ બધા (teo	they all

(continued)

Table 1 (continued)

		badhA)	
तव (tava)	तव,Pro,Gen,-,Sin	તમારામાં (tamArAmAM)	yours
तस्य (tasya)	तस्य,Pro,Gen,Mas,Sin	તેના (tenA)	his
तस्याः (tasyAH)	तस्याः,Pro,Abl,Fem,Sin	તેણીના (teNInA)	her
तु (tu)	तु,Indec	અને (ane)	and, but
ते (te)	ते,Pro,Dat,-,Sin	તેઓ (teo)	they
तेन (tena)	तेन,Pro,Ins,Mas,Sin	ત્યાં (tyAM)	therefore, thus, there
तौ (tau)	तौ,Pro,Acc,Mas,Dua	તેઓ (teo)	they
त्वम् (tvam)	त्वम्,Pro,Nom,-,Sin	તમે (tame)	you
न (na)	न,Indec	ના (nA)	no
नु (nu)	नु,Indec	હવે (have)	at once, now
नो (no)	नो,Indec	નથી (nathI)	not
ननु (nanu)	ननु,Indec	ખરેખર (kharekhara)	indeed
परन्तु (parantu)	परन्तु,Indec	પરંતુ (paraMtu)	but
मम (mama)	मम,Pro,Gen,-,Sin	હું (huM)	my, I
मा (mA)	मा,Indec	નથી (nathI)	no
मे (me)	मे,Pro,Dat,-,Sin	મારા (mArA)	my
य (ya)	य,Indec	ચાલ (chAla)	move, go
यत् (yat)	यत्,Pro,Nom,Neu,Sin	તે (te)	that
यत्र (yatra)	यत्र,Indec	ક્યારે (kyAre)	when, wherever
यथा (yathA)	यथा,Indec	કરતાં (karatAM)	than, as, that
यदा (yadA)	यदा,Indec	ક્યારે (kyAre)	when
यदि (yadi)	यदि,Indec	જો (jo)	if
युयं (yuyaM)	युयं,Pro,Nom,-,Plu	તમે બધા (tame badhA)	you all
येन (yena)	येन,Indec	કારણ કે (kAraNa ke)	as, since, because
वयं (vayaM)	वयं,Pro,Nom,-,Plu	અમે (ame)	we
वा (vA)	वा,Indec	અથવા (athavA)	or
स (sa)	स,Pro,Nom,Mas,Sin	તેમણે (temaNe)	he
सः (saH)	सः,Pro,Nom,Mas,Sin	તેમણે (temaNe)	he
सह (saha)	सह,Indec	એકસાથે (ekasAthe)	together
सा (sA)	सा,Pro,Nom,Fem,Sin	તે (te)	she
स्म (sma)	स्म,Indec	ચોક્કસ (chokkasa)	always, surely
हि (hi)	हि,Indec	તેમણે (temaNe)	because, for

The indeclinables are independent of inflections with respect to any morphological characteristics like case, gender, number, tense, etc. But inflectional affixes for pronouns are subject are mutable.

Most stopwords in any language are classified majorly in article, determiners, and pronouns. And the results in Table 2 exactly reflects the same.

Table 2 Stopword part-of-speech classification through morphological analyzer

No. of stopwords	75	No. of indeclinables	39
		No. of pronouns	36

Table 3 Pronoun bifurcation based on Case markings

Case (Vibhakti)	Count
Nominative	26
Accusative	1
Instrumental	1
Dative	2
Ablative	1
Genitive	3
Locative	0
Vocative	0
Cases unidentified	2
Total pronouns	36

The indeclinables are immutable, while pronouns in Sanskrit are classified into eight cases which are listed in Table 3.

Sanskrit has three numbers namely singular, dual, and plural. Table 4 shows pronoun classified for the same.

Table 5 shows pronouns classified into three genders namely masculine, feminine, and neuter. The unidentified characteristics depicted for case (vibhakti), number, and gender are due to ambiguity in pronoun with multivalued characteristics especially in terms of gender. Few pronoun details were unidentified by analyzer. For example मम, मे,, तव, वयं has gender ambiguity with multivalued characteristics while तनि, तं case were unidentified by analyzer.

Table 4 Pronoun bifurcation based on number

Singular	27
Dual	1
Plural	6
Pronoun number unidentified	2
Total pronouns	36

Table 5 Pronoun bifurcation based on Gender

Masculine	11
Feminine	7
Neuter	9
Pronoun gender unidentified	9
Total pronouns	36

5 Conclusion

The morphological analysis of stopwords is necessary for several unexplored natural language processing tasks. The results can be used to optimize the time complexity factors in processes like morphological parsing, syntax parsing, data caching, etc. As stopwords are common and high-frequency words, it is clear from the results reflecting majorly classified grammatical category into pronouns and indeclinables.

References

1. Raulji J, Saini J (2017) Generating stopword list for Sanskrit language. In: 2017 IEEE 7th international advance computing conference (IACC), Jan 2017. IEEE. pp. 799–802
2. Ghosh K, Bhattacharya A (2017) Stopword removal: Why bother? A case study on verbose queries. In: Proceedings of the 10th annual ACM India compute conference on ZZZ, pp 99–102. ACM
3. Joshi H, Pareek J, Patel R, Chauhan K (2012) To stop or not to stop—Experiments on stopword elimination for information retrieval of Gujarati text documents. In: 2012 Nirma university international conference on engineering (NUiCONE), pp 1–4. IEEE
4. Jha V, Manjunath N, Shenoy PD, Venugopal KR (2016) Hsra: Hindi stopword removal algorithm. In: 2016 international conference on microelectronics, computing and communications (MicroCom), pp 1–5. IEEE
5. Ghag KV, Shah K (2015) Comparative analysis of effect of stopwords removal on sentiment classification. In: 2015 international conference on computer, communication and control (IC4), pp 1–6. IEEE
6. Lazarinis F (2007) Lemmatization and stopword elimination in Greek Web searching. In: Proceedings of the 2007 Euro American conference on Telematics and information systems, p 61. ACM
7. Govilkar S, Bakal JW, Kulkarni SR (2016) Extraction of root words using morphological analyzer for devanagari script. Int J Inf Technol Comput Sci (IJITCS) 8(1): 33
8. Raulji J, Saini J (2019) Sanskrit lemmatizer for improvisation of Morpholological analyzer. In: Presented at international conference on emerging technologies in computer engineering: microservices in big-data analytics, SKIT, Jaipur, India, Feb 2019. ESCI Taylor and Francis
9. Jha GN, Agrawal M, Subhash C, Mishra S, Mani D, Mishra D, Bhadra M, Singh S (2009) Inflectional morphology analyzer for Sanskrit. In: Kulkarni A, Huet G (eds) Sanskrit computational linguistics 1 and 2, LNAI, vol 5402, pp 219–238. Springer
10. Tapaswi N, Jain S (2012) Treebank based deep grammar acquisition and part of speech tagging for Sanskrit sentences. In: CSI 6th international conference on software engineering (CONSEG), Sept 2012. IEEE
11. Bharati A, Kulkarni A, Sheeba V (2006) Building a wide coverage Sanskrit morphological analyzer: a practical approach. The first national symposium on modelling and shallow parsing of Indian languages, IIT Bombay

A Computer Vision Based Approach for Subspace Clustering and Lagrange Multiplier Optimization in High-Dimensional Data

K. R. Radhika, C. N. Pushpa, J. Thriveni and K. R. Venugopal

Abstract In this work, we discuss about the issues raised due to the high-dimensional data in real-life scenario and present a novel approach to overcome the high dimensionality issue. Principal Component Analysis (PCA) based dimension reduction and clustering are considered as promising techniques in this field. Due to computational complexities PCA fails to achieve the desired performance for high-dimensional data whereas, subspace clustering has gained huge attraction from research community due to its nature of handling the high-dimensional data. Here, we present a new approach for subspace clustering for computer vision based applications. According to the proposed approach, first all subspace clustering problem is formulated which is later converted into an optimization problem. This optimization problem is resolved using a diagonal optimization. Further, we present a Lagrange Multiplier based optimization strategy to reduce the error during reconstruction Low-level data from high-dimension input data. Proposed approach is validated through experiments where face clustering and motion segmentation experiments are conducted using MATLAB simulation tool. A comparative analysis is presented shows that the proposed approach achieves better performance when compared with the existing subspace clustering techniques.

Keywords Subspace clustering · Dimension reduction · Computer vision · Optimization

1 Introduction

During the last decade, the demand of data-intensive applications has increased dramatically and an explosive growth is noticed in the data availability. These data are obtained from multiple modalities and sources such as computer vision based surveillance applications where multiple cameras are installed in the desired region and different types of data are captured. This process of data collection has led

K. R. Radhika (✉) · C. N. Pushpa · J. Thriveni · K. R. Venugopal
Department of CSE, UVCE, Bangalore University, Bangalore, India
e-mail: radhika@bmsit.in

S. Fong et al. (eds.), *ICT Analysis and Applications*, Lecture Notes in Networks and Systems 93, https://doi.org/10.1007/978-981-15-0630-7_43

toward various challenges such as data acquisition process, storage, and processing the complex high-dimensional data [1].

These types of high-dimensional data based computations are widely adopted in several areas such as machine learning based applications, image and signal processing, pattern recognition and bioinformatics, etc. For example, let us consider a computer vision based application where image consists of huge amount of pixels and videos containing several frames which generates huge data to process a desired task. This process generates high-dimension data which increases computational complexity and memory requirement of the systems. Moreover, it also affects the performance of applications because of the noise and inaccurate quantity of samples for the given space dimension, known as "curse of dimensionality" [2]. Generally, high-dimensional data are presented in low-dimensional constructions and are not distributed uniformly in the space. Thus, data recovered from the lower dimension structures can help to reduce the computation complexity, memory requirement, and helps to recover the performance of pattern inference, machine learning, and face recognition tasks. However, any image can consist of large number of pixels but computer vision models use fewer pixels and considerations to represent the geometry and changing aspects of the given scene. This process has motivated to develop the robust technique which can be used for representing the high-dimension data into low-dimension structures [3].

Several studies have been presented to address this issue such as dimension reduction and clustering. In this field of dimension reduction, PCA (Principal Component Analysis) is considered a promising technique where high-dimensional space data is obtained using single low-dimension space [4]. These techniques are adopted in several applications as discussed before. However, data points can be extracted from numerous subspaces but the membership subspace data points might be new. For example, in any video sequence, multiple frames are present and multiple moving objects may be presented. Hence, multiple subspaces are required to present these types of scenarios. To overcome these issues, subspace cluster based techniques have gained huge attraction from research communities. This approach of subspace clustering is used for grouping the data into multiple clusters corresponding to the subspace.

Subspace clustering is a branch of clustering which finds cluster in the subspaces in the given dataset. Generally, these techniques are applied in the high-dimensional dataset where irrelevant and multidimensional datasets can result in noisy clusters. However, feature selection techniques help to remove the irrelevant and redundant dimensions from the data whereas subspace clustering techniques analyze the data and identify the relevant data and dimensions and group then into the form of cluster. The subspace clustering techniques are divided into two categories as top-down where initial clusters are formed on the full set and further results are improved in an iterative process, another approach is bottom approach where dense regions are identified in the low-dimensional subspaces and later combined to form the clusters.

This approach facilitates the low-dimension representation from the high dimension data by clustering the data into subspace groups of data points. This process is known as subspace clustering and adopted in computer vision applications for example image segmentation [5], face recognition using clustering [6], video motion segmentation and image processing [7]. However, several techniques of subspace clustering are present in the literature. If the high-dimensional data is clean where less error or negligible errors are present then subspace samples can be represented into subspaces easily [8] and the low-dimension representation can be obtained. If the data suffer from noise and corruptions then it becomes a challenging task to handle the errors which results in poor low-level representation of the data. Hence, a robust model is needed to handle these types of issues in subspace clustering.

1.1 Issues and Challenges in Subspace Clustering

This section presents a brief discussion about issues and challenges present in the subspace clustering approaches.

- Segmentation and model estimation are strongly coupled to each other. If segmentation data are known, then single subspace can be extracted using PCA [9]. On the other hand, if the subspace factors are known, then the data arguments for the finest fit in respective subspace also can be identified. During implementation of these schemes, these parameters are not known hence they need to be solved separately for each problem.
- The dissemination of the data inside the subspace is a crucial parameter which is unknown. If the data is distributed around the cluster and cluster center, then centroid clustering-based approaches can be implemented to solve the clustering problem [10]. On the other hand, if data distribution is arbitrary, then central clustering techniques fails to perform the desired task.
- The position of the identified subspaces is an important parameter [11]. If the subspaces are very close and intersecting, then performing accurate subspace clustering becomes a tedious task.
- Generally, these high-dimensional data may face issues due to the corrupted data where missing data entries, noise, and outlier are the well-known parameters which are responsible for corrupting the data [12].

1.2 Problem Definition

Let us consider that a set of data samples are obtained which may contain different types of errors in linear subspaces. The main aim of subspace clustering is to reduce the errors and rearranging the corresponding segments into their respective subspaces. In this process the "error" represents the deviation between subspaces and the data which can be presented as noisy data, missing entries, outliers, and data corruption. These techniques can be used in different applications such image processing, data mining video summarization, etc.

1.3 Contribution

In this work, a subspace clustering process for dimension reduction for computer vision based applications is proposed. First of all subspace clustering problem is modeled for multiple subspaces and reconstruction of the data is done where noise is present in the original data. In order to perform this task, we formulate an optimization problem which is solved using diagonal matrix based solution. This problem is further solved using Lagrange Multiplier based scheme until the optimal convergence condition is obtained.

1.4 Organization

The complete paper is arranged as follows: A brief discussion about recent studies is discussed in Sect. 2. The proposed model for solving the subspace clustering problem is discussed in Sects. 3 and 4 presents comparative study and finally, Sect. 5 briefs about concluding remarks.

2 Literature Survey

In this section gives a brief discussion about recent studies in the field of subspace clustering.

Peng et al. [13] discussed about the graph-based learning scheme for subspace clustering approach to establish the relation between data points and subspace. However, these techniques suffer from the complexity issues hence, authors developed a new novel approach to represent the error-free data extracted from the data points. In this work, linear intra-subspace projection dominance is performed to compute the coefficients. Later, a sparsity graph is constructed and denoted as L2-graph.

Shao et al. [14] introduced arbitrarily Oriented Synchronized Clusters (ORSC) approach for processing the high-dimensional data that uses synchronization process to handle the complex processes. Mainly this approach is based on the weighted communication model and dynamic clustering process. This technique detects the correlation between arbitrary clusters and arbitrary shaped clusters, initially. Later, outlier detection and noise reduction are applied to improve the performance of subspace clustering.

These techniques are widely adopted in real-time computer vision based applications. Based on this assumption, Javed et al. [15] presented a computer vision based model for background and foreground detection using subspace clustering technique. However, several approaches have been introduced recently. Due to dynamic background, occlusions and jitters, etc., these techniques fail to achieve the desired performance. To overcome these issues, authors developed a novel approach which uses spatial and temporal process of subspace clustering using Robust Principal Component Analysis model. According to this process, motion-aware correlation coefficients are used for computing the spatiotemporal model for the given video sequence and a low-rank model is developed to represent the low-level structures of the video frames.

Tsakiris et al. [16] discussed about the subspace clustering on the missing value data that can be very helpful for reconstruction of data from low-level structures. This technique uses Sparse Subspace Clustering (SSC) for low-dimension representation of the data for pattern recognition and computer vision applications. However, this work gives a theoretical proof that SSC based techniques can be used for subspace clustering where missing data is present.

Xia et al. [17] focused on human motion segmentation strategy using subspace clustering method. In this work, the segmentation problem is transformed into the temporal subspace clustering problem. Later, sparse subspace clustering is considered and geodesic exponential kernel is generated to model the low-level arrangement.

3 Proposed Model

This segment explains the proposed approach for subspace clustering. The complete section is divided as follows: The first section presents a subspace clustering problem formulation scenario which focuses on obtaining the noise-free low-dimension structure data from the high-dimension data. The next phase of the work gives a novel solution for the subspace analysis problem. Finally, implementation scenario using proposed approach for face clustering is explained. Figure 1 shows the complete representation of proposed approach which includes linear subspace creation, aim definition of SSC, decomposition of the subspace clustering problem, optimization solution, and performance analysis.

According to the proposed work, first of all we present linear subspace modeling for the given data and the specific aims are designed for the considered database such as dimension identification, segmentation points and identification of the total number of subspaces, etc. In the next phase, problem definition is defined where PCA is used for decomposition of the problem and later SVD is applied to generate the mean data points. This problem is formulated as optimization problem where dictionary level representation, rank minimization, and subspace memberships are assigned. Finally, we apply proposed solution for different types of databases and the obtained performance is compared with the existing approaches to show the robust performance of proposed model.

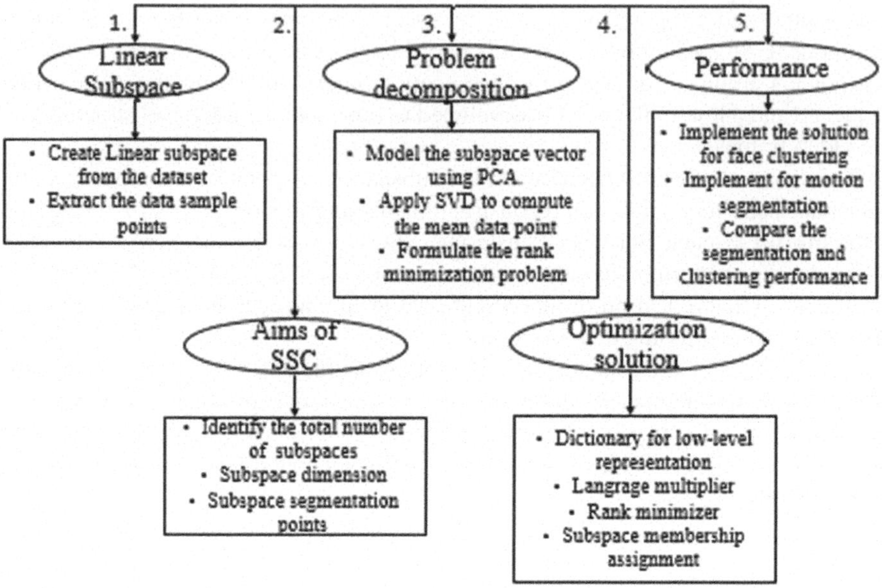

Fig. 1 Proposed work flow chart

3.1 Problem Formulation and Solution

This subsection discusses about the problem formulation for subspace clustering.

Let us consider a data modeling problem where multiple data points are present with the combination of subspaces. A sample representation is depicted in Fig. 2 where two lines and one plane make up a three space union as $\mathbb{S}$.

Let us consider that the extracted points are given as x which are extracted from the unknown union of linear subspace with unknown dimensions, i.e., the extracted

Fig. 2 Sample point set is extracted from $\mathbb{S}^3$ three space union as two line and one plane

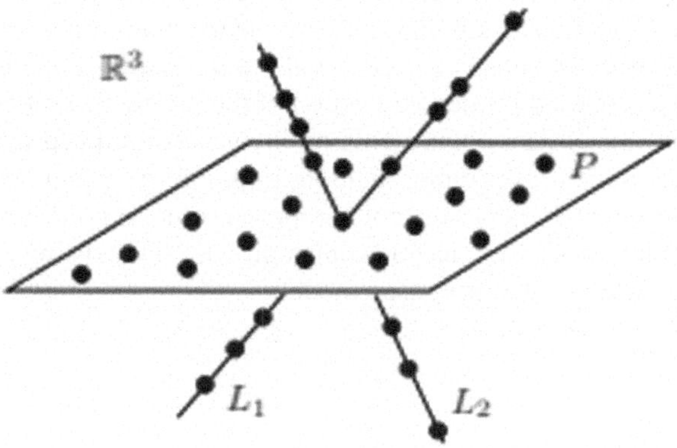

points are given as $\{x_j \in \mathbb{S}^D\}_{j=1}^{N}$ that are having linear subspace as $\{\mathcal{L}_i\}_{i=1}^{n}$ along with unknown dimensions as $d_i = \dim(\mathcal{L}_i), 0 < d_i < D$ where $i = 1, 2, \ldots, n$ where $n \geq 1$. Hence, the subspace can be defined as

$$\mathcal{L}_i = \{x \in \mathbb{S}^D : x = \alpha_i + \mathcal{U}_i y\} \tag{1}$$

where $\alpha_i \in \mathbb{S}^D$ denotes a random point in the considered subspace $\mathcal{L}$, for linear subspaces $\alpha_i = 0$, $\mathcal{U}_i = \mathbb{R}^{D\times d_i}$ denotes the subspace basis and $y \in \mathbb{S}^D$ denotes the low-dimensional structure representation for the data point x. The main aim of this process of subspace clustering is to estimate the total number of subspaces (n), subspace dimension $\{d_i\}_{i=1}^{n}$, basis of subspaces $\{\mathcal{U}_i\}_{i=1,}^{n}$ $\{\alpha_i\}_{i=1}^{n}$ and the segmentation points based on the subspace information.

In this process, if the total number of subspaces is one, then the subspace problem is reduced to computing a vector as $\alpha \in \mathbb{S}^D$, a basis of subspace as $\mathcal{U} \in \mathbb{S}^{D\times d}$ and a low-dimension representation structure given as $\mathcal{W} = [w_1, \ldots., w_N] \in \mathbb{R}^{D\times d_i}$ and the subspace dimension as d. This problem can be modeled into Principal Component Analysis (PCA) model which can be resolved using (Singular Valued Decompositions) SVD. Here, $\alpha = \frac{1}{N}\sum_{j=1}^{N} x_j$ denotes the mean of the data points which are considered and the $(\mathcal{U}, \mathcal{W})$ can be achieved using rank d SVD of the mean subtracted data matrix $X = [x_1 - \alpha, x_2 - \alpha, \ldots, x_N - \alpha] \in \mathbb{R}^{D\times N}$, which can be computed as:

$$\mathcal{U} = U \text{ and } \mathcal{W} = \sum \mathcal{V}^{\mathsf{T}} \text{ where } \mathcal{W} = U \sum \mathcal{V}^{\mathsf{T}} \tag{2}$$

where d can be attained as $d = rank(X)$ with noise-free data. Let X be a r rank SVD given as $X = U\sum \mathcal{V}^{\mathsf{T}}$, $U \in \mathbb{S}^{D\times r}$, $\sum \in \mathbb{S}^{r\times r}$ and $\mathcal{V} \in \mathbb{S}^{N\times r}$. Let us consider that $X_0 \in \mathbb{S}^{D\times n}$ with SVD values, can store a set of n dimensional sections which are extracted from the subspace union $\{\mathcal{L}_i\}_{i=1}^{k}$ with unknown dimensions. Hence, observation vector is generated as:

$$X = X_0 + \mathcal{E}_0 \tag{3}$$

Here, our main aim is to improve the row space data of X_0 to improve the system efficiency.

In order to recover the low-level structure data from the observation vector X, we represent this problem is represented as regularized rank minimization problem which can be given as:

$$\min_{P,\mathcal{E}} rank(D) + \gamma \|\mathcal{E}\|_l \text{ subject to } X = D + \mathcal{E} \tag{4}$$

where $\gamma > 0$ and $._l$ denotes the regularization strategy. However, conventional techniques consider that the extracted data is sampled into single subspace hence union of subspace clustering may be larger and recovery may be inaccurate. In order to deal with this subject, the problem is redefined as general rank minimization:

$$\min_{\mathcal{M},\mathcal{E}} \ rank(\mathcal{M}) + \gamma\|\mathcal{E}\|_l \ \text{ subject to } X = A\mathcal{M} + \gamma\|\mathcal{E}\|_l \ \text{ subject to } X = A\mathcal{M} + \mathcal{E} \tag{5}$$

where A denotes the span dictionary of the given data space, $\mathcal{M}$ denotes the minimizer for low-level representation of data X. Once the optimal solutions are identified for the $\mathcal{M}$ and $\mathcal{E}$ as $\mathcal{M}^*$ and $\mathcal{E}^*$. The original data can be recovered using $A\mathcal{M}^*$.

Let us consider that $A \neq 0$ and $X = A\mathcal{M}$ have feasible solution in the given span of data space as $\mathcal{M}^* = A^\dagger X$. The minimizer problem can be given as $rank(\mathcal{M}^*) = rank(X)$. In this process, dictionary selection plays an important role which displays that the proper selection of dictionary can be helpful to obtain the desired low-level representation. This can be achieved when data columns of A and X are sampled exactly and then minimizer problem can help to achieve the subspace membership for the given data points. Here, we denote that collection of subspaces $\{\mathcal{L}_1, \mathcal{L}_2, \ldots, \mathcal{L}_k\}$ where each subspace has a rank $r_i > 0$. We assume that A_i denotes the total m_i samples of the ith subspace $\mathcal{L}_i$, X_i denotes the sample collection n_i samples. In this case, if the given subspaces are liberated, the optimization problem is denoted in a block diagonal form which is given as:

$$\mathcal{M}^* = \begin{bmatrix} \mathcal{M}_1^* & 0 & 0 & 0 \\ 0 & \mathcal{M}_2^* & 0 & 0 \\ 0 & 0 & \ddots & 0 \\ 0 & 0 & 0 & \mathcal{M}_k^* \end{bmatrix} \tag{6}$$

where $\mathcal{M}^*$ denotes a $m_i \times n_i$ matrix which contains the coefficients with the given rank as $rank(\mathcal{M}_1^*) = rank(X_i)$. This problem is further solved using Lagrange Multiplier approach which can be expressed as follows:

$$\begin{aligned} L = \|\mathcal{H}\|_* + \gamma\|\mathcal{E}\| + tr\left(Y_1^T(X - A\mathcal{M} - \mathcal{E})\right) + tr\left(Y_2^T(\mathcal{M} - \mathcal{H})\right) \\ + \frac{\alpha}{2}(\|X - A\mathcal{M} - \mathcal{E}\|_F^2 + \|\mathcal{M} - \mathcal{H}\|_F^2) \end{aligned} \tag{7}$$

The complete process is given below:

Input: Incomplete data matrix as X, $\mathcal{M}, \mathcal{H}, \mathcal{E}, Y_2, = 0$, $\alpha = 10^{-6}$ Output: reconstructed data matrix, labels and coefficient matrix
Step 1: Initialize the data matrix computation parameters $X_0 \leftarrow initialize$
Step 2: while not convergence
Step 3: estimate the diagonal block
Step 4: fix other parameters and update $\mathcal{H}$ as $\mathcal{H} = \arg min \frac{1}{\alpha} \Vert\mathcal{H}\Vert_* + \frac{1}{2}\left\Vert\mathcal{H} - \left(\mathcal{M} + \frac{Y_2}{\alpha}\right)\right\Vert_F^2$
Step 5: fix other parameters and update $\mathcal{M}$ as $M = \frac{1}{(AA^T+I)}\left((X - \mathcal{E})A^T + \mathcal{H} + \frac{Y_1 A^T - Y_2}{\alpha}\right)$
Step 6: fix other parameters and update error $\mathcal{E}$ $\mathcal{E} = \arg\min \frac{\gamma}{\alpha} \Vert\mathcal{E}\Vert_{2,1} + \frac{1}{2}\left\Vert\mathcal{E} - \left(X - A\mathcal{M} + \frac{Y_1}{\alpha}\right)\right\Vert_F^2$
Step 7: evaluate the convergence stage as $\Vert X - A\mathcal{M} - \mathcal{E}\Vert_\infty < \epsilon$ and $\Vert\mathcal{M} - \mathcal{H}\Vert_\infty < \epsilon$
Step 8: update the diagonal block and rearrange the data according to the obtained updated ranks

4 Results and Discussion

This section presents a complete experimental study using the proposed approach for subspace clustering. The complete experimental study is carried out using MATLAB 2013b running on windows platform. The proposed approach is implemented for face clustering and video motion segmentation purposes. Face clustering is performed using Extended Yale B database [8] and Hopkins 155 database [9]. The proposed approach performance is compared with the several state-of-the art techniques such as LLR [10], SCC [11], SSC [2], LSRC [1], LSA [12], and ARM [6] in terms of misclassification rate which can be computed as

$$Error = \frac{wrong\ clustered\ points}{total\ number\ of\ points} \times 100\% \tag{8}$$

4.1 Face Clustering Comparison

Figure 3 shows a sample outcome of the proposed approach of data reconstruc-

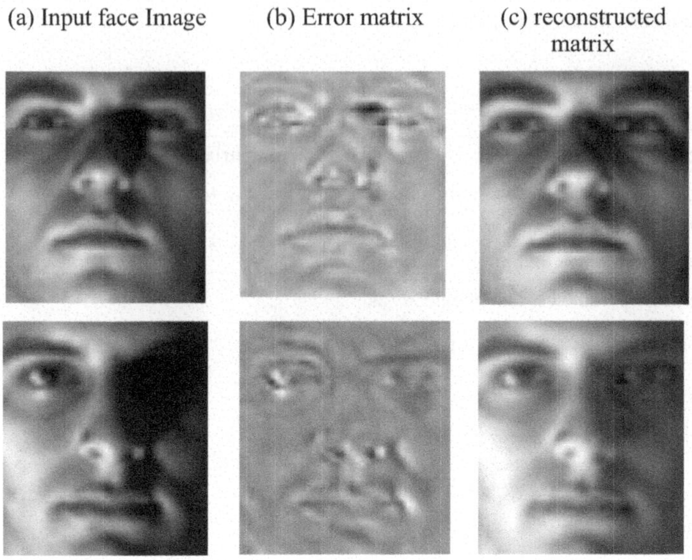

Fig. 3 Sample image, error image and reconstructed image

tion technique where the first column represents the input image, second column represents the error matrix and the outcome reconstructed image is given in third column.

Based on these experiments, we present a comparative study in terms of clustering error rate and equated the performance of the suggested approach with existing techniques of face subspace clustering. To show the robust performance, the proposed approach is compared with a varied number of subjects and evaluated mean and median of the clustering error rates and given in Table 1.

Table 1 shows a comparative analysis for a varied number of subjects. This study shows that the proposed approach accomplishes better performance when compared with the existing techniques, i.e., for 10 subject scenarios, the overall error is 3.81% which is significantly less for existing techniques, due to significant cluster formation with the reduced error when compared with existing techniques. Similarly, error variations are also very less in the proposed approach.

Figure. 4 shows a comparative performance in terms of computation time using LLR, SSC, ARM and proposed approach which is obtained as 25 s, 56 s, 23 s, and 13.2 s. The proposed approach achieves a desired solution by consuming less time and iteration for convergence

Table 1 Clustering error rate comparison

Algorithm	SSC [2]	LLR [10]	LSRSC [1]	LSA [12]	SCC [11]	ARM [6]	PS.
Subject-2							
Median	0.0	0.78	4.69	47.66	7.82	0.78	0.03
Mean	1.86	2.54	5.32	32.80	16.62	1.51	1.42
Subject-3							
Median	1.04	2.60	7.81	50.01	39.05	1.56	0.93
Mean	3.10	4.21	8.47	52.29	38.15	2.26	2.11
Subject 5							
Median	2.51	5.63	12.21	58.08	58.01	3.05	2.11
Mean	4.30	6.90	11.25	59.18	56.81	2.48	2.19
Subject 8							
Median	4.48	10.05	28.03	58.58	59.38	3.32	2.11
Mean	5.84	14.32	23.72	59.18	58.91	3.71	2.59
Subject 10							
Median	5.62	23.58	28.74	57.51	75.77	2.97	2.95
Mean	10.92	22.91	30.31	60.41	73.02	3.85	3.81

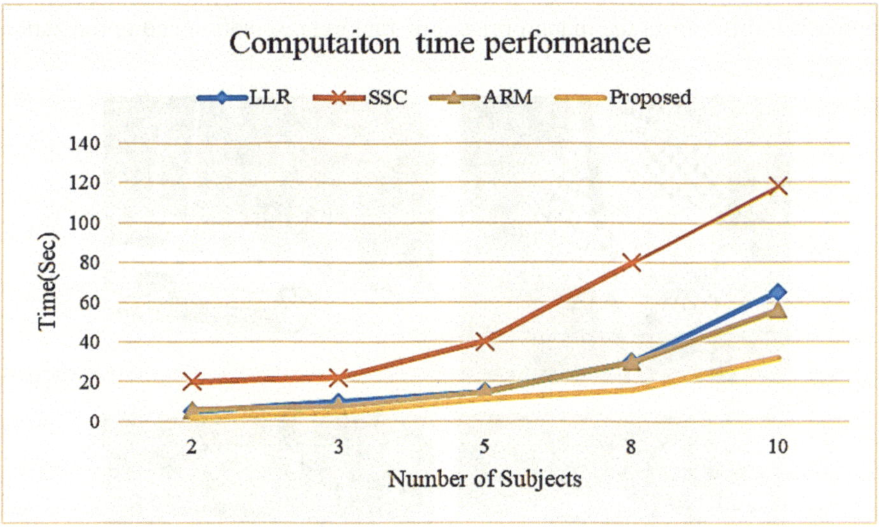

Fig. 4 Computation time performance

4.2 Motion Segmentation Performance Analysis

This section presents another experimental study for computer vision application for motion segmentation. This experimental study is carried out using Hopkins-155 datasets and the performance is evaluated in terms of segmentation error. According to this process of motion segmentation, a video sequence is considered for evaluations and the moving objects are segmented into various spatiotemporal regions which represent a different type of motion.

Figure 5 shows a sample representation of tracing feature during the processing of the input video. This figure represents the background and foreground with different features. The static background maintains similar features during the complete sequence whereas the features of moving objects vary for each frame. These features can be helpful for various types of computer vision applications such as object detection, tracking crowd behavior analysis, etc.

Table 2 shows a comparative performance in terms of motion segmentation error. The experimental study for all motion scenario shows that the overall error for all motion is 1.33%, which is comparatively very low in comparison with the state-of-art techniques. During the process of data reconstruction, data labeling, and coefficient computation, the convergence process also takes place whereas conventional approaches require more iteration and time to converge whereas proposed approach converges fast which in turn helps to reduce the time complexity. Moreover, existing approaches suffer from the inappropriate low-rank representation and optimization

Fig. 5 Video motion segmentation traced features

Table 2 Motion segmentation comparison performance

Algorithm	SSC [2]	LLR [10]	LSRSC [1]	LSA [12]	SCC [11]	ARM [6]	PS.
Motion-2							
Median	0.0	0.0	4.69	4.21	0.0	0.0	0.01
Mean	1.51	2.12	5.32	0.55	2.87	1.45	1.40
Motion-3							
Median	0.55	1.42	3.79	1.46	0.55	0.83	0.02
Mean	4.41	4.02	4.85	7.01	4.39	1.48	1.40
All motion							
Median	0.0	0	0.59	0.88	0	0	0
Mean	2.17	2.55	4.85	4.85	4.11	1.47	1.33

problem which leads to the poor performance of updating the error matrix. Hence, the proposed approach achieves better results when compared with the existing techniques.

5 Conclusions

A novel methodology for a low-level representation scheme for high-dimensional data using subspace clustering technique is proposed. According to this approach, a subspace clustering problem is formulated for a data where multiple number of subspaces are present and data is corrupted with the noise. Here optimization problem is addressed by reconstructing the low-level representation of the data, we formulate an optimization problem which is addressed using a diagonal optimizer solution and Lagrange Multiplier scheme to optimize the error parameters. An extensive experimental study is carried out for facial clustering and human motion segmentation. Results of the proposed approach display significant improvement in the performance of subspace clustering.

References

1. Favaro P, Vidal R, Ravichandran A (2011) A closed form solution to robust subspace estimation and clustering. In: Computer vision and pattern recognition (CVPR), IEEE conference on computer society conference on computer vision and pattern recognition, pp 1801–1807
2. Elhamifar E, Vidal R (2013) Sparse subspace clustering: algorithm, theory, and applications. IEEE Trans Pattern Anal Mach Intell, 2765–2781
3. Von Luxburg U (2007) A tutorial on spectral clustering. Stat Comput, 395–416
4. Arias-Castro E, Lerman G, Zhang T (2017) Spectral clustering based on local PCA. J Mach Learn Res, 253–309

5. Yang A, Wright J, Ma Y, Sastry S (2008) Unsupervised segmentation of natural images via lossy data compression. Comput Vis Image Underst 110(2):212–225
6. Wang W, Yang C, Chen H, Feng X, (2018) Unified discriminative and coherent semi-supervised subspace clustering. IEEE Trans Image Process, 2461–2470
7. Liu G, Lin Z, Yu Y (2010) Robust subspace segmentation by low-rank representation. In: Proceedings of international conference on machine learning, pp 663–677
8. Lee KC, Ho J, Kriegman D (2005) Acquiring linear subspaces for face recognition under variable lighting. IEEE Trans Pattern Anal Mach Intell, 684–698
9. Tron R, Vidal R (2007) A benchmark for the comparison of 3-D motion segmentation algorithms. In: Proceedings of IEEE conference on computer vision and pattern recognition, pp 1–8
10. Liu G, Lin Z, Yan S, Sun J, Yu Y, Ma Y (2013) Robust recovery of subspace structures by low-rank representation. IEEE Trans Pattern Anal Mach Intell, 171–184
11. Chen G, Lerman G (2009) Spectral curvature clustering (SCC). Int J Comput Vis, 317–330
12. Yan J, Pollefeys M (2006) A general framework for motion segmentation: independent, articulated, rigid, non-rigid, degenerate and non-degenerate. In: Computer Vision–ECCV. Springer, pp 94–106
13. Peng X, Yu Z, Yi Z, Tang H (2017) Constructing the L2-graph for robust subspace learing and subspace custering. IEEE Tans Cybern, 1053–1066
14. Shao J, Wang X, Yang Q, Plant C, Böhm C (2017) Synchronization-based scalable subspace clustering of high-dimensional data. Knowl Inf Syst, 83–111
15. Javed S, Mahmood A, Bouwmans T, Jung SK (2017) Background–foreground modeling based on spatiotemporal sparse subspace clustering. IEEE Trans Image Process, 5840–5854
16. Tsakiris MC, Vidal R (2018) Theoretical analysis of sparse subspace clustering with missing entries. arXiv:1801.00393
17. Xia G, Sun H, Feng L, Zhang G, Liu Y (2018) Human motion segmentation via robust kernel sparse subspace clustering. IEEE Trans Image Process, 135–150

Improving Court Efficiency Through ICT Integration: Identifying Essential Areas of Improvement

Rozha K. Ahmed, Khder H. Muhammed, Aleksander Reitsakas, Ingrid Pappel and Dirk Draheim

Abstract Integration of information communication technology (ICT) with the judicial system has recently brought new opportunities in courts toward improved efficiency, quality, and transparency of court cases; better management of cases from registration through case disposal; and extended availability of the judiciary. This paper reflects on the implementation of the e-court system in the Sulaimaniyah Appellate Court in the region of Kurdistan in Iraq. The analysis is based on expert interviews of different stakeholders in the appellate court. The results show significant improvements in terms of the court case management workflow in the following four areas: improved internal daily operations, the security of court cases, concurrent extended access to the judiciary, and transparency. The research aims at extending the body of knowledge for judiciaries, who are on the way to start integrating technology to courts.

Keywords E-court · Court information system · E-services · E-government

R. K. Ahmed (✉) · I. Pappel · D. Draheim
Information Systems Group, Tallinn University of Technology, Tallinn, Estonia
e-mail: rozha.ahmed@taltech.ee; rozha.ahmed@aktors.ee

I. Pappel
e-mail: ingrid.pappel@taltech.ee

D. Draheim
e-mail: dirk.draheim@taltech.ee

K. H. Muhammed
Judicial Council of Kurdistan Region of Iraq, Sulaimaniyah, Iraq
e-mail: khidr.muhammed@sulicourt.com

R. K. Ahmed · A. Reitsakas
Aktors Ltd., Tallinn, Estonia
e-mail: aleksander.reitsakas@aktors.ee

S. Fong et al. (eds.), *ICT Analysis and Applications*, Lecture Notes in Networks and Systems 93, https://doi.org/10.1007/978-981-15-0630-7_44

1 Introduction

Information and communication technology (ICT) can be considered a critical factor in modernizing and reshaping the conventional judiciary. E-court systems are a crucial factor in enhancing the quality of court services with various potential benefits such as online accessibility, timeliness, transparency, efficiency, data accuracy, secured access to files and information, etc. This paper aims at systematically identifying and characterizing these areas of improvement.

In service of the research question, the paper delves into the implementation of the e-court system in the Sulaimaniyah Region Appellate Court as a case study. This project is a pilot for a greater plan to introduce e-government in the Kurdistan region of Iraq. On behalf of the project, the essential daily work routines at court were transformed from *paper to digital* documents.

The analysis is based on structured expert interviews. Further evidence is provided by observations from inside the project. The interviewees have been selected based on their professional background, i.e., all interviewees have long-term experience in the conventional paper-based system and are now active users of the paperless system. We have interviewed five judges, one lawyer, one investigator, and one prosecutor of the Sulaimaniyah Appellate Court, and, furthermore, the head of the IT department of the Sulaimaniyah Appellate Court. Among the interviewees, we had the previous court president, one active and one previous member of the appellate court panel, and, furthermore, a member of the judges' advisory board in the e-court project. Three of the authors have been actively involved as stakeholders in the implementation project, i.e., the project owner and the on-site project leader from the implementing service provider, and the e-court project supervisor from the Sulaimaniyah Appellate Court.

In Sect. 2, we provide a detailed description of the e-court system in the Sulaimaniyah Appellate Court. In Sect. 3, we present the findings of our analysis on the major fields of improvement, based on evidence from the conducted interviews and observation from inside the project. We discuss related work in Sect. 4 and finish with a conclusion in Sect. 5.

2 The E-Court System in the Sulaimaniyah Appellate Court

The aim of this project at Sulaimaniyah Appellate Court was transforming the conventional paper-based system into digital and systematic form, through court case management system, achieving better and more efficient justice delivery. The initial idea of the project emerged in October 2014. The project was then implemented in six stages (planning, system analysis/master plan, prototyping, building infrastructure, piloting, and implementation).

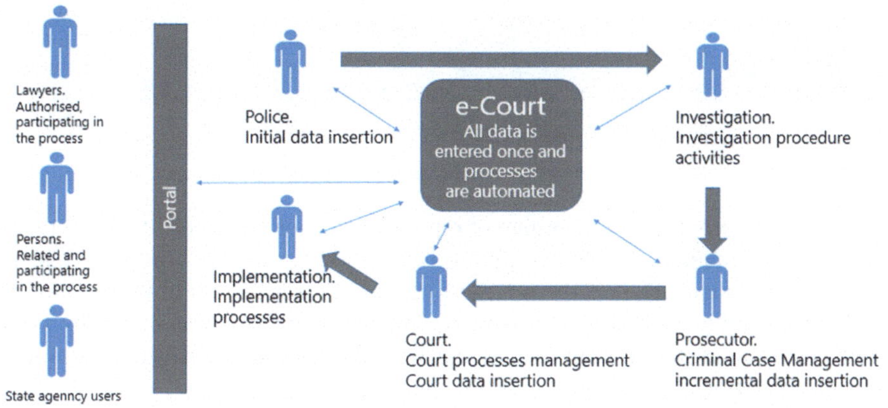

Fig. 1 E-court cooperation

The system has been implemented in all courts under the appellate court of the Sulaimaniyah region.[1]

The system is responsible for case management and, furthermore, for managing several types of certificates at the Civil Primary Court, Personal Status Court, and the Personal Items Court. The procedures of communication in the Sulaimaniyah e-court system are divided into different institutions inside and outside the court. The institutions are the courthouse, three departments in the courts (statistics, notification, and implementation), and external institutions (public prosecution system and police stations). Interdependence between systems and how to communicate includes a group of subsystems that make up the judicial system characterized feature and bonding with each other as an integrated system to perform all the functions and tasks required. There is a linkage between institutions and separate buildings, to facilitate the communication and the electronic transfer of cases between them. Information is entered once, and processes are all automated to allow system users to perform their main function and collaborate efficiently. Moreover, through the implementation of public portal outside court, users have an opportunity to access the system according to their role in the proceedings. The outside users in Sulaimaniyah e-court are categorized into three different types: lawyers, outside agency users, and public users who are case participants can have access to cases according to their relation with the case, see Fig. 1.

Court case management consists of different stages, which all require the user to insert relevant documents and information connected to a specific stage. The system is designed with all required functionalities to assist e-court users to perform all tasks electronically, from the claim submission through generating court decision and case disposal. Figure 2 shows the main cycle of the case proceeding.

---

[1]Civil Primary Court (6 courts), Personal Status Court (5 courts), Personal Items Court, Labour Court, Investigative Court (12 courts), Misdemeanors Court (6 courts), Felony Court (3 courts), and Juvenile Court.

Fig. 2 Case proceeding cycle

The system provides a wide range of reports that cover the needs of the courts in different terms of judicial reference, giving a simplified way to assess the performance of courts and judges. Hence, these reports provide a clear picture of the amount of work achieved by the courts. Knowing the amount of stress experienced by each court helps the administration department to make appropriate decisions towards court organization.

Separate registers with different purposes for cases at each court provide better case management for judges and court officials, with more efficient record registration compared to handwritten registers, which provide a clear overview of the number of cases at different statuses. Also, for better case tracking purpose, the system is implemented with (13) different states[2] of the case. Moreover, old closed paper cases before implementation are all added to a separate archiving system.

Moving on to main functionalities can be described as the digitization of the legal proceedings process, which includes the following:

Case Registration The claim is submitted, and the case is registered electronically in the prime register of every court. The system automatically creates connections to other registers. All case-related parties can access the case simultaneously. This automatic connection results in the reuse of the data which have already been entered to the system.

Fee Management Digital fee management simplifies the work of clerks and auditors at the fee register. It also helps to acquire better statistics of all paid and unpaid fees. Hence, it enables a better overview of fees (and the dates when they are due).

Court Summons Management Digital form of court summons and notifications functionality helps to keep a better look on undelivered/delivered summonses, which in turn helps plan court hearings.

Hearing Management Hearings digitally is more simplified than a paper system. Selecting judges, and participants are all managed systematically.

Documents Registration Documents are all registered digitally. Digital document templates will improve the process of creating documents in several different aspects: quality of documents, readability of documents, easier to copy, consistent design, and the speed of creating documents.

Document Template Management All system-generated letters, decisions, and certificates are added from the template management section. Digital document man-

[2]Draft, pending, registered, in next instance; closed, unified, canceled, in supreme court; suspended, abrogated, dropped out, dropped out date passed; in implementation.

agement registry will make it easier to create system-generated document templates. Furthermore, the automatically generated templates in the system help clerks in saving time and less possibility for error.

Automatic Case Distribution The allocation of cases becomes more transparent (compared to manual allocation) and enables to consider the complexity of cases (in addition to the number of the cases).

User Management The system designed to have four different types of user groups: court users, lawyers, public users, and agency employees. Hence, the user management section provides a better overview of the roles, views, user, and permissions.

Onscreen Hearing Monitor TV screens are installed in the courthouse, next to courtrooms in order to clarify the schedule of courtrooms for all the visitors. These screens help better monitoring of the hearings in every court and decreasing the waiting time by lawyers and litigants, and more systematic manner.
Case-Law Case-law is the summary of previous court judgments, which will become available online. Presenting the judgment summary is an essential functionality for lawyers/barristers and for judges, who will be able to search for summaries of similar cases and refer to them in their ongoing cases. In the purpose of this, the e-court system has implemented an advanced search capability to find similar cases.

Robust Statistics This functionality provides better possibilities for gathering statistics and the analysis of court statistics regarding the number of incoming and disposed cases. Hence, the backlog will be gathered automatically (instead of manual calculation), and the result will be ready in a matter of seconds.

Digital and Physical Case File Connection Connection between digital case file and physical case file implemented through a bar code. If cases are stored digitally and also parts of it are printed out on paper, there is a need for linking the physical and digital case to one another. This bar code is implemented in order to ease the search of digital case file using the physical case file printed bar code.

Electronic Notification Electronic notification sends immediate notification about case updates to all related users. Specific case updates such as case registration, hearings, fees, case participants, and documents are all notified to related case users.

Public View The system provides a separate view for public users, who are connected to cases. Public users can be (case participants, outside court justice agency users, and lawyers). Through a public portal, they can access their case information, get notifications about updates, submit new cases and certificates, and download and upload documents. This functionality results in building trust, confidentiality, and transparency of the public toward the judiciary.

Multilingual System Current system is implemented with three main languages (Kurdish, Arabic, and English). This functionality will aim at decreasing the cost for participant and court. In the paper system, the used language for reading and writing was Kurdish in most of the courts. Hence, case participants who are not their primary language were Kurdish had to get the official court papers and translate it outside the court to the language they use, for example, Arabic or English. The current system is equipped with this functionality to generate all court documents in three languages, and this leads to saving the cost of translation.

3 Findings

3.1 Internal Daily Operations

The interviewed experts pointed out that there are significant improvements in internal daily operation inside the court.

3.1.1 Better Case Management

According to our observations, the whole case process cycle with e-court is more efficient. As an example, in the paper system from the case registration until case disposal, clerks should register case participant information multiple times into different registers. Hence, repetition of the same information with handwritten text gives a possibility for a mistake, in addition to time consumption. In the e-court system, information is inserted to the system and registered once. The registered information will consistently appeared in all registers and can be accessed by all case-related parties. The process starts with entering information once into the main prime register. After the registration, the case will automatically be moved to the fee department, no need to be followed and taken by clerks. Then, the case will be automatically sent for hearing registration. The calendar and timer track the exact date and time set for the hearings, to avoid conflicts. An additional enhancement in the hearings is that hearing details and participant are now shown on the screen and published to the public on the court website. Therefore, case participants and lawyers do not need to be in court for the whole day, waiting for their hearing. After the hearing is set, summonses are generated automatically from the system, with all hearing details and case participant information. Then, notification sent to summons department for dispatching and delivering. In the paper system, summons forms were all filled by clerks with handwritten information that led to many mistakes, in addressee information and all other hearing details such as date and time, court name, case type, and participant names.

The final step is case disposal with closing case by adding the final decision. Currently, the system shows a more organized decision register compared to paper decision register. Moreover, the decision summary is automatically added to every case and shown only by mouse click. That saves the clerk time to rewriting the whole decision in the register.

In addition, a judge[3] agreed the e-court implementation promotes daily internal operations and described that: "*Although, the usage of technology in Kurdistan courts and the project is in the beginning, however, we have seen significant improvements in the case registration process, with less time and more accurate information as compared to the paper system; we, as judges, noticed that e-court enhanced speed in our daily work and case disposal, which at the end benefits the litigants.*"

Another judge[4] added: "*Generally, the main task of the court is restoring rights to the oppressed or investigating in a case or crime, the court should solve and conclude with issuing a fair decision very soon. Also, e-court has helped the process to be achieved in less time, compared to the manual process. There are some situation which time plays an essential key in court cases, for example in some civil cases, even if court despised a case with a fair decision, the actual value of returned right may lose due to the time consumed in the process of finalizing the case. Such cases related to financial issues or properties. Similar issue in criminal cases, there are some cases which take long periods and due to that some traces of crime may fade by time or even disappear.*"

Another operation improvement is case movement from one court to another, when cases are appealed, objected, and referred from one court to another court. In a paper system, bulks of case files were moved from one court to another. A considerable amount of time was needed, for example, days if there are distances between courts, in addition to human resources and effort. Currently, the system takes responsibility of case movement more efficiently, and case data instantly shown in the next court along with notification to all related system users.

In support to that, a judge[5] stated: "*Since, I am using the system from 2016 up to today, I have stayed away from paper and pen (% 90) in my daily work as a judge. Moreover, case movements between different related parties and institutions are done in a matter of minutes. For example situations such as referral, and different types of appeal used to be very slow and consuming days to arrive at the next destination.*"

A prosecutor[6] agreed on that and added: "*One of the significant improvements for prosecution is with case movements between courts and prosecution building in case referral situation. In a conventional system, bulks of the paper file were moved from courts to prosecution and returned to court after prosecution documents have been added. Currently, no movements needed, referred cases instantly appear to prosecutors who are involved in the process, and case data will be available. The system sends notification about case referral.*"

Moreover, a judge[7] added: "*This functionality has saved significant time for appeal judges because, in a paper system, opinion was written on a paper and attached to*

[3]Supreme Court judge, previous court president at the Sulaimaniyah Appellate Court.

[4]Judge at the primary court in the Sulaimaniyah Appellate Court, member of the judges advisory board in the e-court project.

[5]Member of the appellate court panel of the Sulaimaniyah Appellate Court.

[6]Prosecutor at the Sulaimaniyah Appellate Court.

[7]Supreme Court judge, a previous member of the appellate court panel at the Sulaimaniyah Appellate Court.

the case file. The case file was moved from one judge to another multiple time. Each judge in the panel had to read all case details and contents and add the opinion. After that, the case will be moved to the next judge in the panel, with a similar procedure, then finally to the head of the panel. This movement of cases was taking significant time, as every judge in order to read case documents and details, was taking a case file to home with days for revision. While now, the case file is accessible for all panel members, they can review the case and add their opinion simultaneously, without moving case file."

Electronic notification is another key to improving internal daily operation. Announcement of each action regarding the case update is sent instantly to all related parties. As an example, when a case is appealed, instant notification sent to the appellate court that new case is coming in pending status to be registered. Moreover, in courts which are managed by a panel of judges, when each judge adds an opinion, other judges in the panel are notified directly. Another example which already enhanced the case management in terms of notification is the summons delivery. When a summons is delivered, at summons department, and clerks mark it as delivered, notification sent to judges about delivery. In the paper system, many hearings were postponed due to an undelivered summons, which resulted in the non-attendance of participants. Moreover, when a case is referred from investigative courts, notification of referral is sent to prosecutors who are connected to competent courts.

In respect of that, judicial investigator[8] explained the importance of electronic notification by saying: "*Time is an essential factor in the investigation process. The system has saved a considerable amount of time in our journey of investigation. As we get instant updates with electronic notifications about orders by judges along with police activities, and all exchanged data and information.*"

3.1.2 Better Overview of Statistics and Flexible Information Retrieval

The system gives better and more accurate statistics. In a paper system, each court has to count court cases manually and fill statistics information into a paper form, monthly and yearly. Currently, the system gives an accurate and instant case statistics for any required period, and This has resulted in saving time and eliminating errors. Moreover, each register provided with a filter section. Information retrieval can be achieved by a case number or other metadata in the case such as case status, judge name, case types, and many more; this capability is not possible with paper systems.

In support of that, a judge (see footnote 3) said: "*Statistics generally is now faster than before, in a way that is not comparable with a paper system. With more quality and accurate result.*"

[8]Judicial investigator at the police station in the Sulaimaniyah Appeal Court.

3.1.3 Case Distribution

The e-court system implements an automatic *case distribution*. In the paper-based system, case distribution was made on a paper manually. As an example, there are six primary courts, when new cases come, they should be equally distributed over judges. The head judge at primary court (1) is usually responsible for that distribution. He makes a table with handwritten information on how to keep track of equal distribution. Currently, this functionality is controlled by the system. Automatically judges are selected with keeping balances of the distribution. Hence, the time used by a judge for manual distribution can be used for other judiciary tasks.

In this respect, a judge (see footnote 4) supports that: "*In paper system one of the main problems was case distribution among judges, which was not fair and there was a room for participant and lawyers to play with judge selection, in order to get the desired judge. While currently, the system functions automatic distribution equally, systematically, and consequently, the process more transparent.*"

3.2 Security

Electronic systems provide a higher level of security of information than paper systems, due to different levels of authentication and authorization, in addition to functionality of user management rights according to groups and individuals. Restricting access to confidential cases to authorized users—even if from the same court—improves court case security. According to observations by authors, another layer of security provided for viewing documents by case participants and lawyers who are connected to the case, the court will decide on the viewing permission. Moreover, the system also provides an audit trail of every action. Getting the record of log about case history is challenging to maintain with a paper document. Moreover, digital systems have an easy way to backup documents for offsite storage providing safe archives and disaster recovery strategy.

A judge (see footnote 3) supports that by saying: "*More security is achieved through e-court system in two aspects. First; security of case information, only authorized users having access to confidential cases. Second; security of documents from losing and damaging.*"

Moreover, from the IT management perspective, the head of the IT department of the Sulaimaniyah Appellate Court added: "*Securing court cases is one of the essential aspects of justice, even with a conventional system. The current system provides a high level of protection from both sides, software, and hardware. Furthermore, the log of actions is a critical source can be used as a piece of evidence for looking at who and when made any changes in the case. The system provides a set of different roles to be used for each user according to what permissions they have in that court, which gives additional security for accessing case data.*"

A lawyer[9] from the court emphasized the security improvements: "*Security in the system is very strong, as every user accesses through a personal credential with user name and password. Moreover, providing a log section at the end of each case gives more control over the security of the case data, by keeping a record who has accessed and made.*"

3.3 Concurrent Access to the Judiciary

E-court systems extend the concurrent access to the judiciary. Current e-court system gives the possibility to be accessed from outside the courthouse with VPN (Virtual Private Network). On the other hand, this allows judges to not carry out bulks of case files for revision at home.

In support of that, a judge (see footnote 4) stated: "*I, personally do most of the tasks outside the court. This is facilitating judges work. For instance, I have used the system even outside Sulaimaniyah, while I was traveling I still had access to my cases and proceeded some tasks and revised case documents as I usually do inside the courthouse in my room. That makes processing work can be done even during our vacation if necessary.*"

Furthermore, another judge (see footnote 5) added: "*Due to the system, I am not taking back old and dirty case files anymore as before. Currently, I can do all my work remotely from home. No need to move paper files between home and court. Now, everything is electronic and remote access to our work is a significant improvement.*"

3.4 Transparency

The e-court system allows case participants, lawyers, and all other case users possibility to track the status of their case online, and view published hearings and decisions by court authorities, this feature increases transparency. Hence, it can result in fighting corruptions.

On the other hand, the automatic allocation equally distributes cases, which can be publicly seen by all users. That does not leave room for lawyers to play with cases by selecting the desired judges, and all procedure is managed automatically and transparently by the system.

Observant noticed that the system fosters transparency in court case proceeding and justice delivery in general.

All respondents confirmed that transparency with e-court is more visible than with paper system, and some responses have described the benefits of this improvement as the following: "*Transparency is an advantage of the system, specifically in case*

[9]Lawyer at the Sulaimaniyah Appellate Court.

distribution and hearings now become more transparent than before. Transparency delivers justice in the court process and gains public trust toward courts."

Moreover, a judge (see footnote 4) added: "*Transparency is essential in courts, specifically in civil courts hearings are obliged to be held publicly and transparently, while for the criminal court there are some situations which investigation process should be kept secret. E-court has improved transparency in a way that case participants are allowed to access their cases from the public portal and track the progress of their cases.*"

Furthermore, another judge[10] supported that by stating: "*Transparency is a crucial point in our work. All none confidential cases has to be maintained transparently. As an example, investigative case detail view provides updated information about accused status such as; (arrested, escaped, disappeared, not detained, on bail, and deferred fate), which is seen by all involved parties in the investigation cycle, makes following up process much more manageable than in the paper system.*"

4 Related Work

Xu [1] states that e-court systems organize and manage collaboration activities executed by different court institutions and other parties related with the court, in order to make court procedures transparent, faster and efficient; in particular, working from outside the office results in "on-call" and "non-stop" judges. With respect to *concurrent access* to the judiciary, they state that the cooperation in information sharing results in time-saving and more efficient collaboration. Furthermore, they analyze that accessibility of case information allows judges to be more convenient and helps to achieve a higher level of transparency.

In [2], Richard Slowes analyzes, concerning *document template management*, that auto-generated templates help clerks in saving time and decrease the risk of error. Furthermore, they find that *electronic notifications* help in building trust, confidentiality, and transparency [2].

Rungruangpattana and Achalakul [3] note that providing electronic data to the general public is considered as delivering better service by the involved stakeholders. In particular, performing an electronic search is found to be a better service for both court personnel and the public. Also in [3], they deal with the improvement of security by e-court systems. In [4], they explain, how automatic *case registration* can improve the service delivery; i.e., once a case is registered, the system automatically shares the case information with related users. Also, in [4], they analyze the advantages of an improved *user management*, i.e., the categorization of user groups allows for better-organized management of the case workflow.

Singh et al. [5] find that e-court systems help to transfer paper-based court processes into a systematic digital process, in addition to speeding up the process of the judiciary and enhancing transparency and cost-effectiveness.

[10] Judge at the investigative courts in the Sulaimaniyah Appellate Court.

In [6], Ursula Gorham finds that simultaneous access to case information yields to a significant improvement of efficiency. Also, she states that greater security is achieved through more reliable electronic backup copies of court documents.

Luzuriaga and Cechich [7] describe how *electronic notification* saves time and cost, and fosters faster publishing.

Bueno et al. [8] identify essential advantages of e-court systems. E-court systems are accessible 24/7 and provide users authorized remote access. Hence, e-court systems make tasks more efficient. The opportunity to track cases from initiation to the disposal phase enhances transparency and the trust of citizens toward the justice system. Automatic case distribution is an efficient solution, as it and guarantees a homogeneous case distribution over judges.

Mandal et al. [9] describe challenges in the implementation of court case systems, e.g., in measuring case similarities.

Hasan et al. [10] find that managing court cases through an e-court system is more secure and efficient, in particular, due to of reuse of data that has already been entered to the system such as case numbers, participants of proceedings, case details.

Hamin et al. [11] describe how e-court systems enable more efficient daily routines. As an example, *online hearing monitors* help to decrease the waiting times.

5 Conclusion

This paper reflects on the implementation of the e-court system in Sulaimaniyah Appellate Court in the Kurdistan region of Iraq. E-court systems currently play an essential role in courts all around the world. They help to deliver justice services in a more organized and systematic manner. Moreover, they enhance the quality of court services and provide the public with an opportunity to access court services concurrently online. This paper presented an overview of the e-court system in the Sulaimaniyah Appellate Court. The study is based on analyzing expert interviews who are direct users of the system in addition to the inside courthouse observations. The findings show significant improvements in the areas of (i) internal daily operations by judges, prosecutors, lawyers, and clerks at different courts and different levels of roles; (ii) security of court cases; (iii) extended access to the judiciary; and (iv) transparency.

References

1. Xu AL (2017) Chinese judicial justice on the cloud: a future call or a Pandoras box? An analysis of the intelligent court system of China. Inf Commun Technol Law 26(1)

2. Slowes R (2012) Benefits of a modern court case management system. Technical Report L-379393/9-12, Thomson Reuters
3. Rungruangpattana P, Achalakul T (2007) The design framework of the civil court case management system in Thailand. In: IEEE region 10 annual international conference, proceedings/TENCON
4. Rungruangpattana P, Achalakul T (2008) The software prototype of civil court case management in Thailand. In: Proceedings of the 2008 advanced software engineering and its applications, ASEA 2008, pp 221–225
5. Singh M, Nehru M, Dwivedi Y, Rana N, Tamilmani K (2018) Success factors for e-court implementation at Allahabad high-court success factors for e-court implementation at Allahabad high-court. In: Proceedings of PACIS'2018–the 22nd Pacific-Asia conference on information systems, association for information systems
6. Gorham U (2012) State courts, e-filing, and diffusion of innovation: a proposed framework of analysis. In: Proceedings of the 13th annual international conference on digital government research, dg.o'12. ACM, New York, USA, pp 232–239
7. Luzuriaga JM, Cechich A (2011) Electronic notification of court documents: a case study. In: Proceedings of the 5th international conference on theory and practice of electronic governance, ICEGOV'11. ACM, New York, USA, pp 45–50
8. Bueno TCD, Bortolon A, Hoeschl HC, Mattos ES, Santos CS, Theiss I (2003) Using rbc to classify judicial petitions on e-court. In: Proceedings of the 9th international conference on artificial intelligence and law, ICAIL'03. ACM, New York, USA, pp 83–84
9. Mandal A, Chaki R, Saha S, Ghosh K, Pal A, Ghosh S (2017) Measuring similarity among legal court case documents. In: Proceedings of the 10th annual ACM India compute conference, Compute'17. ACM, New York, USA, pp 1–9
10. Hassan KH, Yusoff SSA, Mokhtar MF, Khalid KAT (2016) The use of technology in the transformation of business dispute resolution. Eur J Law Econ 42(2):369–381
11. Hamin Z, Othman MB, Mohamad AM (2012) Benefits and achievements of ICT adoption by the high courts of Malaysia. In: SHUSER 2012-2012 IEEE symposium on humanities, science and engineering research, vol 2, no 100, pp 1233–1238

Bilingual Dictionary for Sanskrit—Gujarati MT Implementation

Jaideepsinh Raulji and Jatinderkumar R. Saini

Abstract Working with cross-linguistic environments where lexico-semantic features are vital, the use of digitized bilingual dictionary cannot be overlooked. Here, Sanskrit-Gujarati bilingual dictionary design, contents and its applicability is discussed. The Sanskrit-Gujarati lemmas are correspondingly mapped so as to facilitate use-cases like machine translation, cross-lingual information retrieval, stemming, lemmatization, and other related task. The dictionary design and implementation is through Comma Separated Verbose (CSV) format and Relational Database Management System (RDBMS), but also convertible to formatted tag-based form for better portability. It is usual to have bilingual dictionary for scarce resourced languages prepared manually as opposed to automated and aligned bilingual corpora method for several Natural Language Processing (NLP) related task.

Keywords Bilingual dictionary · WordNet · Lemma · Semantic · Ontology

1 Introduction

The art of creation and compilation of dictionary is known as Lexicography in the world of linguistics. The idea behind choosing a language pair is realizing the richness of Sanskrit and spread of Gujarati language. Bilingual dictionaries are a pathway to learn and compare the semantic nature of the language for humans as well as machines. In Natural Language Processing (NLP) world it plays a vital role in Cross-Lingual Information Retrieval (CLIR), Machine Translation, and related task. Dictionary, not only serve the purpose of semantics but also gives knowledge of basic grammatical features. Normally, dictionary contains meaning, pronunciation,

J. Raulji (✉)
Ahmedabad University, Ahmedabad, India
e-mail: jaideepraulji@gmail.com

J. Raulji · J. R. Saini
Dr. Babasaheb, Ambedkar Open University, Ahmedabad, India
e-mail: saini_expert@yahoo.com

J. R. Saini
SICSR, Pune, India

S. Fong et al. (eds.), *ICT Analysis and Applications*, Lecture Notes in Networks and Systems 93, https://doi.org/10.1007/978-981-15-0630-7_45

and etymology of a word, but manipulated and modified to suit the needs of NLP task. Analogous to dictionary, wordnets are significant source for NLP-based implementation as they are rich in synonym set and ontological feature coverage. Bilingual dictionaries preferably can be built manually by typing from a printed or similar digital resources. Bilingual parallel aligned corpora also serves great purpose to build such artifact if available. For a resource scarce language, pivot-based approach is most suitable where word of language A is translated to pivot B using bilingual dictionary, similarly B is translated to C. Such implementation leads to noisy data, but can be improvised by intervention of linguistic expert.

2 Related Work

For bilingual dictionary construction constraint approach to pivot-based induction method and framework is implemented by Wushouer et al. [1]. The method is useful for resource scarce languages.

Istvan and Shoichi [2] also proposed a method to build fully automatic pivot-based bilingual dictionary for low resource languages. To maximize recall parameter they used bidirectional selection method with flexible threshold value with a usage of WordNet showing accuracy of 79%. They experimented on Japanese–Hungarian language with English as the pivot language.

In 1995, CE Monier-William's (MW) Sanskrit to English dictionary was digitized under the project—The Cologne Digital Sanskrit Lexicon [3]. The MW dictionary originally was written in transliterated form hence the use of OCR was possible for its digital transformation with a size of 17 megabytes [4]. The representation of text followed three level framework L1 represented Devanagari script with Harvard-Kyoto, ASCII transliteration encodings as level L2, and L3 as sublevel.

Apte's and MacDonell's Sanskrit to English dictionary were digitized under project—Digital Dictionaries of South Asia at University of Chicago [3]. An electronic version of Sanskrit Heritage Sanskrit-French dictionary was developed by Heut [2] in 2000. The dictionary is designed to serve purposes such as lexical database for morphology generation, human readable hypertext encyclopedia, syntax analysis, etc [5].

Dubey et al. [6] developed bilingual English–Hindi dictionary using Wikipedia titles, info boxes, and categories. A comparable corpus was used with pre-created small dictionary, associating words and transliteration.

Sanskrit wordnet is built using Hindi wordnet applying an expansion approach at IIT Mumbai by Kulkarni et al. [7]. Based on the Navya-Nyaya philosophy and Paninian grammar, Sanskrit wordnet is also built by Mohanty et al. [8]. Similarly as a part of IndoWordnet project at IIT Mumbai, Gujarati wordnet with expansion approach using Hindi wordnet is developed by Bhensdadia [9].

3 Approach and Implementation

Looking at the methods used to develop bilingual dictionary as mentioned in the previous section, it can be inferred to be categorized into two ways.

(a) Manually dictionaries can be developed by digitizing from several printed media, if bilingual pair is available. For resource poor languages, a pivot-based method can be implemented for the creation, where language A words are transformed to B, and B to C, where B is pivot. Here alignment, spelling errors, and transliteration factors must be taken care of during pivot transformation otherwise it affects accuracy. Moreover, manual creation of dictionary requires steps to gather information, creation of format for entry, adding information, and verification.

(b) Bilingual dictionary can be developed by processing on parallel aligned corpora, which is considered faster and cheaper. Such automatic methods which are corpus based require steps like preprocessing, alignment, extraction, and filtering. The accuracy solely depends on quality of corpus, alignment, and efficient algorithm.

Here manual approach is used for the development of Sanskrit–Gujarati bilingual dictionary due to unavailability of aligned parallel corpus. To the best of our knowledge there is unavailability of Sanskrit to Gujarati dictionary in digitized form. The printed form of such dictionary is available with description of Sanskrit word in Gujarati language despite synonyms. Moreover such formation does not suffice much for Natural Language Processing (NLP) task like Cross-Lingual Information Retrieval, Machine Translation etc. Hence, digitization by referring few printed Sanskrit–Gujarati dictionary and additionally injecting synonym sets and few grammatical features is planned and implemented. The core dictionary referred was Bruhat Kosh by Naik [10].

The Design of Dictionary

There is no distinction in Sanskrit and Gujarati between homophones and homographs, since the rich orthographic forms reflect phonetics exactly. For etymologically sibling pair languages like Sanskrit and Gujarati, development of dictionary becomes straightforward and faster, but the complexity of orthographical features like richness in diacritic marks in both the languages poses hindrances in typing fluency and fear of spelling errors which was verified and corrected by linguist expert.

The common and widely used text encoding for multilingual digital data is UTF-8. Inputting data for roman script like English is easier due to wide proliferation and usage of QWERTY keyboard, but for non-roman scripts it would be a steep learning curve in terms of typing. But, thanks to Google typing tools which also allowed typed input data through phonetic and inscript methods. The phonetics mode allowed Sanskrit and Gujarati word generation by typing English spellings for non-roman phonetics with IntelliSense and suggestions in Google typing tools which made the task somewhat simpler. For seamless portability across several data transformation format, the dictionary development started in Comma Separated Format (CSV). Later, it was transformed to Relational Database Management System to improve data retrieval (searching), storage, and secured environment, which would facilitate for several application programming interface. The Indian Language Transliteration's

(ITRANS) transliteration remarks for each Sanskrit and Gujarati word is also planned and soon be completed. The record entry is segregated into three broad grammatical categories, viz., indeclinables, pronouns, and remaining POS category.

The POS category consists of words with noun, adjective, and verb category whose design is depicted in Table 1. The Sanskrit head word entry is lemma in sanword1 as per table-1, there can be two others entries of same lemma in a tuple (synonym entry for Sanskrit), viz., sanword2, sanword3. If more than two entries are found, it is added as a separate tuple. The pos1 contains only surface POS information as noun, adjective, and verb, while pos2 contains extra POS information if available. The column gujword1 as described in Table 1 contains description or explanation of the Sanskrit head word in Gujarati, while gujword2 to gujword5 column contains Gujarati lemma synonym. The sample data is shown in Table 7.

Indeclinables are words in Sanskrit which are not inflected by gender, number, and case grammatical characteristics. They are added separately so as to reduce search space. The design format to store the same is depicted in Table 2 and sample data is depicted in Table 5.

Table 1 Design of bilingual dictionary

Name of column	Remarks
Id	To identify word uniquely
sanword1	Sanskrit word
sanword2	Sansrkit word (2) which is synonym of sanword1
sanword3	Sanskrit word (3) which is synonym of sanword1
pos1	Part of Speech tag for the word
pos2	Part of Speech tag for the word (if alternative of pos1)
gujword1	Meaning of Sanskrit word, in description form
gujword2	Meaning (Synonym) of sanword1
gujword3	Meaning (Synonym) of sanword1 (Additional Synonym or near word if available)
gujword4	Meaning (Synonym) of sanword1 (Additional Synonym or near word if available)
gujword5	Meaning (Synonym) of sanword1 (Additional Synonym or near word if available)

Table 2 Format to store indeclinables

Name of column	Remarks
id_dec	Unique Identification integer value for record entry
Indec_word	Sanskrit Indeclinable word
Guj_word	Equivalent Gujarati word synonym for Indec_word

Table 3 Format to store pronouns

Name of column	Remarks
id_pro	Unique Identification integer value for record entry
San_pro_word	Sanskrit Pronoun
Guj_word	Equivalent Gujarati word synonym for San_pro_word
Gender	A gender value (M-Masculine, F-Feminine, N-Neuter)
SDP	A number characteristic for Sanskrit pronoun (S-Singular, D-Dual, P-Plural)
Case	Case marking for the Sanskrit pronoun is an Integer value (1-Nominative, 2-Accusative, 3-Instrumental, 4-Dative, 5-Ablative, 6-Genitive, 7-Locative)

Pronouns are words in Sanskrit which are alternatively used for proper nouns analogous to English language. The inflection pattern based on gender, number, and case markings are drastically complex to identify the part of speech for Sanskrit pronoun simply by pattern matcher algorithm, hence pronouns are separately added to dictionary with Gender-Number-Case (GNC) markings. The design format for storing pronouns is depicted in Table 3 and sample data is shown in Table 6.

Approximately the dictionary stores 29,000 words of which 26,350 are nouns, 2,300 are verbs, 200 are pronouns, and 150 indeclinables. Currently the list is updated manually, but planned to automate the process at a later stage.

4 Results

Several Sanskrit data was collected from Internet sources, segregating data into twelve documents based on availability from web and manually typed from printed resources. Through word matching algorithm implementation, the dictionary was tested against Sanskrit corpus of approximately 7157 words. The results of the experiment is listed in Table 4. The accuracy is calculated as ratio of matched and non-matched words resulting to accuracy of 91.99%. Hence from 7157 words, 628 words were not found in dictionary. Approximately 65% of the not found words were recognized from named entity category and the remaining word match were unsuccessful due to invalid spelling and coalesced words (Sandhi). The named entity example words are काश्मीर (kAshmIra), नेहरूः (neharUH), इलाहाबाद (ilAhAbAda), गुजरात (gujarAta), राजेन्द्रप्रसादः (rAjendraprasAdaH), etc.

5 Conclusion

Manual method of preparing a dictionary is a tedious, costly, and time-consuming job. Moreover, tracking updates due to new words coming up in the language periodically is difficult. Hence, if aligned bilingual corpora for the language pair is available

Table 4 Dictionary search results through automated algorithm

Sr. No.	Document name	Count of total words in corpus	Count of words not found in dictionary	% hit count
1	Text corpus 1	895	84	90.61
2	Text corpus 2	179	38	78.77
3	Text corpus 3	163	15	90.80
4	Text corpus 4	122	23	81.15
5	Text corpus 5	129	9	93.02
6	Text corpus 6	3911	387	90.10
7	Text corpus 7	221	1	99.55
8	Text corpus 8	142	6	95.77
9	Text corpus 9	146	2	98.63
10	Text corpus 10	479	53	88.94
11	Text corpus 11	245	7	97.14
12	Text corpus 12	525	3	99.43
Total		**7157**	**628**	**91.99**

from different domains, then the development of high-precision mapping algorithm would suffice the task of automated dictionary generation. The unavailability of such bilingual corpus forced us for the rationale approach of preparing dictionary. The words which were not found in dictionary can be added to the dictionary manually or an automated algorithm can be designed to add the same but still its description and synonym needs to be added manually. The advantage of the manual method leads to a better accuracy than the automated one.

Appendices

See Tables 5, 6 and 7.

Table 5 Sample table for Indeclinable entries

id_dec	indec_word	gujword
1	अतः	તેથી
2	अत्र	અહીં
3	अथ	પરંતુ
4	अपि	પણ
5	अयं	આ

Table 6 Sample table for pronoun entries [gender(m-masculine, f-feminine, n-neuter, 0-Not Available), sdp(s-singular, d-dual, p-plural), case(1-Nominative, 2-Accusative, 3-Instrumental, 4-Dative, 5-Ablative, 6-Genitive, 7-Locative)]

id_pro	sanword	gujword	gender	sdp	case
1	अयं	આ	m	s	1
2	तस्य	તેના	m	s	6
3	ते	તેઓ	-0	s	4
4	तेन	ત્યાં	m	s	3
5	तौ	તેઓ	m	d	2

Table 7 Sample data for all POS entries

id	San word1	San word2	pos1	pos2	Guj word1	Guj Word2	Guj Word3	Guj Word4
1	अक्ष्		क्री		પહોચી જવું	પહોંચ	વ્યાપ	
2	अती		क्री		ઉલંઘન કરવું	ઉલ્લંઘન	ઓળંગ	
3	अद्		क्री		ખાઈ જવું	ખા	ખાવ	નાશ
4	जगदीश		पुं		જગતનો સ્વામી	જગદીશ	વિષ્ણું	
5	अचिन्त		क्री		વિચારવું તે	વિચાર		

References

1. Wushouer M, Lin D, Ishida T, Hirayama K (2016) A constraint approach to pivot-based bilingual dictionary induction. ACM Trans Asian Low-Res Lang Inf Process 15(1):4
2. István V, Shoichi Y (2009) Bilingual dictionary generation for low-resourced language pairs. In: Proceedings of the 2009 conference on empirical methods in natural language processing, vol. 2-2, 6 Aug 2009. Association for computational linguistics, pp 862–870
3. Goyal P, Huet G, Kulkarni A, Scharf P, Bunker R (2012) A distributed platform for Sanskrit processing. In: Proceedings of COLING 2012, pp 1011–1028
4. Kapp DB, Malten T (1997) Report on the cologne Sanskrit dictionary project. In: 10th international Sanskrit conference, Bangalore, 3 Jan 1997
5. Huet G (2004) Design of a lexical database for Sanskrit. In: Proceedings of the workshop on enhancing and using electronic dictionaries, 29 Aug 2004. Association for computational linguistics, pp 8–14
6. Dubey A, Varma V (2013) Generation of bilingual dictionaries using structural properties. Computacion y Sistemas 17(2): 161–168. ISSN 1405-5546

7. Kulkarni M, Dangarikar C, Kulkarni I, Nanda A, Bhattacharyya P (2010) Introducing Sanskrit wordnet. In: Proceedings on the 5th global wordnet conference (GWC 2010), Narosa, Mumbai, Jan 2010, pp 287–294
8. Mohanty S, Nayak SN (2002) Making of Sanskrit wordnet. In: Proceedings of the international conference on universal knowledge and language Nov 2002, pp 25–29
9. Bhensdadia CK, Bhatt B, Bhattacharyya P (2010) Introduction to Gujarati wordnet. In: Third national workshop on indowordnet proceedings
10. Naik R (2012) Bruhat kosh. Akshara Prakashan, Ahmedabad

Eye Gaze Controlled Head-up Display

Aparna Ramakrishnan, Modiksha Madan, Gowdham Prabhakar, Sachin Deshmukh and Pradipta Biswas

Abstract A myriad of infotainment systems has found its applications in the automobile industry with the burgeoning demand for user comfort and interaction. However, operating such infotainment systems entertain secondary tasks to be carried out at the expense of the primary task of driving. This can increase the cognitive load on the driver and has the potential to keep road safety at stake. This paper presents an intelligent interactive head-up display (HUD) on the windscreen of the driver that does not require them to take eyes off the road while undertaking secondary tasks like playing music, operating vent controls, watching navigation map, and so on. The interactive HUD allows the user to navigate and make selections using eye gaze. The HUD also incorporates provisions to estimate driver's cognitive load and distraction level. User studies show that the system improves driving performance in terms of mean deviation from lane in an ISO 26022 lane changing task compared to touch screen system and participants can undertake ISO 9241 pointing tasks in less than 2 s on average inside a car.

Keywords Automotive UI · Multimodal interaction · Eye gaze controlled system · Virtual touch · Head-up display

1 Introduction

Human–machine interfaces have made incredible headways in recent years, especially in enhancing user experiences while considering human-factor principles and design enhancements. The applications can be found in various domains such as the development of assistive technologies, gaming sectors, automotive industries, etc. In automotive environment, the infotainment systems in use have core stack electronic

A. Ramakrishnan · G. Prabhakar · P. Biswas (✉)
Indian Institute of Science, Bangalore, Karnataka, India
e-mail: pradipta@iisc.ac.in

M. Madan · S. Deshmukh
Faurecia India Pvt Ltd, Pune, India
e-mail: sachin.deshmukh@faurecia.com

S. Fong et al. (eds.), *ICT Analysis and Applications*, Lecture Notes in Networks and Systems 93, https://doi.org/10.1007/978-981-15-0630-7_46

screens or head-up displays which use voice, gesture, or touch as input modalities. More modalities of interactions have gained research interest such as haptic feedback [1], hand gesture interactions [2], customization of instrument displays, predictive models, and personalization of user interface [3, 4]. In automotive applications, the HMIs should be designed in such a way as to provide ease of access and usability to the user. This is very important when the user is the driver himself. Interactive displays should be designed in a way to eliminate an additional cognitive load on the driver which can be brought by his distraction on using the display. In order to mitigate this effect on visual search, Air Gesture system [5] and BullsEye system [6] was introduced. However, this required the user to remember a set of gestures or relative screen item positions. Additionally, in terms of efficiency, touch screen interactions proved to be faster than gesture interactions. This invention focuses on creating a direct manipulation interface by leveraging visual search and eye gaze fixations as a means of controlling electronic displays inside a vehicle.

Introduction of eye-tracking technology has enabled research to explore eye gaze tracking as a modality of interaction. Eye gaze tracking involves the measurement of the eye movement and gaze positions on a screen. However, it is not until recently that eye gaze trackers have entered the market and has been used for analyzing ocular parameters for reading and a variety of human–machine interaction (HMI) tasks. In an automotive environment, a head-up display (HUD) finds better application than conventional displays as it reduces the cognitive load of the driver by eliminating the need to look away from the road. An eye gaze controlled HUD (i) can be placed away from the driver, (ii) can be operated without taking their hands off the steering wheel, and (iii) can be used for people who have restricted abilities of motion. A patent is already filed on the proposed work in the Indian Patent Office with application number 201941009219.

2 Related Work

The eye gaze tracking has been an area of research interest from the eighteenth century. In 1908, the first eye-tracking device was developed by Edmund Huey which could track eye movement during reading process. Since the invention, various research has gone into this domain to increase its efficiency and to bring it to the market level for various applications. The MAGIC system [7] invented in 1999 reduced physical effort and fatigue caused by independent eye gaze pointing and mouse control tasks. This was done by using eye gaze tracking to move the mouse pointer. When encountered with multiple monitor displays, the head of the user must move in a wide range which disrupts eye tracking. Ashdown and colleagues addressed this issue in their paper published in 2005 [8] by incorporating eye tracking and head tracking. The Sideways system [9] introduced a method to allow the users to walk to a display and interact with them without the need of prior calibration. The hMouse [10] and CameraMouse [11] moved the pointer based on head movement. In Fejtova's [12] Magic Key system, the selection was done by eye blinks. Eye gaze controlled interfaces are well investigated for people with severe motor impairment. Penkar [13] and colleagues investigated dwell time settings called Actigaze for the mouse selection for purely gaze-based interactions. Such a dwell time-based selection cannot be implemented in an automotive environment as it requires a tradeoff between performance and error rates and must take driver's safety into concern. A dwell selection system can cause false positives and can trigger wrong selections. Biswas [14] combined gaze-controlled system with single switch scanning system. Mondragon [15] and Poitschke [16] investigated gaze-controlled system in automotive environment but for head-down displays. Our proposed work uses eye gaze for cursor movement in a head-up display (HUD) while selection is done using a wireless switch cascaded with the steering.

3 Problem Statement

The existing head-down displays at center stack entail the need for the drivers to take their eyes off the road while operating them. This could distract the driver and can also pave way as a potential threat to their safety on road. A voice recognition system poses several constraints such as low efficiency natural language processing. The proposed system shows information in front of driver and they can operate it either just by looking at the display or with small finger movements. The system also estimates the drivers' level of distraction and cognitive load [17].

4 Proposed Approach

The proposed system consists of hardware and software parts. Figure 1 shows a picture of the setup inside a car. The system can be deployed in two ways

1. Using a display which reflects the HUD to a tinted windshield (Fig. 1b).
2. A pseudo transparent display which uses video feed on a tablet with controls against it (Fig. 1a).

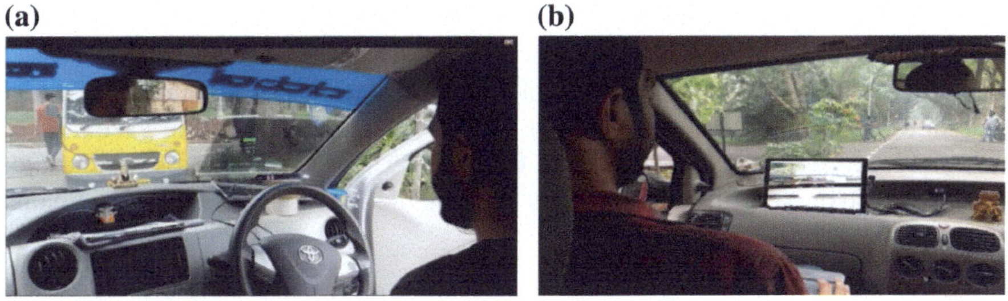

Fig. 1 **a** With reflected screen. **b** Using tablet computer

The assembly consists of two main parts: a rectangular cavity on top to keep the LCD Screen in place and a magnetic eye tracker holder with limited movement allowance; and housing to place a processing unit and other sensors like vibration or light measuring ones just below the screen. The LCD Screen is reflected in the windshield which can be operated by using eye gaze tracking.

The software was selected based on our prior research [18] on gaze-controlled interface and different multimodal fusion strategies. If the user puts his hand on the finger tracking sensor, the pointer stops moving based on eye gaze. When the user removes his hand from the top of the finger tracking sensor, the pointer resumes to move based on eye gaze of the user.

The commercially available eye trackers work on a flat-screen monitor. The eye tracker must be stuck to the monitor in order to give accurate tracking navigation. The system we use recommends placing the eye tracker far away from the windshield and closer to the driver. The curvature of the windshield also prevents the tracker in providing accurate gaze tracking. Therefore, we developed our own calibration software which can track eyes by overcoming the constraints.

We developed a socket program using the API of the eye-tracking vendor and transform the output of the raw gaze coordinates based on screen resolution and zooming level of the rendering device. An example of transformation includes multiplying normalized coordinates by fixed constant and by adding fixed offset if the screen is inverted for mirroring on HUD. Next, we developed a 9-point calibration application that displayed nine squares one at a time on the screen. Each square was of size 50×50 pixels and displayed for a duration of 2.5 s. The cursor control system is implemented in C# and initially runs the 9-points calibration program and uses this to train the neural network. When a target is activated, the target can be selected by pressing a wireless switch attached to the steering wheel.

5 User Studies

5.1 Study 1—Eye-Tracking System in Running Vehicle

Participants: We collected data from 9 participants (5 male, 4 female, average age 29.8 years).

Procedure: We used a Toyota Innova car and a Lenovo Yoga laptop with 12" screen for the study. The car was plying through a motorway while participants were asked to point on regions and make selections. We used a Tobii PCEyeMini gaze tracker with the Tobii SDK to record eye gaze locations. The pointer was moved through eye gaze while the selection was done by clicking the left mouse button. Participants

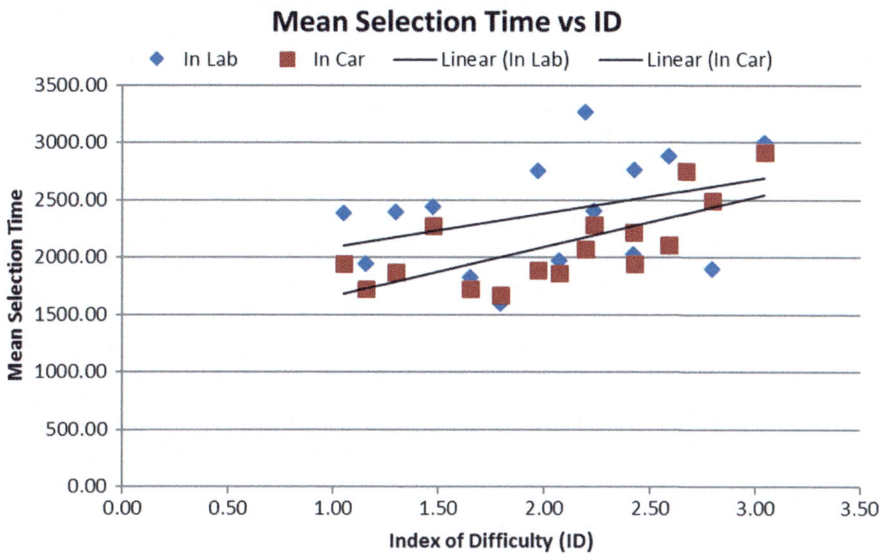

Fig. 2 Comparing selection time of targets using eye tracker inside and outside the car

were instructed to bring the pointer near the target box highlighted in white color and select the box by clicking the left mouse button.

Results: We calculated the mean selection time of each target for each participant and plotted the mean selection time vs index of difficulty (ID) in Fig. 2. We can infer from the graph that the performance of eye tracker inside the car did not deviate much from the performance of eye tracker inside the lab.

5.2 *Study 2—Eye Gaze Controlled HUD for Drivers in Static Vehicle*

Participants: We collected data from 9 participants seating at the driver's seat in the driving position. Among them, there were 6 male and 3 female participants, age range was from 23 to 35 years.

Procedure: We collected data using a standard pointing task displaying 5 targets on screen. Each target was 70 × 70 pixels in size and one of the buttons was rendered differently than other buttons. Participants were instructed first to calibrate using the neural network model to select the designated target as soon as it appears on screen using eye gaze. Figure 3 shows the relative histogram of the response time. The median of selection times was 2.1 s, the average selection time was 1.8 s and standard deviation was 1.1 s.

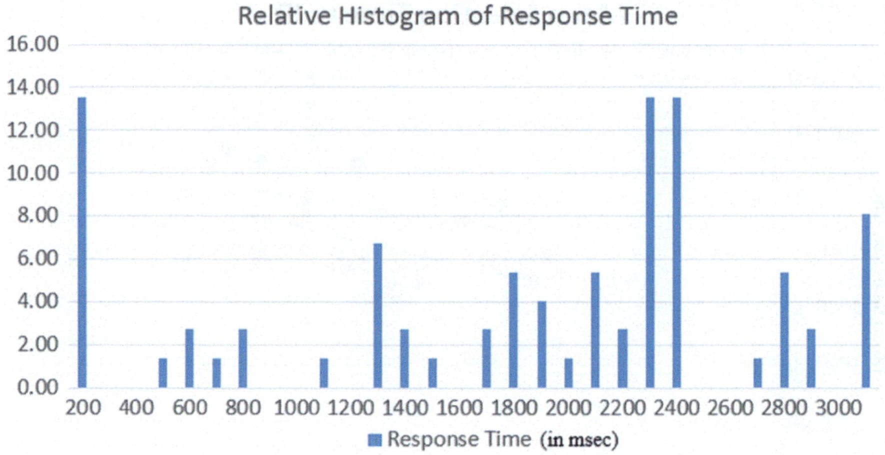

Fig. 3 Histogram of response times

It can be seen from the above results that although the pointing and selection time for both systems are not significantly different, the driving performance is improved in the proposed scheme. In the following paragraphs, we have described the design of the graphical user interface for the gaze-controlled HUD following automotive guidelines.

6 Content Design for Interactive HUD

6.1 Adaptation for Different Ambient Lighting Conditions

The difference in lighting conditions on the interface can originate from a variety of sources in the driving environment and can hinder the driver's visibility of the external environment. It can also induce discomfort and can make the driver weary. This section elaborates on ways to compromise both the reduced legibility and conspicuity of display information due to different lighting conditions of the environment. This can create glare on the display and can cause physical discomfort for the driver.

To mitigate the issues caused by varying lighting conditions, a dynamic contrast display that would adjust automatically to the lighting conditions on the interface was designed. The display would check constantly for the lighting conditions at the surrounding points of an icon and will transform the icon color to a reverse contrast color. The icons change from dark to light and vice versa according to the lighting conditions. The brightness of each icon was calculated using the brightness method in Emgu.CV library. This method gets the hue-saturation-lightness (HSL) value for the corresponding color structure. Any value above the normal lighting condition of

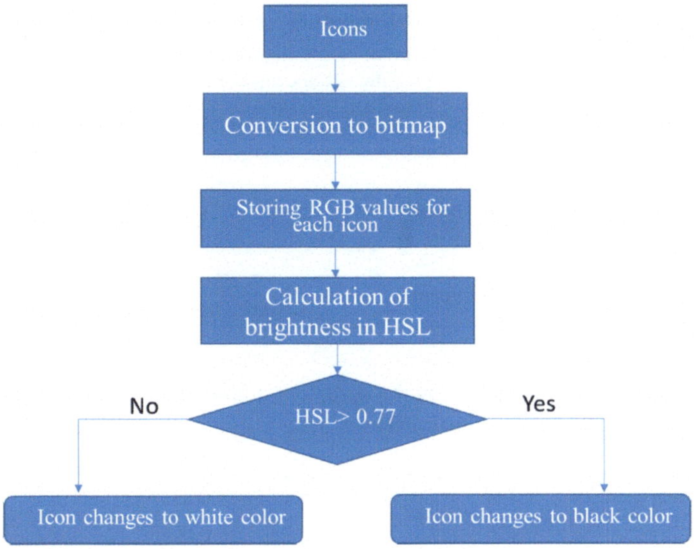

Fig. 4 Algorithm illustration for dynamic reverse contrast display

0.77 [19] renders the icon as dark and anything below this value restores the white color of the icon. Figure 4 illustrates the flowchart of the algorithm.

6.2 Icon Size

Legibility and conspicuity of the icons is vital for visibility and perception. Legibility is not confined to visibility or detection; it is also the capability of being able to discern shapes or characters based on appearance. Both graphical symbols and text which makes up an icon are vital sources in influencing the legibility of the icon. The calculations are made with reference to the existing study [20]. This design goal can be achieved by having icon sizes with optimal visual angle of primary graphical elements at 86 arcminutes (Fig. 5).

Table 1 shows the calculations involved in determining the icon size. The calculations take into consideration the specifications of the tablet used for display. The resolution of the display used is 2736 × 1824 pixels with a DPI (dots per inch) of 267 and a scaling factor of 200%. Table 1 shows the measured and calculated values of the icons.

Figure 6 shows the homepage of the final UI for the HUD by encompassing all the above calculations and considerations for icon size and ambient lighting conditions.

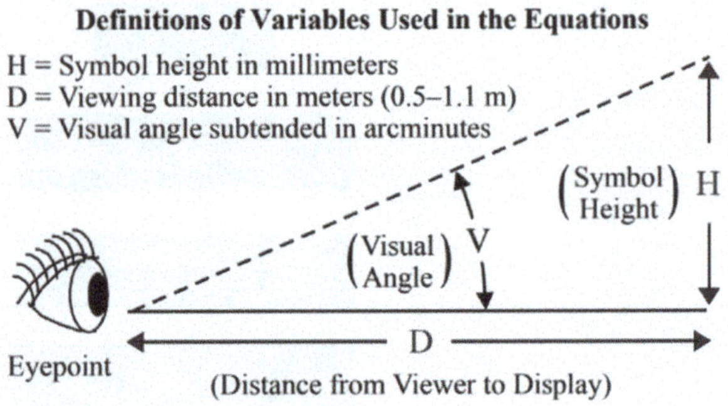

Fig. 5 Visual angle calculation (Adapted from [20])

Table 1 Calculations for icon and text size for the designed car HUD by considering the tablet and reflection resolutions and optimal visual angle

Parameter	Icon	Text
Optimal visual angle (x)	1.433°	0.333°
Tan (x)	0.025	0.0058
Symbol height (H)	D × tan (x) = 120 × 0.025 = 3 cm = 0.275 in	D × tan (x) = 120 × 0.0058 = 0.7 cm = 0.275 in
Symbol height (H)	320.4 px	72.09 px

Fig. 6 Final UI with dynamic reverse contrast display

7 Conclusion

Our proposed system does not require the driver to look off the road while operating the car dashboard and therefore reduces the cognitive load on the driver brought about by the conventional dashboard displays. The system performance was assessed using a dual-task environment incorporating a driving simulator and it is discovered that the pointing and selection times between the suggested scheme and the touch screen display are not substantially distinct, but that driving performance has been enhanced. A second study inside a Toyota Etios car found that participants can undertake pointing and selection tasks at an average time of less than 2 s.

References

1. Chang W, Hwang W, Ji YG (2011) Haptic seat interfaces for driver information and warning systems. Int J Hum Comput Interact 27(12):1119–1132. https://doi.org/10.1080/10447318.2011.555321
2. Ohn-Bar E, Trivedi M (2014) Hand gesture recognition in real-time for automotive interfaces: a multimodal vision-based approach and evaluations. IEEE Trans Intell Trans Syst
3. Feld M, Meixner G, Mahr A, Seissler M, Kalyanasundaram B (2013) Generating a personalized UI for the car: a user-adaptive rendering architecture. In: UMAP 2013, LNCS, vol 7899, pp 344–346
4. Normark CJ (2015) Design and evaluation of a touch-based personalizable in-vehicle user interface. Int J Hum Comput Interact
5. May KR, Walker BN, Gable TM (2014) A multimodal air gesture interface for in vehicle menu navigation. In: Adjunct proceedings of ACM automotive UI 2014
6. Weinberg G, Knowles A, Langer P (2014) BullsEye: an au automotive touch interface that's always on target. In: Adjunct proceedings of ACM automotive UI 2014
7. Zhai S, Morimoto C, Ihde S (1999) Manual and gaze input cascaded (MAGIC) pointing. In: ACM SIGCHI conference on human factors in computing system (CHI)
8. Ashdown M, Oka K, Sato Y (2005) Combining head tracking and mouse in-put for a GUI on multiple monitors. In: CHI Late breaking result
9. Zhang Y, Bulling A, Gellersen H (2013) SideWays: a gaze interface for spontaneous interaction with situated displays. In: ACM SIGCHI conference on human factors in computing system (CHI)
10. Fu Y, Huang TS (2007) hMouse: head tracking driven virtual computer mouse. In: IEEE workshop on applications of computer vision
11. CameraMouse. http://www.cameramouse.com. Accessed 22 Sept 2018
12. Fejtova M et al (2009) Hands-free interaction with a computer and other technologies. Univers Access Inf Soc 8
13. Lutteroth C, Penkar M, Weber G (2015) Gaze vs. Mouse: a fast and accurate gaze-only click alternative. In: Proceedings of ACM UIST 2015
14. Biswas P, Langdon P (2013) A new interaction technique involving eye gaze tracker and scanning systems. In: ACM eye tracking South Africa (ETSA)
15. Mondragon CK, Bleacher B (2015) Eye tracking control of vehicle entertainment systems, Patent No. WO2013036632
16. Poitschke T, Laquai F, Stamboliev S, Rigoll G (2011) Gaze-based interaction on multiple displays in an automotive environment. IEEE Int Conf Syst Man Cybern (SMC), 543–548. ISSN: 1062-922X
17. Biswas P, Prabhakar G (2018) Detecting drivers' cognitive load from saccadic intrusion. Trans Res Part F Traffic Psychol Behav 54 (Elsevier)
18. Biswas P (2016) Exploring the use of eye gaze controlled interfaces in automotive environments. Springer. ISBN: 978-3-319-40708-1
19. Gamgoz N (2000) Effects of hue, saturation, and brightness on attention and preference. Bilkent University
20. Campbell JL, Brown JL, Graving JS, Richard CM, Lichty MG, Bacon LP, Sanquist T (2018) Human factors design guidance for level 2 and level 3 automated driving concepts (Report No. DOT HS 812 555). National Highway Traffic Safety Administration, Washington, DC

Enhanced Gain RMSA with Parasitic Patch Coupled to Non-radiating Edges

Swathy S. Panicker and Sreedevi K. Menon

Abstract In this paper, effect of the mutual and gap coupling on patch antenna characteristics is studied in detail. The study is carried out by placing two patches that are parasitic in nature on both non-radiating sides of Rectangular Microstrip Antenna (RMSA). Gap coupling is examined by altering the distance between the parasitic patch and radiating patch, while mutual coupling can be analysed by altering the width parasitic patch. Mutual coupling and gap coupling contributes to the decrease in overall size of the antenna, increase in gain with a slight decrease in bandwidth. The observations obtained by simulation are confirmed experimentally.

Keywords Rectangular Microstrip Antenna (RMSA) · Voltage Standing Wave Ratio (VSWR) · Parasitic patch · Mutual coupling · Gap coupling

1 Introduction

In high-performance applications like satellite, aircraft, missile, spacecraft etc., where the processes of performing a task, dimensions, weight, cost are constraints, microstrip antennas [1] can be used because they are easy to implement, simple, cheap, accordance with all surfaces (planar and non-planar), low profile, small dimension (size), weight, can be printed directly on the circuit board, etc. But its use is restricted in some applications owing to its narrow bandwidth [2]. Many techniques [3] are there to enhance the bandwidth of MSA. One of these methods includes using stacked patches which enhance the height of the antenna. But when the height increases, surface waves are produced which are not wanted because they take out power from the total obtained for radiation (space waves). Polarisation [4] characteristics and antenna pattern will be lost if the surface waves move in the substrate and scatter at the discontinuities. Another method is using additional resonators which

S. S. Panicker · S. K. Menon (✉)
Department of Electronics and Communication Engineering, Amrita Vishwa Vidyapeetham, Amritapuri, India
e-mail: sreedevikmenon@am.amrita.edu

S. Fong et al. (eds.), *ICT Analysis and Applications*, Lecture Notes in Networks and Systems 93, https://doi.org/10.1007/978-981-15-0630-7_47

can be gap-coupled to any side of a rectangular shaped fed patch [5]. All these methods increase the dimensions of antenna so it cannot be used as array elements.

In enhancing the gain and bandwidth of antenna two equal parasitic patches are introduced on non-radiating edges [6] of RMSA where the dimensions of the parasitic patches, feed length, spacing in between the patch that is feeding and the patch that has parasitic nature plays a crucial role. If the patches, having parasitic nature are placed adjacent (<2 h) then gap coupling occurs. When the resonance frequencies of the two patches, which are parasitic in nature are adjacent then a wider bandwidth [7] is resulted. When the fields of fed and parasitic patches interact with each other coupling occurs. These couplings are of two types: (1) Gap capacitance brings out Gap coupling (2) Inductance of the patches brings out mutual coupling. The aim of this paper is to analyse the outcome of mutual coupling by altering the wideness of the patch with parasitic nature [8]. Inductance comes up with the wideness of the patch which is parasitic in nature. Less width of the patch results in high inductance value. This analysis in mutual coupling helps in reducing the inclusive dimension of antenna and can be used as antenna arrays. There are other techniques which are adapted for gain enhancement in the recent past [9–11]. The proposed method is much simpler than the ones referred.

2 Effects of Mutual and Gap Coupling

There are several ways to upgrade the gain and bandwidth of antenna. Some of them are using parasitic patches, using air as dielectric substrate, incorporating slots on the patch radiator, dual feeding technique, etc. Out of which incorporation of parasitic patches is easier. This parasitic patch can be coplanar and stacked [9]. Stacking increases height so for compactness coplanar techniques are preferred. In coplanar technique the excitation will be given to the main patch. The parasitic patches are electromagnetically coupled with the main patch. When the resonance frequency of feed and parasitic patches come adjacent then broad bandwidth [10] will be the outcome. This parasitic patch is placed on either radiating edge or else the non-radiating edge of patch antenna. In non-radiating edge coupled RMSA the field varies as sinusoidal but in radiating edge coupled RMSA field is varying as linear, so that the gap between the feed and parasitic should be kept small if we are using non-radiating RMSA. This is because the coupling is smaller in non-radiating edge than the radiating edge [11].

A rectangular patch antenna with microstrip line feed is designed on FR4 epoxy substrate with dielectric constant (ϵr) as 4.4, height of the substrate (h) as 1.6, $\tan\delta$ as 0.02, and operating frequency (fr) as 2.15 as in Fig. 1.

This antenna has a VSWR near to 1 at 2.15 GHz as shown in Fig. 2. In the resonance bandwidth gain of the antenna is measured and is as illustrated in Fig. 3.

Since the patch has a low gain due to the high loss tangent, two parasitic patches are added on either side of the main patch. The width (L1) of the parasitic patch

Fig. 1 Patch antenna

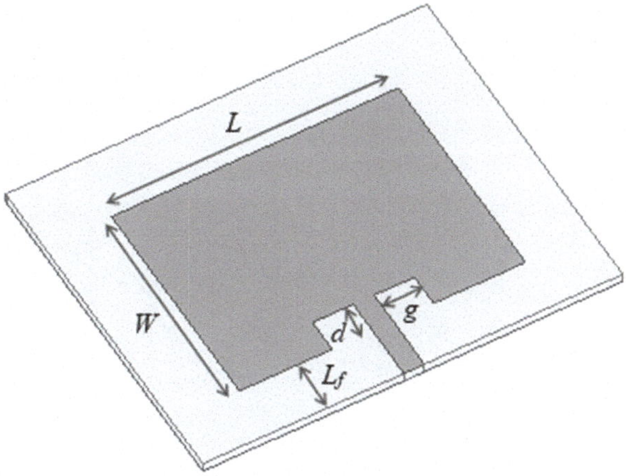

Fig. 2 Reflection characteristics of the patch antenna

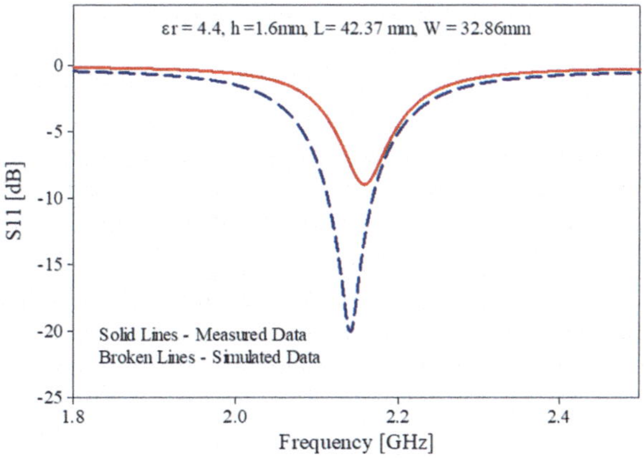

Fig. 3 Gain of the patch antenna

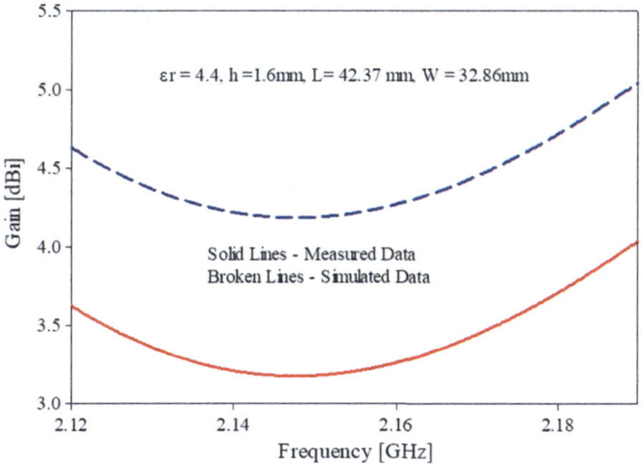

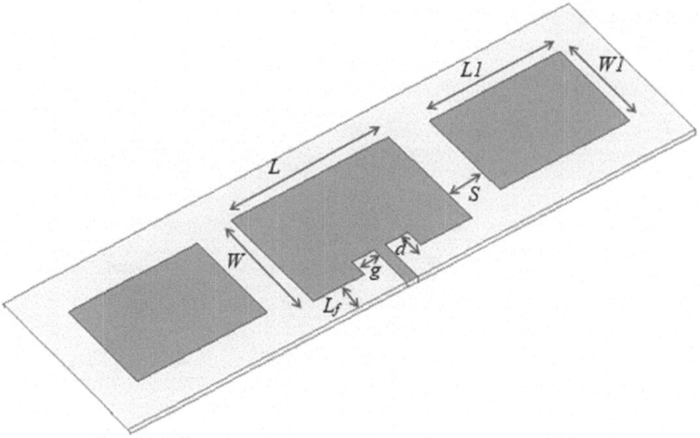

Fig. 4 Patch antenna with parasitic patches

and spacing from the radiating element (S) is optimised for better coupling and gain (Fig. 4).

The width (L1) of the parasitic patch is optimised and the corresponding resonant frequency and gain is presented in Table 1. As the width increases, the gain increases since the surface current increases. But considering compactness, L1 = 35 mm, it is taken as optimum at 2.1 GHz.

For gap coupling analysis the width of parasitic patch L1 is kept as 35 mm and the space between the parasitic and fed patch S, is varied. Coupling capacitance is reduced when the spacing in between the fed patch and parasitic patch increases and produces small bandwidth. The observed results are depicted in Table 2.

As the distance S decreases gain is enhanced. Considering fabrication easiness S = 1 mm is optimised. With the optimisation, parasitic patch of width 35 mm separated by 1 mm from the radiating patch is analysed for its reflection and radiation characteristics.

Table 1 Performance metrics of the antenna—optimisation of mutual coupling

Width L1 (mm)	Frequency (GHz)	Gain (dBi)
35	**2.16**	**4.96**
30	2.16	4.49
25	2.15	4.37
20	2.15	4.14

Table 2 Performance metrics of the antenna—optimisation of gap coupling

Gap S (mm)	Frequency (GHz)	Gain (dBi)
0.5	2.16	5.14
1	**2.16**	**4.93**
1.5	2.15	4.79
2	2.15	4.71
2.5	2.15	4.69
3	2.15	4.61
3.5	2.16	4.43
4	2.15	4.35
4.5	2.15	4.40
5	2.16	4.31

Fig. 5 Prototype of patch antenna with parasitic elements

3 Performance with Parasitic Patch in the Non-radiating Edges

With the parameters discussed in the previous section the proposed antenna as in Fig. 4 is fabricated (Fig. 5) and tested. Initially the antenna reflection is measured as shown in Fig. 6. This antenna performance is compared with the simulated results and is presented in Figs. 7, 8 and 9.

The performance of the antenna by measurement confirms good reflection and radiation characteristics. In both the principle planes cross-polarisation is better than 40 dB ensuring good polarisation purity. The gain in the entire operational bandwidth is improved by ~3 dBi when measured using two-antenna method in the anechoic chamber.

4 Results and Discussions

Inclusion of parasitic patches together with the radiating patch is found to enhance the gain of the antenna system. The width of the parasitic element contributes to the inductance and spacing contributes to the capacitance, providing effective mutual and

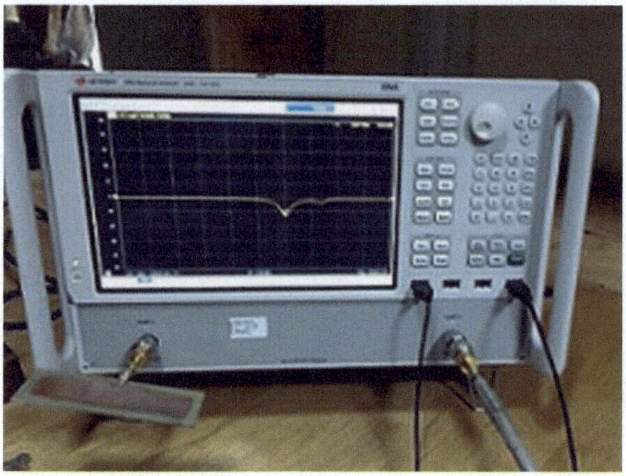

Fig. 6 Measurement of return loss of patch antenna with parasitic elements

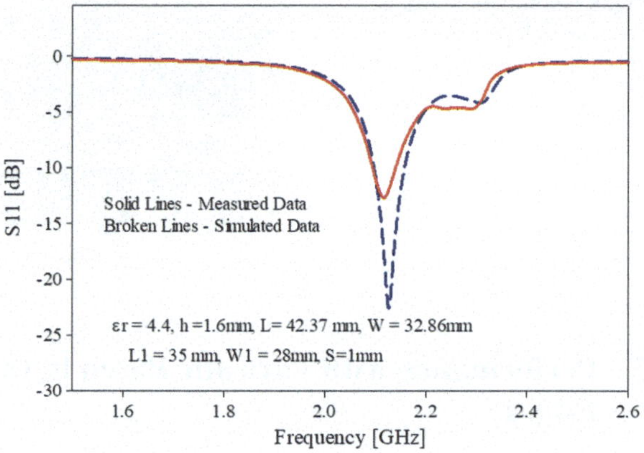

Fig. 7 Return loss of patch antenna with parasitic elements

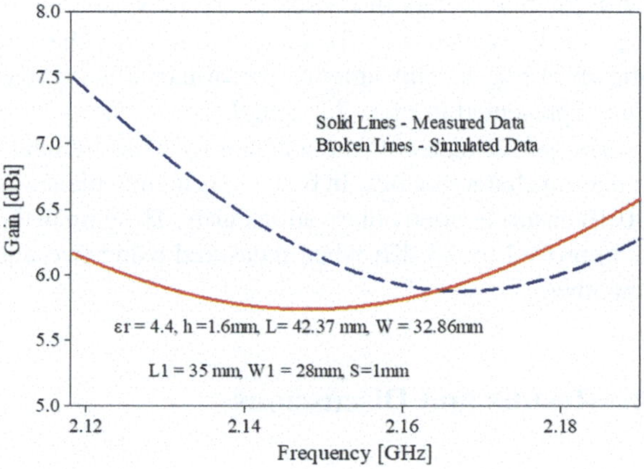

Fig. 8 Gain of patch antenna with parasitic elements

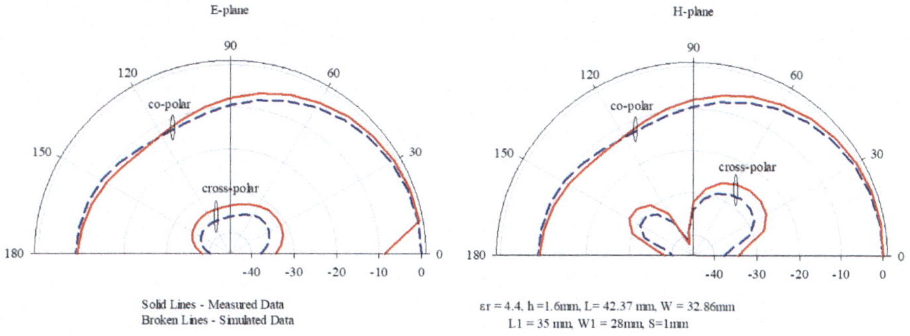

Fig. 9 Radiation pattern of patch antenna with parasitic elements

gap coupling. Optimisation of the system is carried out by HFSS and is validated using network analyser. Usually substrate with high thickness and less dielectric constant are chosen for patch antennas because they can provide enhanced bandwidth, better efficiency, etc. Since FR4 is a lossy material (performance decreases with increase in dielectric constant) gain will be less as shown in the graph of gain of the antenna. Using material of low dielectric constant and loss tangent the proposed technique will provide further enhancement in gain. The proposed solution can be implemented in antenna arrays and systems that require high gain.

References

1. Balanis CA (2016) Antenna theory: analysis and design. Wiley
2. Kumar G, Ray KP (2003) Broadband microstrip antennas. Artech House Inc
3. Serra AA, Nepa P, Manara G, Tribellini G, Cioci S (2007) A wideband dual-polarized stacked patch antenna. IEEE Antennas Wirel Propag Lett 6:141143
4. Lin S, Row J (2008) Monopolar patch antenna with dual-band and wideband operations. IEEE Trans Antennas Propag 56(3):900903
5. Ray KP, Kumar G, Lodwal HC (2003) Hybrid-coupled broadband triangular microstrip antennas. Antennas Propag 51:139–141
6. Mishra PK (2015) Broadband linear and planar microstrip antenna arrays. PhD Thesis, IIT Bombay
7. Chattopadhyay M, Mathur P, Kumar G (2011) Non-radiating edge gap coupled microstrip antennas for 3G applications. In: International conference on microwaves, antenna, propagation and remote sensing (ICMARS)
8. James JR, Hall PS (1989) Handbook of microstrip antennas, vol 1. Peter Peregrinus Ltd
9. Cai Y-M et al (2016) Compact-size low-profile wideband circularly polarized omnidirectional patch antenna with reconfigurable polarizations. IEEE Trans Antennas Propag 64(5)
10. Mak KM, Lai HW, Luk KM (2017) A 5G wideband patch antenna with antisymmetric L-shaped probe feeds. IEEE Trans Antennas Propag 66(2): 957–961
11. Khan QU et al (2016) Higher order modes: a solution for high gain, wide band patch antennas for different vehicular applications. IEEE Trans Veh Technol 66(5):3548–3554

Using Mobile Phones as a Learning Tool in Nature Inspired Furniture Design Process

Upeksha Hettithanthri and Preben Hansen

Abstract Mobile phones are widely used digital devices among young learners. Although it is evident that students use mobile phones for their learning process, until recently it was not recognized as a learning tool, especially in Interior Architecture pedagogy. This article aims to explore the research gap in finding the impacts of using mobile phones as a learning tool in nature inspired furniture design project. The research was conducted using qualitative methodology through observation and interviews which were carried out among 12 selected students in the interior architecture degree program. In this article, we will explore the impacts of using mobile phones, throughout the product development process and how it has affected the students creative design ability. This is an exploratory study conducted within 5 weeks and the design process has been observed as a fly on wall observer throughout the given duration. The research is based on observational notes and open-ended questions given to participants. The results were analyzed through thematic analysis and it reflects on how the exposure of the mobile phones have affected the design ability of the students.

Keywords Design studio · Mobile phones · Design ability

1 Introduction

There are unique ways of teaching and learning Architectural subjects within their study domains [1]. Interior architecture is a specific subject area which deals with cognitive skills and design abilities of the students [2]. Interior architecture pedagogy is developed to inculcate creative design abilities in students and is supported

U. Hettithanthri (✉)
National School of Business Management Green University Town, Pitipana, Homagama, Sri Lanka
e-mail: upeksha@nsbm.lk

P. Hansen
Department of Computer and Systems Sciences/DSV, Stockholm University, 7003, 164 07 Kista, Sweden
e-mail: preben@dsv.su.se

S. Fong et al. (eds.), *ICT Analysis and Applications*, Lecture Notes in Networks and Systems 93, https://doi.org/10.1007/978-981-15-0630-7_48

by student centric teaching methods. Design studio is considered as the core of the design education. The Interior architecture study setup has been made up by wrapping the core educational activities around the pedagogic design studio. Ahmad [3], described design studio as the heart of design education. It is the center of gravity in the interior architectural education, which is structurally similar to architectural education. As described in [4], traditional design studio is a place where the variety of architectural problems were handled through problem based learning. According to D. A. Schon, students tend to learn from reflection while in the design studios and when going through the design processes [5]. Design studios are typically set up to improve face to face interaction and group collaboration of the students. The design studio can be assumed as the core of the interior design curriculum, where students engage in design tasks by analyzing given design problem, synthesizing the solutions and evaluating the given solution [6]. Nicolai and S. Khalid explain that architectural studio education has been formed up into a triangular formation of time and space, instructor or the tutor and physical learning artifacts [7]. It needs to be further elaborated through adding collaborative student activities where students get multiple interaction in learning interior architecture. Casakin and Kreilter [8] highlighted that the traditional design studio practice follows trial and error method in teaching architecture related subjects.

In typical design studio setups, students are engaging in design tasks which contains heavy paper work plus rough models and mock ups [9]. This allows them to test their design ideas before coming into a final design conclusion [10]. Further, it is enriched with practical implications of theories learnt in non-studio-based environments. In typical studio environment, students get the chance to develop their schemes, oversee how the others have solved the given problem through their planning, and further they will be exposed to multiple tutors to get the comments holistically [11].

1.1 Problem Statement

Aim of this study is to understand the impact of using mobile phones as a learning tool in designing a furniture product in interior architectural learning process. With improved ICT facilities, mobile phones have become a multifunctional tool in learning interior architecture [12]. There are many established learning tools such as manual and digital drawing utensils, art accessories and story boards, mood boards, 2D, 3D visualizations and many design communication tools that fall within this category. The use of mobile phones have become common and popular among students, and the aim of this article is to reveal the impact of using mobile phones as a learning tool in different stages of the design process. With that in mind, the following project has been formulated to observe students' learning mechanisms, use of mobile phones and its impact on design ability.

The design project used for this study was developed to allow the students to design a furniture piece after a thorough observation of nature. Students are to reflect their design thinking through the design of their furniture product [10]. The impact of the mobile phone on design process has been observed and measured by the researcher. For this task, students were taken out from the conventional design studio and allowed to be there in a biologically sensitive environment, to feel and explore the true nature and for this a botanical garden in wet zone of Sri Lanka was selected. How students have observed the nature was ascertained by the researcher without disturbing their observation patterns of the nature.

1.2 Research Questions

Observational study was made to understand how students are using mobile phones as a learning tool in different stages of the designing process. The usages of the mobile phones as a learning tool have been observed under 2 major categories: when students are in a natural context and when they are in the design studio context. Research questions were generated as follows.

- What are the impacts of using mobile phones at the observation stage of nature for the students?
- How has the use of mobile phones affected to student's design ability when they are in the design studio context?

2 Methodology

This is a qualitative study carried out in a real world context, based on a fly on wall observation, conducted with a selected group of Interior architecture students from the 1st year completing their 2nd semester [13]. Out of the 33 students in the class only 12 students were selected for the observational study. Ethical clearance was taken from the university to observe the design process of these students. The 12 students were selected by considering the high marks they obtained in design module from their previous semester. Although the researcher has chosen the 1st year 2nd semester design project for the observational study, there was no impact on the non-selected students. Student's design process was observed under three major stages. Stage one of this study is set up to discern the process of observation of nature by students, and stage two is arranged to observe nature recording process of the selected group. The final stage is made to observe the design development process of the students. It is important to state here that the first two stages occurred at a natural context and the final stage took place inside a design studio. The project was undertaken within a period of 5 weeks. Every week, students got 6 hours to engage with this design process according to the semester calendar of the university. Within the very first week of commencement of the semester, the students were taken to a botanical garden. The main purpose of the visit was for the students to be with nature and observe it carefully and to collect inspirations to develop the furniture product. Students were instructed by the module leader to spend 6 hours at the garden and were instructed to observe and record their inspirations or sources for design for their furniture product. The researcher independently and carefully observed students' behavior, how they observed the nature, their creative behavior within that context and how they recorded the nature.

2.1 Data Collection

Random sampling method was used to select students for the observation [14]. The average age of the selected students was 20 years, and there were 6 girls and 6 boys in the selected group. The two main data collection methodologies used were observations and the interviews. Fly on wall observation has been conducted to gather data [13]. The observer used a field dairy in the process of recording data [15]. The notes were taken in chronological order, by allocating individual sheets for each student, with the name of the student, time, location, activity and the tools that they used. The observation was conducted in time intervals of 1 hour. During the field visit of the first hour, behavior of the student, selection of inspirations, nature observation method and recording methods were recorded and at the end of each hour the same were observed. Furthermore, the observer captured photographs during those time intervals of the selected students. Field notes were taken in a sequence of two stages as mentioned earlier. Stage one was when students were in the natural context and Stage two was when they were in the design studio. Every week, 6 hours were allocated for the studio works, and students' weekly progress plus how they were engaged with the design development was recorded. Within the design studio environment, tools used for design development and the methodology followed were observed and recorded weekly. Two main interviews were conducted; one at the end of the nature observation process and the second interview was done at the end of the fourth week. Interviews were recorded and transcribed for the analysis [14].

2.2 Data Analysis

Data analysis was guided by 6 phases of thematic analysis and an inductive approach has been undertaken [16]. Familiarization with the data, creating initial codes, searching for themes, reviewing themes, defining the themes and reporting were the steps followed in generating themes. Initially data has been categorized according to the sources of data generation. Observation was mainly analyzed through field notes taken by the researcher [15]. Data generated through interviews were handled separately. Coding was done in two stages; during the field visit and during the design development process that took place at the studio [15]. Field notes generated codes such as; taking of photographs, searching the internet, observing fine details, sketching, abstractions, and taking samples. Photographs taken by the observer has been categorized according to the codes generated. Design studio field notes generated codes such as, brain storming, prototyping, detail sketches, technical drawings, use of the internet, and to observation of the photographs taken. Those codes generated categories, namely; "supported by mobile phone", "supported by human skills" and "use of mobile phone and human skills". Use of mobile phones and nonuse of mobile phones were the two major themes that resulted in the categorizing of the above data.

Data generated from the interviews were analyzed through a thematic analysis [15]. All the interviews were transcribed and fed into MAXQDA 11. Before feeding interviews into MAXQDA 11, affinity diagram was created by the author to have an initial understanding on codes, categories and themes. Following the first week of activities in observing nature, use of mobile phones and avoiding the use of mobile phones were identified as the two major themes that were discerned by the researcher. Under the theme "use of mobile phones" the following codes were identified as the major reasons extracted from the interview that were conducted at the end of the nature observation process: to take pictures, to zoom in details and browsing of information through the internet. Under the theme of 'avoiding the use of mobile phones', the following codes were noted: Use of sensory perceptions, collecting samples from nature, sketching and drawings.

2.3 Results

The summary of student activities during the nature observation and the design studio activities has been depicted on Table 1.

Thematic analysis resulted in generating 2 main themes; the use of mobile phones and the non-use of mobile phones. Under theme one it was clear that the participants have put great trust on mobile phone as a supportive medium in nature observation. Out of the total number of 12 students, 8 students had used mobile phones to observe the nature, to see fine details which they could not see at a glance, to maintain an easily accessible record and to browse scientific details through the internet. One student out of the 12 had shown no trust on the context he was in and therefore had used the mobile phone 100% to observe the nature. Even when he was inspired by a plant which was not available in the natural context, he continued to carry out the detailing of said observation 100% through his mobile phone by browsing the internet. So, the total percentage of the students who were using mobile phones during the nature observation process is totaled as 66.6%.

Four students were identified during the nature observation process for not using mobile phones. This percentage was calculated to be 33.4%. Three students among them collected samples for further studies and brought them back to the design development process that took place at the design studio. During the design development process that took place at the design studio, all the students (12 out of 12) have used mobile phones under the following categories: to find technical details, to find

Table 1 Nature observation summery

Number of students	Activities engaged with in first 4 h	Use of tools/recourses
3/12	Capturing natural elements of the field, maintained a huge collection of photographs	Mobile phone
4/12	Capturing selected natural elements of the field Used zoom in option of the camera to see tiny details Sketching on the sketch book	Mobile phone Drawing tools
1/12	No much walking, sitting in one place nearly one hour, browsing through internet, saving selected natural elements in a folder in his mobile	Mobile phone to browse internet
3/12	Careful observation of nature, walk on bare foot, spent more than 3 h searching for inspirations, collect samples, sketching selected elements in different angles	Drawing tools
1/12	Collect some samples, draw abstractions with the similar characteristics of the element	Drawing tools

scientific details, to find material properties and share the design among colleagues to get comments.

At the end of the design process, the students' furniture products were assessed by a judging panel established by the module leader with the following criteria: understanding nature, behavior and values, transferring those qualities to the designing of a furniture product, reflections on the inspirations within the product, use of appropriate material and manufacturing technology, skills on communicating the design. Among the students who were using mobile phones during the nature observation process, 4 students have scored between 40 and 45% marks, and 4 students have scored between 45 and 55%. Among students who were not using mobile phones, there is a significant score of higher marks. 3 students among them have scored between 60 and 75%, one student have scored 75–80%. Among 12 students who were using mobile phones through the design development process, 5 students have performed well in providing appropriate manufacturing details, but 3 among them have shown comparatively less skills on transferring the qualities of the natural element selected to the designing of the furniture product and showed poor design communication skills which lead them to score less marks (Table 2).

Students who were not using mobile phones during the nature observation process have scored significantly high marks than students who used mobile phones during the nature observation process. Students who were not using mobile phones have shown better understanding on transferring inner qualities of the natural elements

Table 2 Design studio observation summery

Number of students	Activities engaged in	Use of tools/recourses
4/12	Recording the inspirations, brought back samples to the studio, brainstorming, (week 2 & 3) Developing technical drawings, developing prototypes Testing prototypes, developing final design (week 4 & 5)	Drawing tools-pens Drawing tools-pens, model making materials, mobile phone to browse technical details
4/12	Design ideation, brain storming (w2 & 3) Developing technical drawings, 3D images, perspectives (w4 & 5)	Books, internet, mobile phone, drawing tools Drawing and modeling software
2/12	Changing the source of inspiration, selecting something new, start the process from the beginning (w2 & 3) Test many design forms, develop 2-3 prototypes (w4 & 5)	Mobile phones, drawing tools, Internet
2/12	Trying to understand biological features in depth, deep analysis through engineering and scientific sketches (w2 & 3) Develop prototypes, produce technical drawings, 3D images and try several joinery mechanisms through drafting them technically and later developing prototypes (w4 & 5)	Internet, mobile phone, books related to product development, furniture designing Drawing and modeling software

selected to the furniture piece and it has been reflected through the design products they created.

2.4 Reasons Behind Use of Mobile Phones

In the process of observing the nature, mobile phone was a data recording medium and the main method the students followed for recording is taking photographs. Furthermore, options available in the mobile phone camera, such as zoom-in, has supported them to see tiny details they could not see at a glance. Access to internet through the mobile phone has opened up wide variety of natural recourse to observe and it was the reason for the student who was 100% relying on the mobile phone in making his inspirations. The reasons for not using mobile phones as follows; some students were more enthusiastic on observing the nature through their sensory perceptions and were able to sketch the selected elements very well as a recording medium. Not only that, they also consider mobile phones as a disturbance to their observation process.

3 Discussion

This study was set up to answer the research questions generated as mentioned in Sect. 1.2. Research question one was focusing on finding the impacts of using mobile phones during the nature observation process by Interior architecture students. In answering research question one, the researcher has explored how the students have observed the nature and, how they have used the mobile phones and why they used mobile phones and finally, how it has impacted their design ability. Design ability was measured through the results they undertook for the furniture design they created. The results showed that there was a significantly high level of usage of mobile phones during the nature observation process. Use of mobile phones has limited student's exposure to the nature in different ways. It has made students to rely mostly on the information it gives, rather than discern elements of nature through their own sensory observations. It is clear that the mobile phone has created glasses to observe the nature and it limits sensory perceptions needed in creating novel furniture designs. Students who used their human senses and skills in observing and recording nature were low in number, but they have shown significantly high creative skills in understanding and transferring the inspirations that they gathered through their own observations into product development.

Research question two was created to understand how the impact of the use of mobile phones affects the students' design ability when they are in the design studio. The study observed that during the design studio process, students have used mobile phones to get the support for their design development process which required more technical details of the product to develop its design. It is significant that all of them were using mobile phones at the design studio process, but it was not supportive enough in scoring high marks. Students who have used mobile phones throughout the process from the nature observation to design studio development have scored relatively low marks comparing with the student who have not used mobile phones in the field. Results has shown that the use of mobile phones has limited their design ability. This is due to the over exposure at the early design stage which is difficult to cover up in the latter stages of the design process.

Human cognitive skills are fluctuating at different levels [17]. Designing is a cognitive skill which needs divergent and convergent thinking in a balance. In Interior architectural design process, many manual and digital tools are used, such as drawings, drafting accessories, computer aided design software, to name a few at the top of the list of digital tools. Using tools at different stages of the design process could be effective in increasing design thinking process [10]. It depends on how the user is going to utilize the support of the tools within his or her design process. The mobile phones have come into the play in such scenarios, and due to its multifunctional ability students tends to use it in various ways throughout the design process. However, letting this digital device (mobile phone) take complete control of the design process is hazardous, especially in creative design fields. This study highlights the above point with the one student who had not shown any interest in the context where he was and was more engaged in gathering potential ideas by browsing the net using his mobile device. This could lead the designer to be more introverted and create designers less sensitive to their surroundings and to not appreciate the values in the immediate context, rendering the design process ineffective.

References

1. Afacan Y (2011) Teaching universal design: an empirical research in interior architecture. Proc—Soc Behav Sci 15:3185–3192
2. Afacan Y (2016) Exploring the effectiveness of blended learning in interior design education. Innov Educ Teach Int 53:508–518. https://doi.org/10.1080/14703297.2015.1015595
3. Mutaqi AS (2018) Architecture studio learning: strategy to achieve architects competence. In: SHS web of conferences, vol 41, p 04004. https://doi.org/10.1051/shsconf/20184104004
4. Hill, G.A (2017) The “Tutorless” design studio: a radical experiment in blended learning. J Probl Based Learn Higher Educ 5:111–125
5. Schon, DA (1984) The architectural studio as an exemplar of education for reflection-in-action. 9
6. Demirkan H, Afacan Y (2012) Assessing creativity in design education: analysis of creativity factors in the first-year design studio. Des Stud 33:262–278. https://doi.org/10.1016/j.destud.2011.11.005
7. Steinø N, Khalid MS (2017) The hybrid studio: introducing Google + as a blended learning platform for architectural design studio teaching. J Probl Based Learn Higher Educ 5(1):22–46. https://doi.org/10.5278/ojs.jpblhe.v5i1.1562
8. Casakin H, Kreitler S (2008) Correspondences and divergences between teachers and students in the evaluation of design creativity in the design studio. Environ Plan B Plan Des 35:666–678 (2008). https://doi.org/10.1068/b3405
9. Abdullah NAG, Beh SC, Tahir MM, Ani AIC, Tawil NM (2011) Architecture design studio culture and learning spaces: a holistic approach to the design and planning of learning facilities. Proc Soc Behav Sci 15:27–32 (2011)
10. Cross N (2011) Design thinking: understanding how designers think and work. Berg, Oxford
11. Leathem T, Hillesheim C, Coley A, McGregor S (2019) Student and teacher perspectives on a multi-disciplinary collaborative pedagogy in architecture and construction. High Educ Skills Work Based Learn 121. https://doi.org/10.1108/HESWBL-03-2018-0026
12. Gushchin A, Divakova M (2017) ICT in education of architects. How to strike a balance? Proc Soc Behav Sci 237:1323–1328. https://doi.org/10.1016/j.sbspro.2017.02.217
13. Berry D, Korpan C (2009) Fly on the wall. Collect Essays Learn Teach 2:203–207 (2009)
14. Bansal P(Tima), Smith WK, Vaara E (2018) New ways of seeing through qualitative research. Acad Manag J 61:1189–1195
15. Tracy SJ (2013) Qualitative research methods : collecting evidence, crafting analysis, communicating impact. Wiley-Blackwell, Chichester, UK (2013)
16. Hoskyns S (2016) Thematic analysis. In: Collected work: music therapy research. 3rd edn. Revised Published, Dallas, Barcelona. (AN: 2016-21423). Barcelona, Dallas
17. Dix A (1993) Human-computer interaction. Prentice-Hall, London

Pedestrian Activity Recognition Using 2-D Pose Estimation for Autonomous Vehicles

Pranav Pandey and Jagannath V. Aghav

Abstract Human activity recognition is the task of recognizing activities of any given subject in a scene, from a set of observation over time, taking into consideration the environmental and behavioral factors. It has application in a lot of fields including surveillance, assistance system, threat identification. Human activity recognition plays a vital role in human computer interaction, as it is very important that a computer correctly identifies human activity to really understand the human behavior and learn what the human is trying to convey through their action as more than 50% of communication humans do is through body language. In this paper, we propose a system for identifying activities of pedestrians on road using pose estimation to give the autonomous vehicles a better understanding of the humans' actions and get better at driving and also provide a safer environment for the humans.

Keywords Human activity recognition · Robotics · Human computer interaction · Deep learning

1 Introduction

Human activity recognition is the ability that computer systems must possess to be truly intelligent. We as humans observe and interpret people around us through our senses for understanding and communicating with them efficiently and in doing so, seeing others and understanding their body language, their actions is very important for a better understanding of what others are trying to do and interpret their feelings. Vision is the most important sense of all the senses we have by which we interpret other peoples' body language because about 55% of communication is through body language therefore it is important that the computers should also have this ability to understand humans activity and use it for better communications ad better understanding of human nature. Interpreting human's activity has a varied range of

P. Pandey (✉) · J. V. Aghav
College of Engineering, Wellesley Rd., Shivaji Nagar, Pune, Maharashtra, India
e-mail: pranavp17.comp@coep.ac.in

J. V. Aghav
e-mail: jva.comp@coep.ac.in

S. Fong et al. (eds.), *ICT Analysis and Applications*, Lecture Notes in Networks and Systems 93, https://doi.org/10.1007/978-981-15-0630-7_49

application in the field of artificial intelligence- from surveillance, assistance systems to robotics and autonomous vehicles. It is very important for computers to understand people's activity to be more aware of their actions and provide a safer environment for the them so that both machines and human can work together in a sustainable environment. Autonomous vehicles are inherently designed in such a way that they make it safe for the people to travel in the ego-car and provide a safe environment for the pedestrians and other vehicles other than the ego-car. Recognizing human activity has been a long-researched area, and the literature can be broadly divide into two different classes for the task of activity recognition—One focuses on the sensor based methods to identify the activity by capturing the muscle movement in 3-D space which can be used in wearable devices [1–3]; and the other one is vision based, in which computer vision techniques are used for identifying actions by using several different techniques like depth perception and pose estimation [4]. Although a wide variety of techniques already exists, but there is still space for improvement. In this proposed system, our focus is only on identifying actions related to pedestrians on road and for that we use 2-D human pose estimation to identify their actions. For the task at hand we combine pedestrians pose key-points, their position with reference to the vehicle along with other environmental factors including time of the day, direction of movement, and traffic signal. Using a combination of these attributes, in any given scene we predict if the person present in the scene is crossing or not crossing the road.

2 Dataset

We have made use of two different datasets for realizing our system. Firstly, for the task of pedestrian detection in a frame we used the "OpenImages V4" dataset by Google, it contains images for 600 different classes and is publicly available for download. We have used the "person" class images only from the dataset.

Number of training examples: 10000

Number of test examples: 3000

For the task of training a classifier, which can classify different activities of the pedestrians we used another dataset. The dataset we used in this work was released through a paper in 2016, named, Joint Attention in Autonomous Driving (Rasouli et al.) [5, 6]. The dataset is commonly known as JAAD dataset. It contains 364 video sequences captured from multiple vehicles mounted inside the car and directed outwards towards the pedestrians. The dataset is quite diverse and contains videos from a lot of different types of setting—indoor, outdoor, rain, snow, sunny etc. which makes it fit for our task. The dataset also captures the environmental and behavioral attributes of the pedestrians and the driver as well, which can be used for research in many scenarios like activity identification, intent identification etc. For our work, we have divided all the videos into frames and segregated them manually into two categories—crossing and not crossing and also cropped all the images such that only pedestrian is visible in the image.

Table 1 Weather and geographical attributes present in the dataset

Location	Weather	Time	Road condition
• Indoor—refers to parking • Plaza—refers to outdoor parking (e.g. near mall) • Street—refers to video captured on road	• Cloudy • Clear • Rain • Snow • N/A for Indoor	• Daytime • Nighttime • N/A for Indoor	Whether the road is covered in snow/water or dry • Snow • Rain • Dry

The weather and geographical features that are present in the JAAD dataset are explained in the Table 1.

3 Methodology

For the task of recognizing human activity, we must make a pipeline of multiple tasks that have to be performed to get our expected result. We start by first dividing the input video into frames and feeding each frame into the pedestrians detection module which detects the pedestrians present in the frame, and then the output of this module becomes the output of the pose estimation module, which estimated the body pose of the pedestrians that are present in the frame, and finally the output of pose estimation module is given as input to a classifier along with environmental data and the pedestrians in the frame are classified as crossing or not crossing the road.

3.1 Algorithm

For the task of pedestrian activity recognition, we perform a set of multiple steps on our input data to reach our expected result. The algorithm that we use for our task is given below:

Step 1: Divide the input video into frames.
Step 2: Identify pedestrians in the input frames and draw bounding box around them, also find the center point of the bounding box for further tracking.
Step 3: Estimate body pose key-points of the pedestrians present in the image(frame).
Step 4: Extract body key-points from the JSON files created by the OpenPose library for pose estimation of each frame.
Step 5: Add environmental data to the dataset containing the body key-points for pedestrians and train a support vector machine classifier upon that combined data using 'rbf' kernel.
Step 6: Simultaneously track pedestrians' location and classify their poses to predict their activity.

Result: Pedestrians Activity–Crossing or Not Crossing.

3.2 Frameworks and Libraries Used

For this work, we needed to detect the pedestrian present in the video frames, and for that we made use of TensorFlow framework as backend, and on that we trained Faster RCNN [4] model that we downloaded from TensorFlow model zoo. We have also used OpenPose pose estimation library developed by Carnegie Mellon University for the task. OpenPose is a state-of-the art pose estimation system which estimates the body pose of humans in images, videos, and webcam feeds by imposing a skeleton figure on the people present in the image with all the key-points. We have also used OpenCV library, which has some of the most common computer vision algorithms pre-implemented in it which can be used out-of-the box.

4 Architecture and Implementation

For realizing our system, we created a pipeline of operations that has to be done on each frame of the video, which finally gives us the prediction whether the pedestrian is crossing or not crossing the road. The proposed system architecture is given in the Fig. 1.

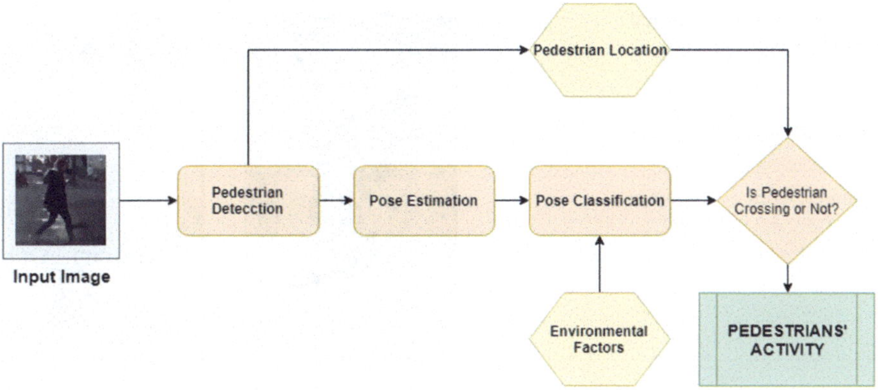

Fig. 1 Flowchart for identifying the pedestrian's activity

4.1 Pedestrian Detection

For the task of pedestrian detection, we trained three different models, among which Faster RCNN worked best and gave the highest accuracy and we were able to identify pedestrians in the frames with high probability, therefore, we have used Faster RCNN in our work. Pedestrian detection [7, 10] is important because we must track the location of the pedestrian in the scene relative to the vehicle which helps in identifying if the person is crossing the road or not. We detect the pedestrian and draw the bounding box around the pedestrian and calculate the center point of the bounding box which acts as a marker for tracking the pedestrians location in the scene (Fig. 2).

Fig. 2 Person detected with 99% probability using our trained faster RCNN model

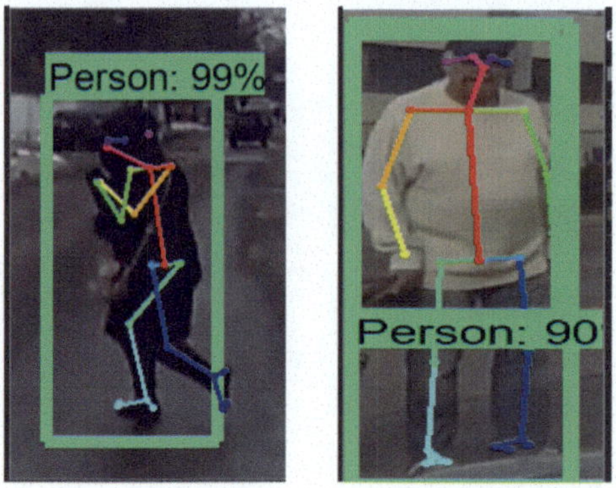

Fig. 3 Body pose estimated using the OpenPose library

4.2 Pose Estimation

Pose estimation is next step in our pipeline for processing the frame. There are a lot of pose estimation algorithms [8, 9] that already exist and provide great accuracy; therefore, we didn't code our own pose estimation system and have used an existing one [14]. We have used OpenPose pose estimation library for the task of estimating the human pose in our work. It is a state-of-the art pose estimation library which provides multiple models like COCO, MPII, Body_25 with different number and location of key-points on the human body. In our work we have done the pose estimation using the BODY_25 model, which identifies 25 different body key-points in the human body. The 25 key-points are as follows:

0—Nose, 1—Neck, 2—RShoulder, 3—RElbow, 4—RWrist, 5—LShoulder, 6—LElbow, 7—LWrist, 8—MidHip, 9—RHip, 10—RKnee, 11—RAnkle, 12—LHip, 13—LKnee, 14—LAnkle, 15—REye, 17—REar, 18—LEar, 19—LBigToe, 20—LSmallToe, 21—LHeel, 22—RBigToe, 23—RSmallToe, 24—RHeel.

We store the detected body key-points in different json files for each frame. The key-points are stored in a entry named 'pose_keypoints_2D' in the json file, and the key-points have a specific format in which they are stored—(x-coordinate, y-coordinate, confidence) for each key-points. These estimated key-points are extracted in the next step for further processing (Fig. 3).

4.3 Pose Classification

There are a lot of pose classification algorithm that already exist [11–13]. But we develop our own pose classification system for our work. After estimating the pose of pedestrians, we extract the key-points for all the frames that are stored in the json file we got as result in the previous step. The key-points are stored in a csv file in the format shown in the Table 2.

Table 2 CSV file structure for the data

Nosex	Nosey	Neckx	Necky	RShoulderx	RShouldery	…
59.6072	23.8137	46.7128	35.4747	49.2132	36.2923	…
72.2265	29.0312	58.7998	40.3501	62.1578	41.6045	…
…	…	…	…	…	…	…

Before training the classifier, we first have to normalize the data because the dataset is prepared manually by cropping the original images and hence the estimated key-points position have different scales. We normalize the data by scaling each value in the range of [0, 1]. The formula for normalizing the data is given below.

$$X_{new} = \frac{X_i - X_{min}}{X_{max} - X_{min}}$$

After normalizing the data, we also must deal with missing values i.e. the key-points which were not identified by the OpenPose system. For filling the missing values, we take mean for each column according to each class and fill those missing values with the mean value. After this, we add the other environmental features, namely, time, direction, signalized, and designated to our data and convert these values into numeric values using one-hot encoding. Finally, we train a SVM classifier on our dataset with 'rbf' kernel and learning rate as 0.001 which we will use to classify the pedestrians poses.

5 Results

Making use of our pipeline, we predicted the pedestrian's activity in the test video which we have selected from the JAAD dataset only, the results are quite satisfactory and the system correctly predicts the pedestrians activity whether they are crossing the road or not for good percentage of the video sequence. There are some inaccuracies in the prediction, like when a person is walking sideways and are near to the car, the system predicts their activity as crossing the road, although they are just walking parallel to the road, we are trying to fix these inaccuracies as we further improve our system. Below we show screenshots of video sequences where our system predicts our two target classes—CROSSING and NOT CROSSING (Fig. 4).

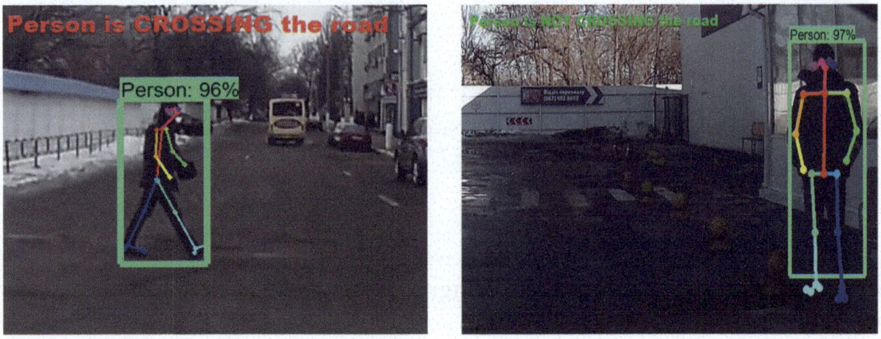

Fig. 4 Final result screenshots showing the pedestrians activity

References

1. Lara OD, Labrador MA (2013) A survey on human activity recognition using wearable sensors. IEEE Commun Surv Tutor 15(3):1192–1209, Third Quarter 2013. https://doi.org/10.1109/surv.2012.110112.00192
2. Kim E, Helal S, Cook D (2010) Human activity recognition and pattern discovery. IEEE Pervasive Comput 9(1):48. https://doi.org/10.1109/mprv.2010.7
3. Anguita D, Ghio A, Oneto L, Parra X, Reyes-Ortiz JL (2012) Human activity recognition on smartphones using a multiclass hardware-friendly support vector machine
4. Hervás BJ, Rodríguez R (eds) Ambient assisted living and home care. In: IWAAL 2012. Lecture Notes in Computer Science, vol 7657. Springer, Berlin. https://doi.org/10.1007/978-3-642-35395-6_30
5. Kotseruba I, Rasouli A, Tsotsos JK (2016) Joint attention in autonomous driving (JAAD). arXiv preprint arXiv:1609.04741
6. Rasouli A, Kotseruba I, Tsotsos JK (2017) Are they going to cross? A benchmark dataset and baseline for pedestrian crosswalk behavior. In: ICCVW
7. Ren S, He K, Girshick R, Sun J Faster R-CNN: towards real-time object detection with region proposal networks
8. Singh A, Agarwal S, Nagrath P, Saxena A, Thakur N (2019) Human pose estimation using convolutional neural networks. In: 2019 Amity international conference on artificial intelligence (AICAI). https://doi.org/10.1109/aicai.2019.8701267
9. Cao Z et al (2017) Realtime multi-person 2D pose estimation using part affinity fields. In: 2017 IEEE conference on computer vision and pattern recognition (CVPR). https://doi.org/10.1109/cvpr.2017.143
10. Benenson R, Omran M, Hosang J, Schiele B (2014) Ten years of pedestrian detection, what have we learned? In: ECCV 2014: Computer Vision—ECCV 2014 Workshops, pp 613–627
11. Ng CW, Ranganath S (n.d.) Gesture recognition via pose classification. In: Proceedings 15th international conference on pattern recognition, ICPR-2000. https://doi.org/10.1109/icpr.2000.903641
12. Gupta JP, Singh N, Dixit P, Semwal VB, Dubey SR (2013) Human activity recognition using gait pattern. Published in IJCVIP .https://doi.org/10.4018/ijcvip.2013070103
13. Zhang S, Wei Z, Nie J, Huang L, Wang S, Li Z (2017) A review on human activity recognition using vision-based method. J Healthcare Eng. Article ID 3090343, 31 pages. https://doi.org/10.1155/2017/3090343
14. Keller C, Gavrila D (2014) Will the pedestrian cross? A study on pedestrian path prediction. IEEE Trans. Intell Transp Syst 15:494–506

Sarcasm Detection Methods in Deep Learning: Literature Review

Shruti Kulkarni and Aparna Biswas

Abstract Sarcasm is rapidly becoming prevalent in all forms of communication today. Each person has a different way of exploiting and understanding sarcasm. It is difficult for even humans to interpret sarcastic texts, as it depends on a lot of things like perspective, context and tone. This makes it a challenging task to train a machine to distinguish sarcastic text from non-sarcastic text. As there are no concrete rules upon which a model to detect sarcasm can be built, we must resort to promising and upcoming techniques to do so. In this paper, we have reviewed the works done in sarcasm detection using deep learning in combination with the natural language processing techniques. We have also proposed an outline of our own system based on current gaps in literature and challenges in the field.

Keywords Sarcasm · Neural network · NLP · Deep learning

1 Introduction

With the latest advances into artificial intelligence, there is an increasing need to widen the boundaries of natural language processing tasks. Sentiment analysis is one of the ways in which machines can achieve better understanding of human speech and make sense out of it. Human language is a symbolic categorical system while we need to train a machine to think like a brain, which requires continuous patterns of activation. Sarcasm detection is an area of NLP which many researchers are trying to find a solution to. Till date, most approaches to sarcasm have treated the task primarily as a text categorization problem. Sarcasm, however, can be expressed in very subtle ways and requires a deeper understanding of human language that standard text categorization techniques cannot grasp.

S. Kulkarni (✉) · A. Biswas
College of Engineering, Pune, India
e-mail: shrutikul22@gmail.com

A. Biswas
e-mail: aparna.comp@coep.ac.in

S. Fong et al. (eds.), *ICT Analysis and Applications*, Lecture Notes in Networks and Systems 93, https://doi.org/10.1007/978-981-15-0630-7_50

In this paper, we have reviewed works dealing with sarcasm detection, concentrating on methods including neural network models and other deep learning techniques. The paper has been arranged as follows.

2 Datasets

There is a surge of data volume due to increase in the usage of social media. This provides us with authentic user-generated data in unstructured format. There are APIs that facilitate smooth scraping of data from Twitter like tweep [1] and Twitter API [2].

For Twitter-based datasets, two approaches to obtain annotations have been used. The first is manual annotation. Ellen Riloff [3] introduced a dataset of tweets which were manually annotated as either sarcastic or not. Maynard and Greenwood [4] conducted a careful study of sentiment classification and used sarcastic tweets as an impact on the aforementioned classification. The second technique to create datasets is the use of hashtag-based supervision. This method is more popular than manually annotating the tweets, as it relies on user perspective more than human identification of sentiment. By providing a hashtag, the author of the tweet is giving additional information that aides in determining the labelling of dataset. There is also the factor of more automation in this category, as it is easier to classify text as sarcastic or non-sarcastic on the basis of well-established sarcasm-indicative hashtags such as #sarcasm, #not, #lol etc. [5] have used this technique to supervise tweets. They have employed a method which ignores such hashtags at the beginning of the tweets, and only include descriptive trailing hashtags for their segregation. For example, "#sarcastic behavior is a form of passive-aggression" will not be labelled as sarcastic.

In case of Reddit-based datasets like SARC [6], there is an inbuilt provision of adding a '/s' at the end of a sarcastic post. This enables the author to provide annotations. So, the labelling of this dataset was largely based on this self-annotated method.

The data from the following figure gives the statistics regarding the amount of data used to train and test the models as reported in previous works. Here, we can see that even though results are satisfactory, the data used may or may not be sufficient to generalize those results on any other model.

3 Approaches

Zhang et al. [2] used a bi-directional gated recurrent neural network for capturing the syntactic and semantic information and to extract contextual features automatically from the tweets and used a pooling neural network. They obtained an accuracy of 87.25% with an F1-score of 77.37 on unbalanced dataset. These results changed after

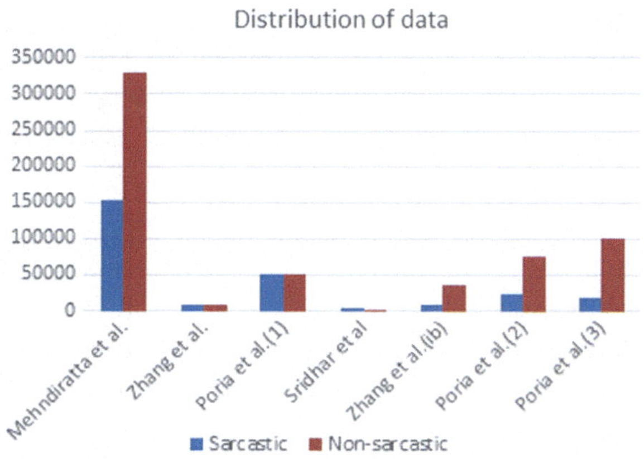

Fig. 1 Data distribution survey

the dataset used was balanced with an accuracy of 79.29% with F1-score of 79.36. This actively illustrates that deep learning models benefit the performance.

Poria et al. [7] have applied pre-trained convolutional neural network models for sentiment, emotion and user personality on three different datasets. These datasets are balances, imbalanced and heavily imbalanced respectively, with figures for the number of tweets provided in Fig. 1 for each dataset. As F1-score is the metric used for natural language processing tasks, results have been evaluated using F1-score for each dataset. They have applied the models both individually and as a combination, with the combined model giving the highest F1-score of 90.34. The third dataset also provided with a high score of 93.30, but there is a possibility of overfitting in such a scenario because dataset is highly imbalanced.

Mehndiratta et al. [1] have used a deep convolutional neural network on Twitter dataset with approximately 1000–10,000 iterations of single hidden layer CNN. They efficiency of their algorithm increased with the number of iterations, reaching a maximum of 89.9 from an average of 81.90 at the highest point. The salient feature of their work is that they have removed certain drawbacks of using Twitter data using text processing algorithms.

Lakshmanan and Anjana [8] have used different machine learning algorithms like Random Forest, Support Vector Machine, Naïve Bayes and Simple CART on Reddit dataset. They have filtered subreddits/pol (politics),/sarcasm for minimizing the data volume. Amongst all the algorithms, simple CART showed best results for detection for sarcasm at an accuracy of 66%.

Table 1 provides a survey of important works in sarcasm detection using deep learning and some baseline methods.

Table 1 Survey of previous works

Authors	Year	Dataset	Approach	Accuracy
Mehndiratta et al.	2017	Twitter	Deep CNN	81.9
Zhang et al.	2016	Twitter	Bi-directional gated Recurrent NN	87.25 F: 77.37
Poria et al.	2017	Twitter	Deep CNN, Pre-trained emotion, sentiment, personality models	F: 90.34
Lakshamanan et al.	2018	Reddit	RF, SVM, NB, Simple CART	66 (max)
Sridhar et al.	2016	Facebook	Hybrid model	82
Zhang et al. (balanced)	2016	Twitter	Bi-directional gated Recurrent NN	79.29 F: 79.36

4 Challenges

Deep learning is a breakthrough technology which has given us state-of-the-art systems. The systems designed using deep learning models provide better results than the previous works including baseline models. The work in deep learning has been made possible due to the integration of faster and stronger processors in the personal computers, adding to the convenience of the average researcher to explore this field. In sarcasm detection too, there is an upsurge in performances of systems using neural network models as a base. A lot of work has been done using commonly available datasets, with more concentration on compiling specific systems which give higher accuracy. However, there is no significant work with the use of context and user modelling which gives us an important research gap. There is also a lack of a generalized model which works for multiple datasets having different schema and post styles.

To sum up everything that has been stated, we have summarized some challenges which the areas in which future work are also can be done:

(1) Larger corpora: Until recently, there was a lack of a standardized dataset which was large and suitable for training a deep learning model effectively. As a result, the extent of work in this area is limited.
(2) Use of contextual information: The inclusion of context embeddings in any language processing task has given better quality of solutions. According to Silvio et al. [9], we can see that the embeddings can be exploited and applied.
(3) Generalized model: According to research, there is a saturation in the performance of any algorithm as datasets get bigger than a certain optimized amount. As larger models tend to overfit and smaller models underfit, there is a need to find a balance between the two by mixing datasets, varying the complexity of models and trial-and-error of the former two as a combination.
(4) User profiling as a feature: Identifying sarcasm in text is a challenging task for humans, making it even more so for machines. Hence, user perspective and history, which form a profile, are helpful in sarcasm detection.

5 Proposed System

Upon analyzing the trends and studying the gaps in the current research, we found that working on large volumes of data requires expertise and specialized algorithms to handle it. This is due to the fact that the main memory, where the data is to be stored, is expensive and hence limited in machine. So, to make our systems cost-effective, we need to perform specialized algorithms on our data which have the capability to use the available memory effectively without losing the results obtained. There is very limited work done in the field of sarcasm detection where large amount of training data has been used. Also, there is a need of a more generalized system to predict sarcasm in long as well as short text posts. While defining our problem statement, we have focused on these two issues.

Our proposed system will be based on the following problem definition: Using sequential deep learning models to effectively detect sarcasm, combined with natural language processing concepts. We are putting forth the idea of a system which includes the GloVe project for word embeddings, recurrent neural network with LSTM. Figure 2 shows the block diagram of our proposed system.

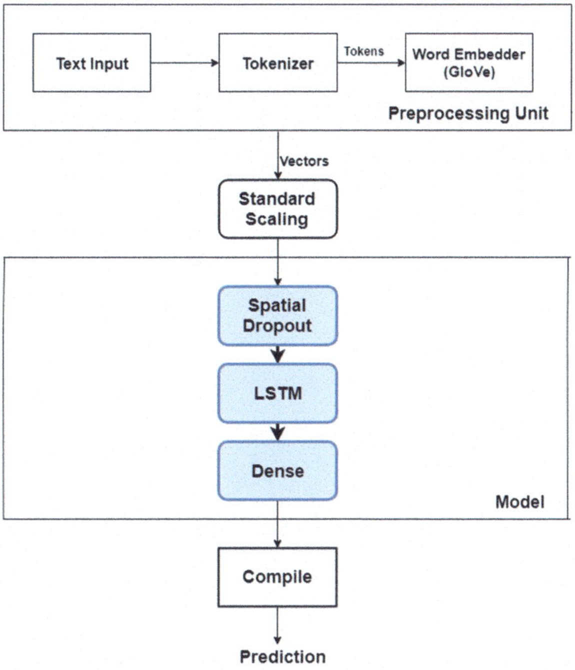

Fig. 2 Block diagram of proposed system

References

1. Mehndiratta P, Sachdeva S, Soni D (2017) Detection of sarcasm in text data using deep convolutional neural networks. Scalable Comput Pract Exper 18(3):219–228
2. Zhang M, Zhang Y, Fu G (2016) Tweet sarcasm detection using deep neural network. In: International conference on computational linguistics: Technical Papers COLING, Osaka, Japan, pp 2449–2460
3. Riloff E, Qadir A (2013) Sarcasm as contrast between positive sentiment and negative situation. EMNLP, pp 704–714
4. Maynard DG, Greenwood MA (2014) Who cares about sarcastic Tweets? Investigating the impact of sarcasm on sentiment analysis. In: LREC, pp 4238–4243
5. Kingma Da (2014) A method for stochastic optimization. arXiv:1412.6980
6. Khodak M, Saunshi N, Vodrahalli K (2018) A large self-annotated corpus for sarcasm. arXiv: 1704.05579v4)
7. Poria S, Cambria E, Hazarika D, Vij P (2016) A deeper look into sarcastic Tweets using deep convolutional neural networks. arxiv.org/abs/1610.08815
8. Lakshmanan B, Anjana A (2018) Sarcasm detection in reddit. Int J Recent Eng Res Dev (IJRERD) ISSN: 2455–8761
9. Joshi A, Tripathi V, Bhattacharyya P, Carman MJ (2016) Harnessing sequence labeling for sarcasm detection in dialogue from TV series 'Friends'. In: CoNLL

AI Based Non-invasive Glucose Detection Using Urine

U. Dhrupad, N. H. Vignesh, Hari Murthy, Chandra Mukherjee and Aynur Unal

Abstract This proposed device uses urine to predict the glucose level present in the patient using non-invasive technique with a high level of accuracy for detection of diabetes. The paper presents a urine glucose level diagnosing and prediction using a computer-based polarimeter held in a portable device, to provide a fast and accurate on-field result. The instrument consists of an LCD screen, optical sensor, Benedict's reagent, a detachable tank, and an embedded system-on-chip (SoC).

Keywords Diabetes mellitus · Benedicts reagent · Non-invasive · Artificial intelligence

1 Introduction

Diabetes is a lifelong disease that is caused by the malfunctioning of the pancreas or when the body is unable to utilize the insulin it produces. An estimated 8.7% of the diabetic population in India lies within the age group of 20–70 years and hence is a growing challenge. It is mostly caused by a combination of factors—tobacco use, unhealthy diets, sedentary lifestyles, rapid urbanization and increasing life expectancy [1]. Although a non-curable disease, it is treatable and prevented by maintaining a strict diet or treated by intake of artificial insulin hormone.

U. Dhrupad (✉) · N. H. Vignesh · H. Murthy · C. Mukherjee
Faculty of Engineering, CHRIST (Deemed to be University), Bangalore, India
e-mail: udhrupad@gmail.com

N. H. Vignesh
e-mail: nh.vignesh7@gmail.com

H. Murthy
e-mail: hari.murthy@christuniversity.in

C. Mukherjee
e-mail: chandra.caluniv@gmail.com

A. Unal
Digital Monozukuri, Palo Alto, CA, USA
e-mail: unalayn@gmail.com

S. Fong et al. (eds.), *ICT Analysis and Applications*, Lecture Notes in Networks and Systems 93, https://doi.org/10.1007/978-981-15-0630-7_51

Blood is the one most widely used body fluid used for glucose measurement in the body. The patient's fingers are pricked to collect the blood which is tested using various chemicals to find out the glucose level in it. The method gives highly accurate results but has its own disadvantages- the process is painful and there is the chance of cross-contamination. Thus, non-invasive techniques which do not require pricking of fingers and blood collection is highly desirable.

Benedict's Reagent is a clinical solution (complex compound of sodium citrate ($Na_3C_6H_5O_7$), copper sulphate pentahydrate ($CuSO_4{\cdot}5H_2O$)) and sodium carbonate (Na_2CO_3) which reacts with urine to form a white precipitate and a color change of the solution, which is dependent on the glucose concentration [2].

The TCS3200 is a color sensor with an array of 16 photodiodes for detecting different wavelengths (red, blue, green) based on appropriate filter's readings. The sensor has a current to frequency converter that generates square wave whose frequency is proportional to the detected light intensity. Analyzing the change in color of the solution, one is able to determine the level of glucose in the urine sample and predict if the patient is diabetic or not.

Urine is one such bodily fluid, for which there is no invasion required. Thus, we propose a non-invasive technique of glucose detection using urine. The device can help provide quick on-field test and diagnosis of the patient's glucose level. In addition, a mobile application is developed which is capable of fetching specific user data from the database and giving the complete picture on the patient's statistics in the past, and also predict future trends.

2 Design

2.1 Components

The device, as shown in Fig. 1 is a small rectangular casing, which houses the embedded system-on-chip (SoC) and an LCD screen to display the content of the result. An automatic Benedict's reagent dispenser in the form of a slender tube is placed. The bottom of the casing houses a color sensor, helping to detect the color of the solution. One end of the casing is equipped with a copper heating element, to heat up the element to its boiling point (50 °C). A detachable urine tank is attached to the device and plugged into a power socket for providing power to help aid the heating process.

2.2 Working

The user, collects the sample of urine (1 ml), into the detachable tank which is hooked to the casing. On switching the device ON, a measured amount of Benedict's reagent

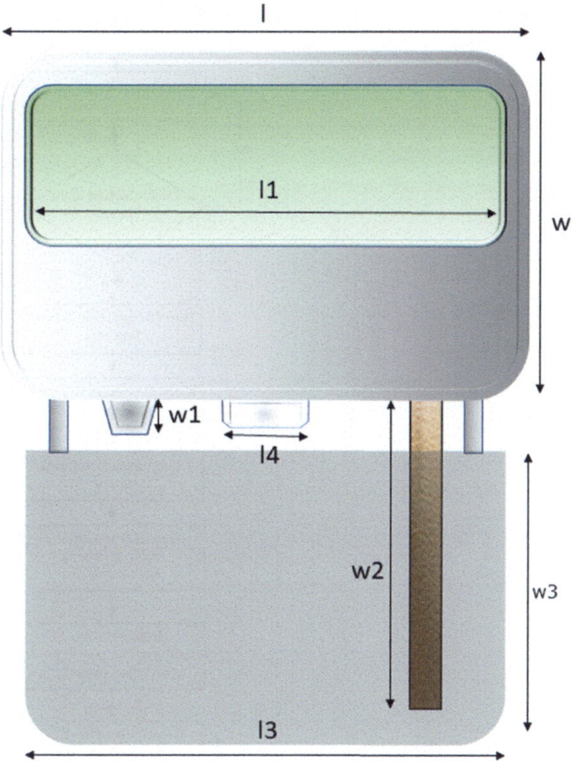

Fig. 1 Geometry of the proposed model ***l*** = 9 cm (rectangular casing), ***l*1** = 7 cm (LCD screen), ***l*3** = 8 cm (detachable urine collecting tank), ***l*4** = 3 cm (color sensor), ***w*** = 8 cm, ***w*1** = 2 cm (reagent dropper), ***w*2** = 10 cm (copper heating rod), ***w*3** = 10 cm

(5 ml) is automatically dropped into the urine sample. The solution is heated up to 50°C, leading to a change its color from light blue to one of the listed colors in Table 1 [3]. The workflow of the same is being described in Fig. 2.

The data, generated by the sensor is being processed on the microcontroller using a predefined program. The workflow of the same as described in Fig. 3.

The processed data is now being uploaded to a cloud database, using Wi-Fi. The data from the database is fetched by the controller, for feeding into the AI predictive algorithm. Further processing of this data is done by the algorithm. The workflow for the same is shown in Fig. 4.

Table 1 Color of solution versus inference on diabetes [3]

Color of solution	Approximate amount of glucose
Blue	Nil
Green	0.5%
Greenish brown	1.0%
Yellow	1.5%
Red brick	2.0% or more

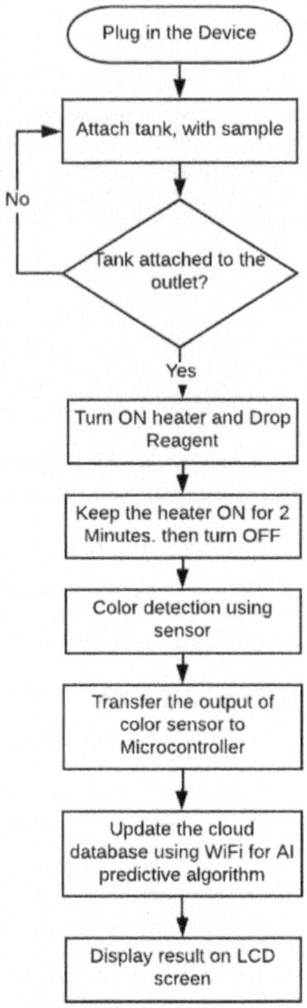

Fig. 2 Flow diagram of working of the proposed model

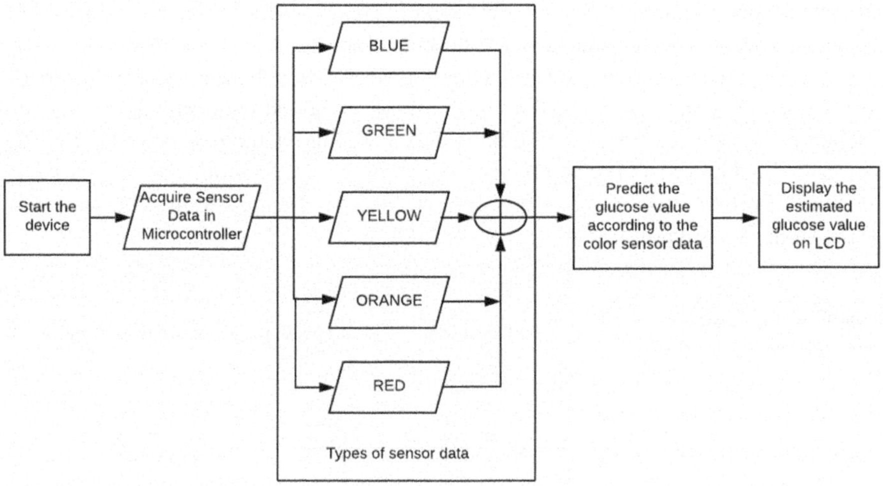

Fig. 3 Flow diagram of program for glucose detection

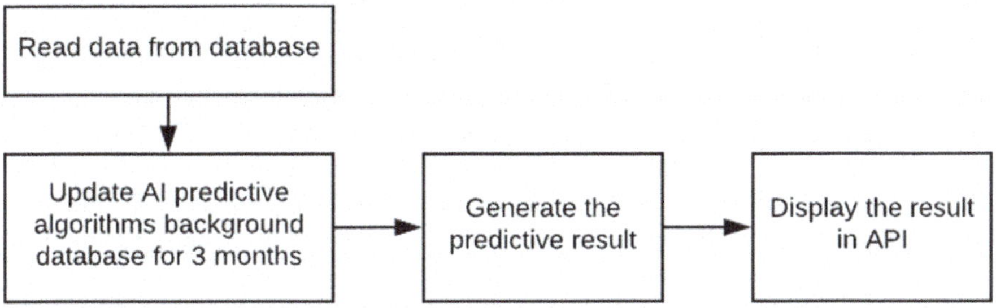

Fig. 4 Flow diagram of AI predictive algorithm

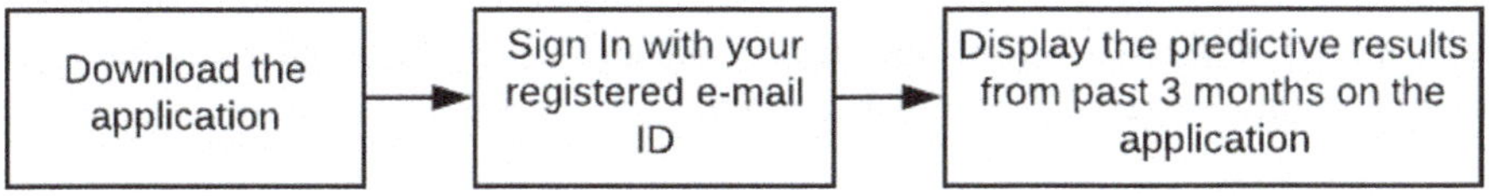

Fig. 5 Flow diagram of the mobile application

The data, from the algorithm, stored on the database, is now accessible using an application developed for mobile devices, for the user to easily accessed using a mobile application to study the entire diabetic history of the patient. The workflow of the application is shown in Fig. 5.

3 Results and Discussion

The color sensor (TCS3200) detects the color change (Fig. 6) and the embedded SoC with Wi-Fi communication determines the amount of glucose present in the given sample.

Fig. 6 Pictorial representation of various color changes

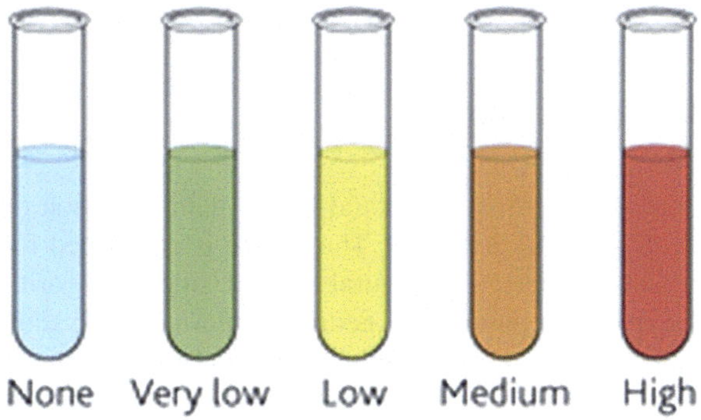

A database is being created, on a cloud service, to keep a track of the patient's frequent tests. The data gathered from frequent tests are uploaded into this database, where every user is allocated a separate profile for tracking his/her diabetic history, which can be accessed via a mobile application provided on multiple OS platforms. A user, with a profile on the database, will only be able to access his/her data through the interface provided on the smartphone application. The data, present on the database is also accessible by local health centers and/or hospitals. Only a registered medical specialist will be able to access and manipulate this database. The specialist now can access the patient's entire diabetic history and prescribe suitable medical aid. The AI algorithm uses the data collected over a period of 3–4 months, to analyze and forecast the health condition of the patient. The algorithm will be able to predict if the patient is going to be high on glucose levels in the coming future or will be normal. The AI model does not produce an accurate result during the initial days of usage, whereas on regular usage of the same, we can see a gradual increase in the accuracy of the same.

As the device functionality depends on the AI algorithm, a huge dataset is necessary to run the algorithm to its full potential. Hence, it is crucial to use the device every day for about 3–4 months. The AI module helps in forecasting the patient's blood sugar levels over a period of time and predict when the patient might suffer from a diabetic attack. This may be communicated to the patient via alerts or reminders onto the smart devices of the user.

Since the results of the day to day tests are being uploaded into the cloud database, the medical representative gets to see a more diversified view of every patient's overall diabetic history and maintain an organized record, without the usage of physical devices and paperwork. The patient benefits from reduced costs of conventional tests done by respective medics, and at less pain and risk of contamination. The functionality and user—operability of the mobile application can be made available in regional languages.

The experimental setup for conducting the test is shown in Fig. 7. The setup showcases a working prototype of the proposed model. The setup shows the color sensor, detecting the change in color of the sample, and displaying the value of the level of glucose on the LCD screen.

4 Conclusion

Here, a non-invasive method for on-field detection of body glucose level has been developed and deployed. The device is developed focusing on the problems faced in diabetes detection in rural areas. Using AI algorithm, it is capable of predicting the risk of a patient for getting diagnosed with diabetes. The device is portable, making it very easy to carry around and henceforth enabling diabetes detection with ease. With further advancements to the mobile application, the interface of the same can be updated to also provide a personalized user news feed, help decide the diet, personalized medication prediction, reminders for consulting a medical professional,

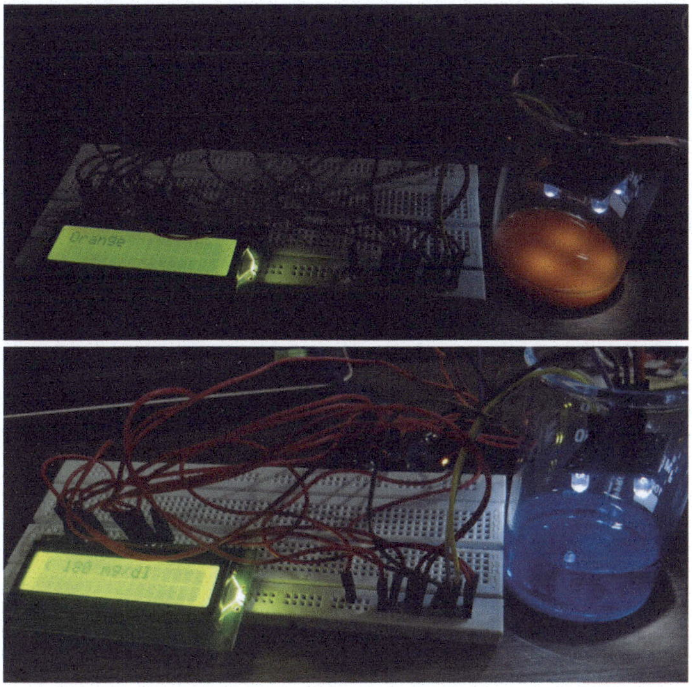

Fig. 7 The experimental setup of working prototype

keeping track of the user's fitness activities and giving necessary feedback on the same.

References

1. www.searo.who.int, World Health Organization (India), 'Diabetes'. http://www.searo.who.int/india/topics/diabetes_mellitus/en/. Accessed 14 June 2019
2. Simoni RD, Hill RL, Vaughan M benedict's solution, a reagent for measuring reducing sugars: the clinical chemistry of Stanley R. Benedict. http://www.jbc.org/content/277/16/e5.full. Accessed 14 June 2019
3. www.himedialabs.com, 'Benedict's qualitative reagent', Revision 1, 2015. http://himedialabs.com/TD/R002.pdf. Accessed 14 June 2019

Improved Ant Colony Optimization in K-Means for Data Clustering

S. S. Bamane, A. J. Umbarkar and M. R. Gaikwad

Abstract Clustering is grouping of similar data points in clusters. Clustering has many applications, particularly in big data analytics. In data mining, traditional algorithm are used for clustering. These algorithms are inefficient in terms quality of cluster. This paper attempts to improvise the traditional K-mean by adding the Ant Colony Optimization algorithm (ACO) for improving the centroid for better clustering. This combination of ACO in K-mean and IACO in K-mean is experimented on iris and skin segmentation supervised datasets. Experimental results show that the performance in terms of F-measure for IACO in K-mean is better than ACO in K-mean and traditional K-means for iris and skin segmentation datasets.

Keywords Evolutionary algorithms · EA · Ant colony optimization · ACO · Clustering · Improved ACO · K-means

1 Introduction

Traditional data clustering algorithms-C-means, K-means and fuzzy C-means are influenced to the centroid of the cluster. These algorithms work with the predefined fixed centroid. This predefined fixed location centroid influence the quality of clustering because of some data points are far away from the predefined centroid. To have quality cluster with better centroid, evolutionary algorithm can be useful.

Ant Colony Optimization (ACO) algorithm is a population based optimization algorithm can be used for varying the centroid and optimizing the cluster formation [1–6]. ACO is an iterative algorithm, which has multiple calculations for fitness function. This quality of ACO with Euclidian distance calculation for cluster formation can be applied in K-mean, to form the quality centroid for a cluster. Centroid position calculation is randomly chosen on the possibility that the cluster should move towards the best arrangement and endeavor to stay away from the bad arrangements. This enhancement in K-mean calculation, improves the quality of clustering. The probability added to ACO algorithm gives improved quality of clusters.

S. S. Bamane · A. J. Umbarkar (✉) · M. R. Gaikwad
Walchand College of Engineering, Sangli, Maharashtra, India
e-mail: anantumbarkar@rediffmail.com

S. Fong et al. (eds.), *ICT Analysis and Applications*, Lecture Notes in Networks and Systems 93, https://doi.org/10.1007/978-981-15-0630-7_52

Fu proposed ACO for three methods of cluster center initialization [1]. Many researchers applied ACO for clustering analysis [2, 3]. Chu and Pan applied constrained ACO for data clustering [4, 7]. Kuo and Cho applied Ant K-Means for Clustering Analysis [5]. Jabbar et al. reviewed the ant-based sorting and ACO-based clustering approaches [6].

Azaryuon and Fakhar applied the ACO for Document Clustering in K-mean [8, 9]. Tao et al. proposed Improved ACO for object clustering [10]. Zhou et al. applied the enhanced firefly algorithm for K-harmonic means clustering [11]. Li and Liu applied Improved Krill Herd Algorithm for finding cluster centers on UCI dataset [12, 13]. Merwe and Engelbrecht worked on the particle swarm optimization algorithm for data clustering to refine the clusters formed by K-means [14]. Hasan and Ramakrishnan reviewed the hybrid evolutionary algorithm for cluster analysis [15]. Tripathi et al. proposed enhanced grey wolf optimization algorithm for data clustering on mapreduce framework for large-scale on data sets [16]. Han et al. applied clustering to document with adaptive dimensional search with different types of datasets [17].

In review of literature, it is seen that, the EAs are used in clustering for finding center or centroid of a cluster. Choosing center or centroid of a cluster plays a vital role in quality of clustering. EAs are the best option to find centroid or centre of the cluster. Many researchers applied EAs for data clustering on a distributed framework such as a mapreduce for improving speedup and quality of clustering.

The paper is organized as follows. Section 2 explains the ACO and IACO Algorithms in short. Section 3 describes datasets, experiential details, results and discussion. Section 4 gives the conclusions.

2 Improved ACO in K-Means

2.1 K-Means Clustering Algorithm

Clustering is collection of articles or data items into relative class of that article or data item. K-means clustering aims to partition *n* articles or data items into *k* clusters in which each articles or data items belongs to the cluster with the nearest mean, serving as a prototype of the cluster [1–6]. Figure 1 is short flowchart of K-mean process.

2.2 ACO and IACO for Data Clustering

ACO is population based metaheuristic that can be utilized for discovering inexact answers to hard optimization problems. ACO is type of optimization algorithms made by the moves of the ant colony. Synthetic 'ant' location most suitable solutions through moving through a parameter space representing all viable answers

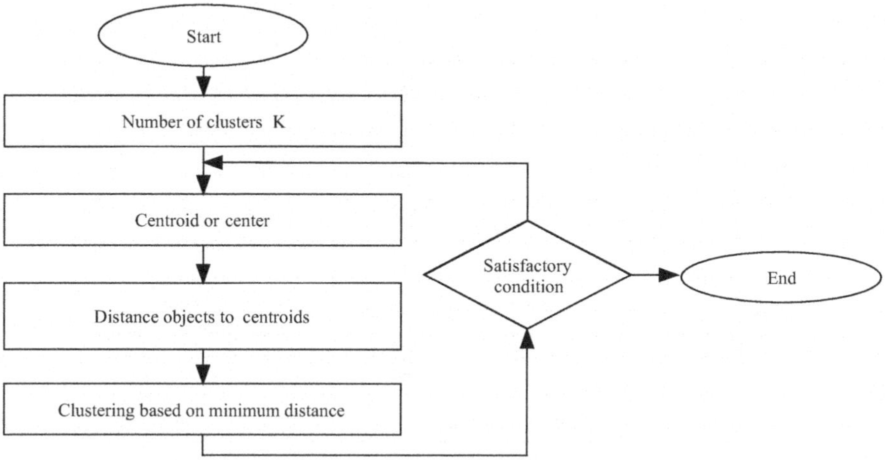

Fig. 1 Flowchart of K-mean [7]

[1–6]. Figure 2 is the flowchart of IACO [1–4, 10]. A step showed in Fig. 2 with bold, is an improvement to ACO [11]. Following are the steps for IACO for clustering.

(1) The data points which are not quality solutions are updated by the Eq. (1). Previous results are abandoned [1–4, 10].

$$s = \left\{ \max\left\{ [t(r,u)]\left[\frac{p_{\mathrm{u}}}{d(r,u).T_{\mathrm{u}}}\right] \right\} \right\} \text{if}\, q \le q_0 \tag{1}$$

where, s is probability of a cluster, q is a random variable uniformly distributed between and 0, 1, ($q \in [0, 1]$), q_0 is a factor which shows a balance factor.

The Euclidean Distance Eq. (2) is the calculation of fitness of each cluster.

$$d(Z_i + Z_j) = \sqrt{\sum\nolimits_{t=1}^{t} (Z_i^t - Z_j^t)^2} \tag{2}$$

where,

$d(Z_i + Z_j)$ is a Euclidean distance, Z_i is a data object cluster i, Z_j is a data object cluster j, t is a dimensions.

(2) The ants which find the quality solutions are to update the pheromone. This is called global update rule. Not all ants are allowed to update the pheromone. Equation (3) is a cluster updations step.

$$T(r,s) \leftarrow (1-p).t(r,s) + (x+a)^{\mathrm{n}} = \sum\nolimits_{\mathrm{k}=1}^{\mathrm{m}} \Delta T_{\mathrm{g}}(r,s) \tag{3}$$

where, r is random number between 0 and 1, $\Delta \mathrm{T}_{\mathrm{g}}$ is given by Eq. (4).

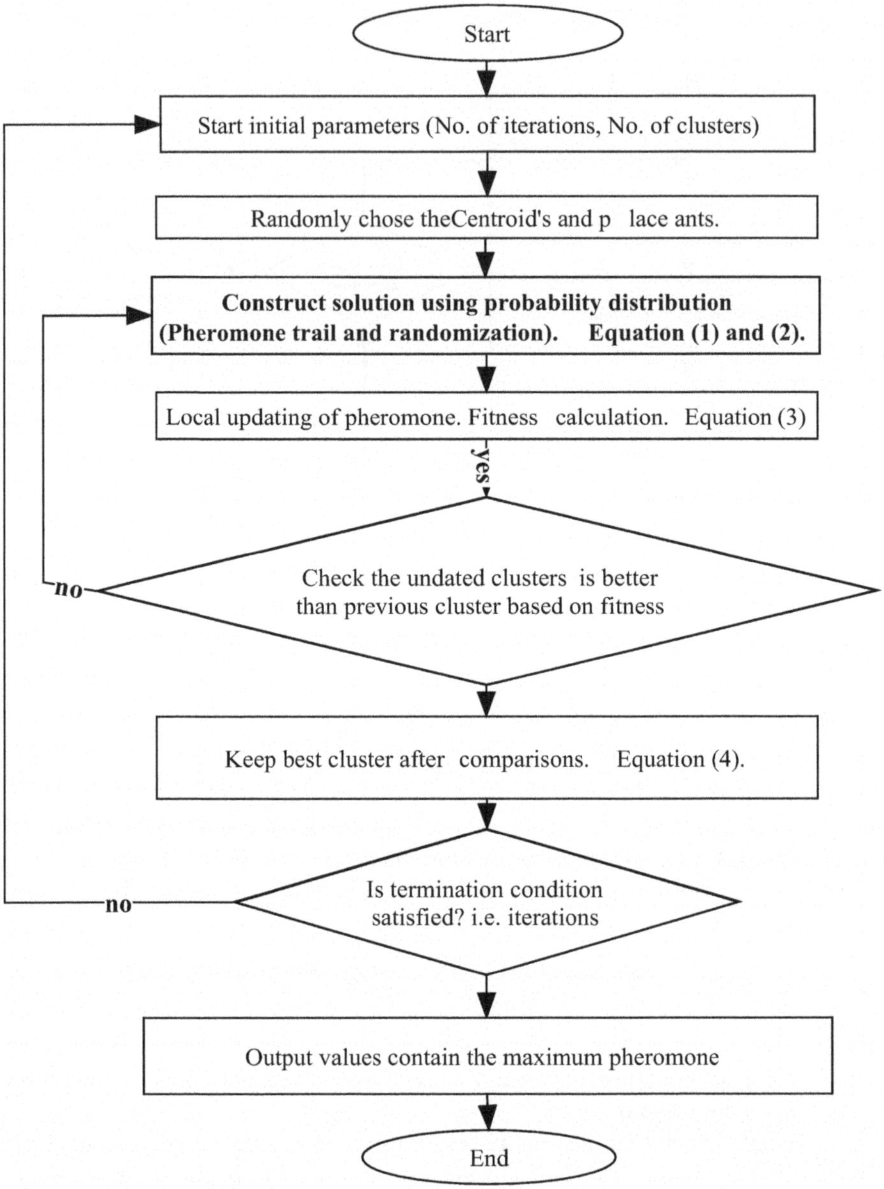

Fig. 2 Flowchart of IACO [1–4, 10]

$$\sum_{k=1}^{m} \Delta T_g(r, s) = \left\{ \begin{array}{r} \frac{1}{T_{gb}}\text{if}(r, s)\ \varepsilon\ \text{global optimal clusters} \\ 0\ \ \text{otherwise} \end{array} \right\} \tag{4}$$

Here, T_{gb} is the shortest time for ant to complete its visit.

(3) The ants are updating the pheromone when the solution is found for the first time. The clusters are updated according to the fitness of clusters. This is called local updated rule given in Eq. (5).

$$T(r, s) \leftarrow (1 - p).t(r, s) + \Delta T_1(r, s) \tag{5}$$

where, $\Delta\ T_1\ (r, s)$ set to the initial pheromone trail concentration T_0.

3 Experimental Results

3.1 UCI Dataset

UCI Iris and Skin segmentation datasets is taken for experimentation in work. The iris dataset contains the sepal length, sepal width, petal length, petal width in centimeters and three classes are Iris Setosa, Iris Versicolor and Iris Virginica. The Skin segmentation dataset is constructed over Blue (B), Green (G), and Red (R) colors space. In the iris dataset, there are of 50 samples for 3 different species of iris measurements [13] (Table 1).

Table 2 depicts the results of the ACO and IACO algorithm for iris dataset. It has the comparison of F-measure along with the k means being compared for iris dataset. It is observed that the proposed method gives the better performance in comparison with the others.

Table 1 Types of datasets: [13]

Sr. no	Name of dataset	Attributes in dataset
1	Iris	13
2	Skin segmentation	4

Table 2 F-measure of ACO in K-mean and IACO in K-mean for iris dataset

Cluster	ACO			IACO		
	Precision	Recall	F-measure	Precision	Recall	F-measure
1	0.86	0.7	0.74	0.97	0.97	0.97
2	0.9	0.8	0.84	0.89	0.68	0.77
3	0.8	0.8	0.8	0.6	0.6	0.6

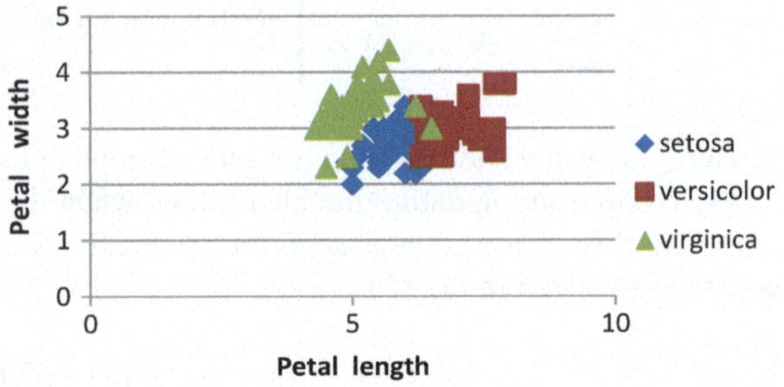

Fig. 3 ACO in K-mean for iris dataset

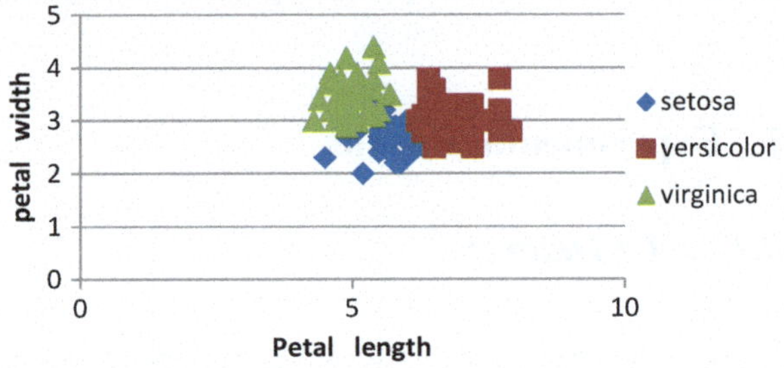

Fig. 4 IACO in K-mean for iris dataset

In Figs. 3 and 4 shows the three cluster points, where in Fig. 4 shows improved cluster over the Fig. 3. Figure 4 shows that, the better cluster data points in three clusters. The iris dataset used for this experimentation is supervised dataset for clustering.

Table 3 shows the results of the ACO algorithm, IACO algorithm and K-mean for iris and skin segmentation datasets. Results tables and plot of IACO shows that, IACO in K-mean method gives the better performance in comparison with ACO and traditional K-mean.

Figures 5 and 6 shows the data points of clusters for ACO in K-mean and IACO in K-mean for skin segmentation. Plots in Figs. 5 and 6 are only plotting 32000 data points for skin segmentation dataset. This is due to the plotting software restriction (Table 4).

Table 3 F-measure of ACO in K-mean and IACO in K-mean for skin segmentation dataset

Cluster	ACO			IACO		
	Precision	Recall	F-measure	Precision	Recall	F-measure
1	0.8	0.8	0.81	0.77	0.9	0.86
2	0.7	0.6	0.64	0.88	0.8	0.85
3	0.8	0.9	0.84	0.55	0.7	0.61

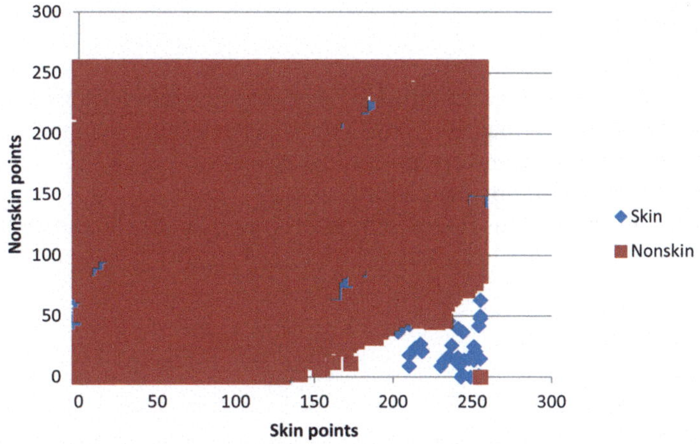

Fig. 5 ACO in K-mean for skin segmentation

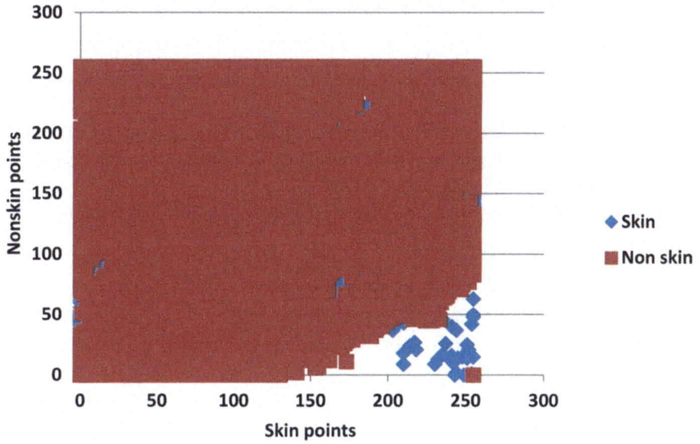

Fig. 6 IACO in K-mean for skin segmentation

Table 4 Compariosn of F-measure for ACO in K-mean, IACO in K-mean and K-means for both the datasets

	ACO	IACO	K-mean
Iris dataset	0.91	0.94	0.75
Skin segmentation	0.79	0.82	0.75

4 Conclusions

In this work, ACO and IACO are applied in K-mean for data clustering. Experimental result show that the performance of IACO in K-mean is better than the ACO in K-mean, and traditional K-means in terms of F-measure. Hence, an IACO gives better performance over the traditional algorithm.

In future scope, the various EAs or hybrid EAs can similarly be experimented for data clustering.

References

1. Fu H (2008) A novel clustering algorithm with ant colony optimization. In: IEEE pacific-Asia workshop on computational intelligence and industrial application, vol 36, pp 66–69, 179
2. Liu X, Fu H (2010) An Effective clustering algorithm with ant colony. J Comput 4:200–210
3. Sinha A, Sahoo G (2007) Ant colony based hybrid optimization for data clustering. Kybernetes 36:175–191
4. Chu S, Pan J (2004) Constrained Ant Colony Optimization for Data Clustering 3157:534–543
5. Kuo R, Cho D (2005) Application of ant K-means on clustering analysis. Comput Math Appl 1709–1724
6. Jabbar AM, Mahamud K, Sagban R (2018) Ant-based sorting and ACO-based clustering approaches: a review. In: International conference on computer applications & industrial electronics. IEEE, pp 217–223
7. Pang C, Hu B, Zhang J, Hu W, Shan Z (2014) Applying data clustering feature to speed up ant colony optimization. Abstract and Applied Analysis 24:8–12
8. Azaryuon K, Fakhar B (2013) A novel document clustering algorithm based on ant colony optimization algorithm. J Math Comput Sci 7:171–180
9. Nagarajan E, Saritha K, MadhuGayathri G (2017) Document clustering using ant colony algorithm. In: International conference on big data analytics and computational intelligence, pp 459–463
10. Tao W, Ma Y, Tian J, Li M, Duan W, Liang Y (2012) An improved ant colony clustering algorithm. In: Zhu R, Ma Y (eds) Information engineering and applications. Lecture notes in electrical engineering, vol 154. Springer, London
11. Zhou Z, Zhu S, Zhang D (2015) A novel K-harmonic means clustering based on enhanced firefly algorithm. In: 5th international conference on intelligence science and big data engineering, big data and machine learning techniques, vol 9243
12. Li Q, Liu B (2017) Clustering using an improved Krill Herd algorithm. Algorithms 10(56):2017
13. UCI repository of machine learning databases (2019) http://archive.ics.uci.edu/m1/index.php. Accessed 21 June 2019
14. Merwe D, Engelbrecht A (2003) Data clustering using particle swarm optimization. In: Proceeding of 2003 congress on evolutionary computation, vol 1. IEEE, pp 215–220
15. Hasan MJA, Ramakrishnan S (2011) A survey: hybrid evolutionary algorithms for cluster analysis. Artif Intell Rev 36:179
16. Tripathi A, Bala M (2018) A novel clustering method using enhanced gray wolf optimizer and mapreduce. Big Data Res 236(3):93–100
17. Han X, Quan L (2017) A novel clustering method using enhanced gravitational search algorithm and mapreduce. Eng Appl Artif Intell 236(3):1–8

Character Segmentation and Recognition of Indian Devanagari Script

Milind S. Khanderao and Sachin Ruikar

Abstract The paper presents Devanagari Character Segmentation and Recognition using neural networks. The hybrid features extraction technique which is the combination of geometric and statistical features is implemented. The geometric feature extraction technique uses directional features of Skeletonized Character image, whereas the statistical feature technique uses distribution of pixel density and Euclid features of the skeletonized character image. For classification, SVM (Support Vector Machine) and MLP (Multi Layer Perceptron) are used as classifiers. The Support Vector Machine has more accuracy as compare to MLP.

Keywords Devanagari script · Character recognition · Statistical feature · Support vector machine

1 Introduction

The Indian Devanagari character segmentation and recognition system, which defines the ability of a machine to analyze and identify the script characters is implemented here. Over the last few decades, machine reading has grown day by day. Recognition Optical character recognition in image processing and artificial intelligence has become one of the most successful applications of technology. Its classification based on two major factors: acquisition of data process and the type of text type (Noise reduced). The goal state is to fetch the character of Devanagari script into digital form after identification. The basic Character set of Devanagari script is called as Aksharas. Samyuktaksharas is the joint character word of Devanagari script. The Samyuktaksharas consists of vowels, consonants and joint characters. The character set of Devanagari script contains of 34 consonants and 12 vowels in addition of 14 vowel modifiers. Besides consonants and vowels, it also contains modifiers called Kana, a slating line placed at the top of character and Matra's which are placed at

M. S. Khanderao · S. Ruikar (✉)
Walchand College of Engineering, Sangli, Maharashtra, India
e-mail: ruikarsachin@gmail.com

M. S. Khanderao
e-mail: milindraok@gmail.com

S. Fong et al. (eds.), *ICT Analysis and Applications*, Lecture Notes in Networks and Systems 93, https://doi.org/10.1007/978-981-15-0630-7_53

left or right part of the character. The half character increases complexity of script and lower modifier too. The Devanagari script writing style is from left to right. The segmentation of character is an operation of decomposing image into sub image of characters. Thereafter several operations like preprocessing, feature extraction, and its respective classification is done. Technology of Devanagari script character recognition had been led to more transform development.

The various researches had been implemented on Devanagari script character recognition. Image of written document in Devanagari script fed as an input image and the editable file taken as output has been implemented. The structure of the script was used in the proposed scheme for segmentation with a homogenous set of features for recognition, which are computationally simple for extracting. Final recognition performed by Support Vector Machine (SVM) classifiers [1]. Patterns orientate segmentation technique for optical character recognition that contributes to document structure analysis. An extended form of pattern oriented segmentation is considered. An efficient and computationally focused method for segmenting character and graphics part of scanned images based on textural cues is used. It segmented using vertical and horizontal projection whereas convex hull technique for feature extraction used [2]. The data set is maintained for preparation and character classification. Data base created by performing standard operations like preprocessing, feature extraction, and distribution of Training set and Testing set. The Deep Convolution Neural Network trains the training set and accuracy evaluated from Testing set [3]. To perform high accuracy for recognition neural network is implemented in the system. The hidden layer and output layer each consist of 33 neurons [4, 5]. The combination classifier used for classification to solve classification problem. The combination of various networks is having more advantage as compare to individual one [6].

2 System Design of Devanagari Optical Character Recognition (DOCR)

Devanagari optical character recognition converts image of text into digital text. It has various applications such as postal address automation, historical document preservation, etc. Character recognition is an interested area in developing systems

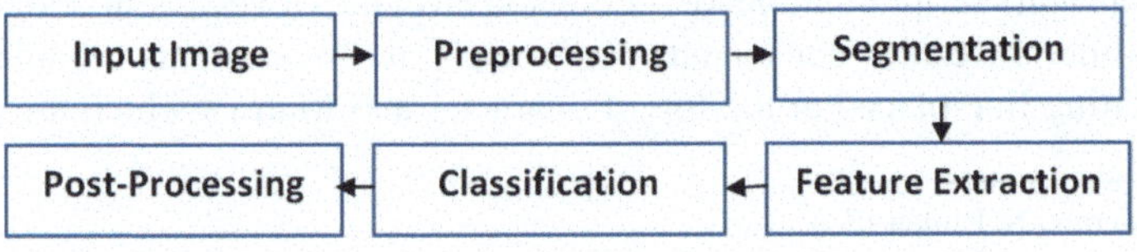

Fig. 1 Steps in DOCR

that works in documents analysis and recognition (see Fig. 1). Input can be taken from sources such as mobile, PDA, digitizer, etc.

2.1 Image Pre-processing

It is the process in which some morphological operations are performed on image, in order to extract useful information from it and convert it into digital form are called as Image pre-processing. Initially the digital images contain various objects with noise and have to remove it without distortion. To enhance the properties of the image, the series of operations performed in pre-processing. To get better results, it is essential for image enhancement. In proposed system two dimensional images is utilized for all the sets of operations. In the preprocessing following operation are performed:

RGB Image to GRAY Image

An RGB image is mainly considered as a True image. Three colors red, green and blue are combined to form any of the color in visible spectrum. Initially raw input image taken is referred as RGB image. Further that is converted into Gray scale to perform necessary operations on it. In gray scale images, only shades of gray are present, which is the required input for all further process.

Noise Removal

Digital images contain various types of noise. In the image acquisition process, it is added in the image. To nullify noise various filters used like high-pass, low-pass, band pass, average filters etc. The median filter is used in proposed system to reduce blurring of character and improve the feature. The median filter avoids edges degradation. We cannot remove it completely but we can reduce it up to certain extent. By applying various filters we can reduce the noise present in the digital image.

Binary Image

An image having only two values for each pixel is nothing but the binary image. Precisely, two, white and black are accustomed for binary. These are also called bi-level or two-level images. Here, each pixel is stored as single bit either 0 or 1. As binary image having only two values (0&1), it reduces the storage requirements as compared with RGB or Gray.

Inversion of Image

Image inversion is a process where dark areas are mapped as light and light areas are mapped as dark. In simple word, black is converted to white and white is converted to black. The resultant black and white image is the replication of original image in white to black and vice versa. The binary image is used as input image for this operation. The inverted image is very helpful for various operations. The objects present in the inverted image more highlighted as compare to binary image. Background of an inverted image is mostly black colored whereas the foreground or the objects are in white colored.

Universe of Discourse

The smallest matrix where entire character skeleton gets fitted is universe of discoure. Features extracted from scanned image of character have positions of different line segments, so universe of discourse has to be selected. After applying it gives the image fitted to required image size and the unwanted background suppressed resulting in enlarging objects of input image.

Image Normalization

It is a process in which range of pixel intensity values is changed. Sometimes, normalization is referred as histogram stretching or contrast stretching. Input image histogram and the processed image histogram show the difference properly. The image normalization enhances the input image, which results in sharpen output image.

Image Skeletonization

It is a process where foreground pixels in a binary image are reduced to a skeletal remnant. For doing the image skeletonization, input image has to undergo some morphological operations such as thinning. It successively erodes away the boundary pixels without affecting the end points of object segment. The morphological thinning process results into skeleton of input image. The alternative method is to first compute distance transform of image. Skeleton then lies along the distinctiveness in distance transform.

2.2 Segmentation

In segmentation process, the pre-processed input image undergoes partitioning an image into multiple parts with extracting the region of interest. It is the basic of image analysis. Quality of image segmentation directly affects the results of concerned image processing operation. Segmentation methods include thresholding, edge detection, graph cut method and machine learning based method. In thresholding objects of interest are separated from its background by some grey-level threshold value.

The $g(i, j)$ is obtained by applying threshold function to $f(i, j)$, then

$$\begin{aligned} g(i, j) &= 1 \quad \text{if } f(i, j) \geq T \\ &= 0 \quad \text{Otherwise} \end{aligned}$$

Threshold is obtained using: $T_r = M[i, j, p(i, j), f(i, j)]$.

In above expression, T is threshold, $f(i, j)$ is gray value of (i, j) & $p\ (i, j)$ is some local property of point such as average gray value of neighborhood centered on point (i, j).

2.3 *Feature Extraction*

Feature extraction is the process where relevant information about different shapes present in pattern is detected and put in a vector to use further for the classification. Depending on the feature vectors system classifies the inputs with significant accuracy. So, features extracted should be highly discriminative with reduced dimensions in order to reduce the computation requirements in classification. If feature extraction is done properly, then it can also reduce the errors such as mean square error or inter-distance differences, etc.

Zoning

In the process of zoning, binary image having Devanagari script characters which is pre-processed and normalized to size of 36×36, is partitioned into n number of equal sized zones. Then features are extracted from individual zone. The advantage of finding features from individual zone over that from whole image is that, it provides more detailed information regarding small and finer details in skeleton of character image. After calculating feature vectors from each zone, they are put in an array to make one feature vector representing the features for a given input image.

Directional Features

These features are extracted from image skeleton, based on the line types forming the character skeleton. To do this, image is zoned or partitioned into 3×3 sub images and features are extracted from each of the zones.

Moment Invariant Features

These are the statistical features. In object recognition moments play significant role. This method measures the intensity function. It gives global character information.

3 Classification and Recognition

After features are extracted, inputs are classified and recognized depending upon the feature values. For this, in this system two classifiers, namely, SVM (Support Vector Machine) and MLP (Multi Layer Perceptron).

3.1 *Support Vector Machines*

The concept of SVM is based on linear separation using hyper plane which is fitted to two classes within multi-dimensional space. SVM is the supervised method of classification in which both target and input data sets are given.

Classification Tasks with Hyper Planes

SVM uses hyper plane to divide the data into some groups of similar elements. Here we consider hyper planes that are linear in nature. In 2-Dimensional space, this would be a straight line and in 3-Dimensional space, it would be a plane.

Maximum Margin

A maximum margin hyper plane, in simple words, is just a line/surface that creates separation between the classes of data. In the given figure, any of the three lines could do the job of separating the data. But the one with the ability to create the greatest separation would take care of some variations in positions of data points near boundary. These variations may lead to miss-classifications.

Support Vectors

Support vectors are those points in each of the classes that are closest to maximum margin hyper plane. Every class should have at least one support vector, but maybe having more than one (see Fig. 2).

In n-dimensional space, the hyper plane is defined as $W \cdot X + B = 0$ where W is a vector of n weights i.e. $[w1, w2 \ldots.. wn]$. B is a single number called the bias.

Using the formula, aim of the algorithm is to find out a set of weights that specify two hyper planes as follows,

$$W \cdot X + B >= 1 \quad W \cdot X + B <= -1$$

These hyper planes should have all the points of the first class fall above the first hyper plane and all points in other class fall beneath second hyper plane. This is possible only when data is linearly separable. The distance between these two planes is 2/||*W*|| · ||*W*|| indicates the Euclidean Norm (the distance from the origin to the vector W). For maximizing distance, ||*W*|| should be minimized.

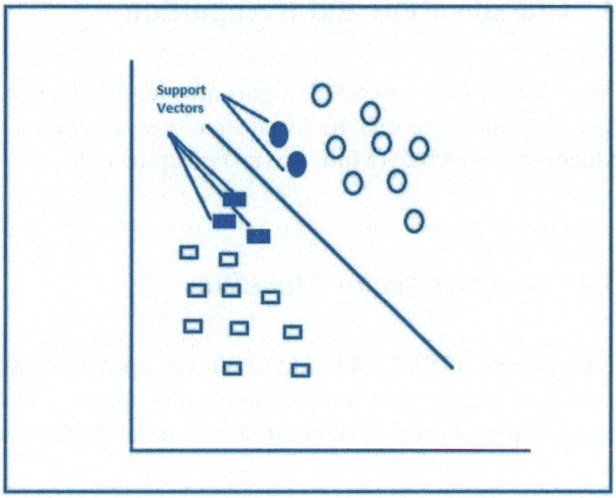

Fig. 2 Support vectors near hyper planes

4 Results

The results produced for Devanagari scripts are shown in Tables 1 and 2. The recognition of Indian Devanagari Script character becomes difficult due to presence of odd characters or similarity in shapes for multiple characters. Scanned image is pre-processed to get a cleaned image and the characters are isolated into individual characters. Preprocessing work is done in which normalization, filtration is performed using processing steps which produce noise free and clean output. Managing our evolution algorithm with proper training, evaluation other step wise process will lead to successful output of system with better efficiency. Use of some statistical features and geometric features through neural network will provided better recognition result of Indian Devanagari Script characters.

Devanagari Character Recognition using neural network performed with hybrid features extraction method. It is the combination of geometric and statistical feature extraction technique. The geometric feature extraction techniques use directional features. The statistical feature technique uses distribution of pixel density and Euclid features of the skeletonized character image. SVM and MLP classifier used for classification of Devanagari Script character. The Support vector machine has more accuracy as compare to MLP. The SVM is slower than the MLP.

Table 1 Results for classification and recognition of Devanagari scripts

Class	SVM	MLP	Class	SVM	MLP
क	73.28	18.4	द	30.01	12.58
ख	23.8	13.84	ध	77.83	30.98
ग	79.34	25.25	न	6.8	6
घ	33.35	14.69	प	54.52	28.12
ङ	84.49	18.67	फ	50.5	27.12
च	17.33	12.71	ब	60.33	13.82
छ	7.8	7.4	भ	7.2	5.1
ज	44.99	16.11	म	63.71	27.25
झ	4.7	6.38	य	56.88	15.91
ञ	87.98	18.42	र	98.5	58.74
ट	75.72	37.04	ल	57.25	16.82
ठ	33.45	20.46	ळ	6.3	6.7
ड	94.01	22.25	व	40.2	12.8
ढ	38.81	36.63	ह	27.73	18.79
ण	25.54	22.39	श	61.92	21.86
त	78.01	16.36	ष	82.27	30.16
थ	78.69	23.44	स	96.5	34.03

Table 2 Results for classification and recognition of Devanagari scripts

Class	SVM	MLP	Class	SVM	MLP
अ	87.5	47.7	०	98.8	97.4
आ	92	59	१	93.5	95.2
इ	75.5	42.6	२	87.8	67
ई	91.7	60.4	३	92.6	65.3
उ	94.3	47.3	४	93.8	65.3
ऊ	94.2	55.5	५	97.3	55.5
ए	99.8	96.1	६	12.3	12.6
ऐ	98.7	67.1	७	31.2	40.3
ओ	79.8	35.9	८	15.4	14.4
औ	96.7	48.5	९	54.3	39.1
अं	92.3	44.3			
अः	32.8	30.9			

5 Conclusion

Devanagari Character Recognition using neural network performed with hybrid features extraction method that is the combination of geometric and statistical feature extraction technique implemented. The geometric feature extraction techniques use directional features. The statistical feature technique uses distribution of pixel density and Euclid features of the skeletonized character image. SVM and MLP classifier used for classification of Devanagari Script character. The Support vector machine has more accuracy as compare to MLP. The SVM is slower than the MLP.

References

1. Sahare P, Dhok SB (2018) Multilingual character segmentation & recognition schemes for Indian document images. IEEE access, vol 6, pp 10603–10617
2. Pachpande S, Chaudhari A (2017) Implementation of Devanagari character recognition system through pattern recognition techniques. In: International conference on trends in electronics and informatics (ICEI), Tirunelveli, pp 717–722
3. M Mathew, AK Singh, CV Jawahar (2016) Multilingual OCR for Indic script. In: IAPR workshop document analysis systems, pp 186–191, Apr 2016
4. Acharya SR (2015) Deep learning based large scale handwritten Devanagari character recognition. In: International conference on software, knowledge, information management and applications, pp 1–6
5. Dhaka VP, Sharma MK (2015) Efficient segmentation technique for Devanagari offline handwritten scripts using feed-forward neural network, neural computer. Application, vol 26. IEEE
6. Neha NKR (2013) An efficient handwritten Devanagari character recognition system using neural network. IEEE

7. Dongre VJ, Mankar VH (2013) Devanagari handwritten numeral recognition using geometric features and statistical combination classifier. Int J Comput Sci Eng 5
8. Quraishi MI, Choudhury JP, De M (2012) Image recognition and processing using artificial neural network, recent advances in information technology. RAIT-IEEE
9. Aggarwal A, Renudhir RR (2012) Handwritten Devanagari character recognition using gradient features. IJARCSSE 2
10. Shelke S, Apte S (2011) A Multistage handwritten marathi compound character recognition scheme using neural networks and wavelet features. Int J Signal Process Image Process Pattern Recogn 4(1)
11. Desai A, Malik L (2011) A modified approach to the thinning of devanagari characters. IEEE
12. Dayashankar Singh AP (2010) Handwritten english character recognition using neural, network. Int J Comput Sci Commun 1(2)
13. Bhattacharya U, Chaudhuri BB (2009) Handwritten numeral databases of indian scripts and multistage recognition of mixed numerals. IEEE Trans Pattern Anal Mach. Intell 31(3)

Application of Machine Learning on Remote Sensing Data for Sugarcane Crop Classification: A Review

Shyamal S. Virnodkar, Vinod K. Pachghare, V. C. Patil and Sunil Kumar Jha

Abstract Sugarcane is a major contributing component in the economy of tropical and subtropical countries like India, Brazil and China. Sugarcane agriculture is empowered with the advancements in the remote sensing technology because of its timely, non invasive, and labor and cost effective capability. Remote sensing data with machine learning algorithms like Support Vector Machine, Artificial Neural Network and Random Forest are proven to be suitable in sugarcane agriculture. The aim of this paper is to present a review of studies that implemented various machine learning algorithms based on remote sensing data in sugarcane crop mapping and classification.

Keywords Machine learning · Remote sensing · Sugarcane crop classification

1 Introduction

Sugarcane is a semi-perennial and one of the most important crops across the world, especially in India, Brazil, and China. Brazil ranked first in sugar production and second in ethanol production, India is the second largest country followed by China. Sugarcane is a raw biological material that produces several products other than sugar, such as ethanol, bagasse, molasses, rum, cachaca. The growth of sugarcane highly depends on the plantation season and area along with other factors like soil type, precipitation, irrigation, and fertilization. There are three plantation seasons in

S. S. Virnodkar (✉) · V. K. Pachghare
College of Engineering, Pune, SPPU, Pune, India
e-mail: ssv18.comp@coep.ac.in

V. K. Pachghare
e-mail: vkp.comp@coep.ac.in

V. C. Patil · S. K. Jha
K. J. Somaiya Institute of Applied Agricultural Research, Karnataka Saidapur, India
e-mail: patil.vc@somaiya.com

S. K. Jha
e-mail: jha.sunilkumar@somaiya.com

S. Fong et al. (eds.), *ICT Analysis and Applications*, Lecture Notes in Networks and Systems 93, https://doi.org/10.1007/978-981-15-0630-7_55

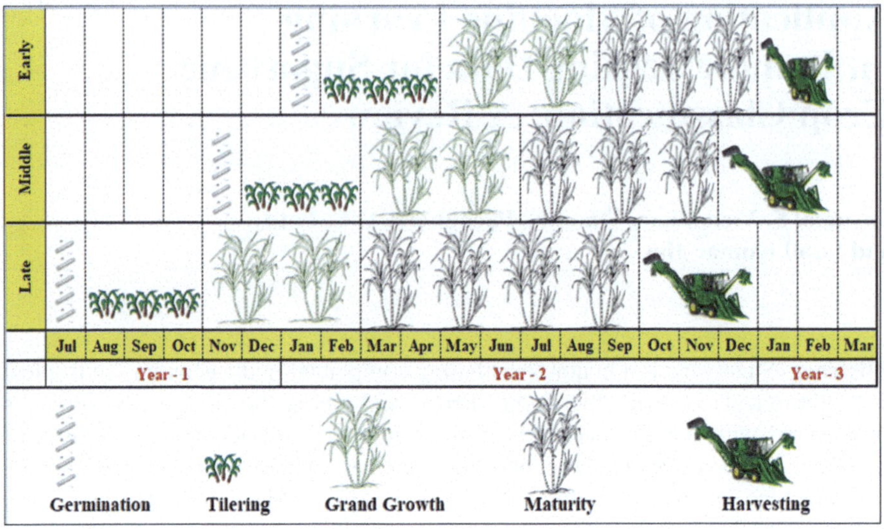

Fig. 1 Sugarcane crop calendar in South India

south India viz. early (Jan–Feb), mid-late (Oct–Nov) and late (Jul–Aug). It undergoes 4 growing phases i.e. germination phase, tillering (formative) phase, grand growth phase, and maturity and ripening phase in south and north India [1], (Fig. 1). Plantation and ratoon are the 2 types of practices to grow sugarcane crops whose plantation cycle is 12–18 months. After harvesting for the first time, it is allowed to grow again without completely ploughed out several times until its quality deteriorate (less no. of shoots), referred to as ratooning. Ratoon crops observed to be low in vigor and mature early than plantation.

Remote sensing (RS) provides an efficacious technique for monitoring crop growth, crop mapping and yield prediction due to its capabilities with regard to its spatial, temporal and spectral resolution [2]. Optical RS data, such as the Landsat-5 Thematic Mapper (TM), Landsat-7 Enhanced Thematic Mapper Plus (ETM+), Moderate Resolution Imaging Spectroradiometer (MODIS), SPOT-5 High Resolution Geometrical (HRG), High Resolution Imaging Camera (CCD) on board of China-Brazil Earth Resources Satellite-2 and—2B (CBERS-2 and—2B), Advanced Spaceborne Thermal Emission and Reflection Radiometer (ASTER), and ENVISAT Advanced Synthetic Aperture Radar (ASAR), have been utilized for mapping sugarcane planting areas, differentiating between sugarcane varieties.

Machine learning, a subsection of Artificial Intelligence (AI), has the power to process abundant data in a nonlinear system. RS generates innumerable data and therefore machine learning algorithms can produce promising results. Numerous machine learning algorithms like decision trees (DT), Support Vector Machine (SVM), Genetic Algorithm (GA), Ensemble Learning (EL) have been successfully applied on RS data in sugarcane crop mapping with good accuracy.

This paper aims to provide a review of the application of machine learning on remote sensing data to classify sugarcane crop. The rest of the paper is shaped as follows. Section 2, presents various image classification approaches and techniques. The literature on sugarcane crop classification is presented in Sect. 3. Section 4 gives a conclusion with future directions for researchers.

2 Remotely Sensed Data Classification Approaches and Techniques

One of the most high interest tasks in RS is to separate the land use and land cover of the earth's surface based on the spectral reflectance of the earth's object through a classification process. Image classification is defined as a process of obtaining extremely useful information from the gigantic satellite imagery and generating maps of classes by designating every pixel of the input satellite image to an information class. Information classes are categorical such as crop type, built up, rock, water body, forest type, tree species, different geological units, etc. Whereas groups of pixels that are nearly similar in their reflectance values in different spectral bands are called spectral classes. RS data classification is executed based on the spectral classes of different features of earth's surface. Different approaches with techniques to classify remotely sensed data are as shown in Fig. 2 [3, 4].

3 Application of Machine Learning for Sugarcane Crop Classification

Over the last 40 years, crop classification has been experimented using RS by many researchers still, it is a challenging task. In 2008, [5] presented a review for sugarcane agriculture which explains RS applications such as variety identification, yield prediction, crop classification, disease detection, health and nutrient status monitoring. The development in machine learning algorithm improves the crop classification task. In this section various ML algorithms in the area sugarcane crop classification based on remotely sensed data are reviewed. Table 1 presents comparative analysis of the research done for sugarcane classification in the form of features evaluated, classifiers used, training samples, input imagery, accuracy achieved, etc.

3.1 Decision Tree

Decision tree is a non parametric method that generates classification tree from training data was implemented by [1] on LISS IV images to identify sugarcane crop

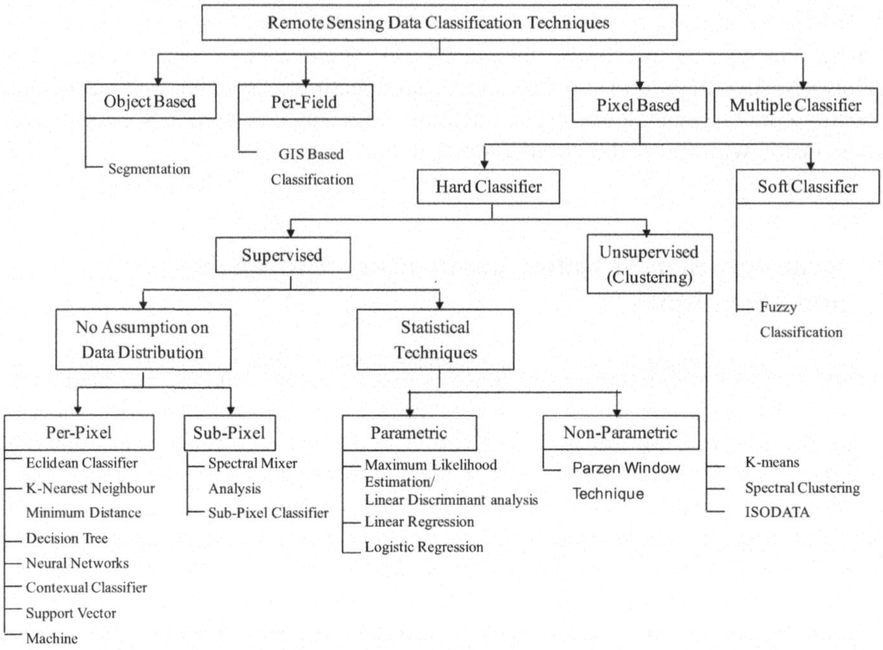

Fig. 2 Image classification approaches in RS

along with other 7 classes for the region of Chhapar village, Muzaffarnagar, India. Classification was implemented using three methods: ISODATA, Maximum Likelihood Classifier (MLC) applied on layered stacked image and DT on vegetation indices. 11 vegetation indices were experimented, among which Normalized Difference Vegetation Index (NDVI), Green Normalized Difference Vegetation Index (GNDVI), Difference Vegetation Index (DVI), Optimized Soil Adjusted Vegetation Index (OSAVI) have a positive effect in the generation of decision tree. Decision tree approach outperformed with 87.93% of overall accuracy.

3.2 *Maximum Likelihood Algorithm*

Maximum probability Algorithm (MLA) is a parametric classifier which doles out a pixel to a class based on its likelihood of having a place with a class whose mean and covariance are demonstrated as forming a normal distribution in multispectral feature space. MLA is a basic classifier which was used widely before development in machine learning algorithms took place.

With the help of MLA, [6] mapped the sugarcane vegetation area in the region of Uttarkhand, India. To classify sugarcane crop and map its variety, authors evaluated the capability of ASTER satellite, by developing the spectral signature of sugarcane

Table 1 Comparative analysis of sugarcane classification studies using machine learning

Ref. No.	Features evaluated	Ground truth data, T-Training V-Validation	Dataset used	Classifier	Method	Overall accuracy	Advantages	Disadvantages	Observations
[1]	11 Vegetation indices	T- 400 V-253	LISS IV	DT	Based on 11 Vegetation indices	87.93	ISODATA and MLC producing acceptable results on single date imagery considering peak growing stage of sugarcane	ISODATA and MLC algorithms unable to classify sugarcane crop areas with acceptable accuracy	Vegetation indices along with high-resolution time series images of the entire crop cycle
				ISODATA	Layer stacking of NIR, Green and Red bands	62.07			
				MLC		76.18			
[6]	NDVI, 3 VNIR bands	T-2500 V-562 Pixels	ASTER (3rd Oct. 2004)	MLA	Using 3 VNIR bands	74.3	ASTER data have proven to be a useful resource to extract the sugarcane information	MLC was unable to separate different varieties of sugarcanes	Three crops are heavily mixed because of water content and crop variety. May not be useful for sugarcane variety mapping. Overall accuracies are low
					3 VNIR bands after NDVI	76.2			
					3 VNIR bands after Atmospheric correction	79.5			
[7]	NDVI, NDWI	T-960 V-320 Data fields	Landsat 8 20 images (Apr 2013–May 2014)	MLC	NDVI layered stacking with MLC	83.8	Landsat NDVI has shown great potential for detecting crop type, for medium sized farms over 1 hector	High resolution datasets are required to map sugarcane fields	NDWI is an effective indicator to detect harvest mode as it achieved high accuracy
					Temporal variations if NDVI and NDWI to detect harvest mode	90			
[8]	B2, B3, B4 bands	249 pixels	IRS LISS IV 2nd Oct. 2014	SVM	Classification on 3 bands	86.04	Good accuracy using spectral features	Paid dataset	Features extracted using GLCM approach outperformed raw band features
					GLCM 8 measures on 3 bands	90.29			

(continued)

Table 1 (continued)

Ref. No.	Features evaluated	Ground truth data, T-Training V-Validation	Dataset used	Classifier	Method	Overall accuracy		Advantages	Disadvantages	Observations
[9]	EVI	Training samples taken from EVI cubic image	Landsat 8 22 images (14th Jan–30th Nov 2015)	SVM	Phenology profile using EVI	90.78		Worked on less training samples	low temporal resolution of imagery creates gaps in data for profile creation	Pairwise multiclass strategy is used by training several SVM's
[10]	G, R, NIR, SWIR bands	500 observations	SPOT 5 (19th May 2016)	MD	Based on class means derived from the training data	74.6		Worked on raw bands, No idices calulation is required	Dataset is paid	Raw bands are useful for classifiation
				MAHD	Similar to MD with covariance matrix used in calculation	81.6				
				MLA	Based on probability of a pixel belonging to a class	87.4				
				SAM	Based on minimum spectral angle	68.2				
				SVM	Kernel based technique	86.4				
[14]	150 bands	2402 pixels and 84 paddocks	EO-1 Hyperion (2nd Apr 2002)			Variety	Cycle/class	Nine sugarcane varieties are classified with satisfactory classification accuracy	Dataset is paid	Machine learning algorithms outperforms the statistical approaches. Paddock level method is useful for sugarcane varieties discrimination
				LDA stepwise	Per pixel classification	76.4	57.4			
				LDA		76.8	57			
				PDA		79.7	62.3			
				RF		87.5	80.4			
				SVM		90	83.9			
				LDA stepwise	Per paddock classification	82.7	63.9			

(continued)

Table 1 (continued)

Ref. No.	Features evaluated	Ground truth data, T-Training V-Validation	Dataset used	Classifier	Method	Overall accuracy			Advantages	Disadvantages	Observations
				LDA		86.9		68.7			
				PDA		85.7		77.1			
				RF		100		97.6			
				SVM		98.8		100			
[15]	NIR, R, G, B bands	Data from field and google earth	Sentinel 2 (19th Feb 2018)	RF	Classification based on a single date layered stacking of NIR, Red, Green, Blue bands	84.22			Freely available dataset. Use of single date imagery	Not much accurate.	RF and SVM worked well on Sentinel raw bands
				SVM		81.85					
[18]	B-G, G, R,NIR,SWIR	No ground data	MSR5	K-nn	k-means clustering	98			No ground survey required	Human intervention required in imagery acquisition	Handheld radiometer was used
[22]	NDVI	Area1(A1)T-182, V-40 Area2(A2)-T-251,V-605 Area3(A3)-T-261,V-606	A1-13 images, A2-13 images, A3-21 images			***A1***	***A2***	***A3***	Significantly higher accuracy using pixel based and object based approach	Mixed results of classification in all three test areas	All classified crops have distinct temporal profile
				TWDTW	PB-NDVI time series	94.8	87.1	74.9			
					OB- NDVI time series	96.2	89.8	78.1			
				RF	PB- NDVI time series	97.1	87.4	88.8			
					OB- NDVI time series	97.6	86.3	88.3			
[28]	NDVI	Toposheets and field data	Landsat 8 12 images 12th Feb 2015–11th Nov 2015	No classifier	NDVI temporal values	90.91			Worked well for ratoon and plant cane discrimination	Covered only 2 phenologies of sugarcane	High accuracy

(continued)

Table 1 (continued)

Ref. No.	Features evaluated	Ground truth data, T-Training V-Validation	Dataset used	Classifier	Method	Overall accuracy	Advantages	Disadvantages	Observations
[30]	Polarimetric data (VV and VH)	Field visit	Sentinel-1 SAR 28 tiles	Knowledge based classifier	Object based crop cycle identification	82.17	Resolved issue of cloudy weather effect	Little low accuracy compared with other techniques	Use of SAR data provides scope to regions having cloud contaminated images
[31]	Objects	T 382, V-500	6 images	AdaBoost with bosted classifier	Time series of spectral, spatial texture and customized attributes	93.6	Good results with cloudy weather and highly mixed crop land having similar spectral reflectance	Need to calculate spatial and texture attribute values	Shown significance of sugarcane crop phenology in classification. Good accuracy in highly fragmented area

and determining the plant features. They worked on 3 bands but NDVI significantly improved the accuracy of the classification.

Reference [7] investigated spatial and temporal contents of Landsat 8 images to map cropping practices viz. sugarcane crop type and harvest mode in the sugarcane-based cropping system. To classify crop type, sugarcane map was generated using time series of NDVI images and then processed in 2 steps i. Firstly this time series was classified using ground survey data and MLC, ii. Recoding and assigning sugarcane classes into one class and other class for other land cover areas. For harvest mode classification, difference in NDVI and Normalized Difference Water Index (NDWI) was used. MLC is a good classifier if proper training samples are provided and giving good accuracy in the range of 75–90% in sugarcane classification.

3.3 Support Vector Machine

Support vector machine is a statistical learning approach to classify heterogeneous data with higher accuracy without assuming input data distribution. SVM learner has the intention of achieving Optimal Separation Hyperplane (OSH) that is a decision boundary between classes which minimizes classification error in training, by having maximum margin and later generalize to unseen data. The margin of the classifier is maximized with the help of support vectors. Support vectors are data points lie closer to the margin mainly contributes to fit the hyperplane. Other data points are discarded because they do not contribute in position and orientation of hyperplane.

Applying SVM on single data of IRS LISS IV with 3 bands (B2, B3, B4), sugarcane and other 6 classes have been classified by [8]. They extracted 8 statistical features using Gray Level Co-occurrence Matrix (GLCM) from 3 bands.

Reference [9] identified sugarcane plantation area and non-sugarcane area by applying SVM using Radial Basis Function (RBF) on time series Landsat data using (Enhanced Vegetation Index) EVI. Instead of ground truth survey data they developed phonology profile from EVI of the crop.

SPOT5 satellite imagery with four spectral bands, Green (G), Red (R), Near InfraRed (NIR), Short Wave InfraRed (SWIR), and 10 m pixel size was assessed by [10] for classification of five crops viz. corn, cotton, sorghum, sugarcane, and non crop. Five supervised techniques such as minimum distance (MD), Mahalanobis distance (MAHD), maximum likelihood, Spectral Angle Mapper (SAM) and SVM, were applied on the original satellite image and on two images generated from this original image. Their results showed that maximum likelihood and SVM performed better than other three classifiers. Original image with MLC achieved the best accuracy of 91.00 and 87.00% for two sites.

Combination of multispectral data from ResourceSat-I (IRS P6) satellite, hyperspectral data from ASTER and microwave data from RADARSAT/RISAT have been explored for classification of crops by [11]. Sub pixel level feature extraction was performed using statistical learning and contextual based algorithms. Additionally,

SVM classifier was implemented for crop mapping, with the use of GA to optimize training set of SVM, and Differential Evolution Algorithms (DE), Comprehensive Learning Swarm Optimization (CLPSO) and Active Learning Algorithms to optimize the kernel parameters. SVM produces above 80% of accuracy in sugarcane crop land mapping (Table 1) and performs well with limited training samples [12] which is the main challenge in RS applications.

3.4 Random Forest

Random Forest (RF) is an ensemble learner [13] that is built by constructing many weak decision trees for classification and regression. It is a nonparametric machine learning algorithm. Bootstrap are randomly selected from an original dataset to construct multitudinous tress with the replacement of samples. The trees are grown in the best possible ways. Each tress is built independently on each other without pruning, based on the two user defined (hyper parameters—Ntree and Mtry) attributes, forms the forest. The majority vote of predictions of all the tress decides the ensemble's final decision. To test new data, it runs through all the produced trees and each tree votes for a class. The class receives maximum votes will be the final selected class.

Variety and crop cycle of sugarcane plants have been predicted using the EO-1 Hyperion hyperspectral data in Australia by [14]. A range of statistical approaches, Linear Discriminant Analysis (LDA) stepwise, LDA and Penalised Discriminant Analysis (PDA), and RF, SVM were implemented for classification firstly on pixel level, later enhanced to paddock level (Table 1). Paddocks usually contain only one sugarcane crop variety of a particular cycle. An object based classifier is good to predict on a per paddock basis using paddock vector boundaries. Paddock based classification method remarkably improved the classification accuracies.

Four bands, NIR, R, G and Blue (B), of a single date Sentinel 2 satellite imagery have been stacked by [15] to classify crops and other land cover. Classification of 11 classes including sugarcane crop was carried out by applying RF and SVM algorithms. The highest classification accuracy (Table 1) was gained by setting penalty parameter to 64, Gamma to 1 for SVM and ntree is 350 and mtry to 1 in case of RF. Feature importance was also computed whose results showed that NIR and B band for RF are of great importance, and NIR and R band contributed more in SVM classifier.

Supervised machine learning classification algorithms need to build training dataset consisting of ground truth observations during satellite overpass period. Such algorithms are unable to operate on other years without new reference data. Classifier generalization or extension [16] is an approach which is used to get rid of the above problem. Reference [17] listed many studies which uses a classifier trained on one reference dataset of a year and use it for another year. They produced three such generalized classifications for plant and ratoon cane classification, based on object based classification with reflectance indices for each object as listed below

- Same year—Training data for same application year

- Multiyear—Several years training data applied on another year
- Multiyear based map updating and change detection—Previous year reference map updating.

3.5 k Nearest Neighbor

k-Nearest Neighbor (k-NN) is a supervised classifier that classifies objects correctly if the dataset is appropriate. The created sample dataset was validated by [18] in their work to classify cotton and sugarcane using k-means. Five spectral bands B, G, R, NIR and SWIR, of Multispectral Radiometer (MSR5) were experimented as features and three clusters were chosen based on three objects bare land, sugarcane and cotton. k-NN was applied for classification on the validated reflectance dataset, resulted in significant increase in accuracy. Authors found all the 5 bands compatible with Landsat imagery, so argued that the same methodology can be experimented on Landsat image bands.

3.6 Dynamic Time Warping

In time series analysis, dynamic time warping (DTW) is one of the algorithms for measuring similarity between two temporal sequences. It was originally invented for speech recognition and recently is used in remote sensing applications. Paper by [19] have used DTW for Sentinel-2 satellite time series image classification to address following challenges—i. Lack of training (ground truth) data due to weather artifacts or labor intensiveness. ii. Unavailability of temporal data due to cloud coverage, iii. Changes in vegetation cycles. The original DTW is extended to multidimensional time series and modified to handle missing temporal samples and cloud covered images. Authors argued that DTW is evenly applicable for SAR images along with optical data. Further, [20] proposed an efficient time series analysis by reducing memory usage by more than a factor of 5, improved temporal classification results with the help of segmentation and reduction in an execution time.

Variation in DTW, time-weighted DTW, for land use and land cover mapping was proposed by [21] to address main issues of having crops phenological cycle more than a year and seasons variation. This TWDTW method gave remarkable results to classify double cropping, forest, pastures, and single cropping from time series pixel based EVI derived from MODIS data. Many existing cropland mapping studies have been focused on time series analysis on pixel based (PB). Reference [22] focused on object based (OB) classification of Sentinel 2 time series data by applying TWDTW on three different test areas and achieved significant accuracy (Table 1).

3.7 Other Techniques

References [23, 24] developed and evaluated a methodology to map sugarcane area ready for harvesting, in a given year using four dates Landsat-5 and Landsat-7 images. Reference [23] mainly focused on merging Object Based Image Analysis (OBIA) and Data Mining (DM) techniques of AI to improve the conventional method of sugarcane mapping from RS through visual interpretation. OBIA was used to represent the acquired knowledge and DM was applied to construct the knowledge model. They implemented the C4.5 algorithm to generate two models, one with the original training set and another by removing important attributes to evaluate the model's efficiency.

Another interesting RS application, the sugarcane field skips mapping was presented by [25]. The procedure is composed of crop rows identification, sugarcane classification, skip extraction and maps creation phases. Crop rows identification was carried out based on NDVI; OBIA was used for classification using ModelBuilder in ArcGIS. In future, they would work on determining the rate of skip increase in ratoons. OBIA also becomes more acceptable than pixel based classification [25] to process UAV images. UAV provides data with very high spatial resolution generates the multi temporal images of AVHRR/NOAA 16 and 17 from April 2001 to March 2010 were analyzed to monitor sugarcane fields by [26]. NDVI values of these images were analyzed using k-means clustering under the Dynamic Time Warping (DTW) function. Five clusters of each season from 2001/2002 to 2009/2010 were produced to monitor the sugarcane fields.

Logistic Regression (LR) was explored by [27] with a gradient descent algorithm to tune the parameters of the model, in order to classify sugarcane land. One of the poorly studied tasks of ratoon sugarcane discrimination and its growth monitoring was explored by [28]. By taking advantage of multi temporal characteristics, NDVIT, temporal values of NDVI, was utilized for ratoon crop identification. RAMiner (Rule Based Associative Classifier Miner), a developed method by [29] created a model using set of associated rules based on NDVI series to identify sugarcane fields with two way classification steps: conviction value and conviction based probability. RAMiner gave highest accuracy 83.4% in comparison with other classifiers.

Time series Sentinel 1 images have been utilized to identify harvested and non harvested sugarcane areas by applying knowledge based crop cycle analysis and segmentation [30]. K-means clustering initially segmented the data into 15 clusters from which crop types were identified based on crop cycle knowledge. Finally classification of three classes, sugarcane harvested area, non harvested area and other crops, was achieved with good accuracy.

Sugarcane cultivation areas in southern China face the challenges of cloudy weather and spectral mixing of crops, makes discrimination of sugarcane crop difficult using RS data [31]. Authors developed a methodology for the mapping of a large sugarcane area using middle resolution satellite data. It includes object oriented based image segmentation with the generation of attribute table followed by building a training set using the Adaboost algorithm and a boosted classifier. The overall training and testing accuracy, proved that middle resolution satellite data is suitable to classify sugarcane crops in the southern China.

Same problem of cloud coverage in images was dealt in [32] by employing time series of images, optical and radar data and cloud gap filling methods to classify crops, pastures and tree plantations at 4 levels, in the heterogeneous region of Sao Paulo, Brazil. Supervised object oriented classification and RF algorithm were used for their study. They concluded that advanced cloud filling methods and Sentinel 1 data did not contribute in overall accuracy. Conversely, time series images were helpful in classification. Reference [33] also observed issue of entire sugarcane growing season in the region of Suixi and Leizhou, South China along with cloud coverage problem. Sentinel 1A SAR data was used for their study along with optical Sentinel 2 data. They devised a technique for early season mapping of sugarcane using RF and XGboost algorithm on time series data. Their framework consists of two procedures:

i. Time series S1A SAR images generate initial sugarcane map which is refined by removing non vegetation area using Sentinel 2 optical data.
ii. S1A based incremental classification to test the framework.

3.8 Deep Learning

A method to extract sugarcane plantation area was proposed by [34] using deep learning with the help of four months multi temporal images of GF-2 and BJ-2. Firstly, non vegetation area was extracted and then temporal processing of sugarcane area was performed, using Deep Convolution Neural Network (DCNN). Input to DCNN were sowing period images, growing period images, matured period images and other data. Authors compared the proposed method with object oriented, DeepLab V3+ and ground data. The overall accuracy of the proposed method is 94.32%.

4 Accucracy Assessment Parameters

4.1 Cohen's Kappa Coefficient

Cohen's Kappa, a robust measure than simple percent calculation, measures the agreement between two raters who each classify number of (N) items into C mutually categories. It is defined by following function

$$k = \frac{P_o - P_e}{1 - P_e} = 1 - \frac{1 - P_o}{1 - P_e} \quad (1)$$

where

Po—Relative observed agreement among raters

Pe—hypothetical probability of chance agreement, using the observed data to calculate the probabilities of each observe randomly seeing each category. Zero value of the Kappa coefficient indicates no agreement between reference data and classified data, whereas value 1 indicates, classified image is totally identical to the reference image.

4.2 Overall Accuracy

The quantitative method of characterizing image classification accuracy is a confusion matrix or error matrix. It is a table that shows correspondence between reference data and classified data. Overall accuracy (OA) is defined as out of all of the reference data what proportion were mapped correctly. It is given as below

$$OA = \frac{\#\text{ of correctly classified sites}}{\#\text{ of reference sites}} \tag{2}$$

The overall accuracy is usually expressed as a percent, with 100% accuracy being a perfect classification where all the reference sites are classified correctly. Figure 3 presents comparative analysis of highest accuracies gained by machine learning and other techniques in the studies of sugarcane classification.

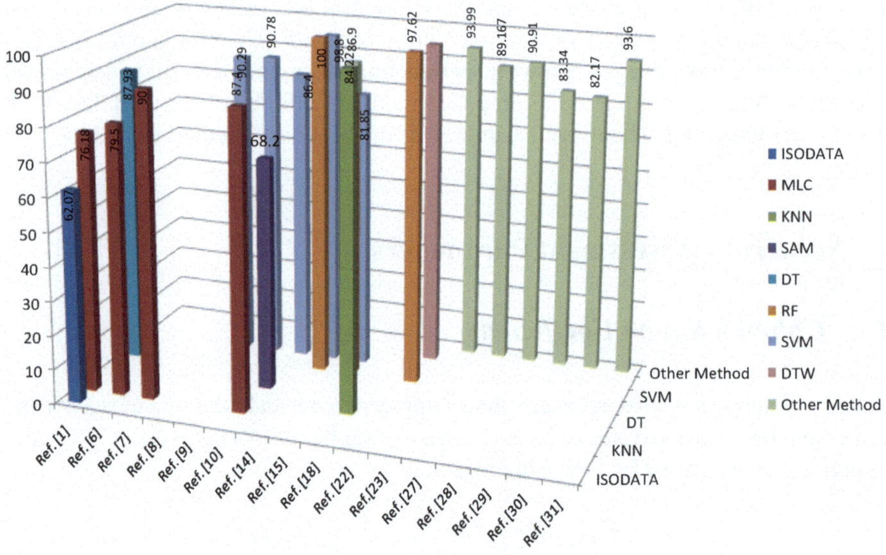

Fig. 3 Comparative analysis of accuracies achieved using machine learning and other algorithms in sugarcane crop classification

4.3 Producer's Accuracy

Producer's Accuracy (PA) is the map accuracy from the producer's point of view. This is how often are real features on the ground correctly shown on the classified map. It is given each classify number of (N) items into C mutually categories. It is defined by following function

$$PA = \frac{\text{\# of reference sites classified accurately of a class}}{\text{Total \# of reference site for that class}} \quad (3)$$

4.4 User's Accuracy

The User's Accuracy (UA) is the accuracy from of a map user's point of view. The User's accuracy indicates how often the class on the map will actually be existing on the ground. The User's Accuracy is defined by following function.

$$UA = \frac{\text{\# of correctly classified sites}}{\text{Total \# of classified sites}} \quad (4)$$

5 Conclusion and Future Directions

This paper presents the significance of machine learning with RS technology in sugarcane crop classification. From the review, it is revealed that object based image analysis outperforms pixel based analysis for sugarcane mapping. The accuracy of all classifiers which are reviewed is ranging from 74.63% to 100%. Classifier such as MLC yields good results when classifying on single date satellite imagery and for sugarcane classification its accuracy is around 74% irrespective of whether it is single date or multi-date imagery. However machine learning with different classification technique i.e. SVM, RF, DTW, AdaBoost, etc. successfully classify sugarcane crop with remarkable accuracy. Considering the fact that sugarcane is highly dynamic crop and its phenological characteristics play an important role in classification; it is concluded that the ML approach using RS data is certified to be applicable, with good accuracy, in sugarcane crop classification. Based on the study of recent work the future directions are identified as follows:

1. Variety Identification—The idea is that varieties have different canopies due to their particular morpho-physical characteristics; hence it might be possible to identify such differences through orbital spectral data. It would help institutions that breed sugarcane varieties for royalties charges, for the propagation of their genetic material.

2. Cloudy images—In southern India generally monsoon starts from June and lasts till September. So 4–5 months of optical satellite data is missing due to heavy cloud coverage over the land. This creates a gap in the data that represents particular growth stage of the sugarcane crop. However assimilation of optical and SAR data helps to cover that phenological stages data but does not produce optimum accuracy. So there is scope to investigate the technique to deal with cloudy images of optical satellite.
3. Satellite Sensors—The sensors in satellite, if can be enhanced enough spatially, temporarily and spectrally then mapping accuracy will definitely increase beyond expectations.

The future work will be to work on sugarcane variety identification, distinguishing between plant cane or ratoon cane and discrimination of sugarcane phenology in the south India area.

References

1. Verma AK, Garg PK, Prasad KSH (2017) Sugarcane crop identification from LISS IV data using ISODATA, MLC, and indices based decision tree approach. Arab J Geosci 10(1):16
2. Xavier AC, Rudorff BFT, Shimabukuro YE, Berka LMS, Moreira MA (2006) Multi-temporal analysis of MODIS data to classify sugarcane crop. Int J Remote Sens 27(4):755–768
3. Lu D, Weng Q (2007) A survey of image classification methods and techniques for improving classification performance. Int J Remote Sens 28(5):823–870
4. https://alan.do/remote-sensing-an-overview-of-common-pixel-classification-techniques-4827c4c3a9ff
5. Abdel-Rahman EM, Ahmed FB (2008) The application of remote sensing techniques to sugarcane (Saccharum spp. hybrid) production: a review of the literature. Int J Remote Sens 29(13):3753–3767
6. Vinod KVK, Kamal J (2010) Development of spectral signatures and classification of sugarcane using aster data. Int J Comput Sci Commun 1:245–251
7. Mulianga B, Begue A, Clouvel P, Todoroff P (2015) Mapping cropping practices of a sugarcane-based cropping system in Kenya using remote sensing. Remote Sens 7(11):14428–14444
8. Dhumal RK et al (2019) A spatial and spectral feature based approach for classification of crops using techniques based on GLCM and SVM. In microelectronics electromagnetics and telecommunications. Springer, Singapore, pp 45–53
9. S. Mulyono and others (2016) Identifying sugarcane plantation using LANDSAT-8 Images with support vector machines. In IOP conference series: earth and environmental science, vol 47, No 1, p 12008
10. Yang C, Everitt JH, Murden D (2011) Evaluating high resolution SPOT 5 satellite imagery for crop identification. Comput Electron Agric 75(2):347–354
11. Khobragade AN, Raghuwanshi (2015) MM Contextual soft classification approaches for crops identification using multi-sensory remote sensing data: machine learning perspective for satellite images. In artificial intelligence perspectives and applications. Springer, pp 333–346
12. Foody GM, Mathur A (2004) Toward intelligent training of supervised image classifications: directing training data acquisition for SVM classification. Remote Sens Environ 93(1–2):107–117
13. Breiman L (2001) Random forests. Mach Learn 45(1):5–32
14. Everingham YL, Lowe KH, Donald DA, Coomans DH, Markley J (2007) Advanced satellite imagery to classify sugarcane crop characteristics. Agron Sustain Dev 27(2):111–117

15. Saini R, Ghosh SK (2018) Crop classification on single date sentinel-2 imagery using random forest and suppor vector machine. Int Arch Photogramm Remote Sens Spat Inf Sci
16. Botkin DB, Estes JE, MacDonald RM, Wilson MV (1984) Studying the earth's vegetation from space. Bioscience, 508–514
17. dos Santos Luciano AC et al (2018) Generalized space-time classifiers for monitoring sugarcane areas in Brazil. Remote Sens Environ 215:438–451
18. Shifa MS, Naweed MS, Omar M, Jhandir MZ, Ahmed T (2011) Classification of cotton and sugarcane plants on the basis of their spectral behavior. Pakistan J Bot 43(4):2119–2125
19. Petitjean F, Inglada J, Gançarski P (2012) Satellite image time series analysis under time warping. IEEE Trans Geosci Remote Sens 50(8):3081–3095
20. Petitjean F, Weber J (2013) Efficient satellite image time series analysis under time warping. IEEE Geosci Remote Sens Lett 11(6):1143–1147
21. Maus V, Câmara G, Cartaxo R, Sanchez A, Ramos FM, de Queiroz GR (2016) A time-weighted dynamic time warping method for land-use and land-cover mapping. IEEE J Sel Top Appl Earth Obs Remote Sens 9(8):3729–3739
22. Belgiu M, Csillik O (2018) Sentinel-2 cropland mapping using pixel-based and object-based time-weighted dynamic time warping analysis. Remote Sens Environ 204:509–523
23. Formaggio AR, Vieira MA, Rennó CD, Aguiar DA, Mello MP (2010) Object-Based image analysis and data mining for mapping sugarcane with landsat imagery in Brazil. Int Arch Photogramm Remote Sens Spat Inf Sci 38:553–562
24. Vieira MA, Formaggio AR, Renno CD, Atzberger C, Aguiar DA, Mello MP (2012) Object based image analysis and data mining applied to a remotely sensed landsat time-series to map sugarcane over large areas. Remote Sens Environ 123:553–562
25. de Souza CHW, Lamparelli RAC, Rocha JV, Magalhães PSG (2017) Mapping skips in sugarcane fields using object-based analysis of unmanned aerial vehicle (UAV) images. Comput Electron Agric 143:49–56
26. do Valle Gonccalves RR, Zullo J, Romani LAS, do Amaral BF, Sousa EPM (2017) Agricultural monitoring using clustering techniques on satellite image time series of low spatial resolution. In: 2017 9th international workshop on the analysis of multitemporal remote sensing images (MultiTemp), pp 1–4
27. Henry F, Herwindiati DE, Mulyono S, Hendryli J (2017) Sugarcane land classification with satellite imagery using logistic regression model. In IOP conference series: materials science and engineering, vol 185, No 11, p 2024
28. Singla SK, Garg RD, Dubey OP (2018) Sugarcane ratoon discrimination using LANDSAT NDVI temporal data. Spat Inf Res, 26(4):415–425. Joao RS, Mpinda STA, Vieira APB, Joao RS, Romani LAS, Ribeiro MX (2018) A new approach to classify sugarcane fields based on association rules, In: information technology-new generations. Springer, pp 475–483
29. Joao RS, Mpinda STA, Vieira APB, Joao RS, Romani LAS, Ribeiro MX (2018) A new approach to classify sugarcane fields based on association rules. In information technology-new generations. Springer, pp 475–483
30. Murugan D, Singh D (2018) Development of an approach for monitoring sugarcane harvested and non-harvested conditions using time series Sentinel-1 data. In IGARSS 2018–2018 IEEE international geoscience and remote sensing symposium, pp 5308–5311
31. Zhou Z et al (2015) Object-oriented classification of sugarcane using time-series middle-resolution remote sensing data based on adaboost. PLoS One 10(11):e0142069
32. de Oliveira Santos C et al (2019) Classification of crops, pastures, and tree plantations along the season with multi-sensor image time series in a subtropical agricultural region. Remote Sens 11(3)
33. Jiang H et al (2019) Early season mapping of sugarcane by applying machine learning algorithms to sentinel-1a/2 time series data: a case study in Zhanjiang City, China. Remote Sens. 11(7):861
34. Zhu M, Yao M, He Y, He Y, Wu B (2019) Studies on high-resolution remote sensing sugarcane field extraction based on deep learning. IOP conference series: earth and environmental science 237(3):32046

Social Media Obsession in Gujarat: An Analytical Study

Priyanka Sharma

Abstract The twenty-first century has given birth to a virtual networking where humans are interconnected via different modes of social media, paying less preference to more active forms of communication. So the people tend to use social media and become more and more obsessed by its use which results in various kinds of cyber attacks as well as many health issues like anxiousness, feeling low, mental instability, etc. This paper focuses on the adverse effects of social media obsession by conducting survey in 5 rural and urban areas of Gujarat. The samples were asked to fill the questionnaire in both the modes, i.e., online as well as offline. On the basis of the results of the outputs obtained from the questionnaire the various aspects of the samples were known and it was found whether they were obsessed with the use of social media. An online system is developed to find the intensity of obsession with social media.

Keywords Social media · Obsession · Data collection

1 Introduction

The twenty-first century has given birth to virtual networking where humans are interconnected via different modes of social media, paying less preference to more active forms of communication. The first social networking site named SixDegrees.com was launched in 1997, which provided users to create profiles, invite friends, organize groups, and also surf their friend list. By 2010, there was an approximate of 1.5 billion users of social networking websites. In the present era there is an alarming hike in the rate of social media users. Social networking has transformed youth—life, society, and shares. It's the social network life that has occupied an important place in our lives, which is clear from our gestures of social networking IDs like Facebook IDs, Twitter handles, and some are also mentioned on our visiting cards. There are lots of benefits on using these websites; youth recognize it and use it for its potential. Evaporation of boundaries between friends and distant family members, staying with

P. Sharma (✉)
Raksha Shakti University, Lavad, Gandhinagar, India
e-mail: ps.it@rsu.ac.in

S. Fong et al. (eds.), *ICT Analysis and Applications*, Lecture Notes in Networks and Systems 93, https://doi.org/10.1007/978-981-15-0630-7_56

the minimum extra value, sharing your opinion and spreading it, (witnessing many successful online statements) and is therefore considered not to be on the social networking site (Social Network PDF). Nowadays the social media is more often used for harmful activities like cyberbullying, misinformation spreads like wildfire, easy prey, isolation of kids with disabilities, and decrease in productivity levels. Due to which there are complaints about anxiousness, feeling low, mental instability, etc. due to obsession of social media applications using the Internet like Facebook, Twitter, Instagram, WhatsApp, and many more. To find out how much people were obsessed with social media a survey was conducted in 5 rural and 5 urban cities of Gujarat and the questionnaire was filled. On the basis of the filled questionnaire the results—how social media affects the people and their mental conditions—were concluded and analyzed.

2 Research Methodology

This research is interdisciplinary in nature carried out from February 1, 2016 to January 31, 2017. The study was taken up in accordance with the guidelines of Institutional ethics committee of the Raksha Shakti University to ensure that the study does not violate any human rights or code of research ethics. The study is titled as the "Risk Analysis of Social Media Obsession in Gujarat—An Analytical Study" which is an attempt to analyze the obsession level of social media across Gujarat.

Social media is the new face of communication in today's environment. It presents new challenges and opportunities for the business community, economy, government, and consumers. The focus of this research is to measure the influence of social media on the consumers of Gujarat. Research involves formulating the problem to be investigated, selecting a suitable research design, choosing and applying appropriate procedures for data collection, and analyzing and communicating the process and findings through a written report. Research methodology is a way to systematically solve the research problem.

2.1 Research Design

All the below mentioned procedures were carried out taking due care of its quality, accuracy, and promptness for maintaining the validity of the work done. The important components of the research methodology are:

Type of Study: The present study is analytical research study. Analytical research is a specific type of research that involves critical thinking skills and the evaluation of facts and information relative to the research being conducted. From analytical research, a person finds out critical details to add new ideas to the material being produced.

Table 1 Age groups of social media user respondents

Age	No. of respondent	Percent
11–20	333	8.3
21–30	1498	37.5
31–40	1093	27.3
41–50	505	12.6
51–60	319	8.0
61–70	187	4.7
71–80	42	1.1
81–90	23	0.6
Total	4000	100.0

Research Design: The present study is qualitative research design. Qualitative research design involves understanding, explaining, exploring, discovering and clarifying situations, feelings, perceptions, attitudes, values, beliefs, and experiences of a group of people.

Locale of the Study: Gujarat was chosen as the locale of the study as per the requirement of the research project. All the samples were collected from 10 districts of Gujarat. The choice of the ten districts was based on the intention to cover both rural and urban setup and also to cover all parts of Gujarat.

Universe of the Study: The study focused on the selected population from five major cities and five major villages from the same district in the state of Gujarat (Table 1). Various categories of people were included in the population selected for the study. The data collected in the form of questionnaires and surveys from different sections of society namely students, housewives, elderly people, government employees, corporate professionals, Laborers, and the ones under unemployment and law enforcement across Gujarat state, covering both rural and urban areas. A structured questionnaire was administered for the representative population both online and offline.

2.2 Data Collection Methods

There are two types of methods for data collection—primary and secondary. Primary data collection method involves gathering fresh information through observation, interviews, or questionnaires. While, secondary data collection method includes both raw data and published sources such as journals, database, transcripts, books, etc.

Target Population

"Internet Users" of Gujarat are considered as a target population for this research because it is prerequisite to have internet connectivity for using social media websites. This is also a reason for having online survey as a method for data collection. Among the internet users, the population narrows down to "Social Media Users" because the research is about studying behavior of "Social Media Users" alone. For that purpose, in questionnaire, there is a checkpoint to take care of this most important factor, which discards the entry of Non-social Media User. Again, the population narrows down to one single state of India, i.e., Gujarat.

Sampling Method

Random and Stratified sampling method was used for the collection of responses on Risk Analysis of Social Media Obsession in Gujarat—An Analytical Study raining need analysis from various professional strata of the society. The participation from various strata was important to generalize the results.

Sample Size

A total of 4000 subjects who volunteered and signed the informed consent to participate in this research were selected for the present study belonging to following eight different categories of occupation and two residential categories of urban or rural belongingness as tabulated:

Processing and Analyzing the Data

In the light of the objectives framed for the study, the collected primary and secondary data were systematically organized. The primary data coded and fed into the excel sheet for processing. The responses were collated and computed in a tabular form. The collected data is analyzed using SPSS (Statistical Package for the Social Sciences).

Screenshot of our web site:

1. **About Project Page** (Fig. 1):

- This is our first page (Home Page) to describe what is our project and its definition, aim of the project, etc.
- First we have About Project, Questionnaires (Form), Ask Query tab.

2. **Questionnaires (Form) Page** (Fig. 2):

- In this page we have a Note section to know why they fill this form like a consent form and some information.

3. **Another Page of Questionnaires** (Fig. 3)

- In questionnaires we ask all questions to choose or select answers to make it easy to fill.

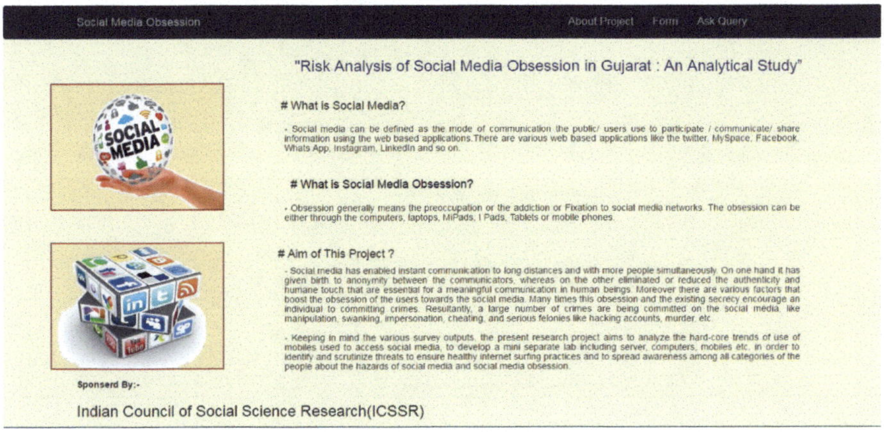

Fig. 1 Screenshot of home page

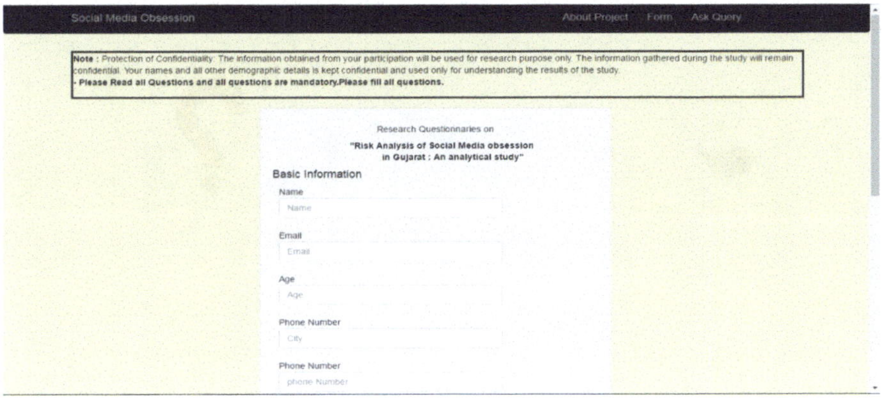

Fig. 2 Screenshot of questionnaires

4. **Ask Query Page** (Fig. 4)

– In the above page, if people have some query about project then they can ask and also we revert in their email.

5. **Thank You Page** (Fig. 5):

– Here is the last page of project that is the thank you page.

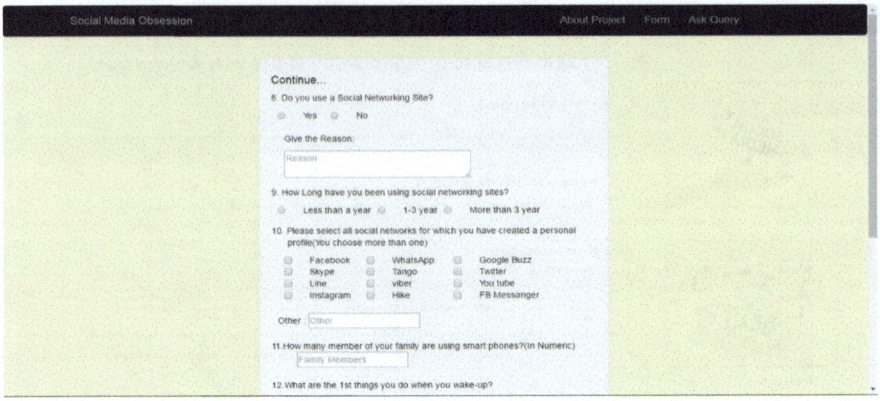

Fig. 3 Screenshot of questionnaires

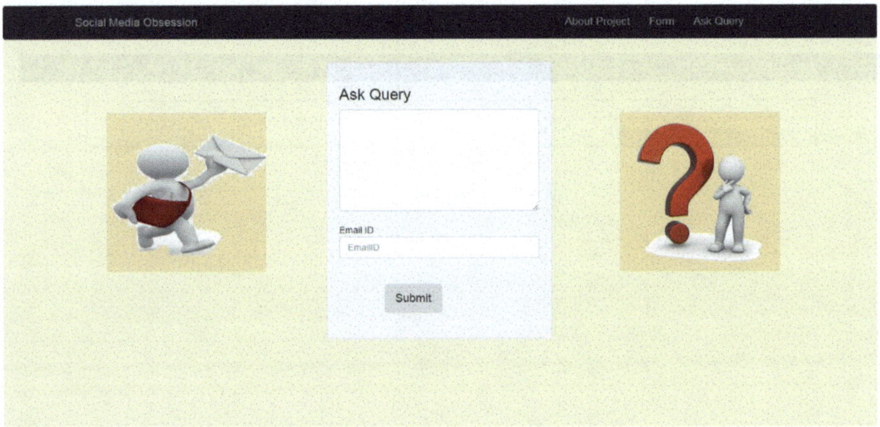

Fig. 4 Screenshot of users query

Fig. 5 Screenshot of thank you

Mobile Application for Social Media Obsession Measure

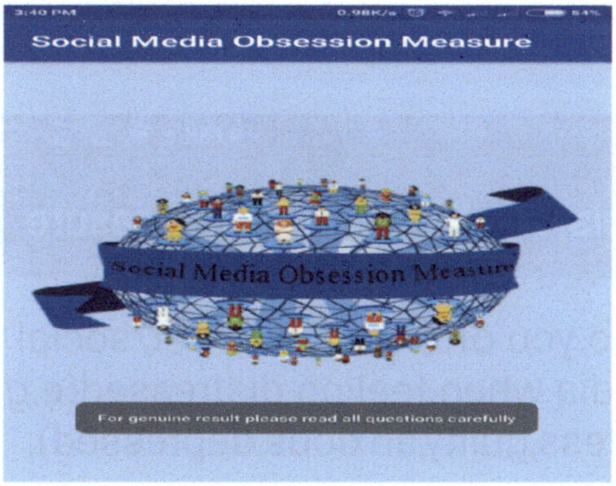

– This is the home page of our mobile application.

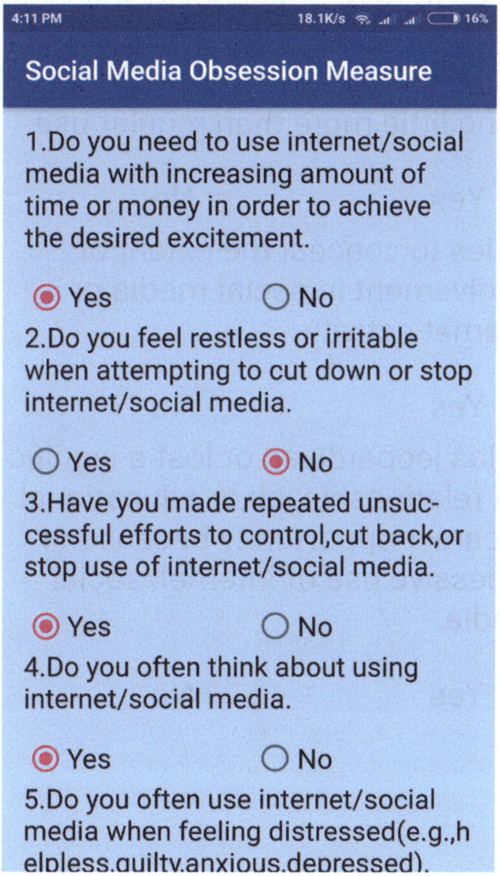

– Here a total of 9 questions and their 4 multiple answers are given. You need to just read all questions carefully and then choose one option among the multiple answers. After that click on the submit button.

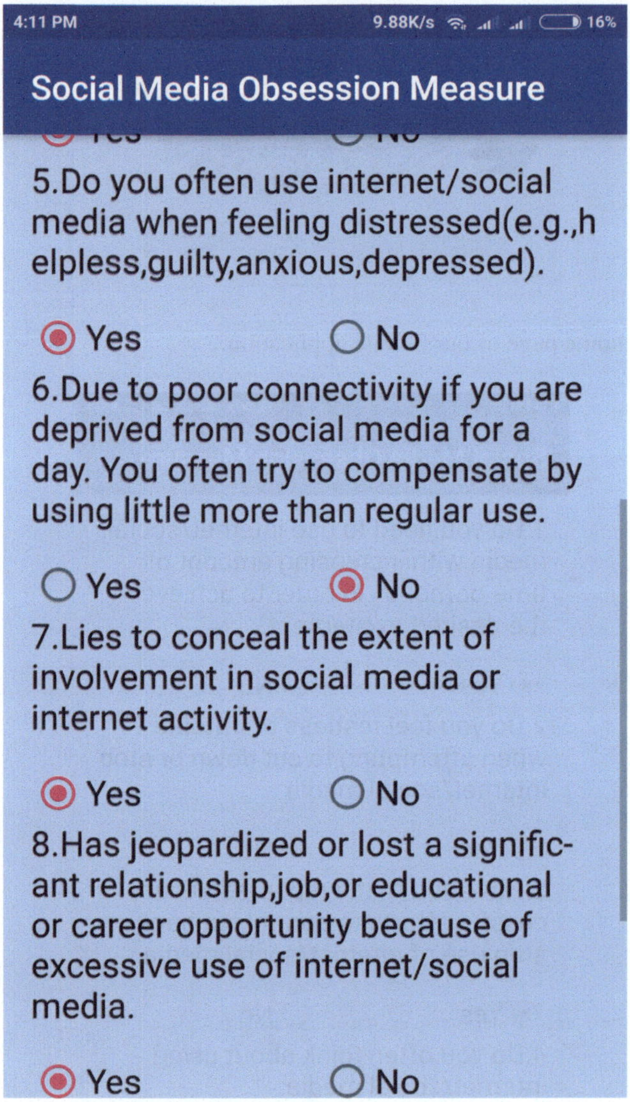

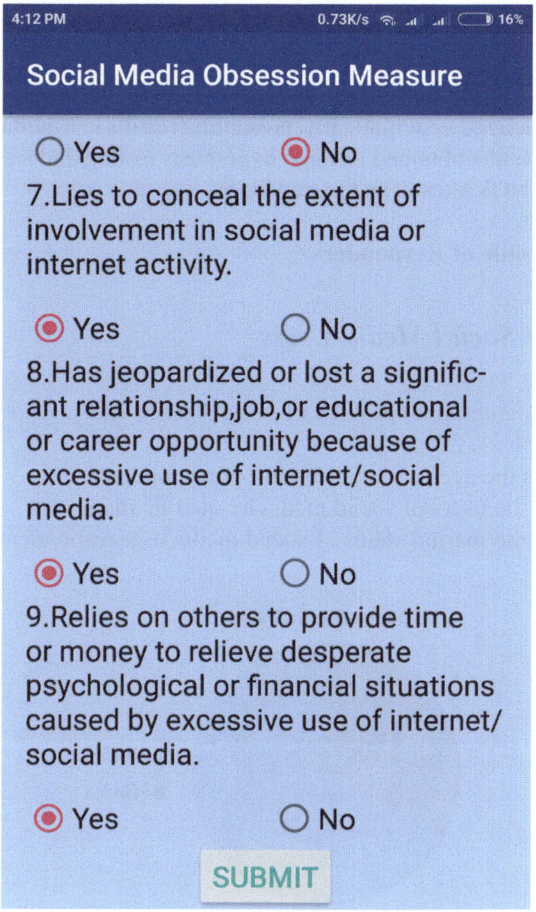

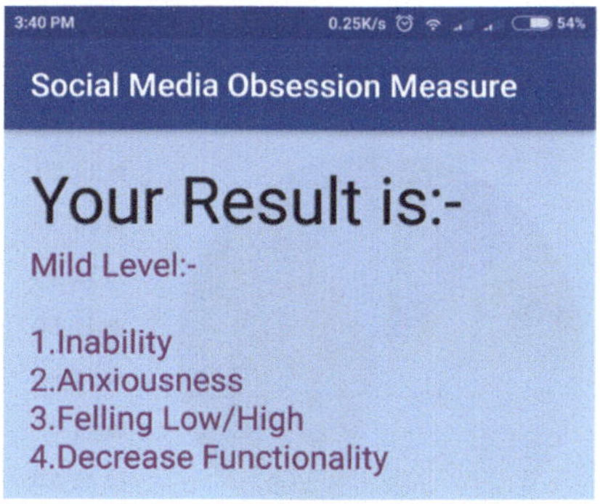

– Here is the result of your answer that you are middle level user of social media so you can't count in obsession.

3 Data Analysis and Interpretation

This chapter presents an analysis of the data collected and results of the research. The chapter is organized systematically, presenting results in a methodical way. Later sections present results obtained through hypothesis testing and statistical tools like factor analysis, ANOVA tests, and paired T-test.

Demographic Profile of Responders

3.1 Profile of Social Media Users

In the following paragraphs, profiles of social media users, who responded, have been shown (Graphs 1, 2, 3 and 4).

Table 1 reflects the age-wise distribution of respondents.

Table 2 shows the users of social media by qualification.

Table 3 shows the marital status of social media user respondents.

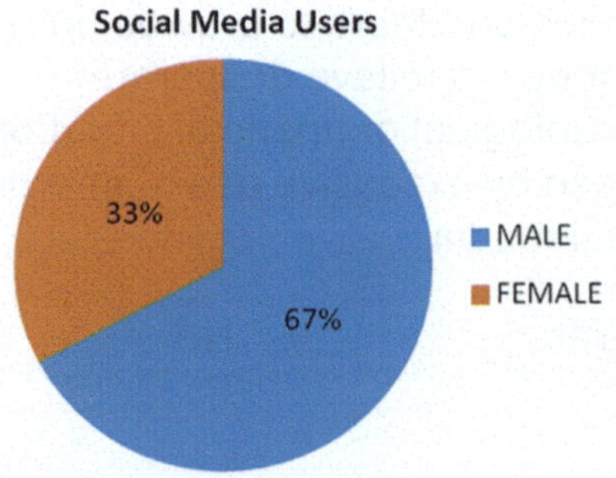

Graph 1 Gender wise social media usage distribution

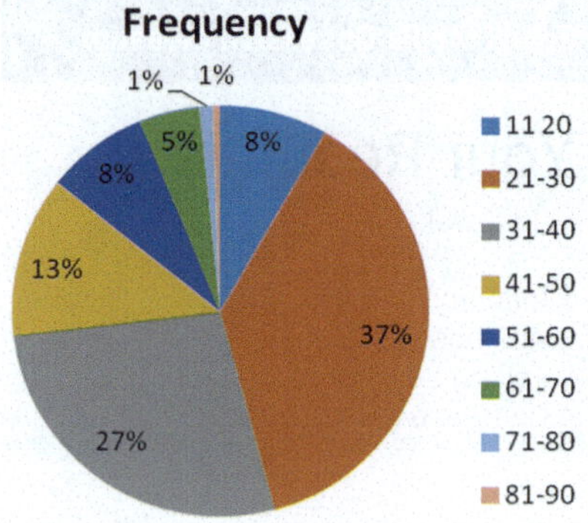

Graph 2 Frequency of social media users by age groups

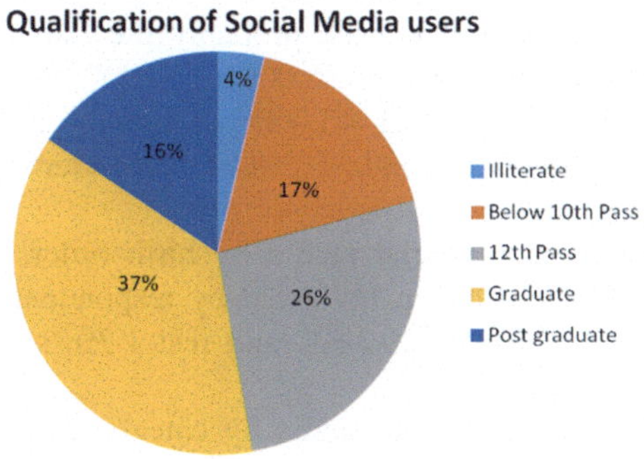

Graph 3 Distribution of social media users based on their qualification

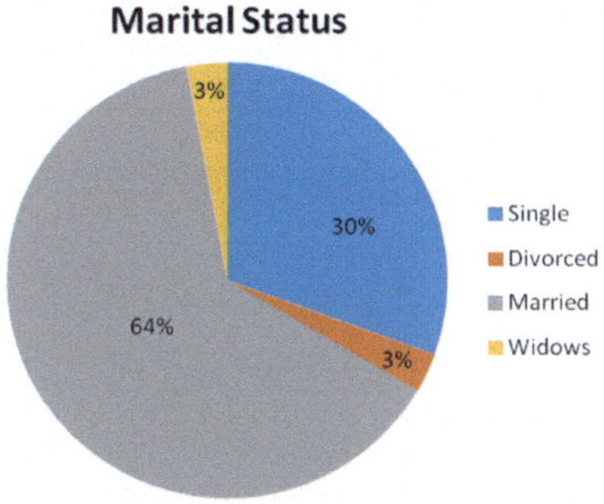

Graph 4 Social media users frequency by their marital status

Table 2 Social media user respondents by qualification

	No. of respondents	Percent
Illiterate	147	3.7
Below 10th Pass	682	17.1
12th Pass	1061	26.5
Graduate	1490	37.3
Post graduate	620	15.5
Total	4000	100.0

Table 3 Marital Status of social media user

	No. of respondents	Percent
Single	1221	30.5
Divorced	116	2.9
Married	2545	63.6
Widows	118	3.0
Total	4000	100.0

Table 4 shows that 3.6% respondents, probably the Labor, belong to "below 5000" category, followed by "5001 to 25,000 (37.6%)". 34.2% of respondents, have claimed about receiving 25,001–45,000 per month, 16.0% of respondents get between 45,001–65,000 per month, 8.7% of respondents receive More than 65,001 Rs. per month (Graph 5).

Table 5 shows the users of social media by birth order. It can be observed that 43.2% respondents are younger. Followed by respondents whose birth order is "Middle one (35.4%)", "Elder (20.3%)" and rest 1.2% respondents belong to other (Graph 6).

Table 6 shows the users of social media by category. All the categories like student, government employee, housewives, elder people, laborers, unemployed

Table 4 Social media user respondents by Income

Monthly income	No. of respondents	Percent
Below 5000	144	3.6
5001–25000	1504	37.6
25001–45000	1366	34.2
45001–65000	638	16.0
More than 65001	348	8.7
Total	4000	100.0

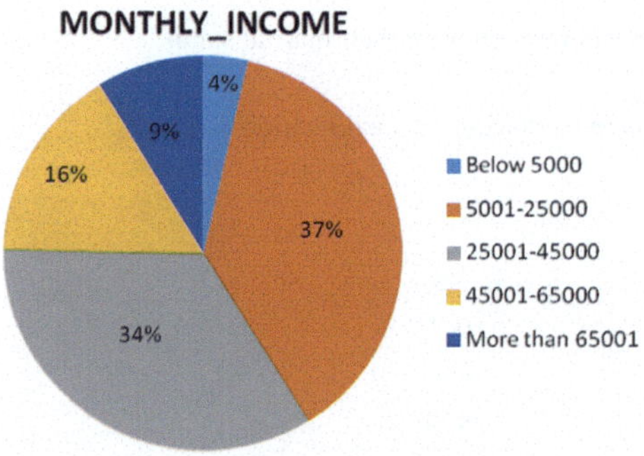

Graph 5 Social media users based on their monthly income

Table 5 Social media user respondents by Birth Order

	No. of respondents	Percent
Younger	1726	43.2
Middle One	1416	35.4
Elder	811	20.3
Other	47	1.2
Total	4000	100.0

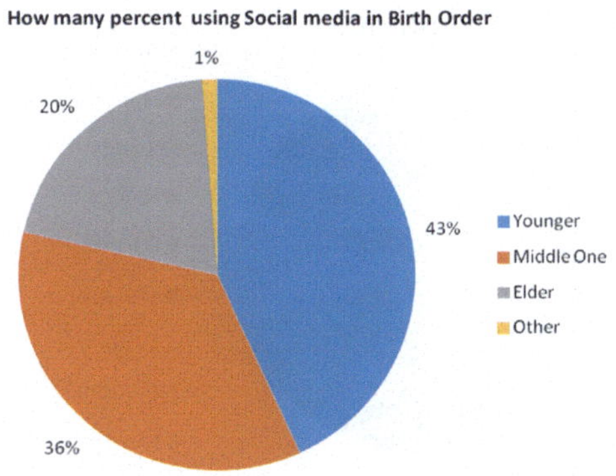

Graph 6 Percentage of social media users by birth order

Table 6 Social media user respondents by Category

Category		No. of respondents	Percent
Valid	Student	500	12.5
	Gov. Employee	500	12.5
	Housewife	500	12.5
	Elder People	500	12.5
	Laborer	500	12.5
	Unemployed	500	12.5
	Law Enforce	500	12.5
	Corporate	500	12.5
	Total	4000	100.0

people, law enforcement individuals, corporate professional were same as 12.5% (Graphs 7 and 8).

Table 7 shows the use of social media respondents. It can be observed that 68.2%

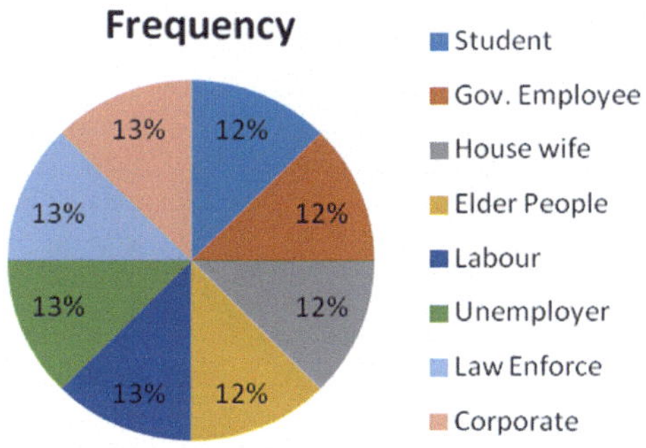

Graph 7 Frequency of social media users by their job profile

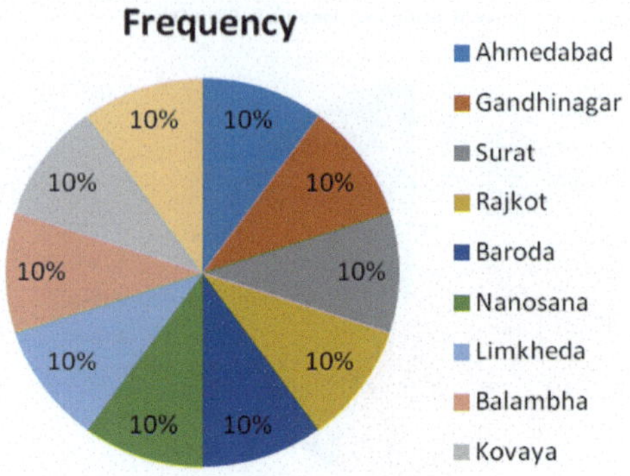

Graph 8 District wise social media users frequency

Table 7 Social media user respondents by Use of social media

	No. of respondents	Percent
YES	2727	68.2
NO	1273	31.8
Total	4000	100.0

respondents use social media sites. And the other 31.8% respondents donot use social media sites.

The other detail about respondents like their use of social media(in year),use of social media daily (in hour) and which social networking sites they use the most and for which reasons it is as follows (Graphs 9, 10, 11 and Tables 8, 9, 10).

From the above frequency table it can be noted that among the respondents in the study are Facebook, Skype, Line, Instagram, Tango, Hike, WhatsApp, Twitter, Viber, IMO, FB Messenger, Hangout, Other. On the first five positions with 61.7% (Whatsapp), 58.8% (Facebook), 49.3% (Hike), 44.4% (Instagram), 37.0% (FB Messenger) of users, respectively are using social networking sites. On first five

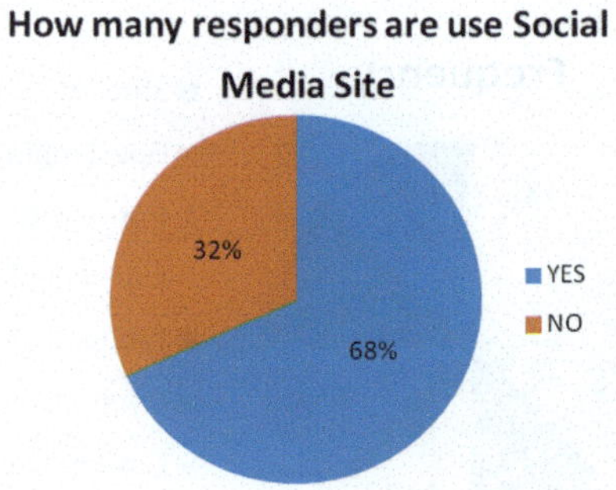

Graph 9 Percentage of social media users among the respondents

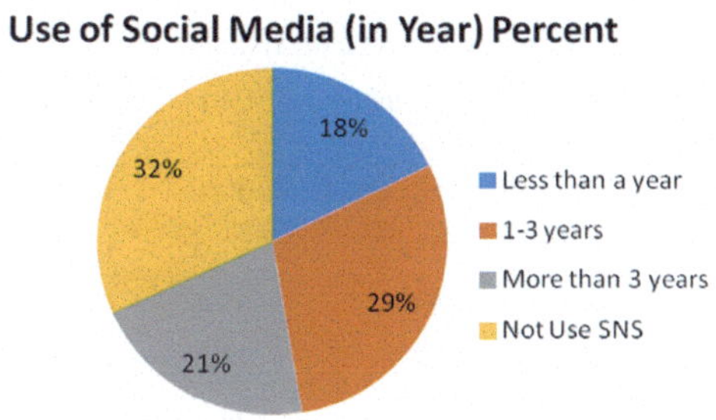

Graph 10 Percentage social media use (in year)

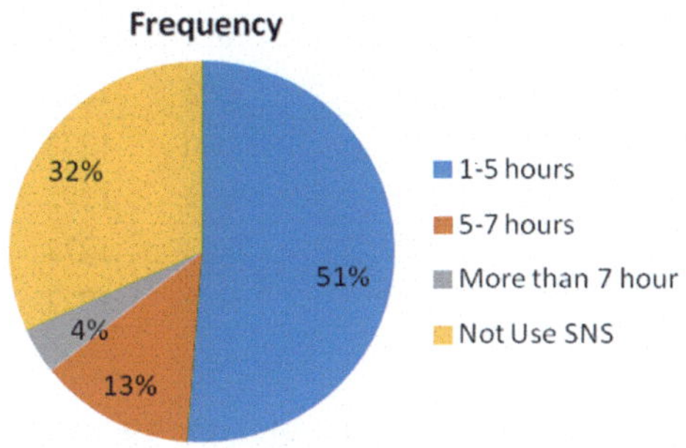

Graph 11 Amount of time spent by the respondent on social media

Table 8 Usage of social media respondents

	No. of respondents	Percent
Less than a year	710	17.8
1–3 years	1180	29.5
More than 3 years	837	20.9
Not Use SNS	1273	31.8
Total	4000	100.0

Table 9 Frequency of social media usage among respondents

	No. of respondents	Percent
1–5 h	2050	51.3
5–7 h	520	13.0
More than 7 h	159	4.0
Not Use SNS	1273	31.8
Total	4000	100.0

Table 10 Statistical usage of SNS statistical usage of SNS

SNS	Using Freq	(%)	Not using Freq	(%)	Not use SNS Freq	(%)
Face Book	2352	58.8	376	9.4	1272	31.8
Skype	450	11.3	2278	57.0	1272	31.8
Line	192	4.8	2536	63.4	1272	31.8
Instagram	1777	44.4	951	23.8	1272	
Tango	139	3.5	2589	64.7	1272	31.8
Hike	1970	49.3	758	19.0	1272	31.8
WhatsApp	2467	61.7	261	6.5	1272	31.8
Twitter	471	11.8	2257	56.4	1272	31.8
Viber	240	6.0	2488	62.2	1272	31.8
IMO	530	13.3	2198	55.0	1272	31.8
FB Messenger	1479	37.0	1249	31.2	1272	31.8
Hangout	195	4.9	2533	63.3	1272	31.8
Other	77	1.9	2651	66.3	1272	31.8

negative with 6.5% (WhatsApp), 9.4% (Facebook), 19.0% (Hike), 23.8% (Instagram), 31.2% (FB Messenger) of users, respectively not using social networking sites.

The below frequency table explains the usage of SNS according to the priority of the respondents. The priority list asked in the self-reporting inventory is to share your experience, to get opinions, to make professional contacts, to make new friends, to keep in touch with friends and family, to play games, to share videos, to find information. The respondents were asked to rate the usage according to the priority from 1 to 8 (Graph 12).

SNS Usage According to the Priority

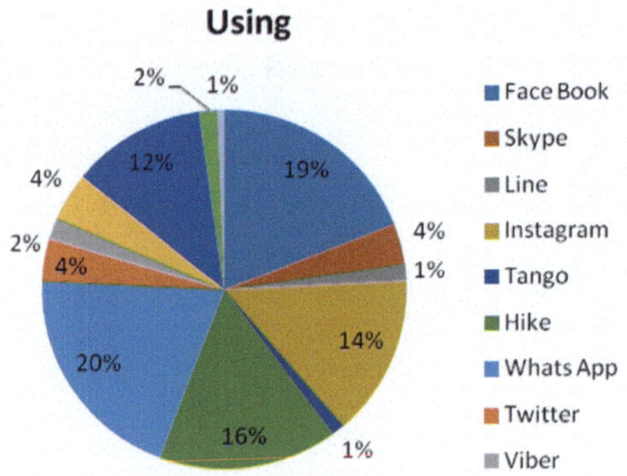

Graph 12 Priority of the SNS applications by the respondents

SNS	1	%	2	%	3	%	4	%	5	%	6	%	7	%	8	(%)
Share exp	576	14.4	348	8.7	358	9	349	8.7	322	8.1	249	6.2	200	5	326	8.2
Get opinion	268	6.7	469	11.7	379	9.5	445	11.1	377	9.4	302	7.6	356	8.9	132	3.3
Make cont	263	6.6	348	8.7	526	13.2	365	9.1	408	10.2	370	9.3	232	5.8	216	5.3
Make new friend	275	6.9	438	11	343	8.6	453	11.3	429	10.7	284	7.1	271	6.8	235	5.9
To keep touch with friends and family	533	13.3	401	10	310	7.8	314	7.9	360	9	270	6.8	267	6.7	273	6.8
Play games	205	5.1	261	6.5	241	6	212	5.3	273	6.8	415	10.4	453	11.3	668	16.7
Share video	202	5.1	264	6.6	320	8	287	7.2	246	6.2	47	13.4	393	9.8	1	0
Find info	416	10.4	232	5.8	253	6.3	334	8.4	310	7.8	331	8.3	367	9.2	476	11.9

General Findings on Demographic Factors of Respondents:

- Male users are 67.3% and females are 32.8%. Male users outweigh the female users, but it is not a significant difference. Female users seem to be in lines, with the male users on the usage of social media platforms in Gujarat.
- Respondents from 31 to 40 age group, are most actively involved in using social media platforms, followed by the age group of 21–30 year olds. It is thus possible to assume that these two age groups are majorly involved in social media platform's usage.
- Social media usage by education, is highest among the "Graduates" (37.3%) followed by users who are "Post graduates" (15.5%) and higher secondary students (26.5%). Thus, it seems that youth are the majority users of social media
- Social media usage by marital status, showed that major users belong to the "Married" category. They are the highest users of social media (63.6%), followed by "Single" category (30.5%).
- According to the survey result of Table 6 shows that 3.6% respondents, probably the Labor, belong to "below 5000" category, followed by "5001 to 25,000 (37.6%)". 34.2% of respondents, have claimed about receiving 25,001–45,000 per month, 16.0% of respondents get between 45,001 to 65,000 per month, 8.7% of respondents receive more than 65,001Rs. per month.
- The users of social media by category.All the categories like student, government employee, housewives, elder people, laborers, unemployeed, law enforcement individuals, corporate professionals were same as 12.5%.

Social Media Consumption Practice Among Respondents

- The survey results show that social media is accessed from various places ranging from cybercafés to mobile phones, at home and work place. The interesting fact is that according to the survey results 36.6% of people are visiting cybercafé, which is highest among all, for accessing social media platforms. Hence, it can be safely deducted out of this fact that social media attracts its users to use it even if they do not have internet access at home or on-the-go devices, and have to visit cybercafés for the same. Informal interviews with cybercafe owners, revealed the fact that students who belong to teenage group, specially visit cybercafes to play games on social media and surf on social media websites.
- The survey results depicts that more number of people are visiting social media platforms in recent past, not more than a year's time, cumulatively 17.8%, as against users with longer tenure (1–3 years)—31.8. The research result suggests that majority people in Gujarat spend "1–5 h" per day (51.3%) on social media platforms. It was also found that 4% respondents spend more than 7 h on social media. hence it can be inferred that these 2.3% respondents may possess either smartphones or other on-the-go devices, which keeps them connected with social media platforms for more than 7 h.

Social Media Consumption Pattern Among Respondents

- It was found that social media is consumed in various proportions (less, average and high usage) by different categories of people based on their 152 demographic factors. With the help of Chi-Square analysis, it was found that there exist associations between demographic factors of respondents and the social media usage of respondents. That means social media is being consumed by every mix of consumers, i.e., gender wise, age group wise, city wise, occupation, income, education, and internet access wise.
- Be it male or female, any individual ranging from the age of 13–55 and above, be it from any city of Gujarat, belonging to any income category from no income to more than 60,000 per month and be it a student, housewife, or entrepreneur—they all consume social media according to their preferred use. 5.1.4 Connections of Respondents on Social Media
- Research results concerning connections of respondents on social media found that respondents majorly connect with their families, online and offline friends, and classmates. They also prefer maintaining connections with their colleagues, customers, and other business contacts. Thus, according to survey results, it can be said that social media are not just networks to keep up the social life active, but also to build professional relationships and to carry out job hunting activities. The social graph is expanding. The idea of "Friends" is changing. Everybody who the person is communicating with on social media are categorized as friends, whether they are family or professional contacts.
- The decision to join social media platforms is majorly done based on a friend's recommendation. Social media are networks to connect people. Hence, it was observed that the influence of "friends" holds a major influence in the decision to join any particular social media network.
- Social media helps in making new relations. This statement was checked in this research. It was found that the respondents majorly are connected with their existing offline connections on social media. But, there are also large number of respondents who connect with new people on social media. Hence, social media seems to be an extended platform for making new relations and maintaining existing relationships.
- Majority of the respondents (74.4%) believe that they have not lost any of their relations because of social media networks. This finding seems to show the healthy usage of social media. It was observed that only 11.4% respondents claimed to have lost their relations due to social media. There may be various personal or professional reasons behind it. A person normally loses a connection due to indifference in opinions, nature, behavior, etc. in an offline world. The actions of offline world seem to have a direct impact on online behavior too. Hence, it may result in losing a connection on social media. The other remaining respondents are totally not aware of whether they have lost relations due to social media or not. This category of people might not be too much involved to analyze their online connections.
- It was found that 38.3% of respondents feel that they did not locate lost relations through social media networks. There may be several reasons for the same. One

might not have an intention to rebond; the opposite person might have blocked them; they might have canceled or deleted their social media accounts, or simply a person might not have a social media account at all, etc. However, on the other end, 36.6% of respondents claim to have located their lost relations through social media. Locating lost relations on social media is among the major reasons why people join a social media network—to connect with people who are far away, or to connect with long lost friends who used to be their school mates or college mates, etc. There may be various reasons behind it.

4 Conclusion

Thus we find that society is socially obsessed irrespective of gender, age, demographic location, and other important factors. This should be taken into consideration for a healthy and better society.

References

1. Uses of Social Networking Sites on Young Adults to Infer on the Different Types of Users
2. Social Media Research and Its Effect on Our Society
3. The Harmful Effects of Social Networks and Other Electronic Media
4. 'Internet addiction' and its consequences
5. "Mass Media can Effect Socio-Political Changes: An Analysis"
6. Online Social Networking and Addiction- a Review of the Psychological literature
7. Social Networking Sites—A Critical Analysis of Its Impact on Personal and Social Life
8. The Global Cyber Crime Industry
9. Risk society Towards a New Modernity
10. Networking-A Study of Indian Youth by Shivani Arora
11. Facebook Addiction: Factors Influencing an Individuals Addiction
12. Social Media and Academic Performance of Students
13. The Negative Effects Of Social Networking Sites Such As Facebook Among Asia Pacific University Scholars In Malaysia
14. "Exploring The Effects Of Social Medisa Use On The Mental Health Of Young Adults"
15. Impact of Social Networking Sites (SNS) on The Youth of India
16. Joseph AL (2009, April 23) Man arrested for posting ob-scene profile of ex-boss on Orkut. The Indian express
17. Krishnan S (2009, March 21) Mumbai Police tie up with Orkut to nail offenders. The Indian Express
18. Kumar A (2008, May 19) Orkut used in credit card scam to buy airline tickets. The Deccan Chronicle
19. https://www.schooliseasy.com/2014/02/social-media-in-theclassroom/
20. http://www.edudemic.com/how-students-benefit-from-usingsocialmedia/

ICT and Sustainability Development in India

Sanjay Gaur, Leena Sharma, Vaishali Singh and Pallavi Saini

Abstract Development and sustainability both are crucial parts of any economy. Developing countries are working for sustainability development whereas developed countries are focused on growth and developments. The Indian economy stands in stable economies worldwide. There are various sectors that provide their contribution to the GDP growth. The IT sector in India plays a significant role in development as well as sustainability development during the slowdown in economy. At present the Indian IT industries and support services are leading industries. With the slogan of "Digital India" rapid development in this sector take place now. At present Indian IT industries' participation in GDP is 7.7% approx and till 2025 it is expected to reach 10%. The present paper is an effort to depict a view of Indian IT industries and its role and presence in national development and sustainability. The study focuses on the current slight economy slowdown and position of IT industries.

Keywords Economy · IT industries · Slowdown · Development · Sustainability · Software

1 Introduction

As we know that India is one of the growing economies of the world. It also secured a place in the top 10 economies of the world, even at present we are at the number five economy of the world. Still it is noticeable that this is at a time of economical slowdown worldwide. The Indian GDP for the first half of current year 2019 also went down, which was not expected. It may have a lot of factors including environmental to political. But the slowdowns in economy generate challenge for sustainability. Different revenue sectors are very much affected by the growth of economy. Sustainability is one of the key points in every aspect for survival. At present the Indian IT industries

---

S. Gaur (✉)
Jaipur Engineering College and Research Centre, Jaipur, India
e-mail: sanjay.since@gmail.com

L. Sharma · V. Singh · P. Saini
St. Xavier's College, Jaipur, India
e-mail: leenasharma@stxaviersjaipur.org

S. Fong et al. (eds.), *ICT Analysis and Applications*, Lecture Notes in Networks and Systems 93, https://doi.org/10.1007/978-981-15-0630-7_57

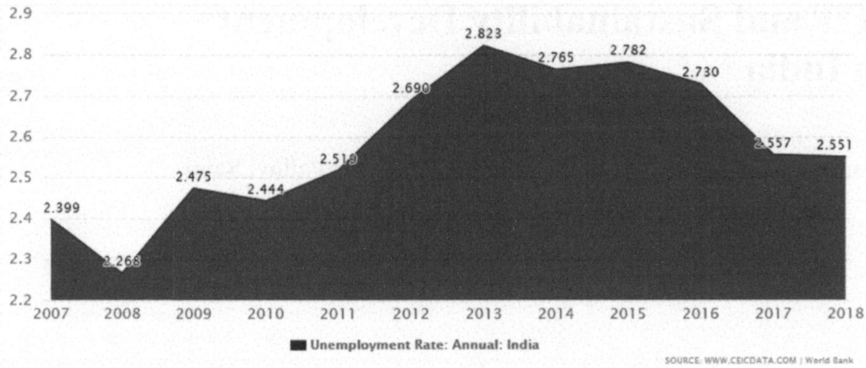

Fig. 1 Unemployment rate in India (www.cdcdata.com|World Bank)

and support services are leading industries. With the slogan of "Digital India," rapid development in this sector takes place now. At present Indian IT industries' participation in GDP is 7.7% approx and till 2025 it is expected to reach 10%. This study is an effort to depict Indian IT industries and their role in national development and sustainability. The study focuses on the economical slowdown and position of IT industries.

2 Unemployment Rate in India

It is a considerable truth that the population of India was 1,332.00 million in March 2019, and labor force participation rate went down to 51.93% in December 2018. As per data in December 2018 the unemployment rate in India was recorded 2.55%, which slightly dropped in comparison of December 2017 which was 2.56. During December 1991 to December 2019 the average unemployment rate was 2.67 percent. It was recorded that in December 2003 the unemployment rate was 3.18%, which was the highest one and 2.27% was the lowest which was recorded for December 2008. These data were produced by World Bank (Fig. 1).

3 Current GDP Scenario

According to the economic survey for the year 2018–19, this was publicized by Union Minister of finance and Corporate Affairs Ms. Nirmala Sitharaman on July 4, 2019. In the announcement, 7% growth rate was forecast for the financial year 2020 as compared to the existing growth rate 6.8%. In the upcoming years Micro, Small, and Medium Enterprises (MSMEs) are targeted to earn expected incomes, to generate employment and to enhance productivity. To become an economy of US$ 5

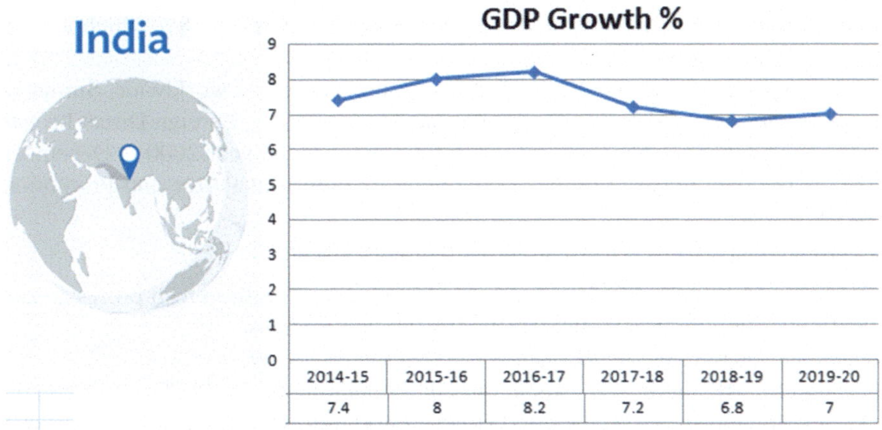

Fig. 2 GDP growth rate

trillion by the financial year 2024–2025, the sustainable actual 8% GDP growth rate is required (Fig. 2).

The GDP Growth may be understood by facts that, the estimated GDP growth rate is 7% for the financial year 2019–20. The GDP for the financial yare 2018–19 was about 6.8% compared to 7.2% for the financial year 2017–18. The GVA growth at basic prices is expected to be 6.1 per cent in 2017–18. The Fiscal Deficit in the financial year 2019 ended with 3.4% of GDP and debt to GDP ratio of 44.5% (Provisional). On the other hand expenses of the Central Government cut down by 0.3% of GDP, in the financial year 2018–19 over 2017–18.

4 ICT and Support Market

The information communication technology (ICT) for the Indian market is a big sector in respect of business and employment. During worldwide slow down including India, the ICT sector of India is doing well, which plays an important role in sustainable development.

India is a worldwide giant for the sourcing market and also rapidly growing, accounting for just about 55% market share of the US$ 185–190 billion worldwide services sourcing business in 2017–18. Indian IT and supportive companies have set up over 1,000 universal liberation centers in about 80 nations worldwide. There is capability to become worldwide digital power and digital hub by means of 75% of universal digital talent available in the nation.

In the year 2018–19 Indian IT and support industry grew to US$ 181 billion. In the financial year exports concern to IT and support industry increased to US$ 137 billion whereas domestic income developed to US$ 44 billion. Investments in ICT by Indian industries in the year 2018 grew over 9% to which reaches US$ 87.1 billion.

It is forecasted that revenue from the digital services may reach 38% which will be US$ 350 billion by the year 2025.

It is the potency of Indian ICT industries to attract attention worldwide. Similarly domestic hardware and software sector attracted cumulative "Foreign Direct Investment" (FDI) inflows US$ 35.82 billion during the financial year 2000 to December 2018, as per data released by the Department of Industrial Policy and Promotion (DIPP).

Following are the development in the Indian ICT sector:

- NASSCOM launched a platform to up-skilling over 2 million tech professionals and same for 2 million competent employees and students.
- Income growth in the BFSI vertical stood at 6.80% yearly.
- In March 2018, there were over 1,140 GICs operating beyond India.
- PE investments in the sector stood at US$ 2,400 million in Q4 2018.
- Venture Capital (VC) investments in the IT & ITeS sector stood at US$ 53.0 million during Q4 2018.

5 Achievements: Sustainability Remarks

Despites the slowdown in the economy, it was significant that about 200 Indian IT firms were presented in more than 80 countries. The Indian IT export reached its record height with the amount of US$ 126 billion for the year 2017–18. In that sequence in the year 2018–19, the revenue generated by India IT firms was US$ 181 billion, which is the highest one.

For the future point of view, India is the highest offshoring destination for IT industries worldwide. India encompasses its potential for onshore and offshore services to universal clients; up-and-coming technologies at the present proffer all new-fangled gamut of prospect for top IT firms in India. The export income of such industries is estimated to grow up 7–9% y-o-y to US$ 135–137 billion in FY19. Also industry is expected to cultivate US$ 350 billion by the year 2025 and BPM is expected for US$ 50–55 billion out of the total.

6 Sustainability of ICT Sector

In the scenario of the world, India is a leading sourcing destination, it is almost 55 percent of US$ 185–190 billion market in the financial year 2017–18. This is due to a world-class technically sound and potential manpower. Also living and survival cost in India is 6–7 times inexpensive with respect to European countries as well as the United States.

In the latest, IT sector, Cloud service, and support scenario in India are probable to rise three times to US$ 7.1 billion by the year 2022 in association with the development of artificial intelligence, Machine learning, big data, Internet of things, and data analytics. It is predicted in the Cloud Next Wave of Growth in India report. The IT industry of India contributes almost 7.7% in the GDP of the nation and it is predicted that it would be approximately 10% by the year 2025. Also it holds 20% share in the total of export the Nation.

By the year 20125, it is expected that IT-IBM sector may grow to US$ 350 billion and similarly BPM may be for US$ 50–55 billion out of the entire revenue. There is strong possibility that the digital support area will grow 37–38 of the total industry revenue.

The IT industry generates approx 3.97 million jobs in the country in which 105,000 were generated in the financial year 2018 and it is expected that at the end of the year 2019 there would be 250,000 new jobs. Even the hardware and software area attracts a big part of "Foreign Direct Investment" inflows of about US$ 35.82 billion during April 2000 and December 2018 and ranks second in inflow of FDI, as per data public by the "Department of Industrial Policy and Promotion" (DIPP). The Indian government also provides support for better sustainable development through Special Economic Zones (SEZs), software technology parks of India (STPI), by making ease in procedure for IT investment and single window clearance for setting up facilities.

7 The Indian ICT Sector

In general, the information communication sector of India holds major sections which are (1) Information Technology Industries that is directly concerned with computer and related areas of activities and second one is (2) Telecomm Industry which is focused with digital network, communication, cell phone, internet facilities, etc.

We can also classify the ICT sector according to software, IT-enabled services and business domain of the IT segment with support services. The software section is concerned with the software development, application and app development, software engineering and research and development in the area.

The IT-enabled service comprises IT projects, consultancy, and approach for customization of task, IT management, and solution support. Outsourcing is one of the big sectors, hardware support and development, network and infrastructure management. In this section we can also include training and support for operation of the services.

The third major section is IT enables services and business process outsourcing. This section is fully dedicated to develop business from all the support vectors of information technology. With the cumulative efforts of theses section of ICT, overall more than 7.8% GDP contribution is provided by IT sector which is one of the major contributions and provides sustainability in modern age.

Figure 3 shows the status of IT sectors in India for the financial year 2018–19. Here it is clearly shown that the IT sector is one of major and fast developing sector of the Indian economy and plays a major role in sustainable development.

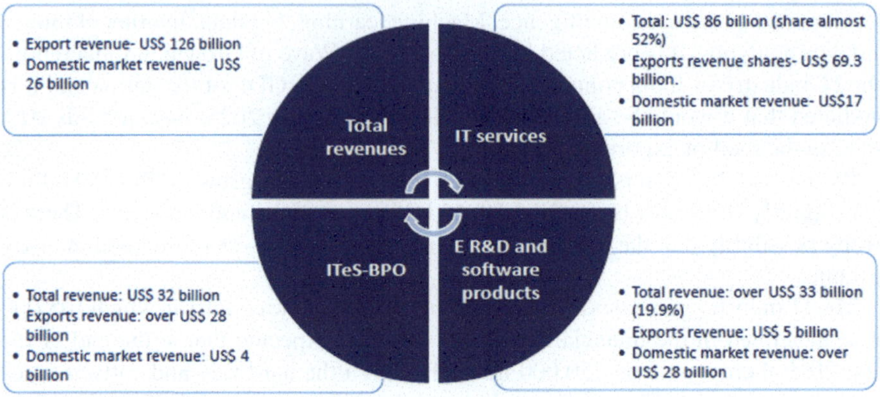

Fig. 3 Market share of IT (financial year 2018–19)

Fig. 4 Market Leaders

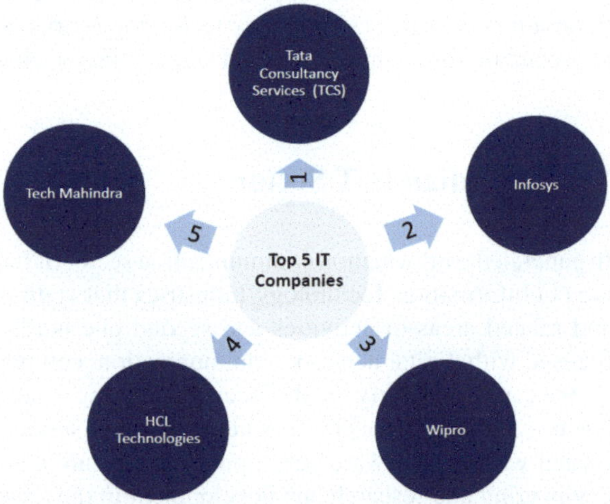

The major players of the IT sector in the Indian market are Tata Consultancy Services (TCS), Infosys, Tech-Mahindra, Wipro and HCL. TCS is a universal boss in the IT sector and its 70% income is earned by TataSons. Infosys is a big Indian brand, it generated a sum of Rs 60,878 crores (~7.6 billioneuros) in 2018. Tech-Mahindra's revenue was Rs 23,562 crores (~3 billioneuros) in 2018. The HCL's revenue was Rs. 21,000 crores (~2.5 billioneuros) and Wipro's revenue was Rs. 44,902 crores (~5.6 billioneuros) in 2018.

The growth drivers of IT sectors are helpful to drive the nation's sustainability. The availability of potential pool at lesser rates helps in cost-cutting about 60–70 percent. Advanced technologies such as digital technologies, Data analytics, artificial intelligence, social media marketing, cloud technologies, and Machine learning are creating novel paradigms. The research in field of digitization and automation is the

Fig. 5 Growth Drivers

backbone for Indian IT sector. Now Indian IT industries are in full demand by Europe and the rest of the world (Figs. 4 and 5).

8 Conclusion

The present paper is a blend of GDP scenario of the Indian economy especially for the financial years from 2014 to 2019, the unemployment situation of the nation are also clubbed with the economy. The focus of the study is on Indian IT sector and sustainability development in slowdown economy. The role and contribution of IT industries are shown by the help of possible statistics. The role and responsibility figure of the leaders are also discussed in the precise formats along with the growth drivers. Prediction for future also gives upcoming picture and roadmap of sustainability development for the nation. So overall this is an attempt to show an effort of Indian IT industries and its contribution to the development of the nation.

References

1. Hilty LM, Hercheui MD (2010) ICT and sustainable development. In: What kind of information society? governance, virtuality, surveillance, sustainability, resilience. IFIP advances in information and communication technology, 2010, Volume 328/2010, edited by Jacques Berleur, Magda D. Hercheui, Lorenz M. Hilty. Berlin Heidelberg New York: Springer ISBN: 978-3-642-15478-2, pp 227–235
2. Hilty LM (2008) Information technology and sustainability. Essays on the Relationship between ICT and Sustainability. Books on Demand, Norderstedt. http://www.amazon.de/gp/product/3837019705/

3. India's unemployment rate hit four-decade high of 6.1% in 2017–18, says NSSO survey. *Businesstoday.in*. Retrieved 14 March 2019
4. By hiding unemployment data, the government will not help the economy—or itself. *The Telegraph. Calcutta. 17 January 2019*. Retrieved 20 February 2019
5. *Tandon, Rajguru.* India Needs 10 Million Jobs Per Annum Till 2030 To Counter Unemployment. *BW Businessworld*. Retrieved 25 February 2019
6. Media Reports, Press Information Bureau (PIB), Department of Industrial Policy and Promotion (DIPP) statistics, Department of Information and Technology, Union Budget 2017–18, https://www.ibef.org/research/reports, (Exchange Rate Used: INR 1 = US$ 0.0139 as of Q3 FY19)
7. https://www.ibef.org/research/reports, (Exchange Rate Used: INR 1 = US$ 0.0139 as of Q3 FY19)Jobs data not finalised: Government after NSSO 'Report'. *The Economic Times. 1 February 2019*. Retrieved 14 March 2019
8. India Unemployment Rate [1991–2019] [Data & Charts] –CEIC, https://www.ceicdata.com/en/indicator/india/unemployment-rate
9. Key Reforms to Sustain India's Growth Momentum, https://www.adb.org/news/key-reforms-sustain-indias-growth-momentum, © 2019 Asian Development Bank
10. https://www.researchgate.net/publication/220939797_ICT_and_Sustainable_Development
11. https://sustainabledevelopment.un.org/content/documents/15836India.pdf
12. https://link.springer.com/content/pdf/10.1007/978-3-642-15479-9_22.pdf
13. https://link.springer.com/content/pdf/10.1007/978-3-642-15479-9_22.pdf*Media Reports, Press Information Bureau (PIB), Department of Industrial Policy and Promotion (DIPP) statistics, Department of Information and Technology, Union Budget 2017-18,* https://www.ibef.org/research/reports, (Exchange Rate Used: INR 1 = US$ 0.0139 as of Q3 FY19)

The manufacturer's authorised representative in the EU is Springer Nature Customer Service Centre GmbH, Europaplatz 3, 69115 Heidelberg, Germany. If you have any concerns regarding our products, please contact ProductSafety@springernature.com

Printed and bound by CPI Group (UK) Ltd, Croydon, CR0 4YY
30/09/2026
02225448-0019